McGraw-Hill My Math

Welcome to *My Math* — your very own math book! You can write in it – in fact, you are encouraged to write, draw, circle, explain, and color as you explore the exciting world of mathematics. Let's get started. Grab a pencil and finish each sentence.

My name is ______________________________.

My favorite color is ______________________________.

My favorite hobby or sport is ______________________________.

My favorite TV program or video game is

______________________________.

My favorite class is ______________________________.

Math, of course!

Bothell, WA • Chicago, IL • Columbus, OH • New York, NY

connectED.mcgraw-hill.com

STEM McGraw-Hill is committed to providing instructional materials in Science, Technology, Engineering, and Mathematics (STEM) that give all students a solid foundation, one that prepares them for college and careers in the 21st century.

Send all inquiries to:
McGraw-Hill Education
STEM Learning Solutions Center
8787 Orion Place
Columbus, OH 43240

ISBN: 978-0-02-116195-9 *(Volume 2)*
MHID: 0-02-116195-X

Printed in the United States of America.

21 22 23 24 25 QSX 22 21 20 19 18

Our mission is to provide educational resources that enable students to become the problem solvers of the 21st century and inspire them to explore careers within Science, Technology, Engineering, and Mathematics (STEM) related fields.

Meet The Artists!

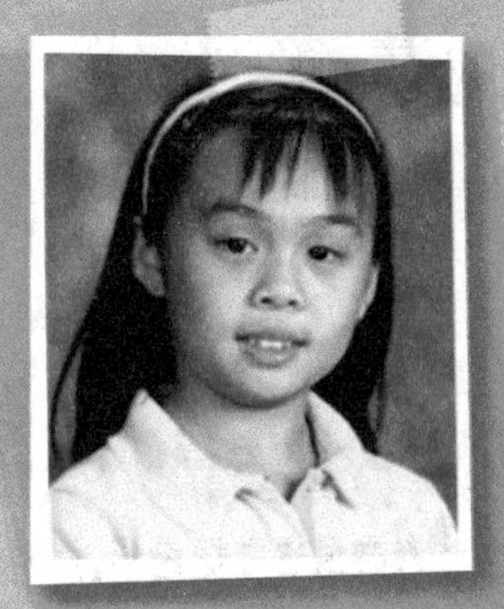

Carolyn Phung

Invertebrate Art My class was studying invertebrates, and we discovered that insects have their own symmetry. Then we talked about radial symmetry which is like a wheel. I put these ideas together to form my artwork. *Volume 1*

Grace Kramer

Math is Awesome 4 Math challenges me to learn new things. When I think about math problems they are sometimes hard to solve just like puzzles. Math is puzzles that are all around me. *Volume 2*

Other Finalists

Jesus Pallares
Math Roller Coaster

Isidro Tavares
Righteous Math

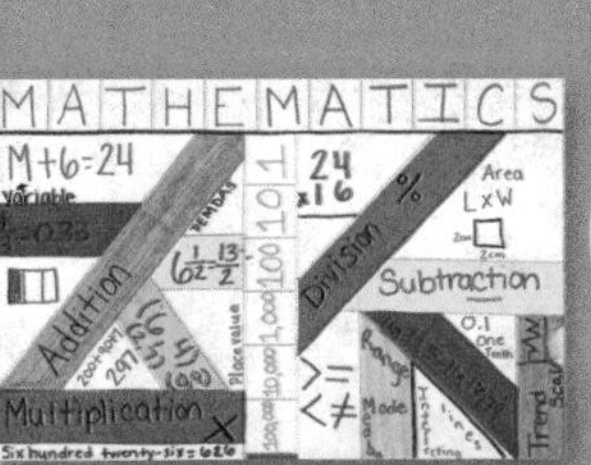

Jaquise Hickman
Roadmap of Math

Avi Sanan
Jumping Fish

Isa Weiss
Sailing Through Math

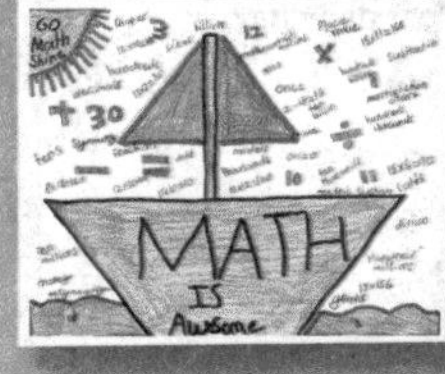

Luis Rodriguez
Math is Awesome 12

Mikayla Pilgrim
My Many Colored Numbers

Kayley Spiller
The Part Tree

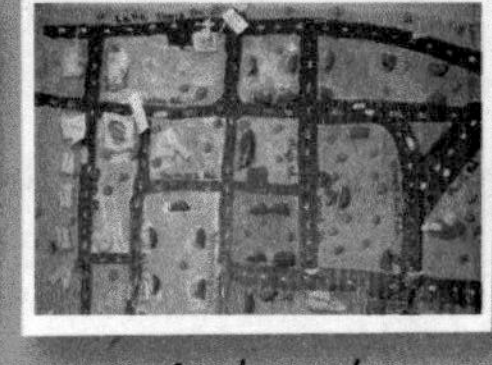

Carl Zent
Math Means a Working Community

Calista Smith
Skinny Jeans

Find out more about the winners and other finalists at www.MHEonline.com.

We wish to congratulate all of the entries in the 2011 *McGraw-Hill My Math* "What Math Means To Me" cover art contest. With over 2,400 entries and more than 20,000 community votes cast, the names mentioned above represent the two winners and ten finalists for this grade.

GO digital

it's all at
connectED.mcgraw-hill.com

Go to the Student Center for your eBook, Resources, Homework, and Messages.

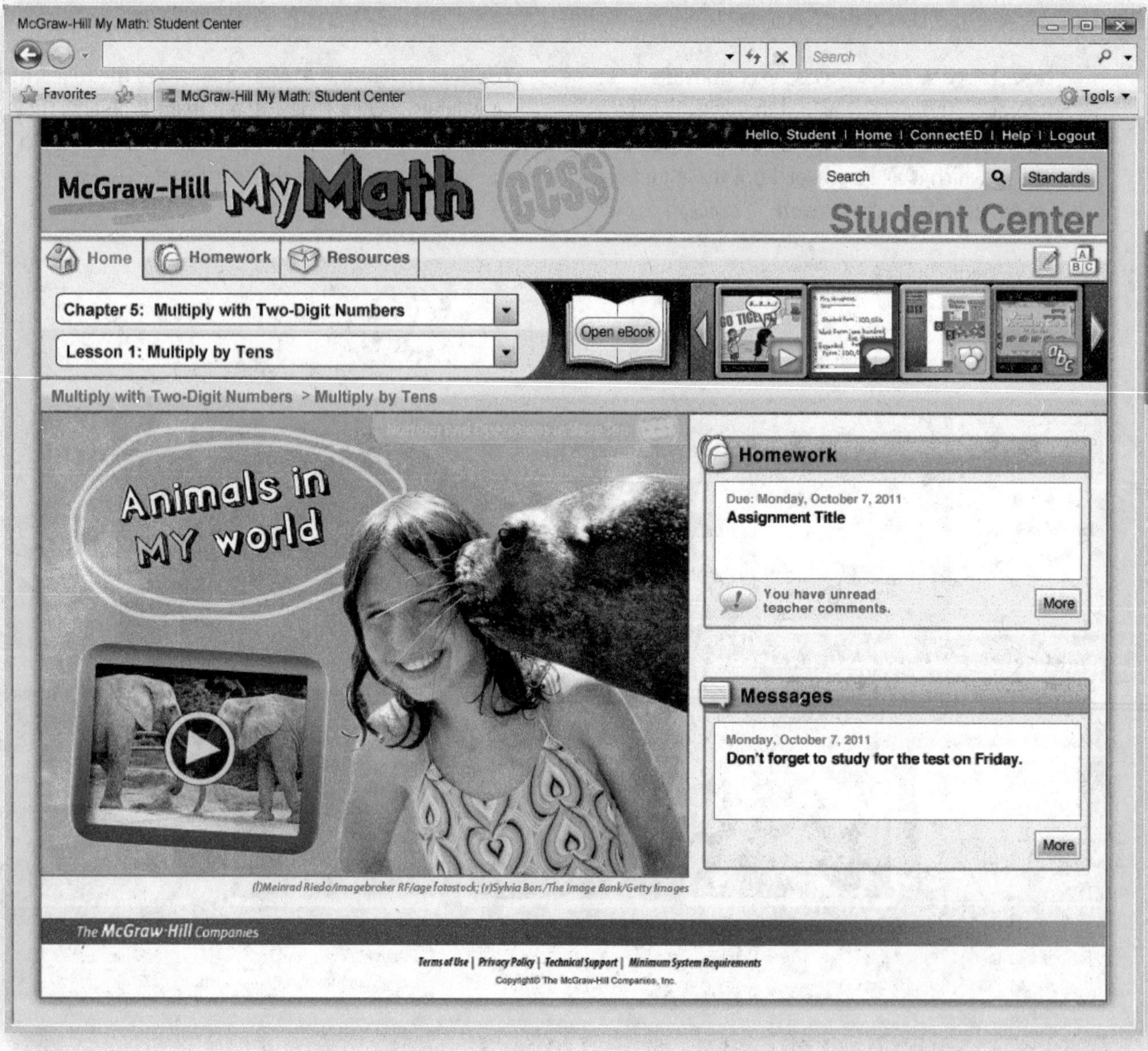

McGraw-Hill My Math: Student Center

Hello, Student | Home | ConnectED | Help | Logout

McGraw-Hill My Math

Student Center

Home | Homework | Resources

Chapter 5: Multiply with Two-Digit Numbers

Lesson 1: Multiply by Tens

Open eBook

Multiply with Two-Digit Numbers > Multiply by Tens

Animals in MY world

Homework

Due: Monday, October 7, 2011

Assignment Title

You have unread teacher comments.

More

Messages

Monday, October 7, 2011

Don't forget to study for the test on Friday.

More

(l)Meinrad Riedo/imagebroker RF/age fotostock; (r)Sylvia Bors/The Image Bank/Getty Images

The McGraw-Hill Companies

Terms of Use | Privacy Policy | Technical Support | Minimum System Requirements

Write your Username ______ Password ______

Get your resources online to help you in class and at home.

Vocab	Watch	Tools	Check

Find activities for building vocabulary.	Watch animations of key concepts.	Explore concepts with virtual manipulatives.	Self-assess your progress.

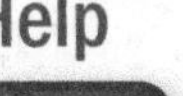

eHelp	Games	Tutor

Get targeted homework help.	Reinforce with games and apps.	See a teacher illustrate examples and problems.

Scan this QR code with your smart phone* or visit mheonline.com/stem_apps.

*May require quick response code reader app.

Contents in Brief

Organized by Domain

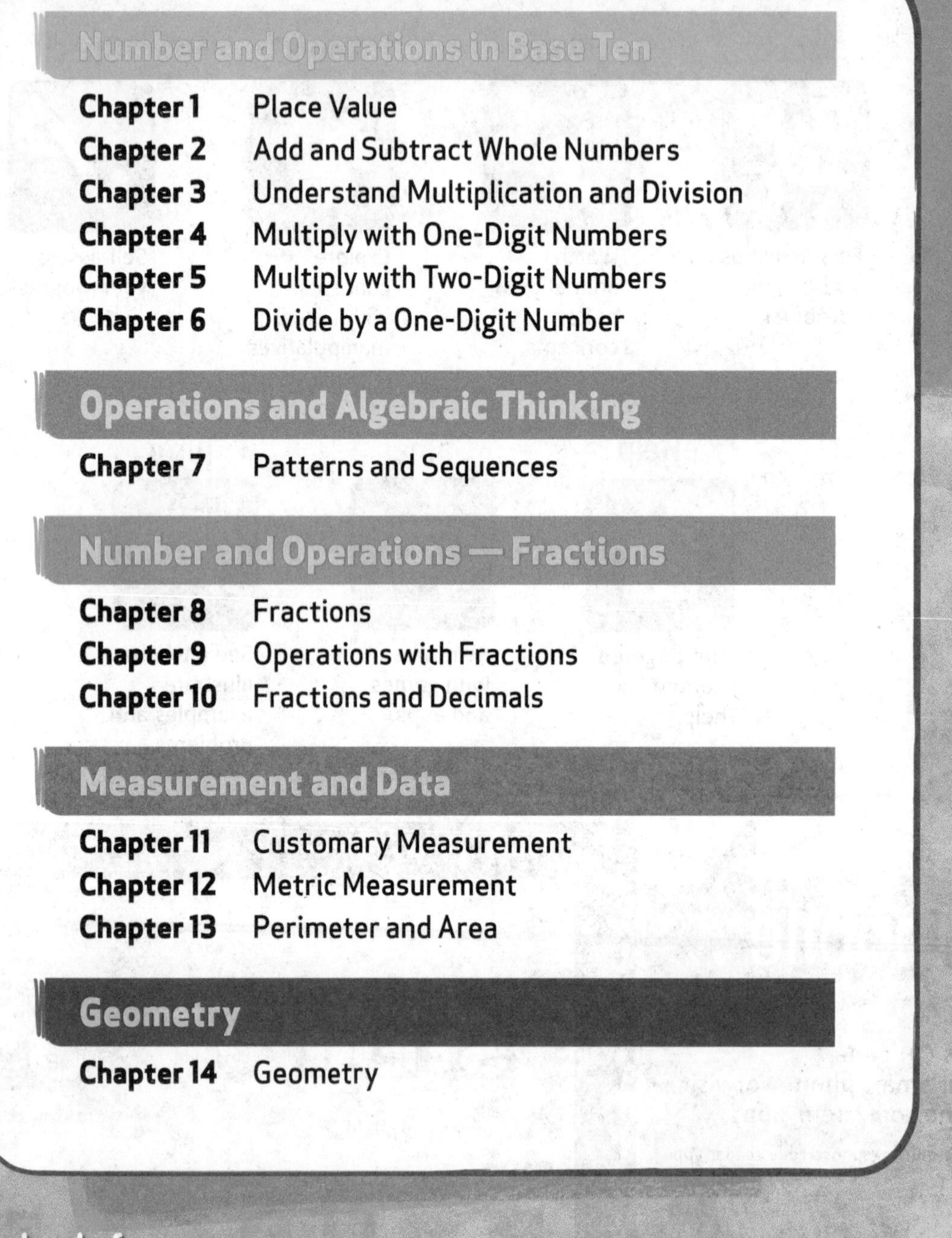

Chapter 1 Place Value

ESSENTIAL QUESTION
How does place value help represent the value of numbers?

Are you ready for the great outdoors?

Getting Started

Lessons and Homework

Wrap Up

Look for this!
Click online and you can watch videos that will help you learn the lessons.
Watch

connectED.mcgraw-hill.com

Chapter 2 Add and Subtract Whole Numbers

ESSENTIAL QUESTION
What strategies can I use to add or subtract?

connectED.mcgraw-hill.com

Chapter 3 Understand Multiplication and Division

ESSENTIAL QUESTION
How are multiplication and division related?

Getting Started

Lessons and Homework

Wrap Up

eHelp

Look for this!
Click online and you can get more help while doing your homework.

Chapter 4 Multiply with One-Digit Numbers

ESSENTIAL QUESTION
How can I communicate multiplication?

Getting Started

Lessons and Homework

Wrap Up

I hope I can multiply my savings!

connectED.mcgraw-hill.com

Chapter 5 Multiply with Two-Digit Numbers

ESSENTIAL QUESTION
How can I multiply by a two-digit number?

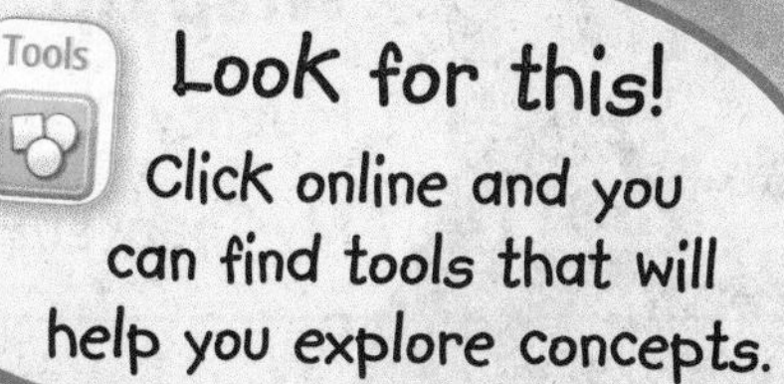

Chapter 6

Divide by a One-Digit Number

ESSENTIAL QUESTION
How does division affect numbers?

Getting Started

Lessons and Homework

Wrap Up

connectED.mcgraw-hill.com

Chapter 7 Patterns and Sequences

ESSENTIAL QUESTION
How are patterns used in mathematics?

Tutor

Look for this!
Click online and you can watch a teacher solving problems.

Chapter 8 Fractions

ESSENTIAL QUESTION
How can different fractions name the same amount?

Getting Started

Lessons and Homework

Wrap Up

Chapter 9 Operations with Fractions

ESSENTIAL QUESTION
How can I use operations to model real-world fractions?

Getting Started

Lessons and Homework

Wrap Up

Look for this! Vocab Click online and you can find activities to help build your vocabulary.

Chapter 10

Fractions and Decimals

ESSENTIAL QUESTION
How are fractions and decimals related?

Getting Started

Lessons and Homework

Wrap Up

connectED.mcgraw-hill.com

Chapter 11 Customary Measurement

ESSENTIAL QUESTION
Why do we convert measurements?

Check

Look for this!
Click online and you can check your progress.

Chapter 12 Metric Measurement

ESSENTIAL QUESTION
How can conversion of measurements help me solve real-world problems?

Getting Started

Lessons and Homework

Wrap Up

Chapter 13 Perimeter and Area

ESSENTIAL QUESTION
Why is it important to measure perimeter and area?

Getting Started

Lessons and Homework

Wrap Up

Chapter 14 Geometry

ESSENTIAL QUESTION
How are different ideas about geometry connected?

Getting Started

Lessons and Homework

Wrap Up

connectED.mcgraw-hill.com

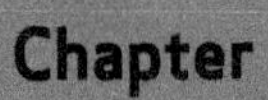

Chapter 8 Fractions

ESSENTIAL QUESTION

How can different fractions name the same amount?

Now We're Cooking!

Watch a video!

Watch

MY Common Core State Standards

Number and Operations – Fractions

4.NF.1 Explain why a fraction $\frac{a}{b}$ is equivalent to a fraction $\frac{(n \times a)}{(n \times b)}$ by using visual fraction models, with attention to how the number and size of the parts differ even though the two fractions themselves are the same size. Use this principle to recognize and generate equivalent fractions.

4.NF.2 Compare two fractions with different numerators and different denominators, e.g., by creating common denominators or numerators, or by comparing to a benchmark fraction such as $\frac{1}{2}$. Recognize that comparisons are valid only when the two fractions refer to the same whole. Record the results of comparisons with symbols $>$, $=$, or $<$, and justify the conclusions, e.g., by using a visual fraction model.

4.NF.3 Understand a fraction $\frac{a}{b}$ with $a > 1$ as a sum of fractions $\frac{1}{b}$.

4.NF.3b Decompose a fraction into a sum of fractions with the same denominator in more than one way, recording each decomposition by an equation. Justify decompositions, e.g., by using a visual fraction model.

4.NF.5 Express a fraction with denominator 10 as an equivalent fraction with denominator 100, and use this technique to add two fractions with respective denominators 10 and 100.

Operations and Algebraic Thinking *This chapter also addresses this standard:*

4.OA.4 Find all factor pairs for a whole number in the range of 1–100. Recognize that a whole number is a multiple of each of its factors. Determine whether a given whole number in the range 1–100 is a multiple of a given one-digit number. Determine whether a given whole number in the range 1–100 is prime or composite.

It's a lot of stuff, but I'll be able to figure it out!

Standards for Mathematical PRACTICE

1. Make sense of problems and persevere in solving them.
2. Reason abstractly and quantitatively.
3. Construct viable arguments and critique the reasoning of others.
4. Model with mathematics.
5. Use appropriate tools strategically.
6. Attend to precision.
7. Look for and make use of structure.
8. Look for and express regularity in repeated reasoning.

= focused on in this chapter

Name ....

Check ← Go online to take the Readiness Quiz

Graph each fraction on a number line.

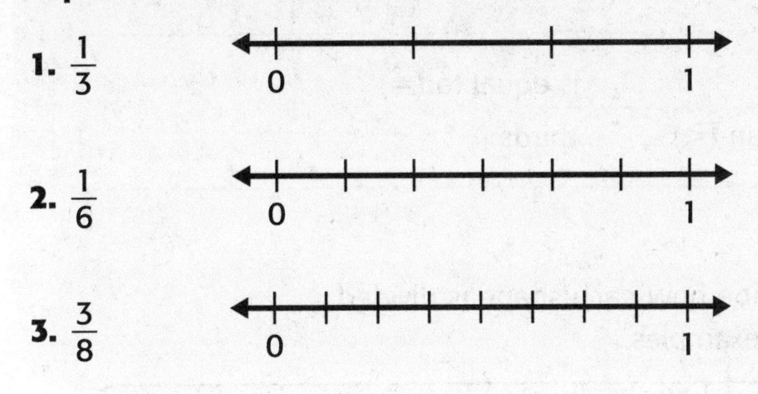

1. $\frac{1}{3}$ 0 1

2. $\frac{1}{6}$ 0 1

3. $\frac{3}{8}$ 0 1

Use the number line to determine whether the two fractions are equivalent. Write yes or no.

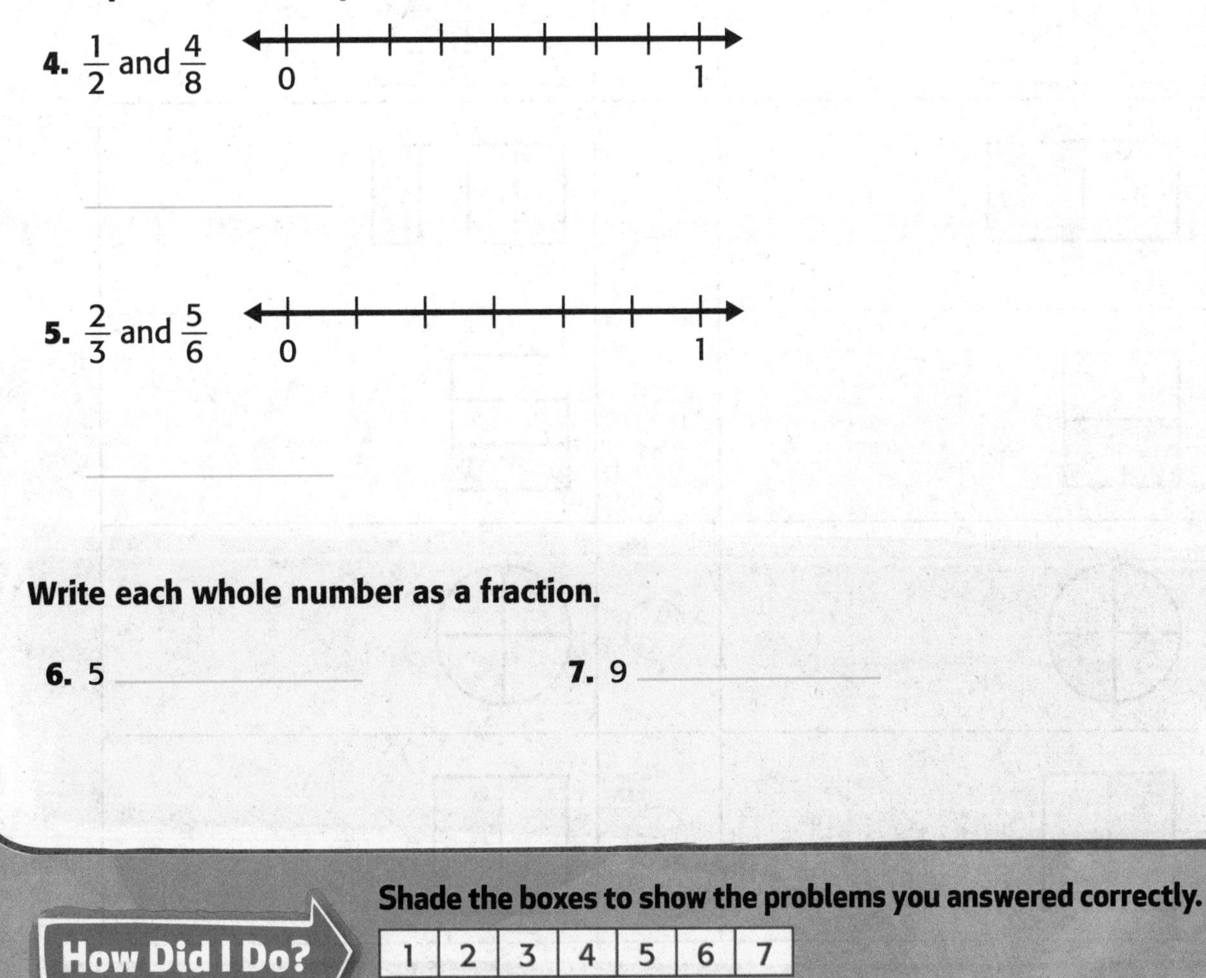

4. $\frac{1}{2}$ and $\frac{4}{8}$ 0 1

____

5. $\frac{2}{3}$ and $\frac{5}{6}$ 0 1

____

Write each whole number as a fraction.

6. 5 ______________ **7.** 9 ______________

How Did I Do?

Shade the boxes to show the problems you answered correctly.

1	2	3	4	5	6	7

Name

Review Vocabulary

fourths | halves | is equal to (=)
is greater than (>) | is less than (<) | thirds

Making Connections

Use the review vocabulary to describe how each shape is divided. Then use symbols to compare the examples.

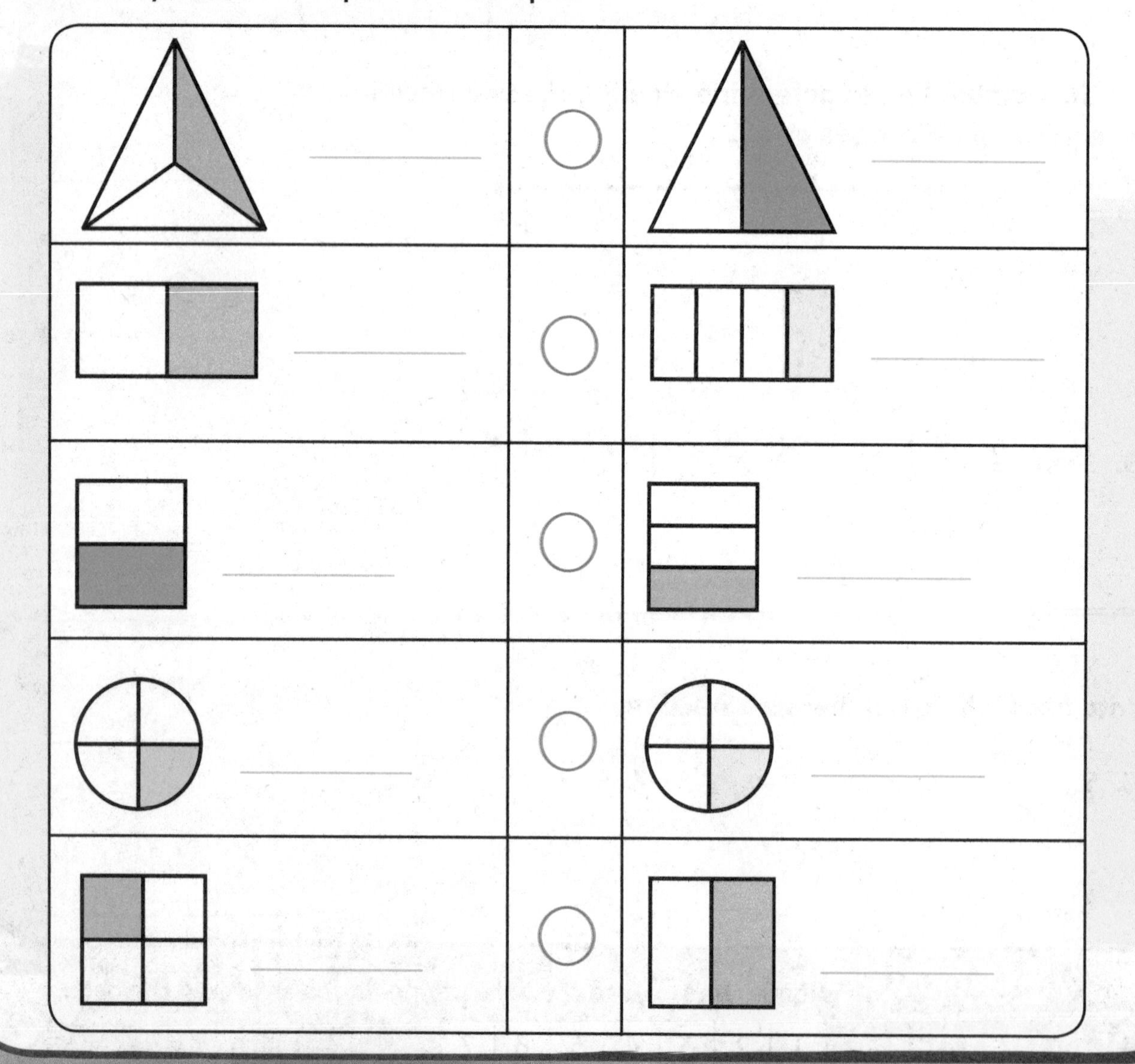

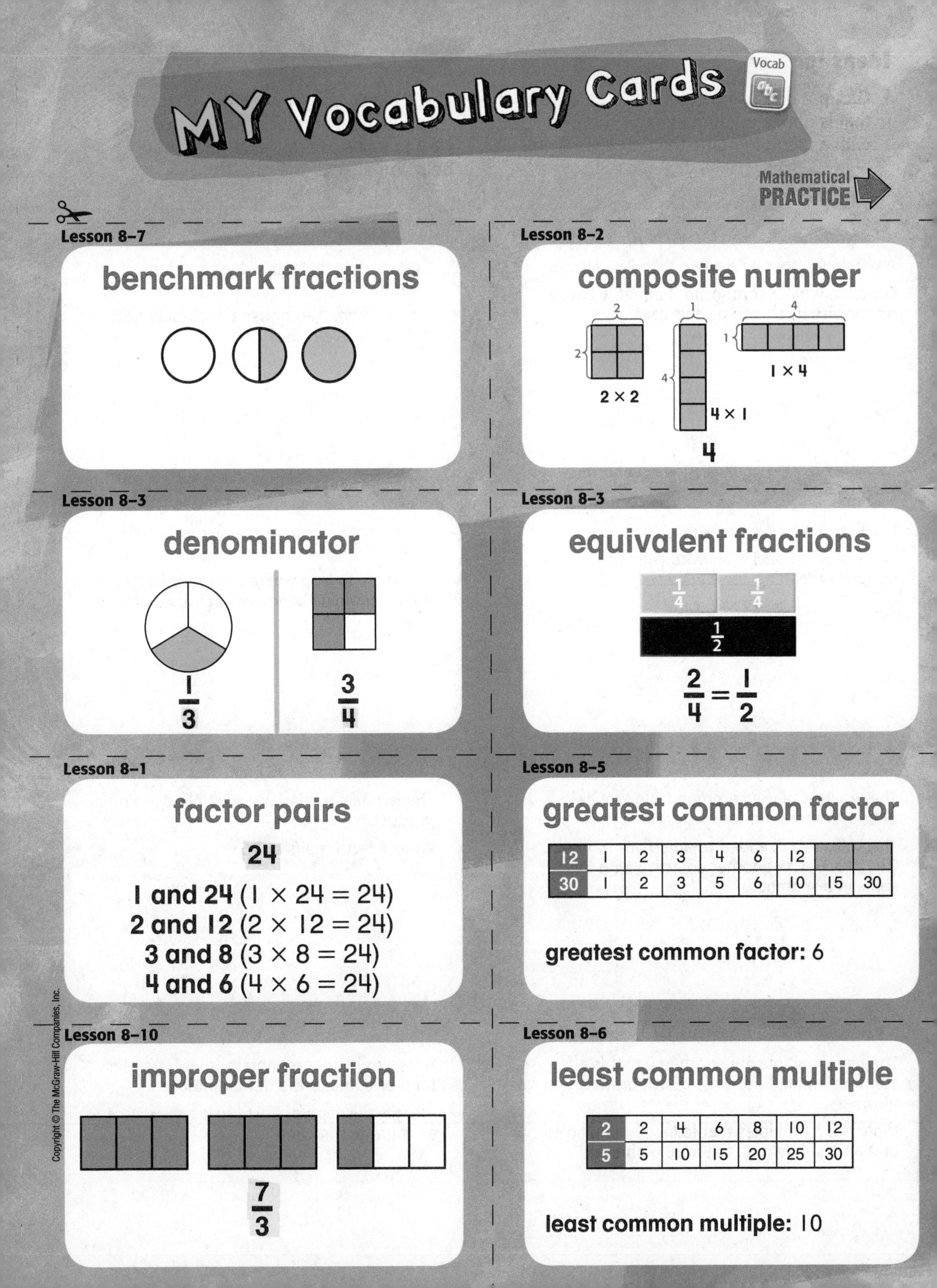
MY Vocabulary Cards
Vocab
abc
Mathematical PRACTICE
Lesson 8–7
benchmark fractions
Lesson 8–2
composite number
2
2
2 × 2
1
4
4 × 1
4
1
1 × 4
4
Lesson 8–3
denominator
1/3
3/4
Lesson 8–3
equivalent fractions
1/4
1/4
1/2
2/4 = 1/2
Lesson 8–1
factor pairs
24
1 and 24 (1 × 24 = 24)
2 and 12 (2 × 12 = 24)
3 and 8 (3 × 8 = 24)
4 and 6 (4 × 6 = 24)
Lesson 8–5
greatest common factor
12 1 2 3 4 6 12
30 1 2 3 5 6 10 15 30
greatest common factor: 6
Lesson 8–10
improper fraction
7/3
Lesson 8–6
least common multiple
2 2 4 6 8 10 12
5 5 10 15 20 25 30
least common multiple: 10
Copyright © The McGraw-Hill Companies, Inc.

Ideas for Use

- Group 2 or 3 common words. Add a word that is unrelated to the group. Then work with a friend to name the unrelated word.
- Use a blank card to write this chapter's essential question. Use the back of the card to write or draw examples that help you answer the question.

A whole number with more than two factors.

Describe what a composite number is based on the visuals shown on this card.

Common fractions that are used for estimation.

When do you see or use benchmark fractions in your everyday life?

Fractions that have the same value.

What is an antonym, or word with an opposite meaning, for *equivalent*?

The bottom number in a fraction. It tells the total number of equal parts.

Write a tip to help you remember which number is the denominator and which is the numerator.

The greatest of the common factors of each of two or more numbers.

The suffix *-est* is added to adjectives to mean "the most." Use another adjective with this suffix in a sentence.

The two factors that are multiplied to find a product.

Write 3 factor pairs of 36.

The least multiple greater than 0 that is a common multiple of each of two or more numbers.

Write the 3 multiples that follow the multiples of 2 and 5 shown on the front of the card.

A fraction with a numerator that is greater than or equal to the denominator.

Identify this type of fraction: $2\frac{4}{5}$. Rewrite it as an improper fraction.

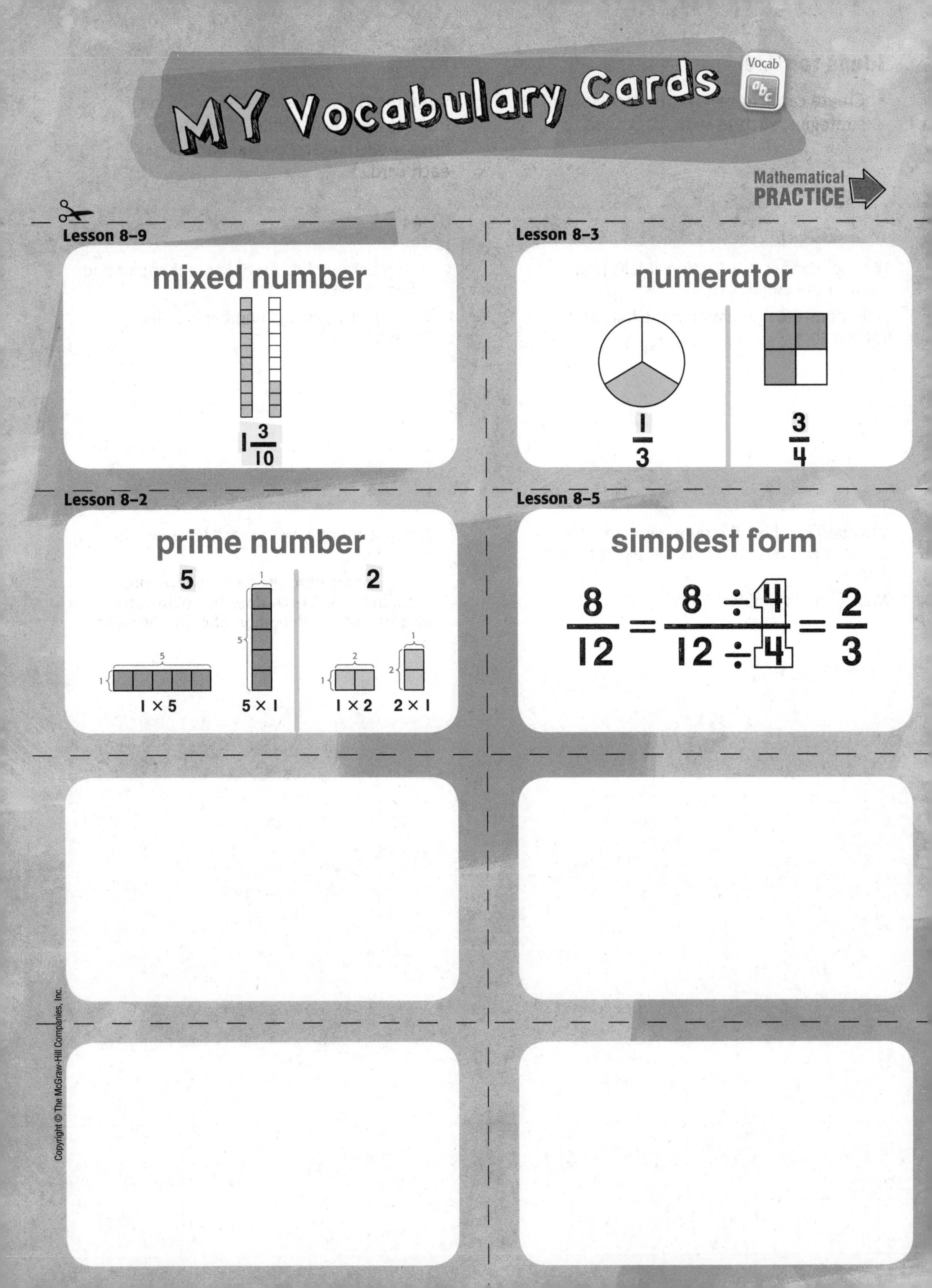
MY Vocabulary Cards
Vocab
abc
Mathematical PRACTICE
Lesson 8-9
mixed number
1 3/10
Lesson 8-3
numerator
1/3
3/4
Lesson 8-2
prime number
5
2
1 × 5
5 × 1
1 × 2
2 × 1
Lesson 8-5
simplest form
8/12 = (8 ÷ 4)/(12 ÷ 4) = 2/3

Ideas for Use

- Create cards to review problem-solving strategies, such as work backward.
- Write the name of each lesson on the front of each card. Write a few study tips for each lesson on the back of each card.

The top number in a fraction. It tells how many of the equal parts are being used.

What does the numerator in the following fraction show? $\frac{9}{20}$

A number that has a whole number part and a fraction part.

Explain why a mixed number is called "mixed."

A fraction in which the numerator and the denominator have no common factor greater than 1.

Write $\frac{15}{45}$ in its simplest form.

A whole number with exactly two factors—1 and itself.

Prime comes from the Latin root *primus*, meaning "first." How does this help you understand the definition of *prime number*?

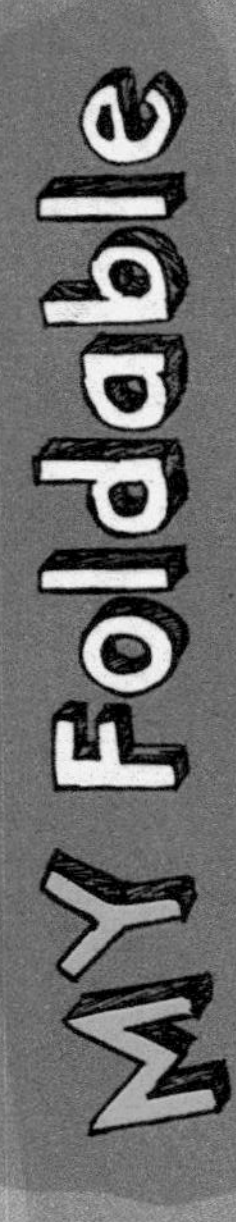

FOLDABLES® Follow the steps on the back to make your Foldable.

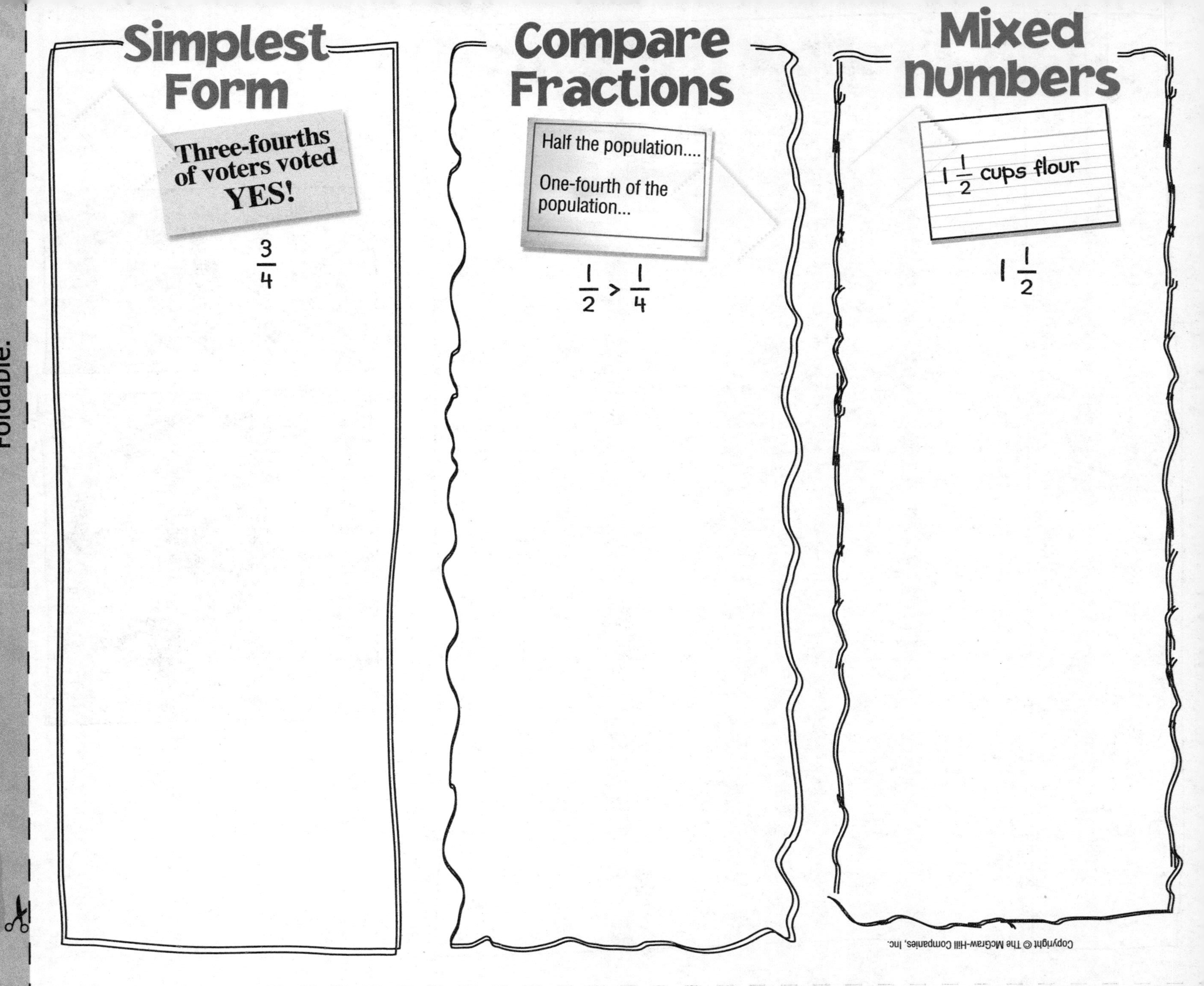

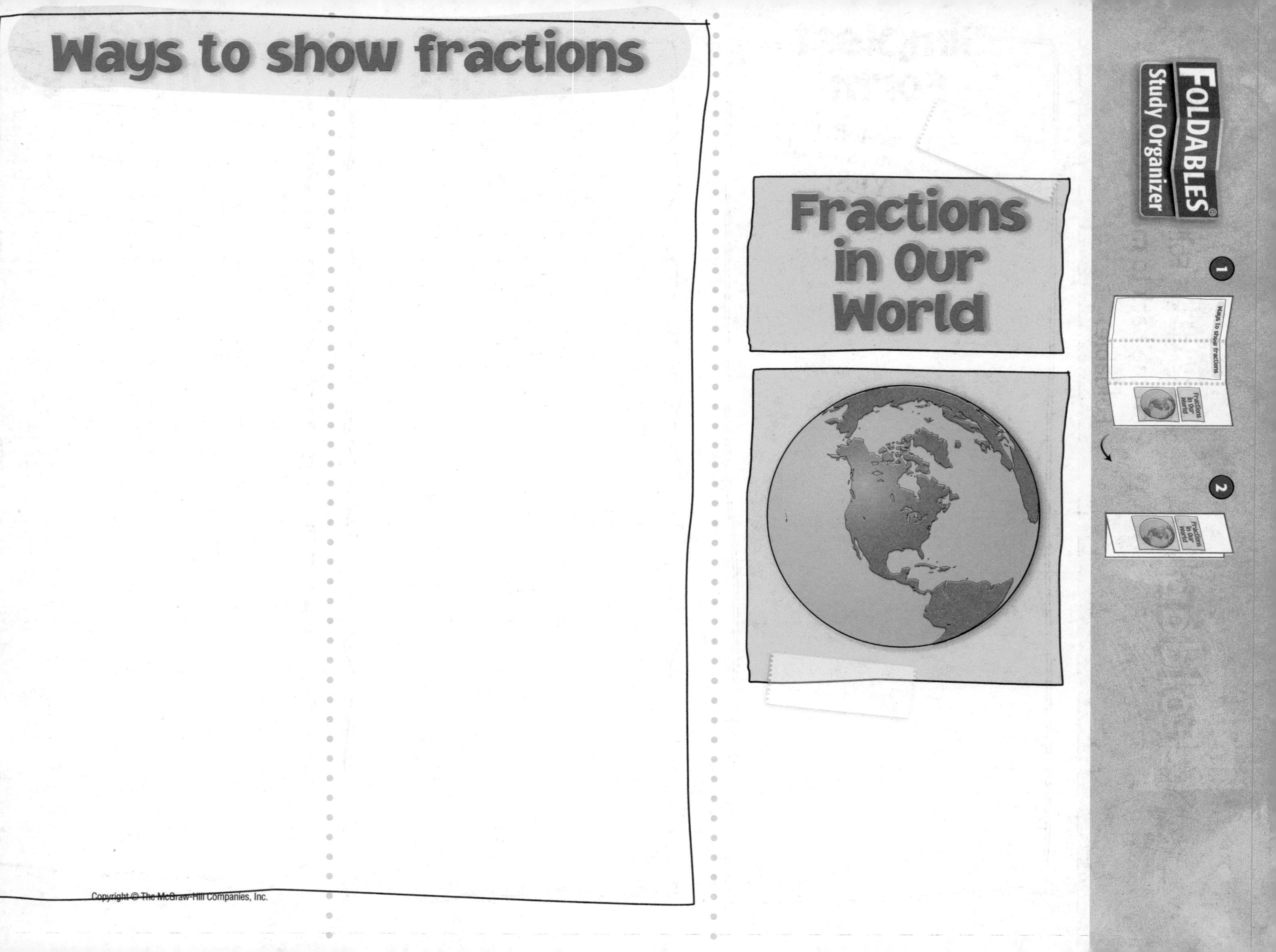
Ways to show fractions
Fractions in Our World
FOLDABLES
Study Organizer
1
Ways to show fractions
Fractions in Our World
2
Fractions in Our World
Copyright © The McGraw-Hill Companies, Inc.

Name ....

Factors and Multiples

Lesson 1

ESSENTIAL QUESTION
How can different fractions name the same amount?

Factors are numbers that are multiplied together to form a product. The two factors that are multiplied together are the **factor pairs** of the product.

Example 1

A chef is arranging 48 strawberries in an array on a tray. Find all the factor pairs of 48 to see what kind of arrays can be made. Draw and label one of the arrays.

Think of multiplication equations that result in a product of 48.

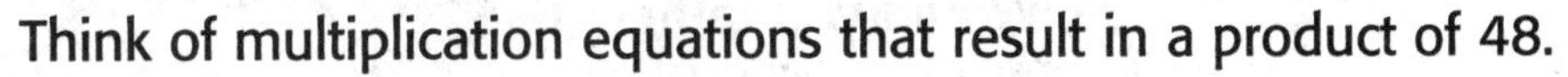

$1 \times 48 = 48$ 1 and 48 are a factor pair of 48.

$2 \times 24 = 48$ 2 and 24 are a factor pair of 48.

$3 \times$ _______ $= 48$ 3 and _______ are a factor pair of 48.

$4 \times$ _______ $= 48$ 4 and _______ are a factor pair of 48.

$6 \times$ _______ $= 48$ 6 and _______ are a factor pair of 48.

48 is a multiple of each of the factors above.

So, the following arrays can be made: _______ × _______, _______ × _______,

_______ × _______, _______ × _______, and _______ × _______.

My Drawing!

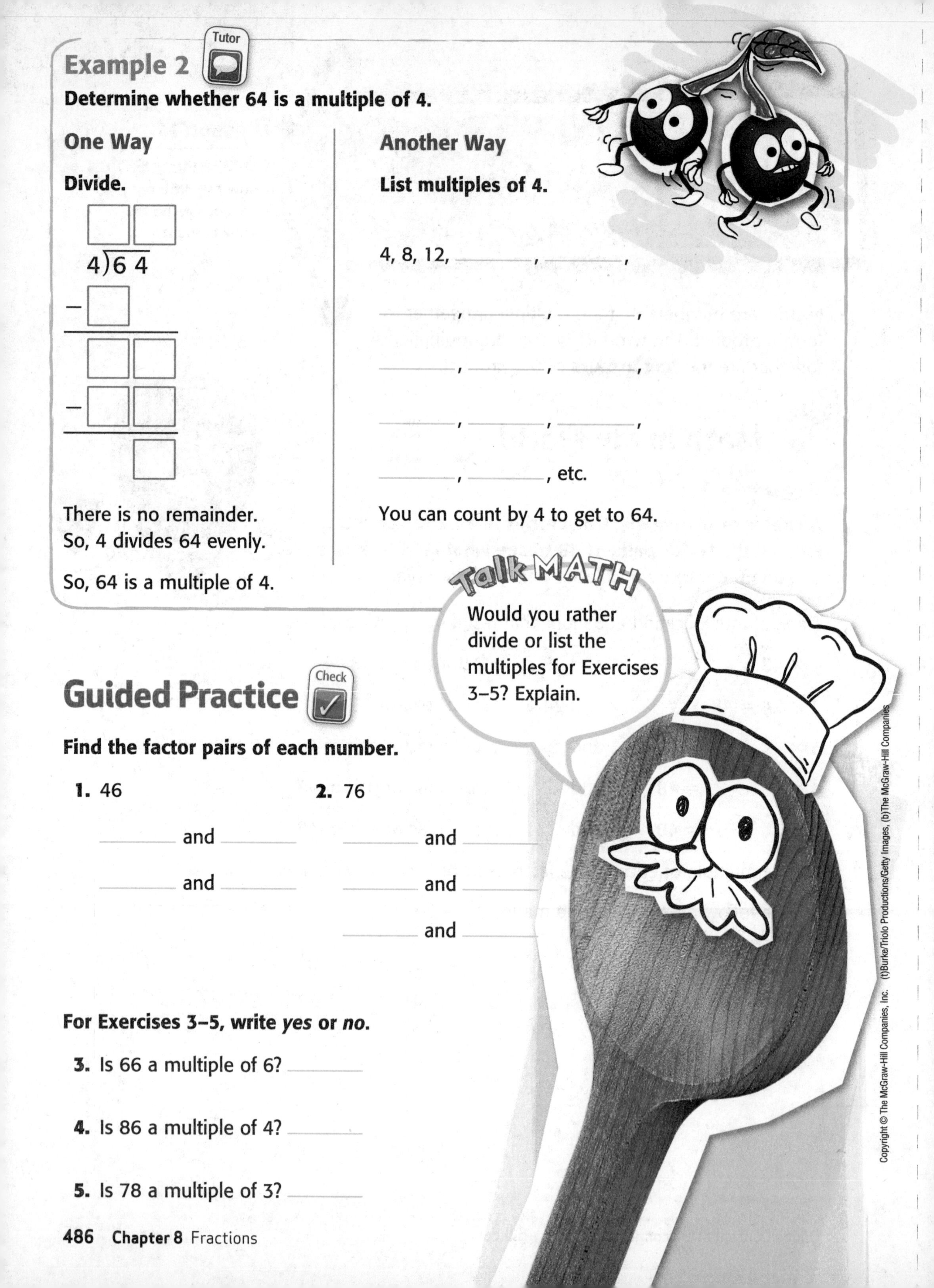

Example 2

Determine whether 64 is a multiple of 4.

One Way

Divide.

$$\begin{array}{r} \square\square \\ 4\overline{)64} \\ -\square \\ \hline \square\square \\ -\square\square \\ \hline \square \end{array}$$

There is no remainder.
So, 4 divides 64 evenly.

Another Way

List multiples of 4.

4, 8, 12, ______, ______,

______, ______, ______,

______, ______, ______,

______, ______, ______,

______, ______, etc.

You can count by 4 to get to 64.

So, 64 is a multiple of 4.

Guided Practice

Find the factor pairs of each number.

1. 46

______ and ______

______ and ______

2. 76

______ and ______

______ and ______

______ and ______

For Exercises 3–5, write *yes* or *no*.

3. Is 66 a multiple of 6? ______

4. Is 86 a multiple of 4? ______

5. Is 78 a multiple of 3? ______

Name ....

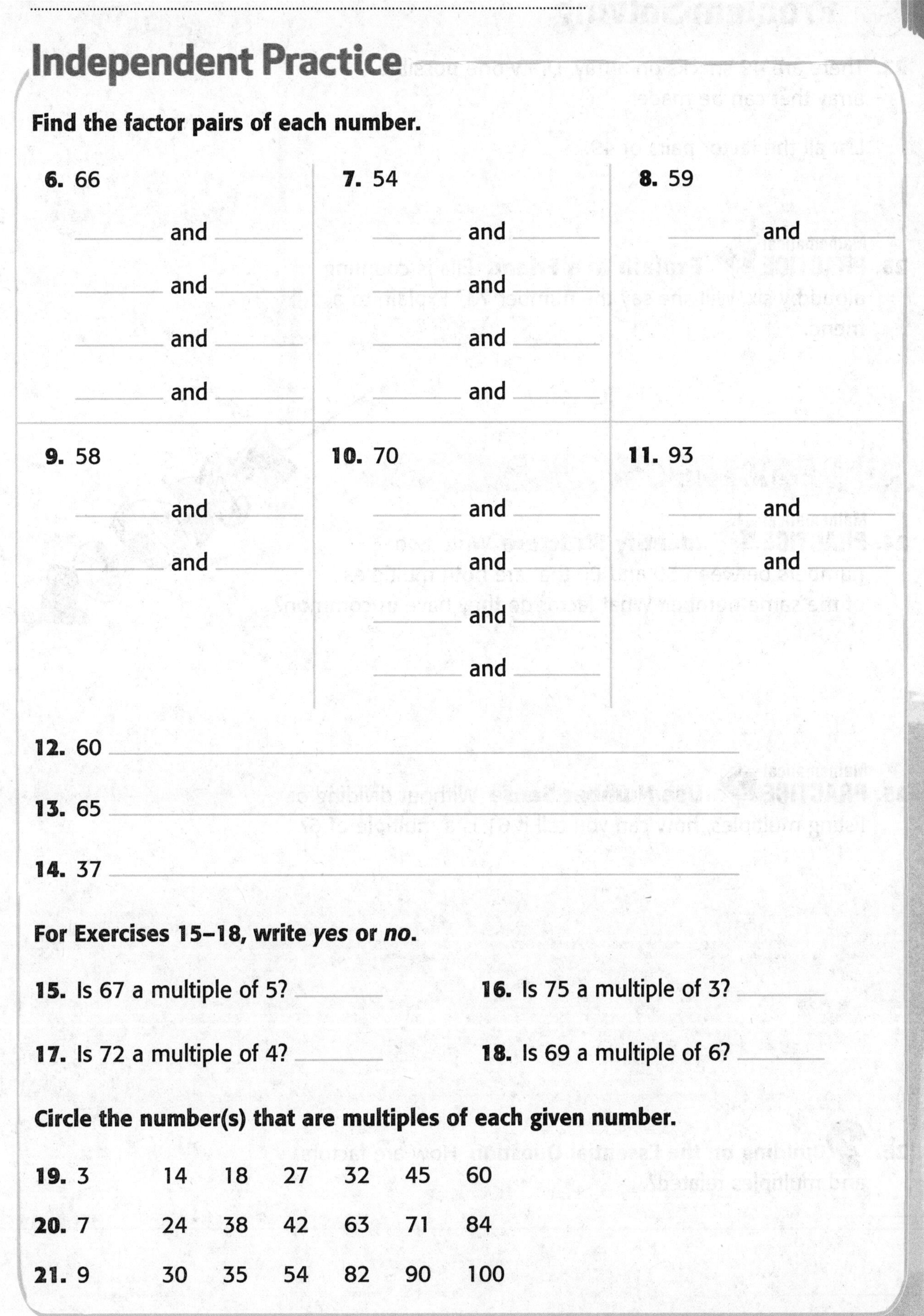

Independent Practice

Find the factor pairs of each number.

6. 66

_____ and _____

_____ and _____

_____ and _____

_____ and _____

7. 54

_____ and _____

_____ and _____

_____ and _____

_____ and _____

8. 59

_____ and _____

9. 58

_____ and _____

_____ and _____

10. 70

_____ and _____

_____ and _____

_____ and _____

_____ and _____

11. 93

_____ and _____

_____ and _____

12. 60 ______________________________

13. 65 ______________________________

14. 37 ______________________________

For Exercises 15–18, write *yes* or *no*.

15. Is 67 a multiple of 5? _____

16. Is 75 a multiple of 3? _____

17. Is 72 a multiple of 4? _____

18. Is 69 a multiple of 6? _____

Circle the number(s) that are multiples of each given number.

19. 3   14   18   27   32   45   60

20. 7   24   38   42   63   71   84

21. 9   30   35   54   82   90   100

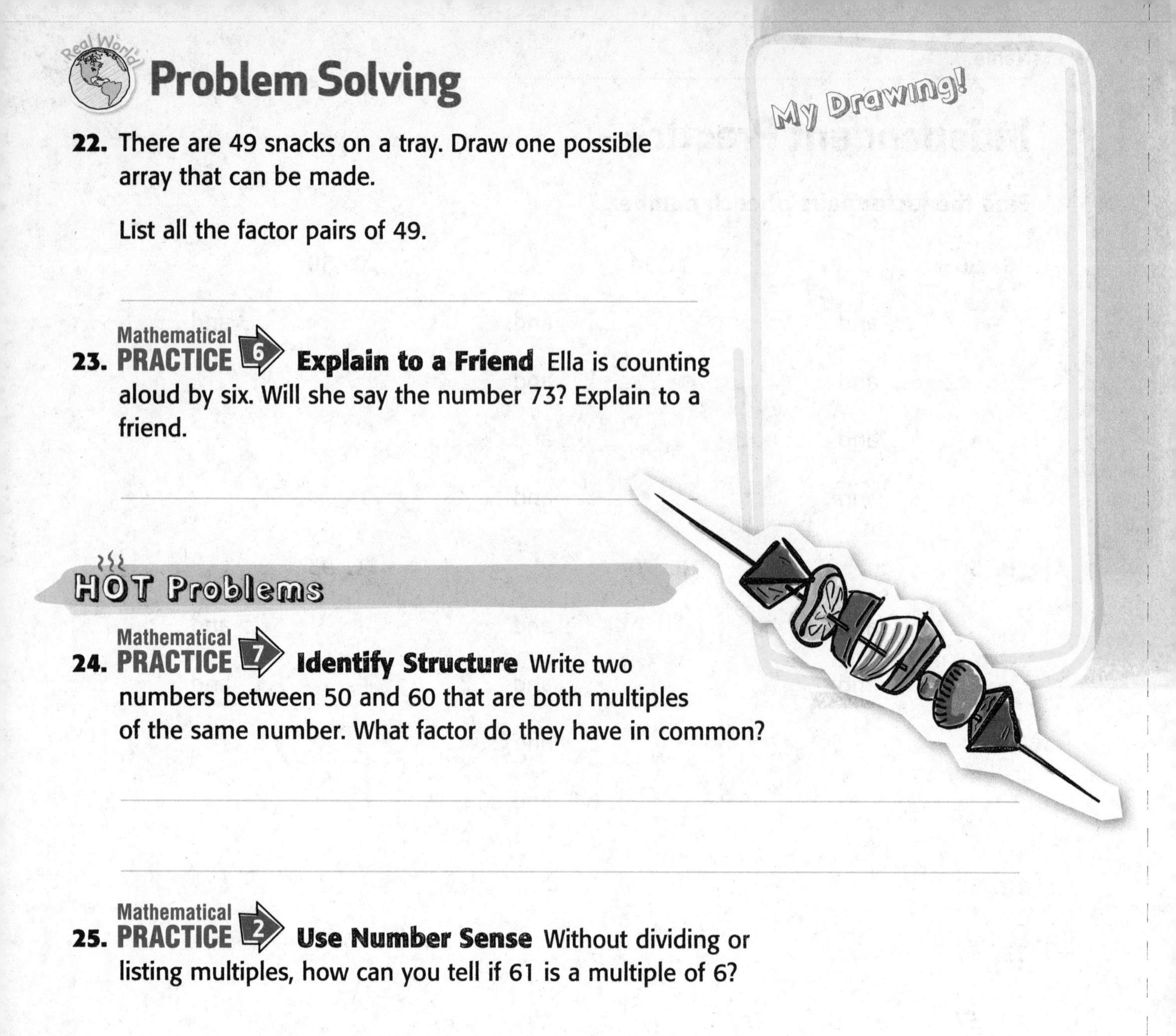

Problem Solving

22. There are 49 snacks on a tray. Draw one possible array that can be made.

List all the factor pairs of 49.

23. Mathematical PRACTICE 6 **Explain to a Friend** Ella is counting aloud by six. Will she say the number 73? Explain to a friend.

HOT Problems

24. Mathematical PRACTICE 7 **Identify Structure** Write two numbers between 50 and 60 that are both multiples of the same number. What factor do they have in common?

25. Mathematical PRACTICE 2 **Use Number Sense** Without dividing or listing multiples, how can you tell if 61 is a multiple of 6?

26. **Building on the Essential Question** How are factors and multiples related?

Name ....

MY Homework

Lesson 1

Factors and Multiples

Homework Helper

eHelp

Need help? connectED.mcgraw-hill.com

Mr. Carlton has 12 pictures to put on the bulletin board in his classroom. List the factor pairs to find how many different ways Mr. Carlton can arrange the pictures in an array.

Think of multiplication equations that result in a product of 12.

$1 \times 12 = 12$ 1 and 12 are a factor pair of 12.

$2 \times 6 = 12$ 2 and 6 are a factor pair of 12.

$3 \times 4 = 12$ 3 and 4 are a factor pair of 12.

So, Mr. Carlton could arrange the 12 pictures in the following arrays: 1×12, 2×6, or 3×4.

Check

Use models to show possible arrays. The arrays can also be turned the other way.

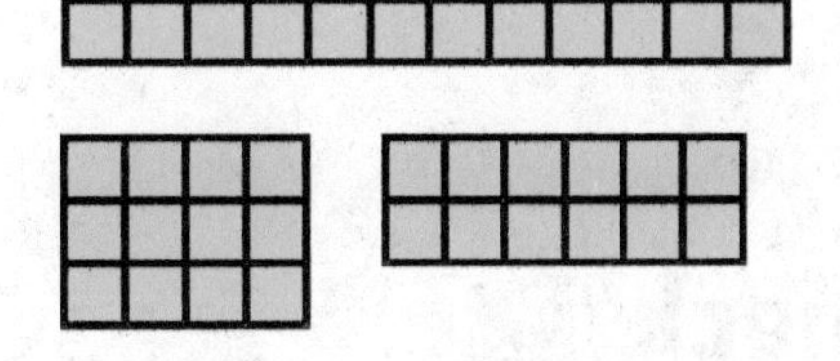

Practice

Find the factor pairs of each number.

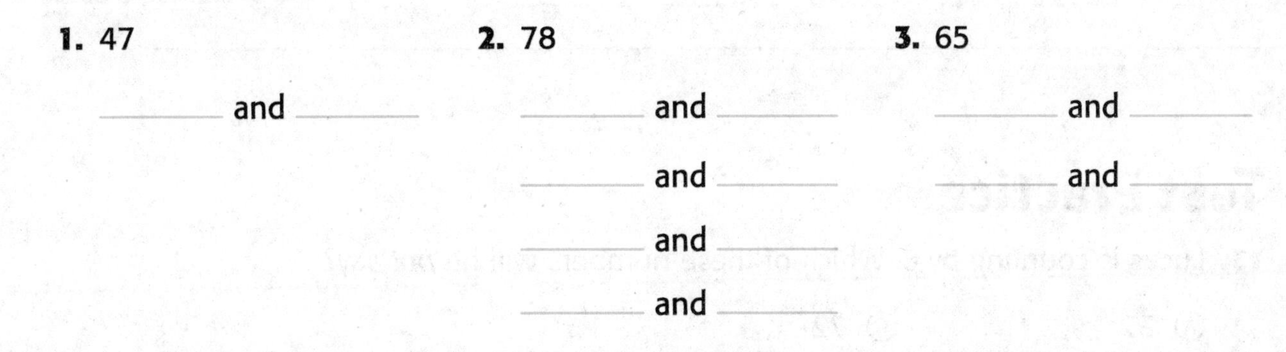

1. 47

_____ and _____

2. 78

_____ and _____

_____ and _____

_____ and _____

_____ and _____

3. 65

_____ and _____

_____ and _____

Find the factor pairs of each number.

4. 56

_____ and _____

_____ and _____

_____ and _____

_____ and _____

5. 30

_____ and _____

_____ and _____

_____ and _____

_____ and _____

6. 71

_____ and _____

For Exercises 7–10, write *yes* or *no*.

7. Is 43 a multiple of 7? _____

8. Is 56 a multiple of 6? _____

9. Is 80 a multiple of 4? _____

10. Is 42 a multiple of 3? _____

Problem Solving

11. Mathematical PRACTICE 4 **Model Math** Janelle is selling lemonade. She has poured 36 cups of lemonade to display. Draw one possible array Janelle could use to display the cups.

My Drawing!

Vocabulary Check

Vocab

12. Write a definition for factor pairs. Then give an example.

_____

_____

Test Practice

13. Lucas is counting by 8. Which of these numbers will he *not* say?

Ⓐ 32

Ⓑ 56

Ⓒ 72

Ⓓ 84

Name ....

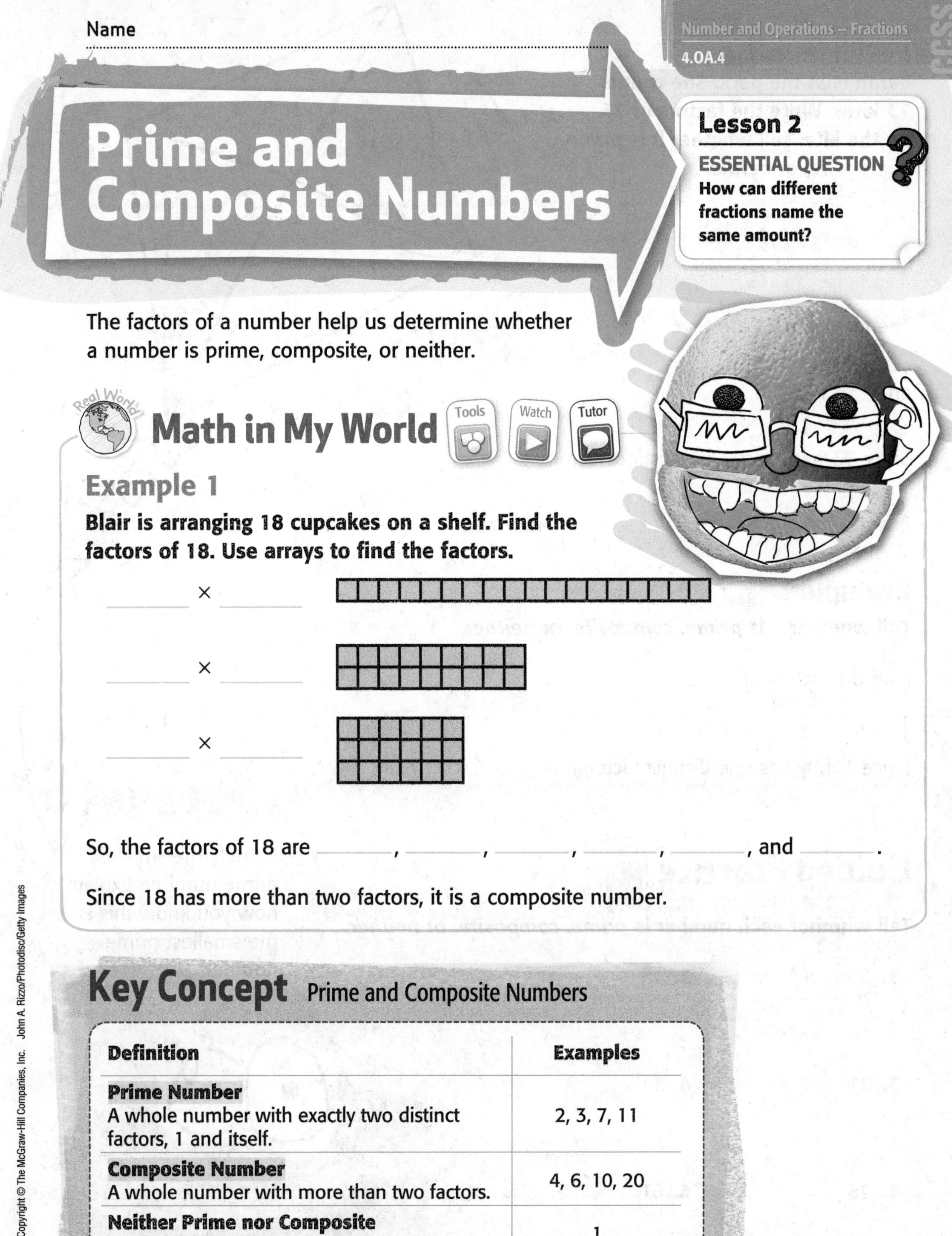

Prime and Composite Numbers

Lesson 2

ESSENTIAL QUESTION
How can different fractions name the same amount?

The factors of a number help us determine whether a number is prime, composite, or neither.

Math in My World

Example 1

Blair is arranging 18 cupcakes on a shelf. Find the factors of 18. Use arrays to find the factors.

_______ × _______

_______ × _______

_______ × _______

So, the factors of 18 are _______, _______, _______, _______, _______, and _______.

Since 18 has more than two factors, it is a composite number.

Key Concept Prime and Composite Numbers

Definition	Examples
Prime Number A whole number with exactly two distinct factors, 1 and itself.	2, 3, 7, 11
Composite Number A whole number with more than two factors.	4, 6, 10, 20
Neither Prime nor Composite 1 only has one distinct factor.	1

Example 2

Sarah is at the park. She counts 73 kites. Write the factors of 73 on the kite. Tell whether it is *prime, composite,* or *neither*.

The factors of 73 are ______

and ______.

Since 73 has exactly two distinct factors,

it is a ______________ number.

Example 3

Tell whether 1 is *prime, composite,* or *neither*.

Find the factors of 1.

1×1

Since 1 only has one distinct factor, it is ______________.

Talk MATH

Identify the smallest prime number. Explain how you know this is the smallest prime number.

Guided Practice

Tell whether each number is *prime, composite,* or *neither*.

1. 5 ______

2. 15 ______

3. 21 ______

4. 31 ______

5. 26 ______

6. 61 ______

Name

Independent Practice

Tell whether each number is *prime, composite,* or *neither.*

7. 1 ____

8. 3 ____

9. 4 ____

10. 14 ____

11. 29 ____

12. 41 ____

13. 50 ____

14. 63 ____

15. 65 ____

16. 79 ____

17. 84 ____

18. 97 ____

19. Circle the prime numbers. Cross out the numbers that are composite or neither.

1	2	3	4	5	6	7	8	9	10
11	12	13	14	15	16	17	18	19	20
21	22	23	24	25	26	27	28	29	30
31	32	33	34	35	36	37	38	39	40
41	42	43	44	45	46	47	48	49	50

20. Write a prime number that is greater than 50.

____

21. Write a composite number that is greater than 70.

____

Problem Solving

22. What prime number is greater than 88 and less than 95?

23. Ken is planting vegetables in his garden. He has 20 seeds. Determine whether 20 is a prime or composite number. If it is composite, list all of the ways Ken can arrange the seeds in even rows.

24. Mathematical PRACTICE 4 **Model Math** Susana is making a quilt by sewing together square pieces of fabric. She has 36 fabric squares. Draw an array to show how she can create a quilt that has the same number of squares in each row and each column.

My Drawing!

Is 36 prime or composite? ____________

HOT Problems

25. Mathematical PRACTICE 3 **Draw a Conclusion** The numbers 17, 31, and 37 are prime numbers. Reversing the order of the digits to make 71, 13, and 73 also result in prime numbers. Does reversing the order of the digits of a 2-digit prime number always result in a prime number? Explain.

26. **Building on the Essential Question** How are factors related to prime numbers?

MY Homework

Lesson 2
Prime and Composite Numbers

Homework Helper

eHelp Need help? connectED.mcgraw-hill.com

Patrice is having a tea party. There will be 13 people at the tea party altogether. Can Patrice divide the chairs evenly among more than 1 table? Explain.

Find the factors of 13 and decide if 13 is a prime number, a composite number, or neither.

The factors of 13 are 1 and 13. So, 13 is a prime number.

Patrice cannot divide the chairs evenly among more than 1 table because 13 is a prime number.

1

13

Type of Number	Definition
prime number	a whole number with exactly two factors, 1 and itself (Examples: 17, 29, 41)
composite number	a whole number with more than two factors (Examples: 8, 30, 56)
neither prime nor composite	a number that has only one distinct factor (Example: 1)

Practice

Tell whether each number is *prime, composite,* or *neither*.

1. 16 ________

2. 37 ________

3. 50 ________

4. 41 ________

5. 1 ________

6. 81 ________

Tell whether each number is *prime, composite,* or *neither*.

7. 0 ______ **8.** 11 ______ **9.** 90 ______

10. 75 ______ **11.** 53 ______ **12.** 23 ______

Problem Solving

13. Colby has 16 jars of spices. He wants to arrange them in arrays. What arrays could he use to arrange them?

______

14. Winnie has 7 soccer trophies she wants to display in an array. How many different arrays are possible? Explain.

______

15. Mathematical PRACTICE 1 **Keep Trying** Identify the two prime numbers that are greater than 25 and less than 35.

______

16. Identify two composite numbers that each have 8 as a factor.

______

Vocabulary Check

Vocab

Draw a line to match the vocabulary term with its example.

17. prime number • 61

18. composite number • 21

Test Practice

19. Which of the following is a prime number?

Ⓐ 67 Ⓒ 63

Ⓑ 65 Ⓓ 60

Check My Progress

Vocabulary Check

Vocab

1. Circle the **prime** numbers. Draw an X on the **composite** numbers.

4	7	5	9	11	8	6	14

For Exercises 2–5, write factor, multiple, or factor pairs.

2. 35 is a ____________ of 7.

3. 24 and 3 are ____________ of 72.

4. 6 is a ____________ of 66.

5. 90 is a ____________ of 9.

Concept Check

Check

6. Find the factor pairs of 80.

____

For Exercises 7–9, write *yes* or *no*.

7. Is 50 a multiple of 5? ____________

8. Is 63 a multiple of 4? ____________

9. Is 72 a multiple of 3? ____________

Tell whether each number is *prime*, *composite*, or *neither*.

10. 4 ____________

11. 3 ____________

12. 8 ____________

13. 1 ____________

Real World!

Problem Solving

14. Mandy is thinking of a prime number that is greater than 41 and less than 47. What is Mandy's number?

15. There are ten desks in a room. Determine whether 10 is prime or composite. If it is composite, list all of the ways the desks can be arranged if they are in even rows.

16. Is there a way to place 29 books on shelves so that each shelf has the same number of books, with more than one book on each shelf? Explain.

17. There are 48 cans of soup. Can an array be made with 5 equal rows? Explain.

My Work!

Test Practice

18. Which number is prime?

Ⓐ 4

Ⓑ 5

Ⓒ 6

Ⓓ 9

Name ....

Hands On
Model Equivalent Fractions

Lesson 3

ESSENTIAL QUESTION
How can different fractions name the same amount?

The top number on a fraction is the **numerator**.
The bottom number on a fraction is the **denominator**.
Fractions that represent the same part of a number are **equivalent fractions**.

Build It

Generate two fractions that are equivalent to $\frac{1}{3}$.

1 Model $\frac{1}{3}$.

Place a $\frac{1}{3}$ - tile.

2 Find a fraction equivalent to $\frac{1}{3}$.

Place $\frac{1}{6}$ - tiles below the $\frac{1}{3}$ - tile to equal the length of the $\frac{1}{3}$ - tile.

How many $\frac{1}{6}$ - tiles did you place? ______

So, $\frac{1}{3}$ and $\frac{2}{6}$ are equivalent fractions.

3 Find another fraction equivalent to $\frac{1}{3}$.

Place $\frac{1}{12}$ - tiles below the $\frac{1}{6}$ - tiles to equal the length of the $\frac{1}{3}$ - tile.

How many $\frac{1}{12}$ - tiles did you place? ______

So, $\frac{1}{3}$ and $\frac{4}{12}$ are equivalent fractions.

So, $\frac{1}{3}$, $\frac{\square}{6}$, and $\frac{\square}{12}$ are equivalent fractions.

Try It

Generate two fractions that are equivalent to $\frac{1}{4}$.

1 The first number line is divided into fourths. Plot $\frac{1}{4}$ on the number line.

2 The second number line is divided into eighths.

What fraction is at the same location as $\frac{1}{4}$? $\frac{\square}{\square}$

Plot this fraction on the number line.

3 The third number line is divided into twelfths.

What fraction is at the same location as $\frac{1}{4}$? $\frac{\square}{\square}$

Plot this fraction on the number line.

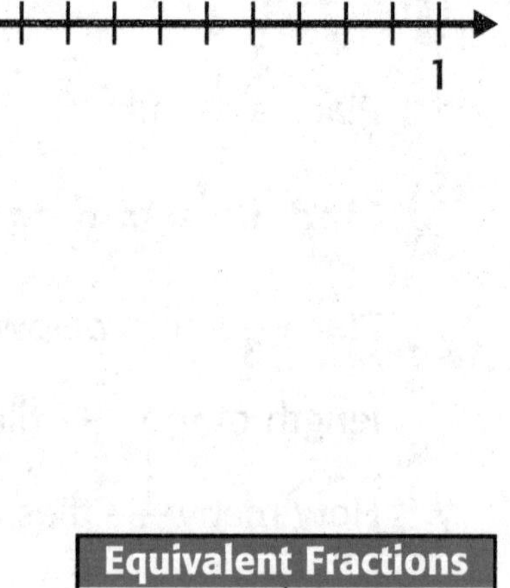

So, two fractions that are equivalent to $\frac{1}{4}$ are $\frac{\square}{\square}$ and $\frac{\square}{\square}$.

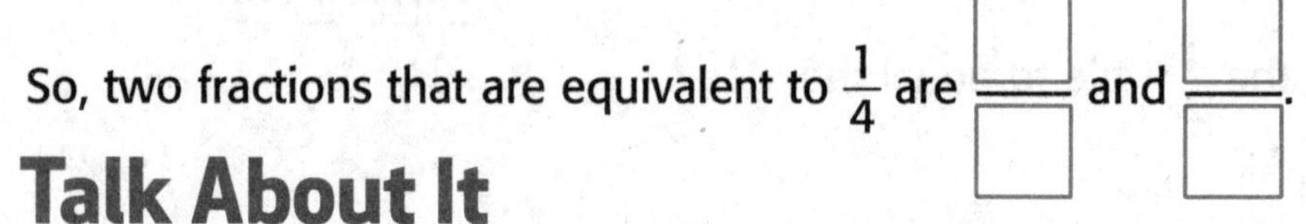

Talk About It

1. **Mathematical PRACTICE 8 Look for a Pattern** The table shows some equivalent fractions. Study the table. Describe the pattern between the numerators and denominators of two equivalent fractions.

Equivalent Fractions	
$\frac{1}{3}$	$\frac{2}{6}$
$\frac{1}{3}$	$\frac{4}{12}$
$\frac{1}{4}$	$\frac{2}{8}$
$\frac{1}{4}$	$\frac{3}{12}$

2. **Mathematical PRACTICE 3 Draw a Conclusion** Without using fraction tiles or number lines, determine whether $\frac{1}{2}$ and $\frac{3}{6}$ are equivalent fractions. Explain.

Name ....

Practice It

Recognize whether the fractions are equivalent. Write *yes* or *no*. Use fraction tiles or number lines.

3. $\frac{2}{4}$ and $\frac{6}{12}$ ________

4. $\frac{6}{8}$ and $\frac{5}{10}$ ________

5. $\frac{2}{3}$ and $\frac{3}{5}$ ________

6. $\frac{9}{12}$ and $\frac{3}{4}$ ________

7. $\frac{4}{6}$ and $\frac{8}{12}$ ________

8. $\frac{2}{3}$ and $\frac{6}{10}$ ________

Generate two equivalent fractions for each fraction. Use fraction tiles or number lines.

9. $\frac{2}{4}$ ____________

10. $\frac{2}{6}$ ____________

11. $\frac{4}{8}$ ____________

12. $\frac{5}{10}$ ____________

13. $\frac{1}{3}$ ____________

14. $\frac{2}{3}$ ____________

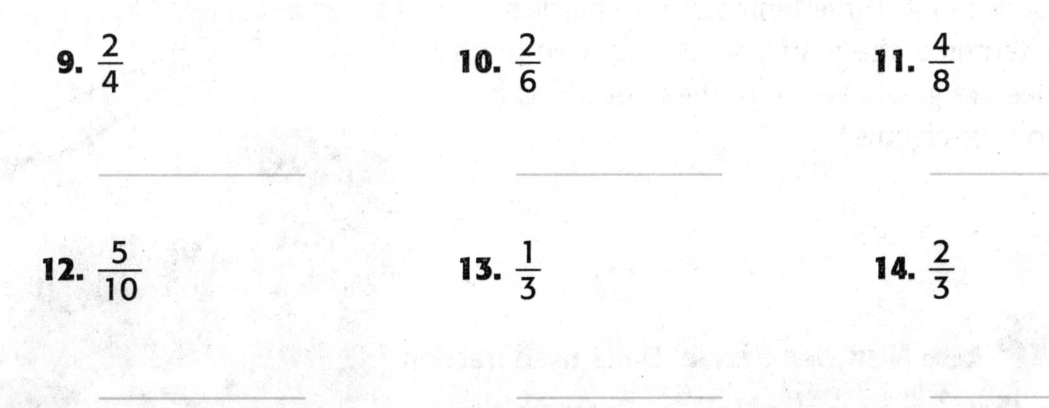

My Work!

Apply It

15. Mathematical PRACTICE 4 **Model Math** There were 10 baked goods in a basket. Four of them were sold. Write a fraction to show the part of the baked goods that were not sold. Then write an equivalent fraction to this number.

My Work!

16. Two-thirds of a jar of peanut butter has been used. Write an equivalent fraction.

17. A jar has marbles in it. Three-tenths of the marbles are red. Five-tenths of the marbles are blue. Two-tenths of the marbles are green. Which of these fractions is equivalent to four-eighths?

18. Mathematical PRACTICE 2 **Use Number Sense** Daria used fraction tiles to show that $\frac{3}{5}$ is equivalent to $\frac{6}{10}$. Compare the number and size of fraction tiles needed to model each fraction.

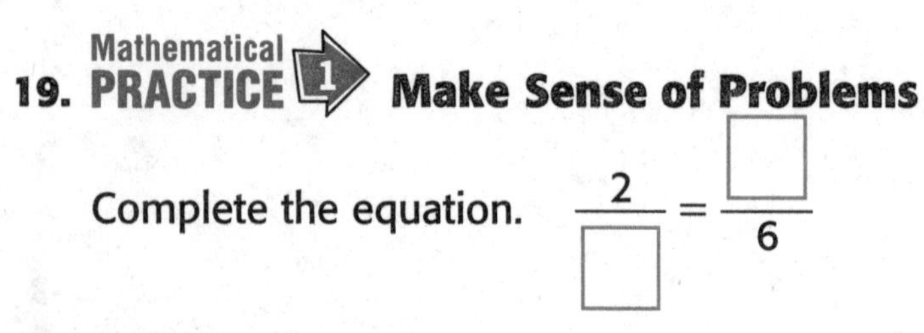

19. Mathematical PRACTICE 1 **Make Sense of Problems**

Complete the equation. $\frac{2}{\square} = \frac{\square}{6}$

Write About It

20. Write a real-world example of equivalent fractions.

Name ....

Lesson 3

Hands On: Model Equivalent Fractions

Homework Helper

eHelp Need help? connectED.mcgraw-hill.com

Determine whether $\frac{1}{2}$ is equivalent to $\frac{3}{6}$.

One Way **Use fraction tiles.**

1. **Model $\frac{1}{2}$.**

2. **Model $\frac{3}{6}$.**

   Line up three $\frac{1}{6}$-fraction tiles below the $\frac{1}{2}$-fraction tile.

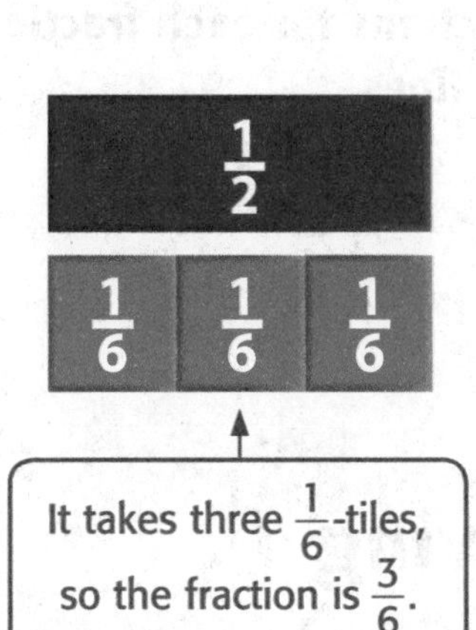

It takes three $\frac{1}{6}$-tiles, so the fraction is $\frac{3}{6}$.

Since they are the same length, the fractions are equivalent.

So, $\frac{1}{2} = \frac{3}{6}$.

Another Way **Use number lines.**

1. Divide the first number line into halves.

2. Divide the second number line into sixths.

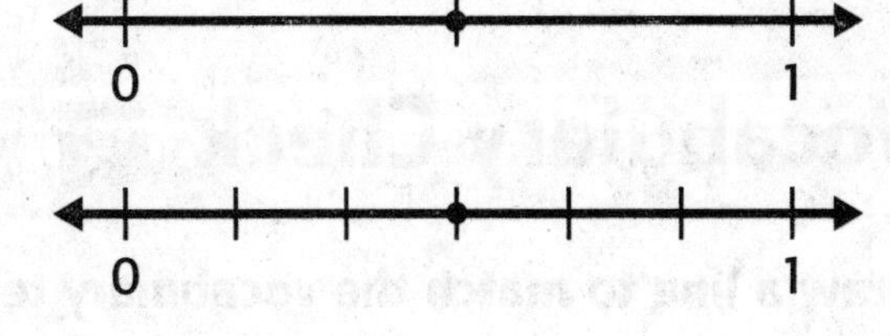

3. Count the number of sixths that are in one half.

The number lines show that $\frac{1}{2}$ and $\frac{3}{6}$ are at the same point.

So, they are equivalent fractions.

Practice

Recognize whether the fractions are equivalent. Write *yes* or *no*. Use fraction tiles or number lines.

1. $\frac{3}{5}$ and $\frac{6}{8}$ ________

2. $\frac{4}{5}$ and $\frac{5}{6}$ ________

3. $\frac{2}{4}$ and $\frac{6}{12}$ ________

4. $\frac{2}{3}$ and $\frac{4}{6}$ ________

5. $\frac{8}{12}$ and $\frac{4}{6}$ ________

6. $\frac{5}{6}$ and $\frac{9}{10}$ ________

Generate two equivalent fractions for each fraction. Use fraction tiles or number lines.

7. $\frac{1}{3}$ ________

8. $\frac{8}{12}$ ________

9. $\frac{3}{4}$ ________

Problem Solving

10. Mathematical PRACTICE 3 **Justify Conclusions** Francie lives $\frac{1}{5}$ mile from the school. Jake lives $\frac{2}{10}$ mile from the school. Do they live the same distance from the school? Explain.

____

____

Vocabulary Check

Vocab

Draw a line to match the vocabulary term with its example.

11. numerator

12. denominator

13. equivalent fractions

- $\frac{6}{10}$ and $\frac{3}{5}$
- the number 1 in $\frac{1}{4}$
- the number 4 in $\frac{1}{4}$

Equivalent Fractions

Lesson 4

ESSENTIAL QUESTION
How can different fractions name the same amount?

You have used models and number lines to find equivalent fractions. Multiplication can also be used to find equivalent fractions. Fractions that have the same numerator and denominator are equivalent to one whole.

Math in My World

Tools Watch Tutor

Example 1

A recipe for spaghetti and meatballs calls for $\frac{3}{4}$ pound of ground beef. Find two fractions that are equivalent to $\frac{3}{4}$.

One Way Use models.

Model $\frac{3}{4}$.

Place six $\frac{1}{8}$ - tiles below the $\frac{1}{4}$ - tiles. Place nine $\frac{1}{12}$ - tiles below the $\frac{1}{8}$ - tiles.

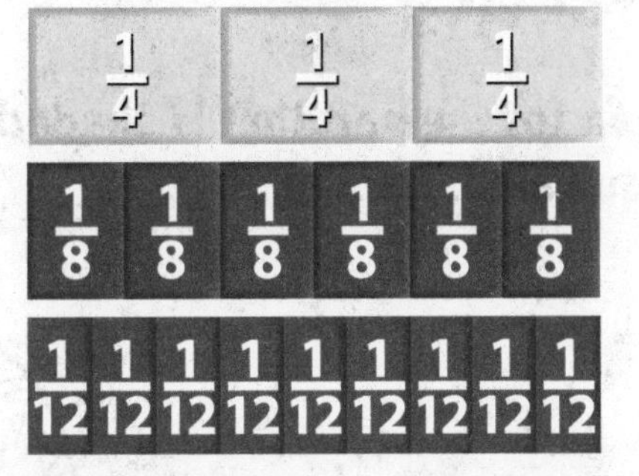

The $\frac{1}{8}$ - tiles are smaller than the $\frac{1}{4}$ - tiles, so there are more of them.

The $\frac{1}{12}$ - tiles are smaller than the $\frac{1}{4}$ - tiles and $\frac{1}{8}$ - tiles, so there are more of them.

Another Way Use multiplication.

Multiply the numerator and denominator by the same number.

$\frac{3}{4} = \frac{3 \times 2}{4 \times 2} = \frac{6}{8}$ $\frac{3}{4} = \frac{3 \times 3}{4 \times 3} = \frac{9}{12}$

So, two fractions that are equivalent to $\frac{3}{4}$ are $\frac{6}{8}$ and $\frac{9}{12}$.

You can use multiplication to write a fraction with a denominator of 10 as an equivalent fraction with a denominator of 100.

Example 2

Liliana has a container of dry beans. The table shows the fraction of each type of bean. Which type of bean shows a fraction that is equivalent to $\frac{3}{10}$ of the container?

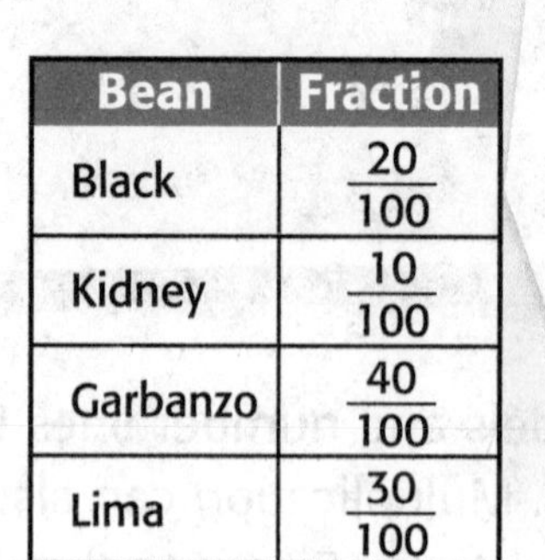

Bean	Fraction
Black	$\frac{20}{100}$
Kidney	$\frac{10}{100}$
Garbanzo	$\frac{40}{100}$
Lima	$\frac{30}{100}$

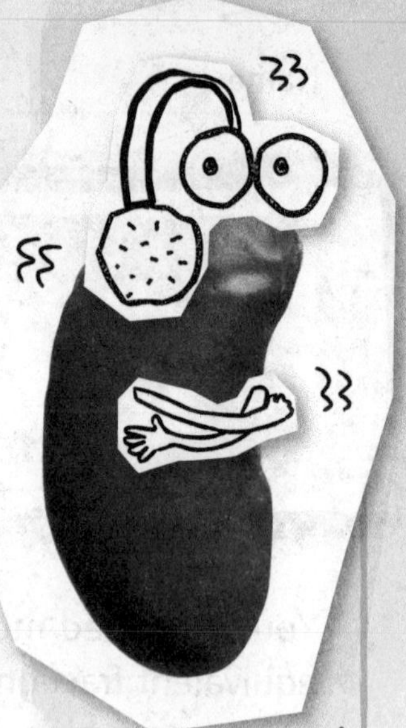

Each fraction in the table has a denominator of 100. Use multiplication to find which fraction is equivalent to $\frac{3}{10}$.

Multiply.

$$\frac{3}{10} = \frac{3 \times 10}{10 \times 10} = \frac{\square}{\square}$$

So, $\frac{3}{10}$ of the container has ______ beans.

Talk MATH

Tell why $\frac{3}{4}$, $\frac{6}{8}$, and $\frac{9}{12}$ are equivalent fractions. Give an example of another set of three equivalent fractions.

Guided Practice

Write the fraction for the part that is shaded. Then find an equivalent fraction.

1.

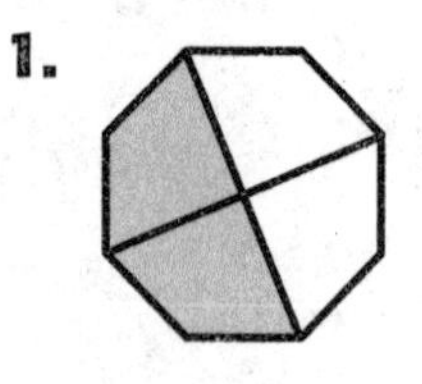

______

______

2.

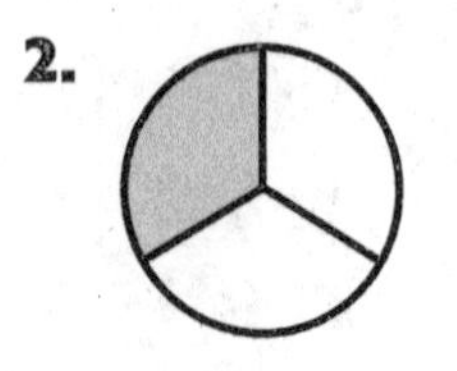

______

______

Name

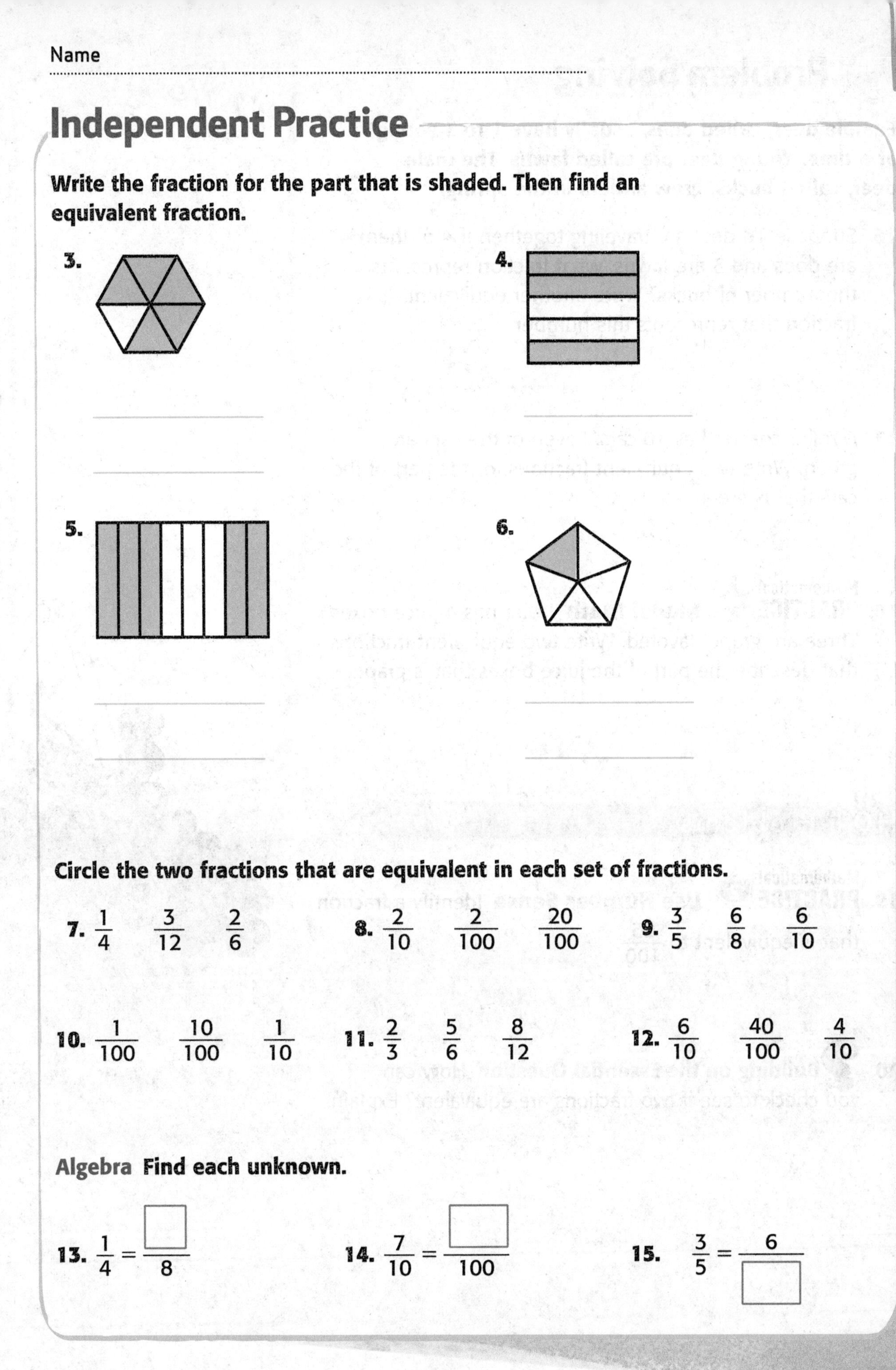

Independent Practice

Write the fraction for the part that is shaded. Then find an equivalent fraction.

3.

4.

5.

6.

Circle the two fractions that are equivalent in each set of fractions.

7. $\frac{1}{4}$ $\frac{3}{12}$ $\frac{2}{6}$

8. $\frac{2}{10}$ $\frac{2}{100}$ $\frac{20}{100}$

9. $\frac{3}{5}$ $\frac{6}{8}$ $\frac{6}{10}$

10. $\frac{1}{100}$ $\frac{10}{100}$ $\frac{1}{10}$

11. $\frac{2}{3}$ $\frac{5}{6}$ $\frac{8}{12}$

12. $\frac{6}{10}$ $\frac{40}{100}$ $\frac{4}{10}$

Algebra **Find each unknown.**

13. $\frac{1}{4} = \frac{\square}{8}$

14. $\frac{7}{10} = \frac{\square}{100}$

15. $\frac{3}{5} = \frac{6}{\square}$

Problem Solving

Female deer, called does, usually have 1 to 3 young at a time. Young deer are called fawns. The male deer, called bucks, grow antlers in the spring.

16. Suppose 10 deer are traveling together. If 4 of them are does and 3 are fawns, what fraction represents the number of bucks? Write another equivalent fraction that represents this number.

17. A roller coaster has 10 cars. Seven of the cars are green. Write two equivalent fractions for the part of the cars that is green.

18. Mathematical PRACTICE 4 **Model Math** Javier has 4 juice boxes. Three are grape flavored. Write two equivalent fractions that describe the part of the juice boxes that is grape.

My Work!

HOT Problems

19. Mathematical PRACTICE 2 **Use Number Sense** Identify a fraction that is equivalent to $\frac{25}{100}$.

20. **Building on the Essential Question** How can you check to see if two fractions are equivalent? Explain.

Name

MY Homework

Lesson 4
Equivalent Fractions

Homework Helper

eHelp

Need help? connectED.mcgraw-hill.com

Write the fraction for the part that is shaded. Then find two equivalent fractions.

Find the fraction that represents the shaded part.

4 ← number of shaded parts

8 ← total number of parts

Find equivalent fractions.

Multiply the numerator and denominator by the same number, for example, 2.

$$\frac{4 \times 2}{8 \times 2} = \frac{8}{16}$$

Multiply the numerator and denominator by another number, for example, 3.

$$\frac{4 \times 3}{8 \times 3} = \frac{12}{24}$$

So, the fraction represented by the circle is $\frac{4}{8}$.

Two equivalent fractions are $\frac{8}{16}$ and $\frac{12}{24}$.

Practice

Write the fraction for the part that is shaded. Then find an equivalent fraction.

1. 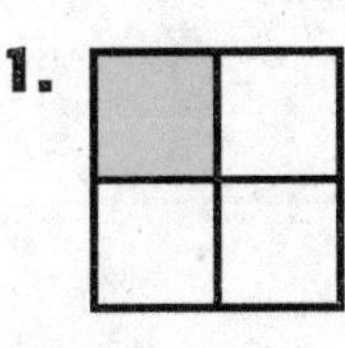

______

______

2. 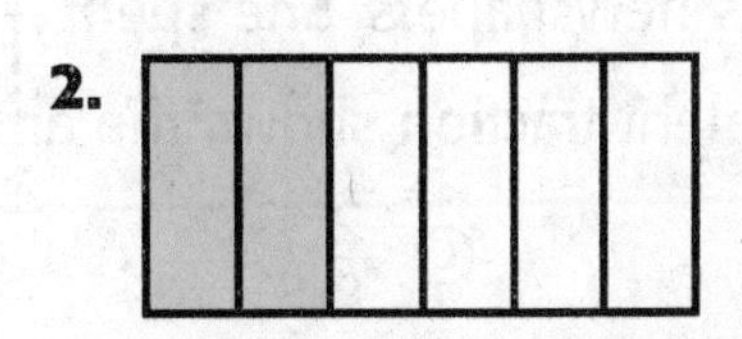

______

______

Write the fraction for the part that is shaded. Then find an equivalent fraction.

3.

4.

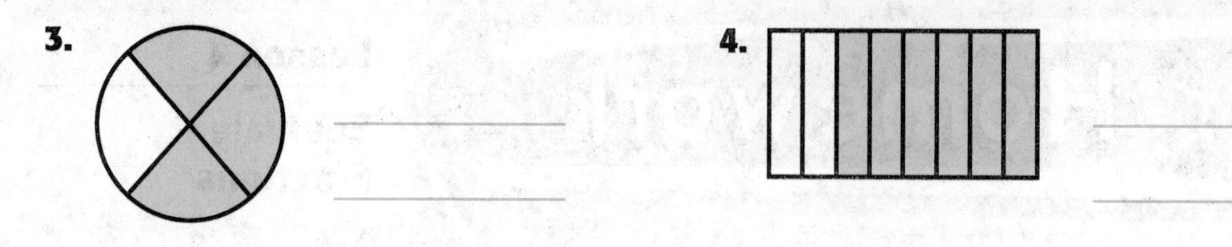

Find an equivalent fraction for each fraction.

5. $\frac{20}{100}$ ______ 6. $\frac{2}{8}$ ______ 7. $\frac{90}{100}$ ______

Algebra Find each unknown.

8. $\frac{6}{12} = \frac{x}{2}$

$x =$ ______

9. $\frac{3}{10} = \frac{x}{100}$

$x =$ ______

10. $\frac{5}{8} = \frac{10}{x}$

$x =$ ______

Problem Solving

11. **Mathematical PRACTICE 2 Use Number Sense** Janie has 4 pieces of fruit. Two of the pieces of fruit are bananas. Write two fractions that describe the fraction of fruit that are bananas.

______

12. A box contains 4 red pencils and 6 black pencils. What fraction of the pencils are red? Write two equivalent fractions.

______

Test Practice

13. Laura delivers newspapers. She spent $\frac{4}{12}$ of her savings on a new CD. Which equivalent fraction shows the amount Laura spent?

Ⓐ $\frac{1}{9}$

Ⓑ $\frac{1}{3}$

Ⓒ $\frac{2}{8}$

Ⓓ $\frac{2}{3}$

Simplest Form

Lesson 5

ESSENTIAL QUESTION
How can different fractions name the same amount?

A fraction is in **simplest form** when its numerator and denominator have no common factor other than 1. The simplest form of a fraction is equivalent to the fraction.

Math in My World

Watch Tutor

Example 1

There are 12 ingredients in a bread recipe. Esteban needs to buy 8 out of 12, or $\frac{8}{12}$, of the ingredients. Write $\frac{8}{12}$ in simplest form.

To write a fraction in simplest form, divide both the numerator and the denominator by the greatest common factor. The **greatest common factor** is the greatest of the common factors of two or more numbers.

1 Find the common factors of 8 and 12.

Factors of 8: 1, 2, 4, 8

Factors of 12: 1, 2, 3, 4, 6, 12

Common factors: ___, ___, ___

Circle the greatest common factor.

2 Divide by the greatest common factor.

$$\frac{8 \div 4}{12 \div 4} = \frac{\square}{\square}$$

The numbers 2 and 3 have no common factor other than 1.

So, $\frac{8}{12}$ in simplest form is $\frac{\square}{\square}$.

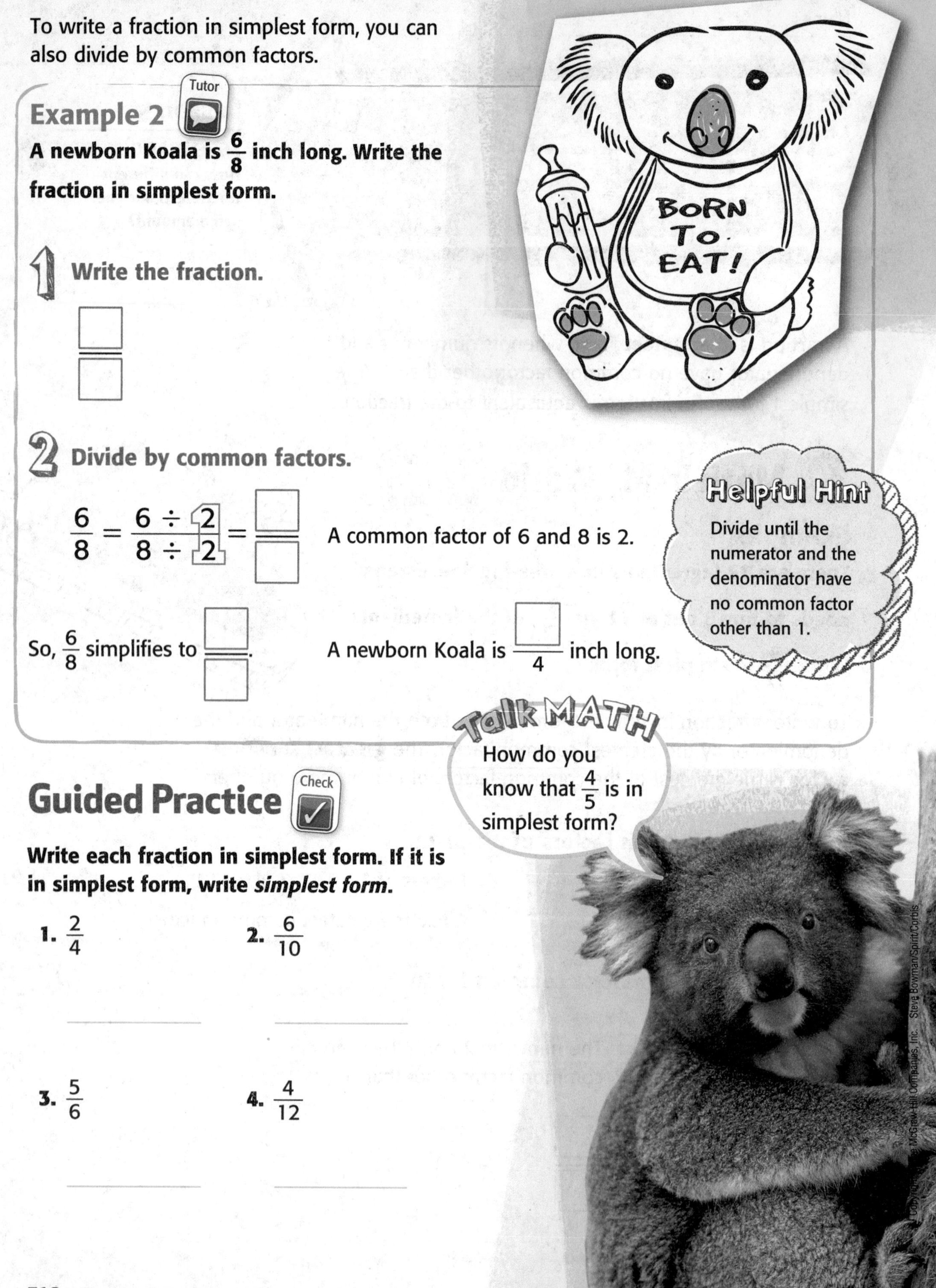

To write a fraction in simplest form, you can also divide by common factors.

Example 2

Tutor

A newborn Koala is $\frac{6}{8}$ inch long. Write the fraction in simplest form.

1 Write the fraction.

$$\frac{\square}{\square}$$

2 Divide by common factors.

$$\frac{6}{8} = \frac{6 \div 2}{8 \div 2} = \frac{\square}{\square}$$

A common factor of 6 and 8 is 2.

So, $\frac{6}{8}$ simplifies to $\frac{\square}{\square}$.

A newborn Koala is $\frac{\square}{4}$ inch long.

Helpful Hint

Divide until the numerator and the denominator have no common factor other than 1.

Talk MATH

How do you know that $\frac{4}{5}$ is in simplest form?

Guided Practice

Check

Write each fraction in simplest form. If it is in simplest form, write *simplest form*.

1. $\frac{2}{4}$ ________

2. $\frac{6}{10}$ ________

3. $\frac{5}{6}$ ________

4. $\frac{4}{12}$ ________

Name

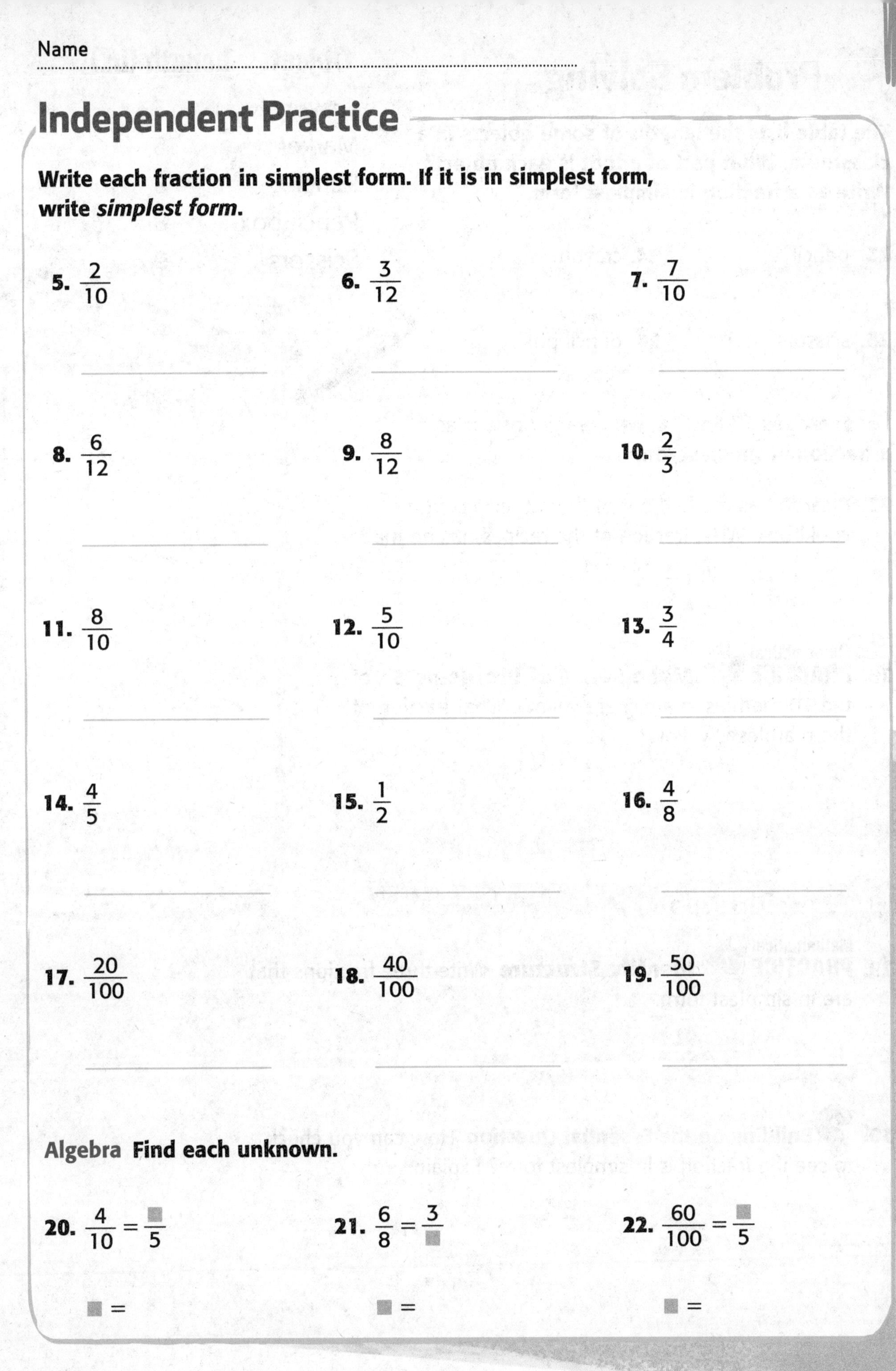

Independent Practice

Write each fraction in simplest form. If it is in simplest form, write *simplest form*.

5. $\frac{2}{10}$ ______

6. $\frac{3}{12}$ ______

7. $\frac{7}{10}$ ______

8. $\frac{6}{12}$ ______

9. $\frac{8}{12}$ ______

10. $\frac{2}{3}$ ______

11. $\frac{8}{10}$ ______

12. $\frac{5}{10}$ ______

13. $\frac{3}{4}$ ______

14. $\frac{4}{5}$ ______

15. $\frac{1}{2}$ ______

16. $\frac{4}{8}$ ______

17. $\frac{20}{100}$ ______

18. $\frac{40}{100}$ ______

19. $\frac{50}{100}$ ______

Algebra **Find each unknown.**

20. $\frac{4}{10} = \frac{■}{5}$

■ =

21. $\frac{6}{8} = \frac{3}{■}$

■ =

22. $\frac{60}{100} = \frac{■}{5}$

■ =

Problem Solving

The table lists the lengths of some objects in a classroom. What part of a foot is each object? Write as a fraction in simplest form.

Object	Length (in.)
Crayon	3
Marker	5
Pencil	6
Pencil box	8
Scissors	9

23. pencil ________ **24.** crayon ________

25. scissors ________ **26.** pencil box ________

My Work!

For Exercises 27 and 28, write each answer as a fraction in simplest form.

27. Ricardo has made 4 out of the 12 recipes in his cookbook. What fraction of the recipes has he made?

____

28. Mathematical PRACTICE 1 **Make Sense of Problems** Six of the 10 marbles in a bag are yellow. What fraction of the marbles is yellow?

____

HOT Problems

29. Mathematical PRACTICE 7 **Identify Structure** Write three fractions that are in simplest form.

____

30. **Building on the Essential Question** How can you check to see if a fraction is in simplest form? Explain.

____

____

____

Name ....

MY Homework

Lesson 5
Simplest Form

Homework Helper

Need help? connectED.mcgraw-hill.com

Sophie found 12 golf balls. Six of them are yellow. What fraction of the golf balls are yellow? Write the fraction in its simplest form.

1 Write the fraction.

6 out of 12 golf balls are yellow.

The fraction is $\frac{6}{12}$.

Helpful Hint

A fraction is in its simplest form when its numerator and denominator have no common factor other than 1.

2 Find the common factors of 6 and 12.
Circle the greatest common factor.
Factors of 6: 1, 2, 3, ⑥
Factors of 12: 1, 2, 3, 4, ⑥, 12

3 Divide by the greatest common factor.

$$\frac{6 \div 6}{12 \div 6} = \frac{1}{2}$$

So, $\frac{1}{2}$ of the golf balls are yellow.

Practice

Write each fraction in simplest form. If it is in simplest form, write *simplest form.*

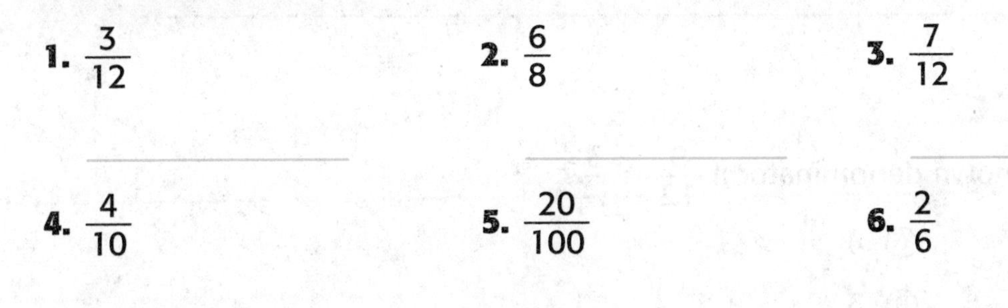

1. $\frac{3}{12}$ ________

2. $\frac{6}{8}$ ________

3. $\frac{7}{12}$ ________

4. $\frac{4}{10}$ ________

5. $\frac{20}{100}$ ________

6. $\frac{2}{6}$ ________

Write each fraction in simplest form. If it is in simplest form, write *simplest form*.

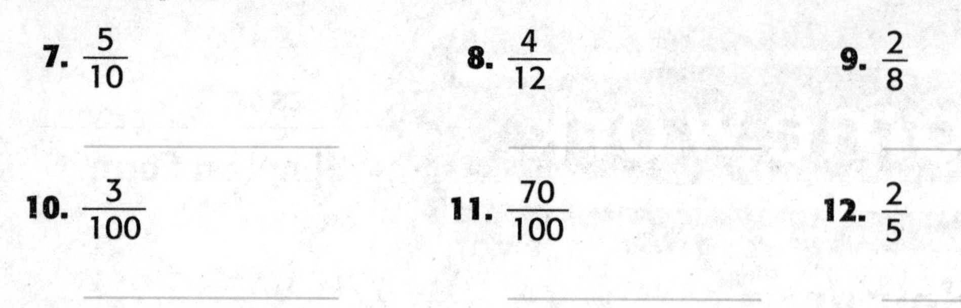

7. $\frac{5}{10}$ ________

8. $\frac{4}{12}$ ________

9. $\frac{2}{8}$ ________

10. $\frac{3}{100}$ ________

11. $\frac{70}{100}$ ________

12. $\frac{2}{5}$ ________

Problem Solving

13. Mathematical PRACTICE 1 **Make Sense of Problems** Latitia had 12 marbles. After she gave 2 marbles to Emilia, she has $\frac{10}{12}$ of her marbles left. What fraction of the marbles did Latitia give away? Write in simplest form.

____

14. Ryan has 8 kittens. Two are white. What fraction of the kittens are *not* white? Write in simplest form.

____

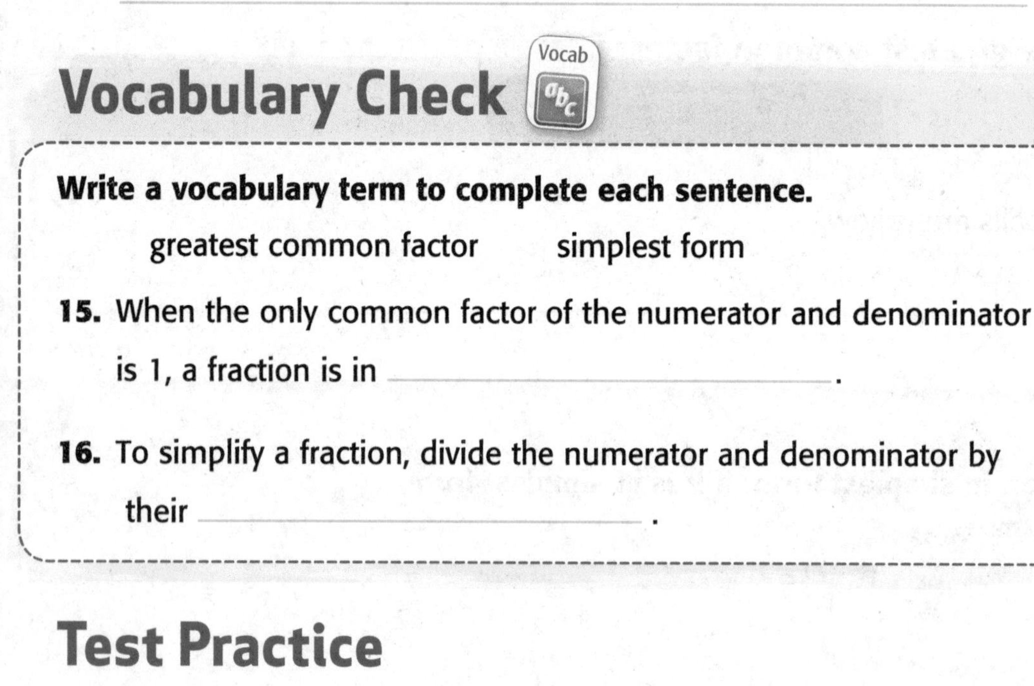

Vocabulary Check

Write a vocabulary term to complete each sentence.

greatest common factor   simplest form

15. When the only common factor of the numerator and denominator is 1, a fraction is in ________.

16. To simplify a fraction, divide the numerator and denominator by their ________.

Test Practice

17. What is the unknown denominator if $\frac{9}{12} = \frac{3}{■}$?

Ⓐ 8   Ⓒ 4

Ⓑ 6   Ⓓ 3

Name ....

Compare and Order Fractions

Lesson 6

ESSENTIAL QUESTION
How can different fractions name the same amount?

To compare fractions, create equivalent fractions with the same denominators or the same numerators. Use the **least common multiple**, or the least multiple common to sets of multiples.

Example 1

Ramon has an insect collection. The table shows the lengths of four insects in his collection. Which is longer, a mosquito or a whirligig beetle?

Insect	Length (in.)
Mosquito	$\frac{1}{4}$
Field cricket	$\frac{5}{8}$
Whirligig beetle	$\frac{3}{8}$
Lightning bug	$\frac{1}{2}$

1 Find the least common multiple of the denominators.

Multiples of 4: 4, **8**, 12, 16, ... ← mosquito

Multiples of 8: **8**, 16, 24, ... ← whirligig beetle

The least common multiple of the denominators is ______.

2 Create equivalent fractions that use 8 as the denominator.

Mosquito

$$\frac{1}{4} = \frac{1 \times 2}{4 \times 2} = \frac{2}{8}$$

Whirligig Beetle

$$\frac{3}{8} = \frac{3 \times 1}{8 \times 1} = \frac{3}{8}$$

3 Compare the numerators.
When the denominators are the same, the fraction with the greater numerator is the greater fraction.

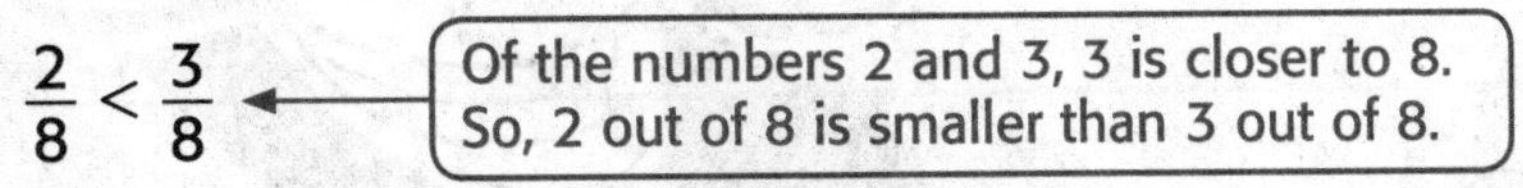

$\frac{2}{8} < \frac{3}{8}$ ← Of the numbers 2 and 3, 3 is closer to 8. So, 2 out of 8 is smaller than 3 out of 8.

So, a ______________ is longer.

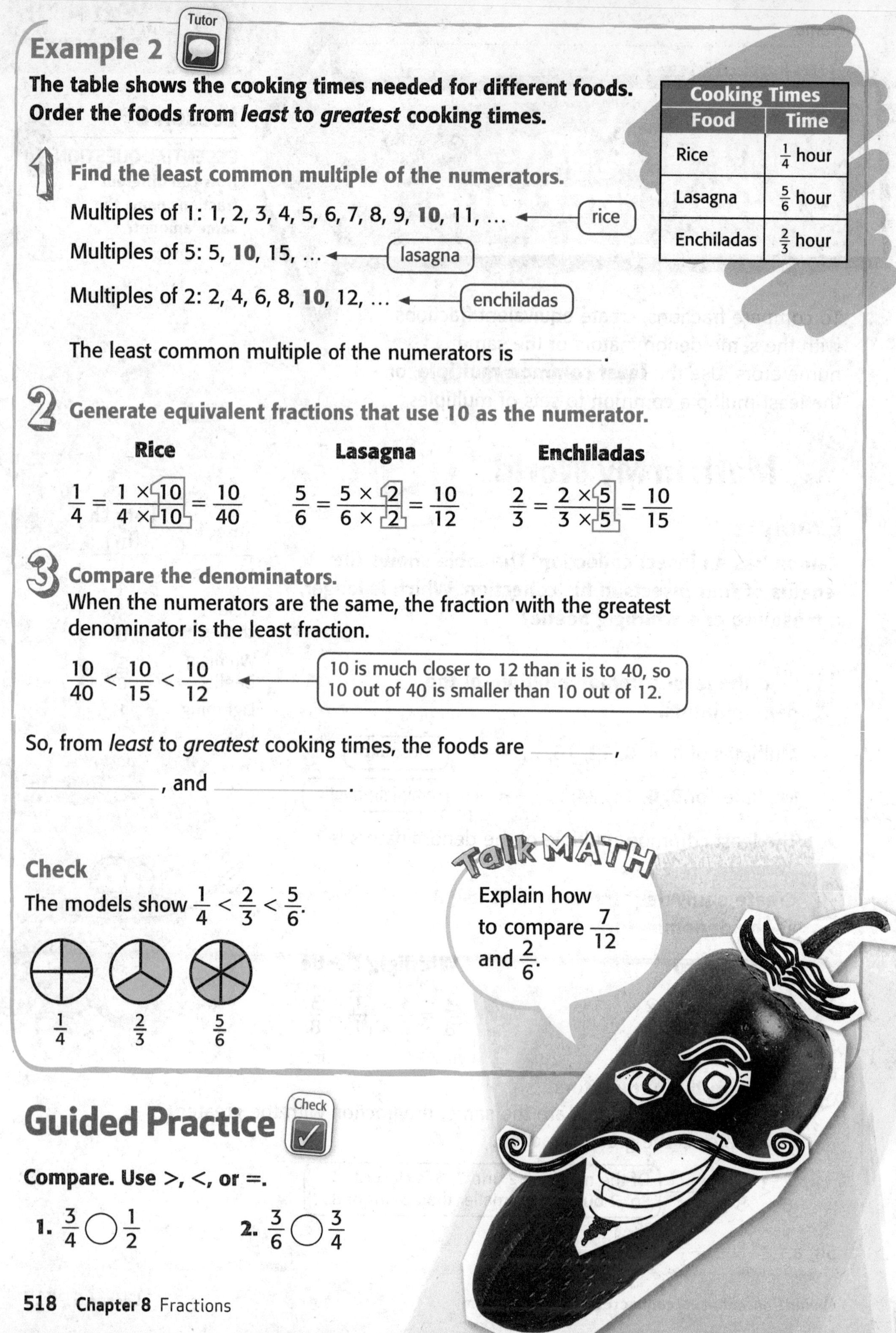

Example 2

Tutor

The table shows the cooking times needed for different foods. Order the foods from *least* to *greatest* cooking times.

Cooking Times	
Food	**Time**
Rice	$\frac{1}{4}$ hour
Lasagna	$\frac{5}{6}$ hour
Enchiladas	$\frac{2}{3}$ hour

1 **Find the least common multiple of the numerators.**

Multiples of 1: 1, 2, 3, 4, 5, 6, 7, 8, 9, **10**, 11, ... ← rice

Multiples of 5: 5, **10**, 15, ... ← lasagna

Multiples of 2: 2, 4, 6, 8, **10**, 12, ... ← enchiladas

The least common multiple of the numerators is ______.

2 **Generate equivalent fractions that use 10 as the numerator.**

Rice

$\frac{1}{4} = \frac{1 \times 10}{4 \times 10} = \frac{10}{40}$

Lasagna

$\frac{5}{6} = \frac{5 \times 2}{6 \times 2} = \frac{10}{12}$

Enchiladas

$\frac{2}{3} = \frac{2 \times 5}{3 \times 5} = \frac{10}{15}$

3 **Compare the denominators.**

When the numerators are the same, the fraction with the greatest denominator is the least fraction.

$\frac{10}{40} < \frac{10}{15} < \frac{10}{12}$ ← 10 is much closer to 12 than it is to 40, so 10 out of 40 is smaller than 10 out of 12.

So, from *least* to *greatest* cooking times, the foods are ______, ______, and ______.

Check

The models show $\frac{1}{4} < \frac{2}{3} < \frac{5}{6}$.

$\frac{1}{4}$ $\frac{2}{3}$ $\frac{5}{6}$

Guided Practice

Check

Compare. Use >, <, or =.

1. $\frac{3}{4}$ ◯ $\frac{1}{2}$

2. $\frac{3}{6}$ ◯ $\frac{3}{4}$

Name ....

CCSS

Independent Practice

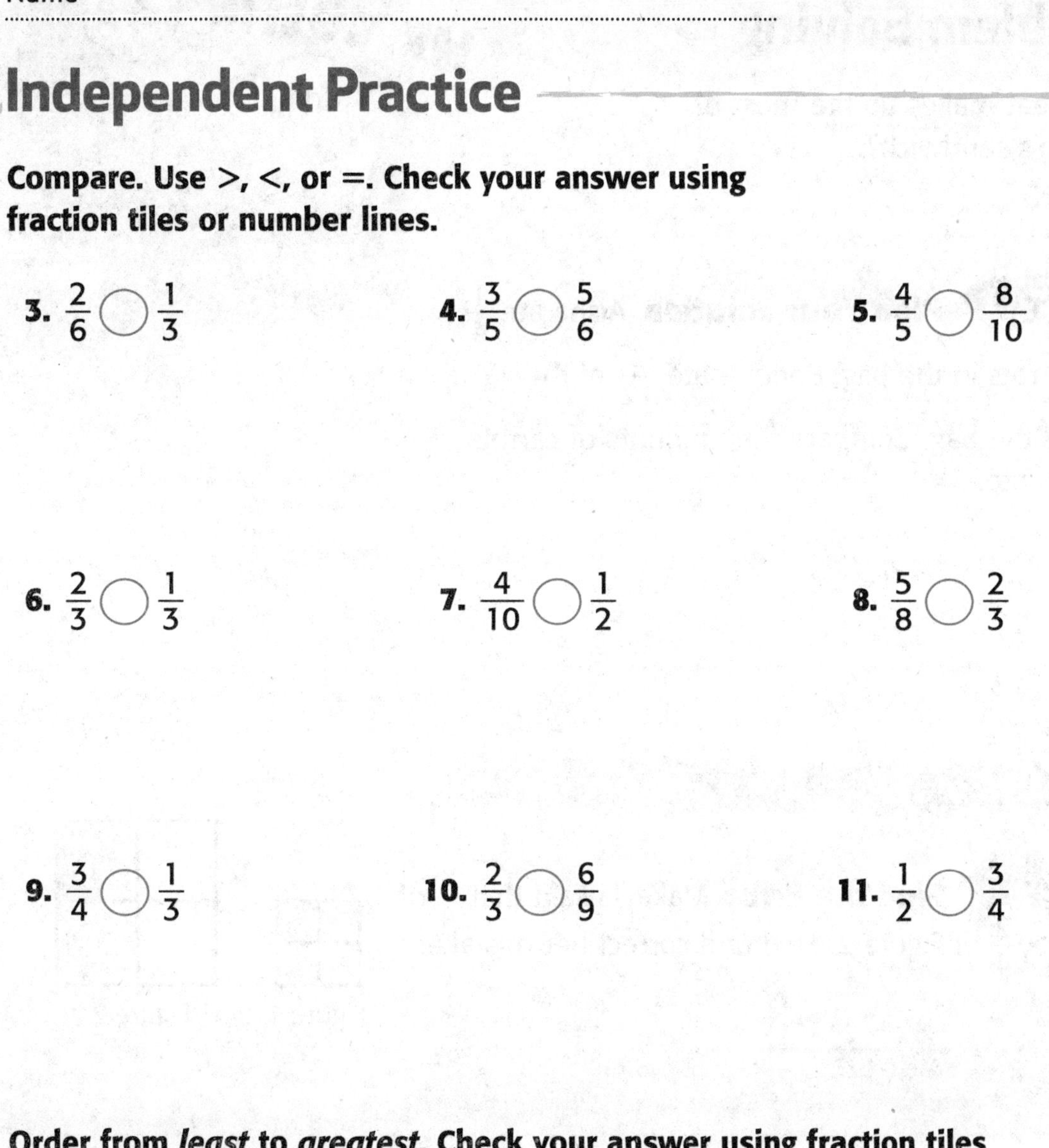

Compare. Use >, <, or =. Check your answer using fraction tiles or number lines.

3. $\frac{2}{6} \bigcirc \frac{1}{3}$

4. $\frac{3}{5} \bigcirc \frac{5}{6}$

5. $\frac{4}{5} \bigcirc \frac{8}{10}$

6. $\frac{2}{3} \bigcirc \frac{1}{3}$

7. $\frac{4}{10} \bigcirc \frac{1}{2}$

8. $\frac{5}{8} \bigcirc \frac{2}{3}$

9. $\frac{3}{4} \bigcirc \frac{1}{3}$

10. $\frac{2}{3} \bigcirc \frac{6}{9}$

11. $\frac{1}{2} \bigcirc \frac{3}{4}$

Order from *least* to *greatest*. Check your answer using fraction tiles or number lines.

12. $\frac{4}{6}, \frac{1}{3}, \frac{3}{3}$

13. $\frac{3}{4}, \frac{2}{3}, \frac{7}{8}$

14. $\frac{3}{10}, \frac{3}{5}, \frac{3}{4}$

______________ ______________ ______________

15. $\frac{1}{6}, \frac{2}{5}, \frac{3}{4}$

16. $\frac{3}{8}, \frac{2}{5}, \frac{2}{10}$

17. $\frac{3}{5}, \frac{2}{4}, \frac{1}{3}$

______________ ______________ ______________

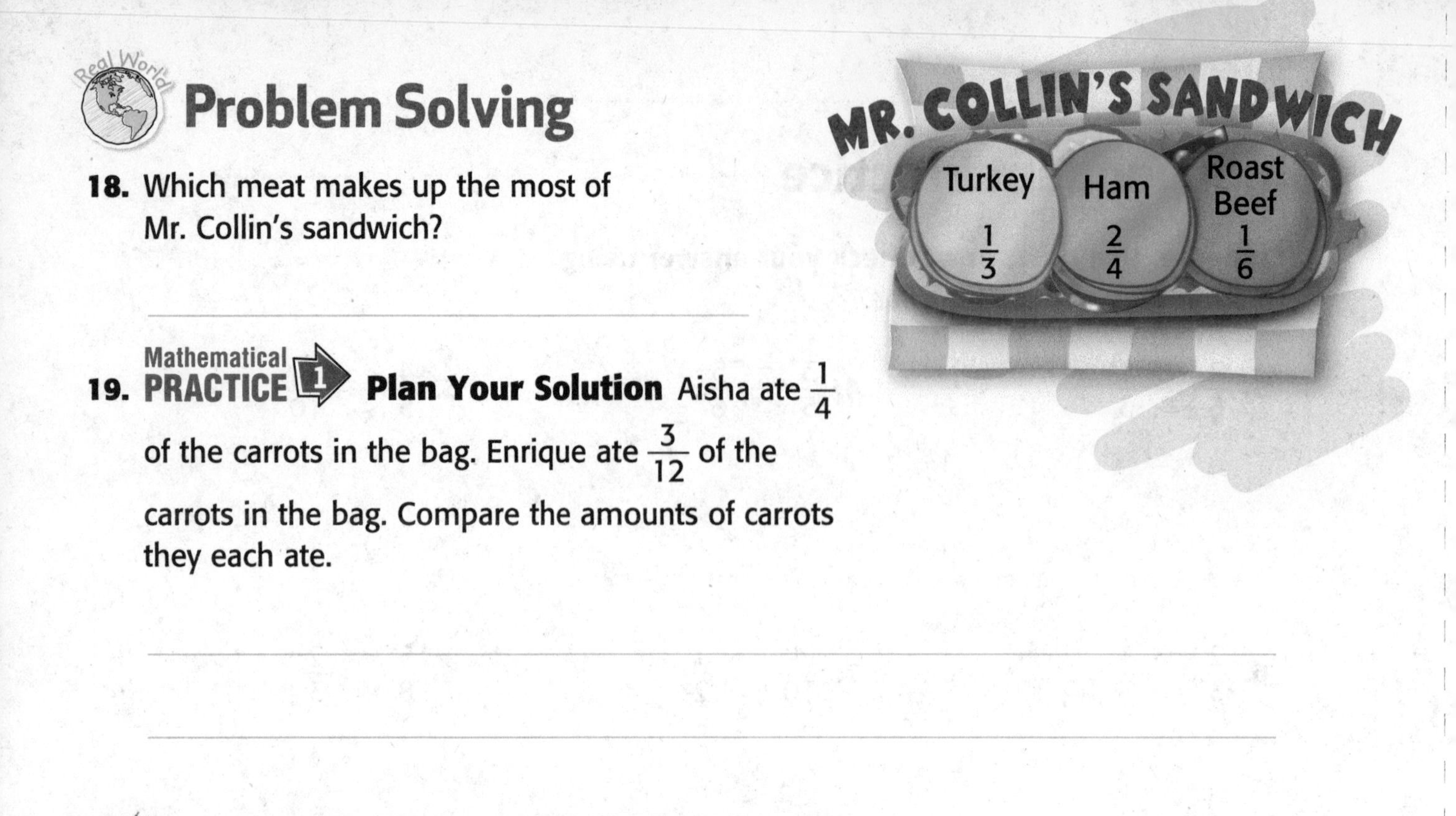

Problem Solving

18. Which meat makes up the most of Mr. Collin's sandwich?

19. Mathematical PRACTICE 1 **Plan Your Solution** Aisha ate $\frac{1}{4}$ of the carrots in the bag. Enrique ate $\frac{3}{12}$ of the carrots in the bag. Compare the amounts of carrots they each ate.

HOT Problems

20. Mathematical PRACTICE 3 **Find the Error** Makayla said that $\frac{3}{4}$ of Figure 1 > $\frac{2}{4}$ of Figure 2. Find and correct her mistake.

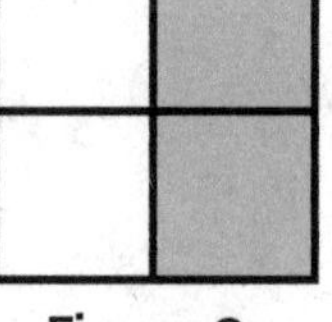

Figure 1 Figure 2

21. Mathematical PRACTICE 2 **Reason** Write three fractions that are *not* greater than $\frac{1}{2}$.

22. **Building on the Essential Question** How can I compare two fractions with the same numerator?

Name

MY Homework

Lesson 6

Compare and Order Fractions

Homework Helper

eHelp Need help? connectED.mcgraw-hill.com

Ellen has three cans of paint that are the same size. The can of blue paint is $\frac{2}{3}$ full. The can of green paint is $\frac{3}{4}$ full, and the can of yellow paint is $\frac{1}{2}$ full. Order the paint colors from least to greatest amounts.

Compare $\frac{2}{3}$, $\frac{3}{4}$, and $\frac{1}{2}$.

1 Find the least common multiple of the denominators.

Circle the least common multiple.

Multiples of 2: 2, 4, 6, 8, 10, (12), 14

Multiples of 3: 3, 6, 9, (12), 15

Multiples of 4: 4, 8, (12), 16

The least common multiple is 12.

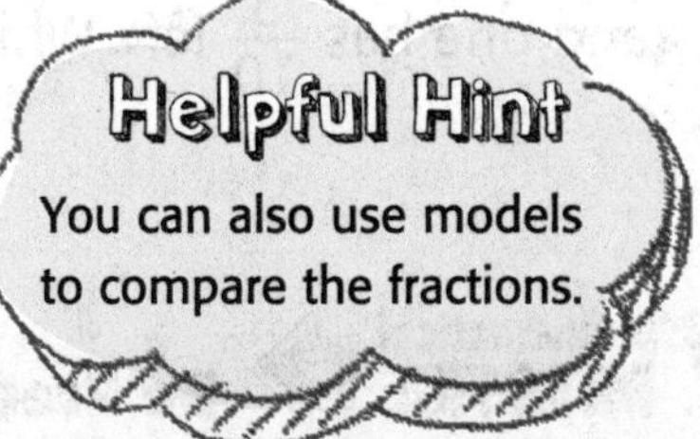

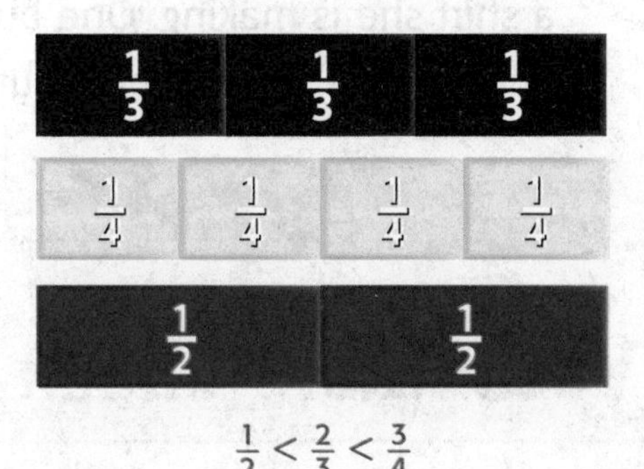

$\frac{1}{2} < \frac{2}{3} < \frac{3}{4}$

2 Create equivalent fractions.

Multiply to find equivalent fractions with 12 as the denominator.

$\frac{2 \times 4}{3 \times 4} = \frac{8}{12}$ $\frac{3 \times 3}{4 \times 3} = \frac{9}{12}$ $\frac{1 \times 6}{2 \times 6} = \frac{6}{12}$

3 Compare and order.

$\frac{6}{12} < \frac{8}{12} < \frac{9}{12}$

So, $\frac{1}{2} < \frac{2}{3} < \frac{3}{4}$.

From *least* to *greatest,* the amount of each color of paint is yellow, blue, and green.

Practice

Compare. Use >, <, or =.

1. $\frac{1}{2}$ ◯ $\frac{1}{3}$
2. $\frac{5}{12}$ ◯ $\frac{1}{4}$
3. $\frac{4}{5}$ ◯ $\frac{8}{10}$
4. $\frac{7}{10}$ ◯ $\frac{4}{5}$
5. $\frac{1}{5}$ ◯ $\frac{2}{10}$
6. $\frac{2}{5}$ ◯ $\frac{2}{8}$
7. $\frac{9}{10}$ ◯ $\frac{7}{8}$
8. $\frac{3}{10}$ ◯ $\frac{4}{8}$
9. $\frac{1}{4}$ ◯ $\frac{6}{12}$

Order from *least* to *greatest*.

10. $\frac{4}{8}, \frac{1}{3}, \frac{2}{3}$ ________
11. $\frac{5}{6}, \frac{7}{12}, \frac{3}{4}$ ________
12. $\frac{1}{2}, \frac{7}{8}, \frac{2}{8}$ ________
13. $\frac{1}{3}, \frac{1}{4}, \frac{5}{6}$ ________

Real World

Problem Solving

14. Patti has two glue sticks that are partially used. One has $\frac{1}{5}$ left, and one has $\frac{3}{10}$ left. Which glue stick has more glue?

____

15. Mathematical PRACTICE 6 **Be Precise** Lola measures three buttons for a shirt she is making. One button is $\frac{1}{8}$ inch, one is $\frac{3}{8}$ inch, and one is $\frac{1}{4}$ inch. Which button is smallest? Which is largest?

____

Vocabulary Check

Vocab

16. What is the least common multiple of 3 and 8? ________

Test Practice

17. Which fraction is *not* greater than $\frac{1}{2}$?

Ⓐ $\frac{7}{8}$ Ⓑ $\frac{4}{6}$ Ⓒ $\frac{3}{5}$ Ⓓ $\frac{2}{5}$

Name ....

Use Benchmark Fractions to Compare and Order

Lesson 7

ESSENTIAL QUESTION
How can different fractions name the same amount?

Benchmark fractions are common fractions, such as $\frac{1}{2}$, that are often used to compare and order fractions.

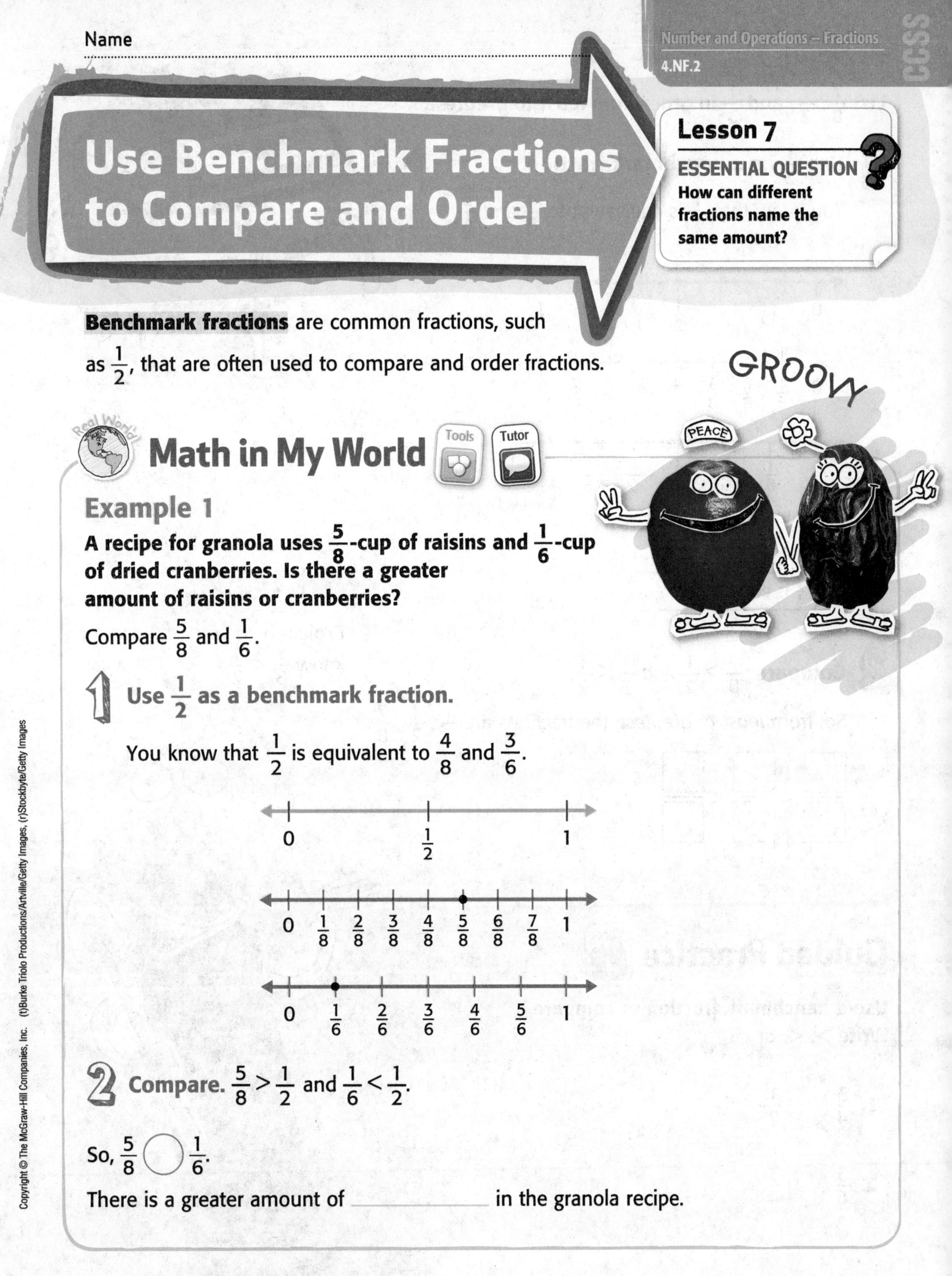

Real World Math in My World

Tools Tutor

Example 1

A recipe for granola uses $\frac{5}{8}$-cup of raisins and $\frac{1}{6}$-cup of dried cranberries. Is there a greater amount of raisins or cranberries?

Compare $\frac{5}{8}$ and $\frac{1}{6}$.

1 **Use $\frac{1}{2}$ as a benchmark fraction.**

You know that $\frac{1}{2}$ is equivalent to $\frac{4}{8}$ and $\frac{3}{6}$.

0, $\frac{1}{2}$, 1

0, $\frac{1}{8}$, $\frac{2}{8}$, $\frac{3}{8}$, $\frac{4}{8}$, $\frac{5}{8}$, $\frac{6}{8}$, $\frac{7}{8}$, 1

0, $\frac{1}{6}$, $\frac{2}{6}$, $\frac{3}{6}$, $\frac{4}{6}$, $\frac{5}{6}$, 1

2 **Compare.** $\frac{5}{8} > \frac{1}{2}$ and $\frac{1}{6} < \frac{1}{2}$.

So, $\frac{5}{8}$ ◯ $\frac{1}{6}$.

There is a greater amount of ____________ in the granola recipe.

Example 2

List $\frac{7}{8}$, $\frac{1}{2}$, and $\frac{1}{3}$ in order from *least* to *greatest*.

1 Use $\frac{1}{2}$ as a benchmark fraction.

You know that $\frac{1}{2}$ is equivalent to $\frac{4}{8}$ and $\frac{3}{6}$.

0, $\frac{1}{2}$, 1

0, $\frac{1}{8}$, $\frac{2}{8}$, $\frac{3}{8}$, $\frac{4}{8}$, $\frac{5}{8}$, $\frac{6}{8}$, $\frac{7}{8}$, 1

0, $\frac{1}{6}$, $\frac{2}{6}$, $\frac{3}{6}$, $\frac{4}{6}$, $\frac{5}{6}$, 1

$\frac{1}{3}$ is equivalent to $\frac{2}{6}$.

2 Compare. $\frac{7}{8} > \frac{1}{2}$ and $\frac{1}{3} < \frac{1}{2}$.

So, from *least* to *greatest*, the fractions are $\frac{1}{3}$, $\frac{1}{2}$, $\frac{7}{8}$.

$\frac{\square}{\square} < \frac{\square}{\square} < \frac{\square}{\square}$

Talk MATH

Explain how you know $\frac{1}{8} < \frac{1}{2}$.

Guided Practice

Check

Use a benchmark fraction to compare. Write >, <, or =.

1. $\frac{3}{4} \bigcirc \frac{1}{2}$

2. $\frac{3}{6} \bigcirc \frac{3}{4}$

Name

Independent Practice

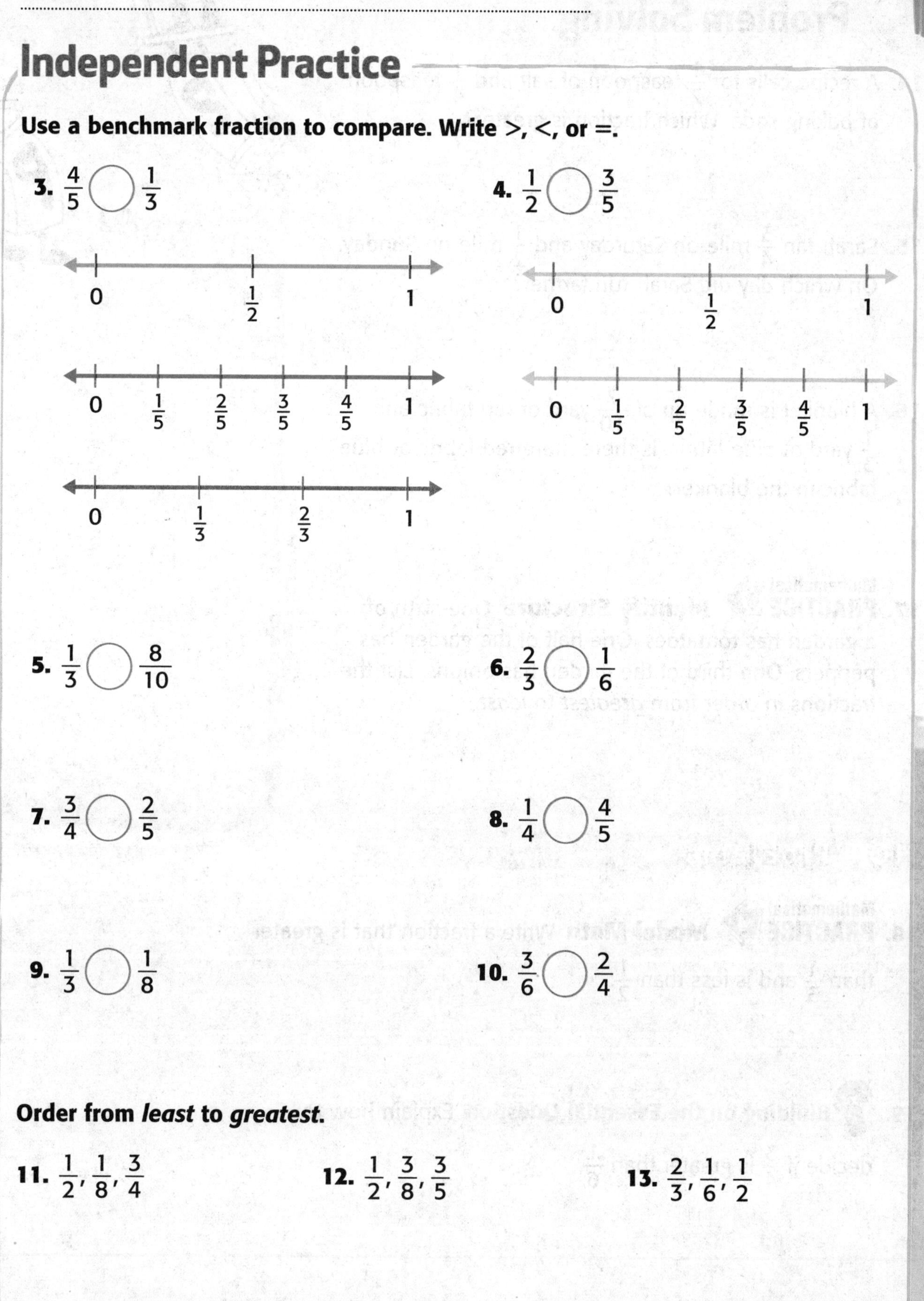

Use a benchmark fraction to compare. Write >, <, or =.

3. $\frac{4}{5}$ ◯ $\frac{1}{3}$

4. $\frac{1}{2}$ ◯ $\frac{3}{5}$

5. $\frac{1}{3}$ ◯ $\frac{8}{10}$

6. $\frac{2}{3}$ ◯ $\frac{1}{6}$

7. $\frac{3}{4}$ ◯ $\frac{2}{5}$

8. $\frac{1}{4}$ ◯ $\frac{4}{5}$

9. $\frac{1}{3}$ ◯ $\frac{1}{8}$

10. $\frac{3}{6}$ ◯ $\frac{2}{4}$

Order from *least* to *greatest*.

11. $\frac{1}{2}, \frac{1}{8}, \frac{3}{4}$

12. $\frac{1}{2}, \frac{3}{8}, \frac{3}{5}$

13. $\frac{2}{3}, \frac{1}{6}, \frac{1}{2}$

Real World

Problem Solving

14. A recipe calls for $\frac{1}{2}$ teaspoon of salt and $\frac{1}{4}$ teaspoon of baking soda. Which fraction is greater?

HUH?!?

My Work!

15. Sarah ran $\frac{3}{4}$ mile on Saturday and $\frac{1}{2}$ mile on Sunday. On which day did Sarah run farther?

16. A blanket is made up of $\frac{7}{10}$ yard of red fabric and $\frac{1}{3}$ yard of blue fabric. Is there more red fabric or blue fabric in the blanket?

17. Mathematical PRACTICE 7 **Identify Structure** One-sixth of a garden has tomatoes. One-half of the garden has peppers. One-third of the garden has onions. List the fractions in order from *greatest* to *least*.

HOT Problems

18. Mathematical PRACTICE 4 **Model Math** Write a fraction that is greater than $\frac{1}{3}$ and is less than $\frac{1}{2}$.

19. **Building on the Essential Question** Explain how to decide if $\frac{3}{4}$ is greater than $\frac{1}{6}$.

Name ....

MY Homework

Lesson 7
Use Benchmark Fractions to Compare and Order

Homework Helper

eHelp Need help? connectED.mcgraw-hill.com

Connor has completed $\frac{3}{8}$ of his homework. Aidan has finished $\frac{5}{6}$ of his homework. Who is closer to being finished with his homework?

Use benchmark fractions, or common fractions, to compare $\frac{3}{8}$ and $\frac{5}{6}$.

1 Use $\frac{1}{2}$ as a benchmark fraction.
You know that $\frac{1}{2}$ is equivalent to $\frac{4}{8}$ and $\frac{3}{6}$.

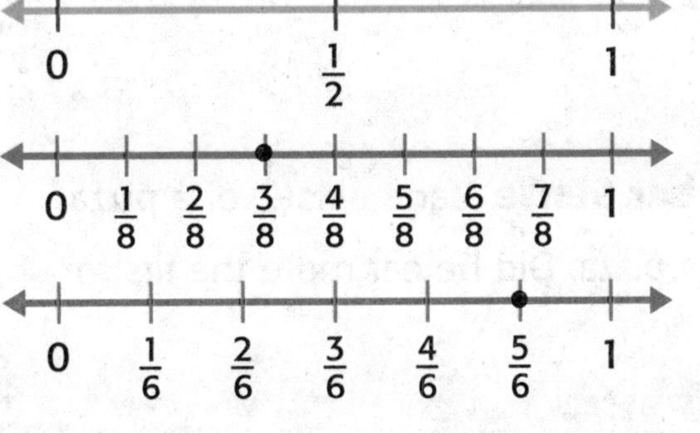

2 Compare.
$\frac{3}{8}$ is less than $\frac{1}{2}$, and $\frac{5}{6}$ is greater than $\frac{1}{2}$.

So, Aidan is closer to finishing his homework than Connor.

Practice

Use a benchmark fraction to compare. Use >, <, or =.

1. $\frac{3}{8}$ ◯ $\frac{3}{4}$

2. $\frac{4}{6}$ ◯ $\frac{3}{8}$

3. $\frac{1}{8}$ ◯ $\frac{5}{6}$

Use a benchmark fraction to compare. Use >, <, or =.

4. $\frac{1}{3} \bigcirc \frac{3}{4}$

5. $\frac{3}{5} \bigcirc \frac{2}{6}$

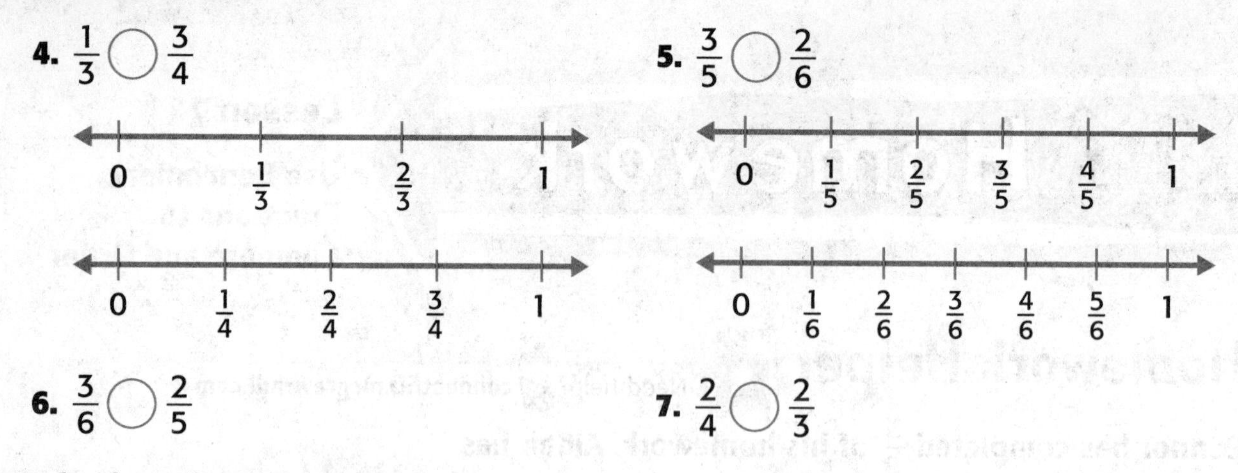

6. $\frac{3}{6} \bigcirc \frac{2}{5}$

7. $\frac{2}{4} \bigcirc \frac{2}{3}$

Order from *least* to *greatest*.

8. $\frac{1}{2}, \frac{1}{4}, \frac{2}{3}$ ____________

9. $\frac{3}{6}, \frac{3}{4}, \frac{2}{8}$ ____________

Problem Solving

10. Bailey has an hour to get ready for school. She spends $\frac{3}{6}$ of that time getting dressed. She spends $\frac{1}{4}$ of that time eating breakfast. Does Bailey spend more time getting dressed or eating breakfast?

____

11. **Mathematical PRACTICE 2** **Use Number Sense** Zane eats $\frac{1}{8}$ of a pizza. Later he eats $\frac{1}{2}$ of the same pizza. Did he eat more the first or second time?

____

Vocabulary Check

Vocab

12. How do I use benchmark fractions?

____

Test Practice

13. Which fraction is not greater than $\frac{2}{5}$?

Ⓐ $\frac{3}{10}$ Ⓑ $\frac{7}{10}$ Ⓒ $\frac{2}{4}$ Ⓓ $\frac{3}{6}$

Check My Progress

Vocabulary Check

Vocab

1. Write two **equivalent fractions**. ______

2. Write the fraction that has a 4 in the **numerator** and a 6 in the **denominator**.

______

3. Write the fraction from Exercise 2 in **simplest form**. ______

4. Explain the difference between a **greatest common factor** and a **least common multiple**.

______

______

______

Concept Check

Check

Recognize whether the fractions are equivalent. Write *yes* or *no*. Use fraction tiles or number lines.

5. $\frac{1}{4}$ and $\frac{3}{12}$ ______

6. $\frac{4}{6}$ and $\frac{6}{12}$ ______

7. $\frac{5}{10}$ and $\frac{3}{6}$ ______

Generate two equivalent fractions for each fraction. Use fraction tiles or number lines.

8. $\frac{2}{8}$ ______

9. $\frac{1}{3}$ ______

10. $\frac{2}{4}$ ______

Problem Solving

11. Adina read 60 out of 100 pages in a comic book. Write the fraction of the pages she read in simplest form.

______

12. The table shows how much time each student needs to finish an art project. Does Simón need more or less time than Phil? Explain.

Student Time	
Simón	$\frac{1}{3}$ hour
Phil	$\frac{3}{4}$ hour

______

______

______

13. Write the fraction for the part that is shaded. Then find an equivalent fraction.

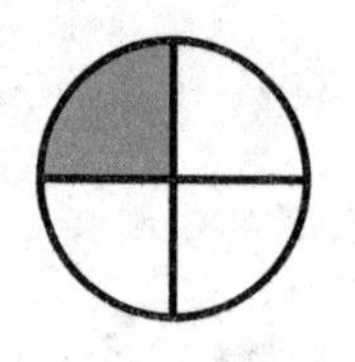

______

My Work!

Test Practice

14. Which fraction is in simplest form?

Ⓐ $\frac{2}{10}$

Ⓑ $\frac{3}{12}$

Ⓒ $\frac{4}{12}$

Ⓓ $\frac{3}{8}$

Name ....

Real World

Problem-Solving Investigation

STRATEGY: Use Logical Reasoning

Lesson 8

ESSENTIAL QUESTION
How can different fractions name the same amount?

Learn the Strategy

Watch Tutor

Marla used flour, sugar, and brown sugar to make a treat. Use the clues below to find the amount of each ingredient she used.

- **The amounts were $\frac{3}{4}$ cup, $\frac{1}{4}$ cup, and $\frac{2}{3}$ cup.**
- **She used more flour than sugar.**
- **She used more sugar than brown sugar.**

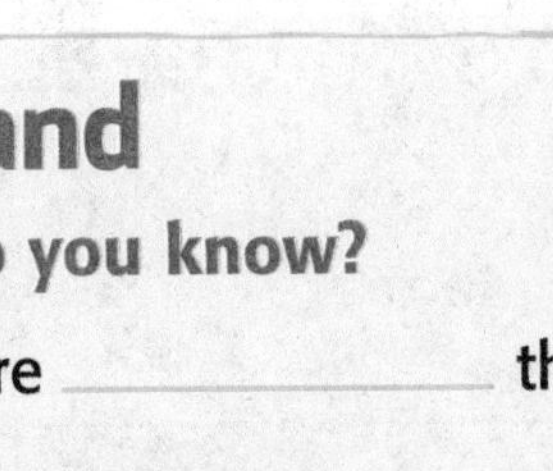

1 Understand

What facts do you know?

Marla used more __________ than sugar and more sugar than __________.

What do you need to find?

the amount of each __________ she used

2 Plan

I can use logical reasoning to solve the problem.

3 Solve

The order of the ingredients from greatest to least is __________, __________, and __________.

The greatest fraction is $\frac{3}{4}$. So, Marla used $\frac{3}{4}$ cup of flour. $\frac{2}{3} > \frac{1}{4}$

So, she used $\frac{2}{3}$ cup of sugar and $\frac{1}{4}$ cup of brown sugar.

4 Check

Does your answer make sense? Explain.

____

Practice the Strategy

Ameer is making salsa using tomatoes, onions, and black beans. Use the clues below to find the amount of each ingredient he used.

- **The amounts are: $\frac{1}{2}$ cup, $\frac{3}{4}$ cup, and $\frac{1}{3}$ cup.**
- **He is using more tomatoes than black beans and more black beans than onions.**

1 Understand

What facts do you know?

____

____

What do you need to find?

____

Plan

____

3 Solve

4 Check

Does your answer make sense? Explain.

____

Name

Apply the Strategy

Solve each problem by using logical reasoning.

My Work!

1. Sophia is making a salad with tomatoes, cucumbers, and mozzarella cheese. Use the clues to find the amount of each ingredient.

- The amounts are $\frac{3}{6}$ cup, $\frac{2}{5}$ cup, and $\frac{3}{4}$ cup.
- There is a less amount of tomatoes than cucumbers.
- There is a less amount of cheese than tomatoes.

2. Mason walked on Monday, Wednesday, and Friday. Use the clues to find how far he walked each day.

- The distances were $\frac{6}{8}$ mile, $\frac{1}{4}$ mile, and $\frac{1}{6}$ mile.
- He did not walk the farthest on Monday.
- He walked less on Friday than Monday.

3. Mathematical PRACTICE 1 **Plan Your Solution** Emma saw birds, fish, and bears at the zoo. One-third of the animals did not have feathers or fur. One-half of the animals have four legs. There were one-sixth of the animals left. What fraction of the animals were birds?

Review the Strategies

Use any strategy to solve each problem.

- Work backward.
- Make a table.
- Make a model.
- Look for a pattern.

My Work!

4. Emil bought his mom a dozen roses. Some of the roses are shown below. The rest are white.

Of which color were there the most?

____

What fraction of the roses were that color? Write in simplest form.

____

Write an equivalent fraction.

____

5. Dirk ran eight-tenths of a mile at track practice.

Leslie ran $\frac{80}{100}$ of a mile. Who ran farther? Explain.

____

____

6. Mrs. Keys has 36 pens. Yesterday she had half that amount plus 2. How many pens did she have?

____

7. Mathematical PRACTICE 3 **Draw a Conclusion** There are bananas, pears, and peaches in a bag. One-eighth of the fruits are pears. One half are bananas. Are there more pears or bananas?

____

Name ....

MY Homework

Lesson 8

Problem Solving: Use Logical Reasoning

Homework Helper

eHelp Need help? connectED.mcgraw-hill.com

Hailey made raisin bread. She used flour, raisins, and water. The amounts of these ingredients were $\frac{1}{3}$ cup, $\frac{1}{4}$ cup, and $\frac{2}{3}$ cup. She used more flour than water. She used more water than raisins. How much of each ingredient did Hailey use?

1 Understand

What facts do you know?

Hailey used flour, raisins, and water to make bread. The amounts of the ingredients were $\frac{1}{3}$ cup, $\frac{1}{4}$ cup, and $\frac{2}{3}$ cup.

What do you need to find?

I need to find how much of each ingredient Hailey used.

2 Plan

I will use logical reasoning to solve the problem.

3 Solve

The order of the ingredients from greatest to least amounts is flour, water, and raisins. The order of the amounts from greatest to least is $\frac{2}{3}$ cup, $\frac{1}{3}$ cup, and $\frac{1}{4}$ cup.

So, Hailey used $\frac{2}{3}$ cup of flour, $\frac{1}{3}$ cup of water, and $\frac{1}{4}$ cup of raisins.

4 Check

Does the answer make sense?

Yes. The clues match the answer.

Real World

Problem Solving

Solve each problem by using logical reasoning.

My Work!

1. Mathematical PRACTICE 2 **Stop and Reflect** Ryan has his artwork displayed at the library, the mall, and the bank. Use the clues to find the fraction of his art that is displayed at each place.
   - $\frac{1}{4}$ of the art is at one location, $\frac{1}{8}$ of the art is at the second location, and $\frac{5}{8}$ of the art is at the third location.
   - There is more of Ryan's art at the library than the mall.
   - There is less of Ryan's art at the bank than at the mall.

____

____

2. Benjamin made a fruit salad with strawberries, blueberries, and kiwi. Use the clues to find the amounts of each ingredient.
   - The amounts were $\frac{3}{4}$ cup, $\frac{2}{8}$ cup, and $\frac{1}{2}$ cup.
   - Benjamin used more blueberries than strawberries.
   - Benjamin used more strawberries than kiwi.

____

____

3. Layla wrote a report about insects. She listed the lengths of tiger beetles, carpenter ants, and aphids. The lengths were $\frac{1}{2}$ inch, $\frac{5}{8}$ inch, and $\frac{1}{8}$ inch. A tiger beetle is bigger than a carpenter ant. A carpenter ant is bigger than an aphid. List the sizes of each insect.

____

____

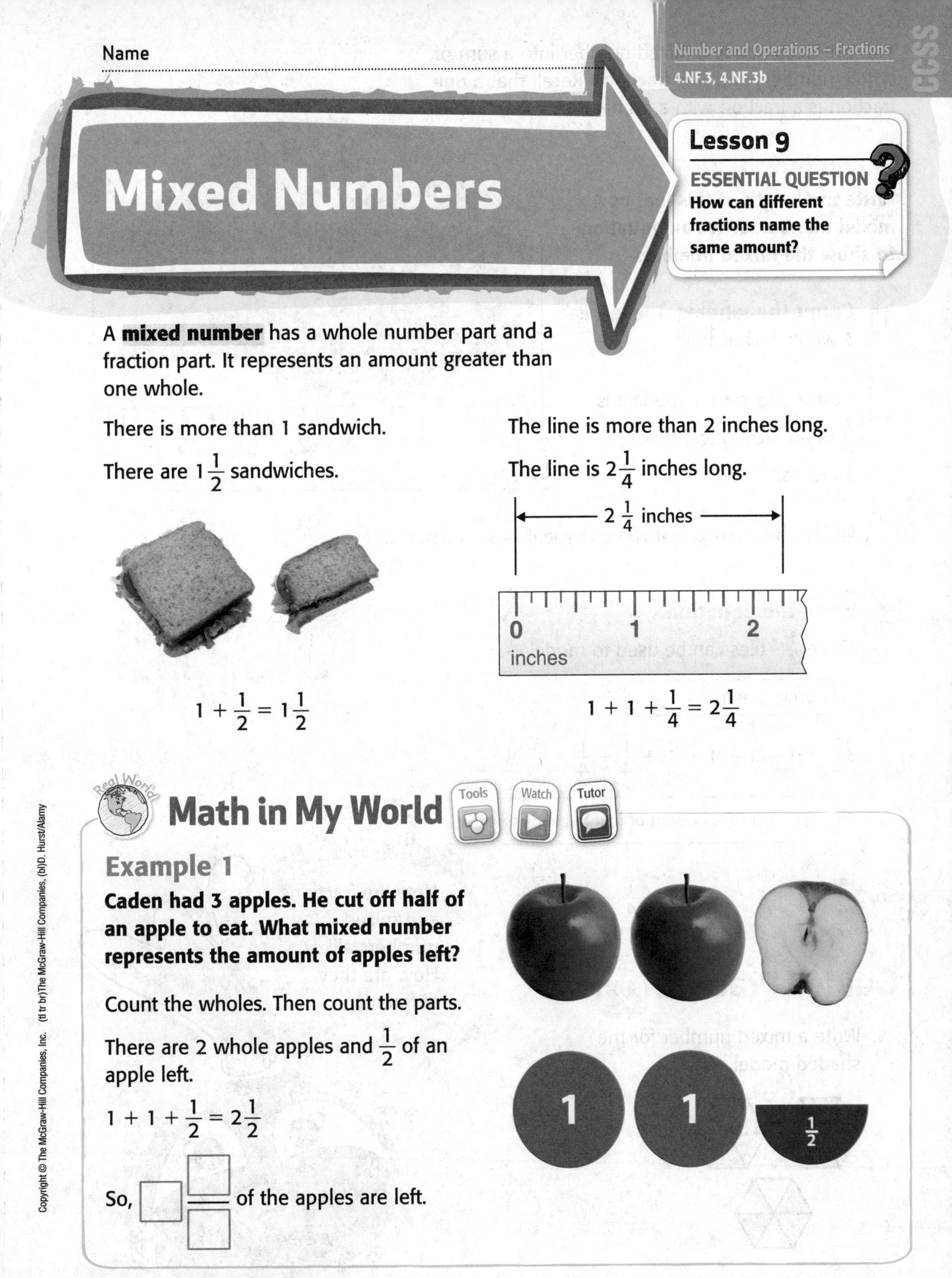

Mixed Numbers

Lesson 9

ESSENTIAL QUESTION
How can different fractions name the same amount?

A **mixed number** has a whole number part and a fraction part. It represents an amount greater than one whole.

There is more than 1 sandwich.

There are $1\frac{1}{2}$ sandwiches.

$$1 + \frac{1}{2} = 1\frac{1}{2}$$

The line is more than 2 inches long.

The line is $2\frac{1}{4}$ inches long.

$$1 + 1 + \frac{1}{4} = 2\frac{1}{4}$$

Math in My World

Example 1

Caden had 3 apples. He cut off half of an apple to eat. What mixed number represents the amount of apples left?

Count the wholes. Then count the parts.

There are 2 whole apples and $\frac{1}{2}$ of an apple left.

$1 + 1 + \frac{1}{2} = 2\frac{1}{2}$

So, $\square\frac{\square}{\square}$ of the apples are left.

You can decompose a mixed number into a sum of whole numbers and unit fractions. Recall that a unit fraction is a fraction with a numerator of 1.

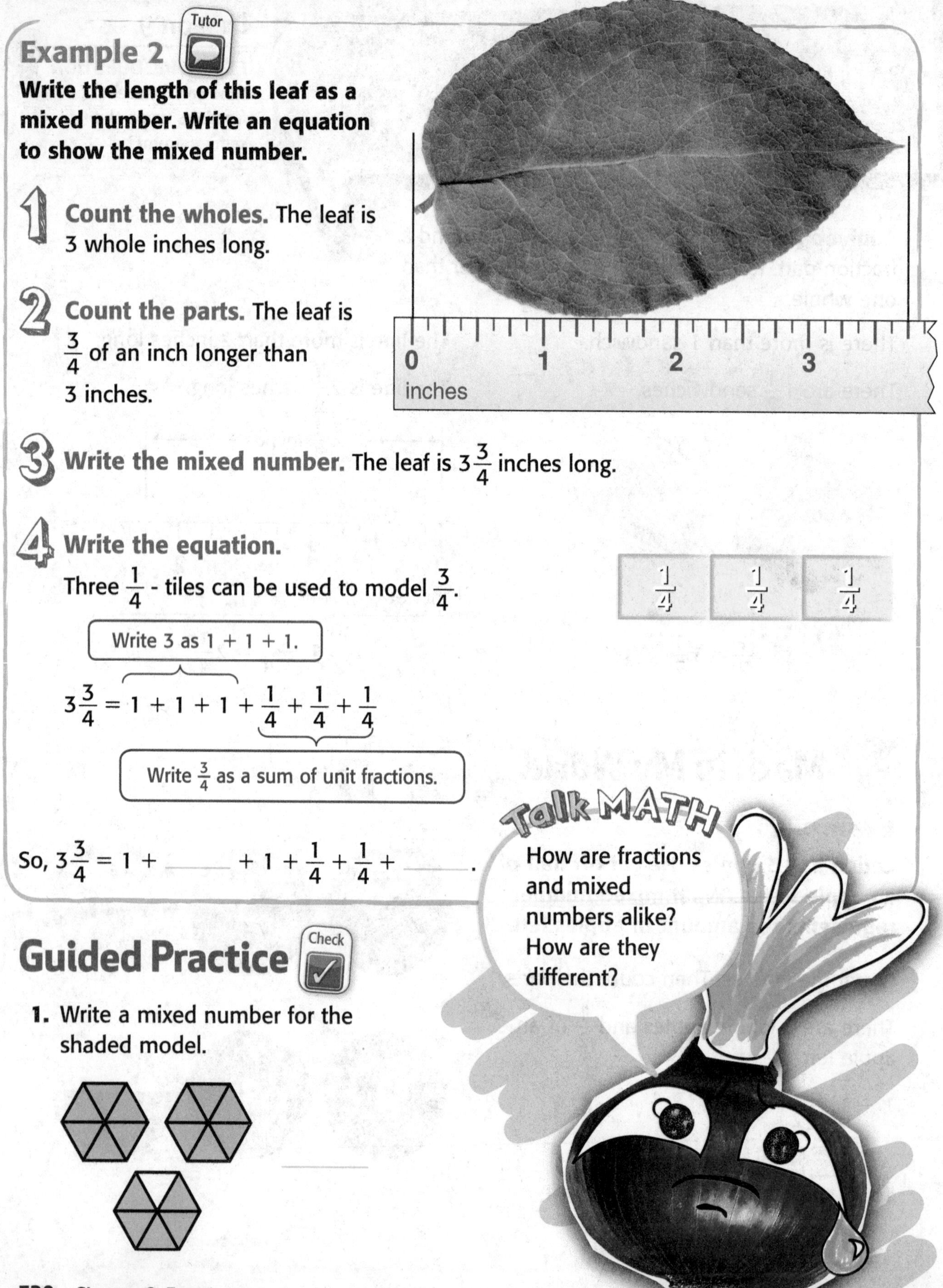

Example 2

Write the length of this leaf as a mixed number. Write an equation to show the mixed number.

1. **Count the wholes.** The leaf is 3 whole inches long.

2. **Count the parts.** The leaf is $\frac{3}{4}$ of an inch longer than 3 inches.

3. **Write the mixed number.** The leaf is $3\frac{3}{4}$ inches long.

4. **Write the equation.**

Three $\frac{1}{4}$ - tiles can be used to model $\frac{3}{4}$.

Write 3 as 1 + 1 + 1.

$$3\frac{3}{4} = 1 + 1 + 1 + \frac{1}{4} + \frac{1}{4} + \frac{1}{4}$$

Write $\frac{3}{4}$ as a sum of unit fractions.

So, $3\frac{3}{4} = 1 +$ _____ $+ 1 + \frac{1}{4} + \frac{1}{4} +$ _____.

Talk MATH

How are fractions and mixed numbers alike? How are they different?

Guided Practice

1. Write a mixed number for the shaded model.

_____

Name

Independent Practice

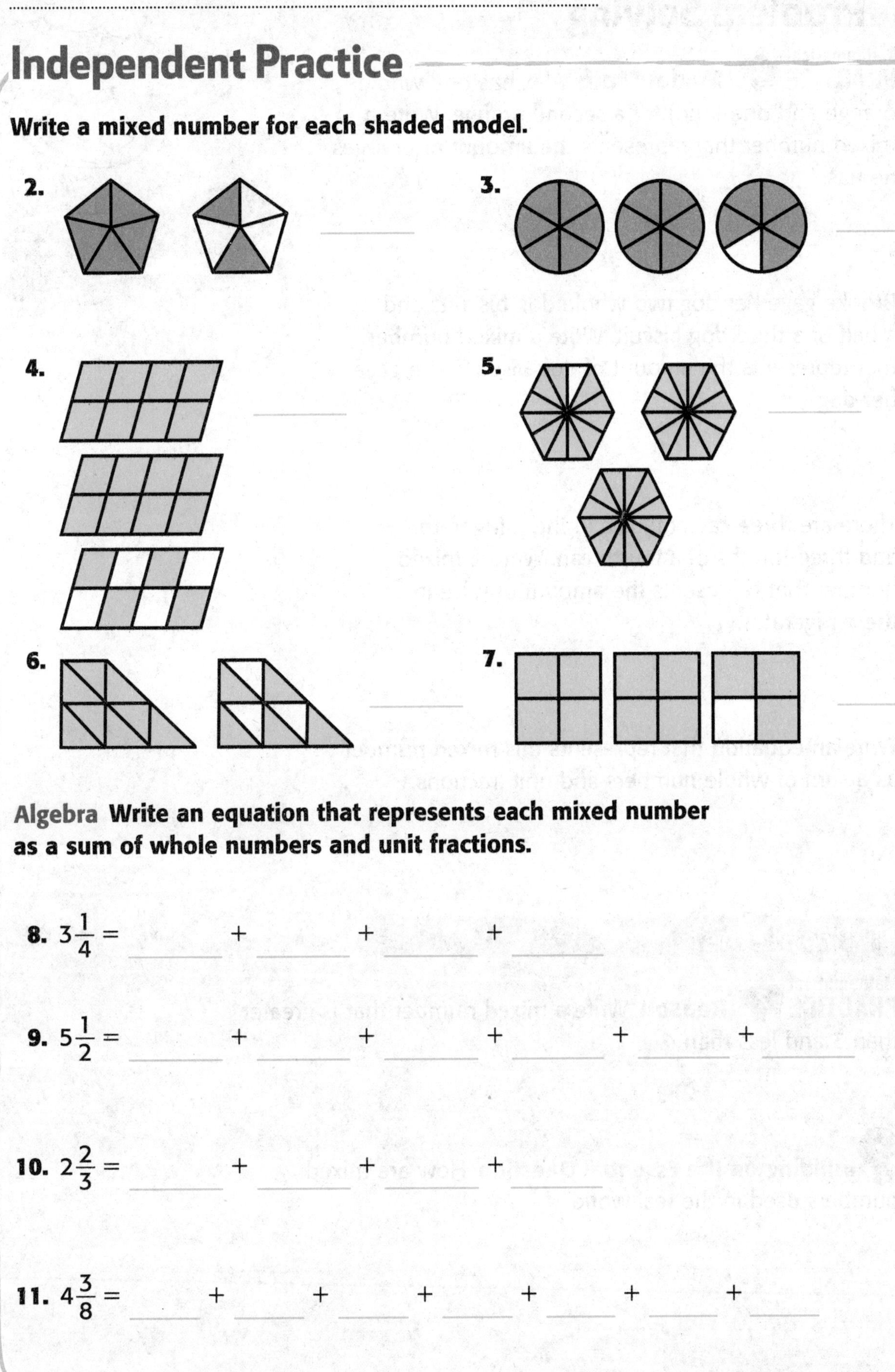

Write a mixed number for each shaded model.

2.

3.

4.

5.

6.

7.

Algebra Write an equation that represents each mixed number as a sum of whole numbers and unit fractions.

8. $3\frac{1}{4}$ = ______ + ______ + ______ + ______

9. $5\frac{1}{2}$ = ______ + ______ + ______ + ______ + ______ + ______

10. $2\frac{2}{3}$ = ______ + ______ + ______ + ______

11. $4\frac{3}{8}$ = ______ + ______ + ______ + ______ + ______ + ______ + ______

Problem Solving

12. Mathematical PRACTICE 4 **Model Math** Alex has one whole orange and one-fourth of a second orange. Write a mixed number that represents the amount of oranges he has.

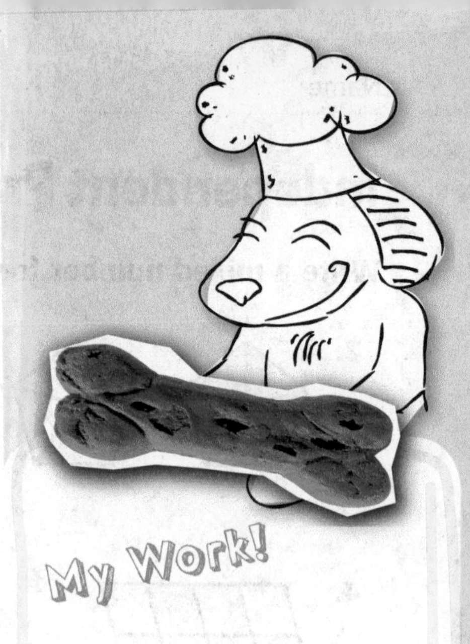

13. Brooke gave her dog two whole dog biscuits and a half of a third dog biscuit. Write a mixed number that represents the amount of dog biscuits she gave her dog.

14. There are three cans of juice in the refrigerator and three-fourths of a fourth can. Write a mixed number that represents the amount of juice in the refrigerator.

Write an equation that represents this mixed number as a sum of whole numbers and unit fractions.

HOT Problems

15. Mathematical PRACTICE 2 **Reason** Write a mixed number that is greater than 3 and less than 4.

16. **Building on the Essential Question** How are mixed numbers used in the real world?

MY Homework

Lesson 9
Mixed Numbers

Homework Helper

eHelp Need help? connectED.mcgraw-hill.com

Walter and his classmates are taking notes for a group report. Their note cards are shown. What is the total amount of notes Walter's group has taken? Write an equation to show the mixed number.

1 Count the wholes.
There are 2 cards full of notes. $1 + 1 = 2$

2 Count the parts.
There are two cards that are $\frac{1}{3}$ full of notes. $\frac{1}{3} + \frac{1}{3} = \frac{2}{3}$.

3 Add the wholes and the parts. $2 + \frac{2}{3} = 2\frac{2}{3}$

So, $1 + 1 + \frac{1}{3} + \frac{1}{3} = 2\frac{2}{3}$.

Walter's group has taken notes on $2\frac{2}{3}$ cards for their report.

Practice

Write a mixed number for each shaded model.

1. 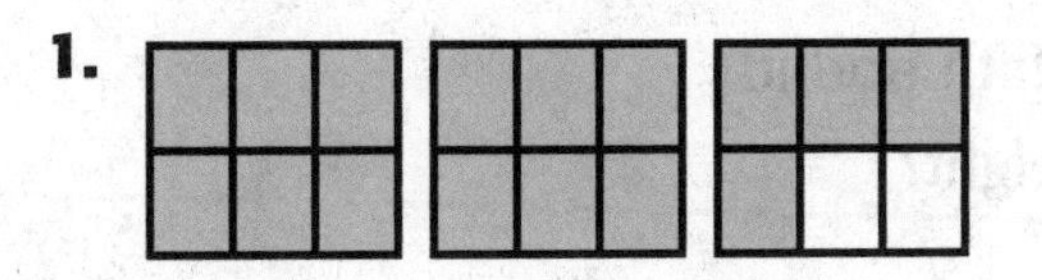

____

2. 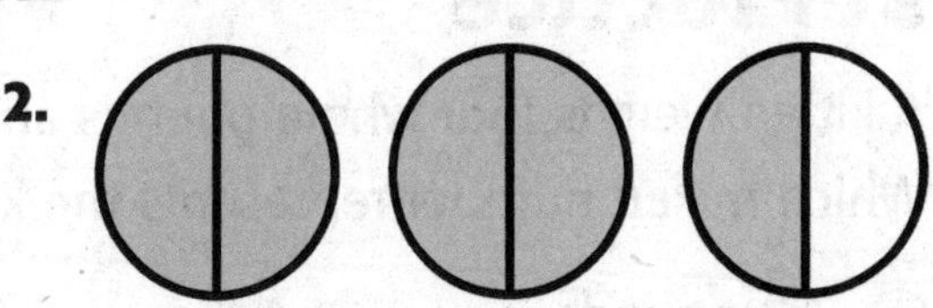

____

Write a mixed number for each model.

3.

4.

Algebra **Write an equation that represents each mixed number as a sum of whole numbers and unit fractions.**

5. $4\frac{1}{4}$

6. $1\frac{5}{6}$

Problem Solving

7. Mathematical **PRACTICE** 2 **Use Number Sense** There are 2 whole bagels and three-fourths of a third bagel. Write a mixed number that represents the amount of bagels.

8. There are 3 whole packages of pasta and one-third of a fourth package of pasta. Write a mixed number that represents the amount of packages.

Vocabulary Check

Vocab

9. Write an example of a mixed number.

Test Practice

10. A kitten weighs four whole pounds and $\frac{2}{3}$ of a fifth pound. Which mixed number represents the kitten's weight?

Ⓐ $3\frac{2}{3}$ pounds

Ⓑ $4\frac{1}{3}$ pounds

Ⓒ $4\frac{2}{3}$ pounds

Ⓓ $5\frac{2}{3}$ pounds

Name

Mixed Numbers and Improper Fractions

Lesson 10

ESSENTIAL QUESTION
How can different fractions name the same amount?

An **improper fraction** has a numerator that is greater than or equal to its denominator. Mixed numbers can be written as improper fractions.

Mixed Numbers	Improper Fractions
$1\frac{1}{2}$ $2\frac{3}{4}$ $3\frac{5}{6}$	$\frac{3}{2}$ $\frac{11}{4}$ $\frac{23}{6}$

Math in My World

Tools Tutor

Example 1

Nyoko is selling pies at a bake sale. Each pie has 5 slices. There are 7 slices left. What fraction of the pies is left?

One Way

Count the wholes and the parts.

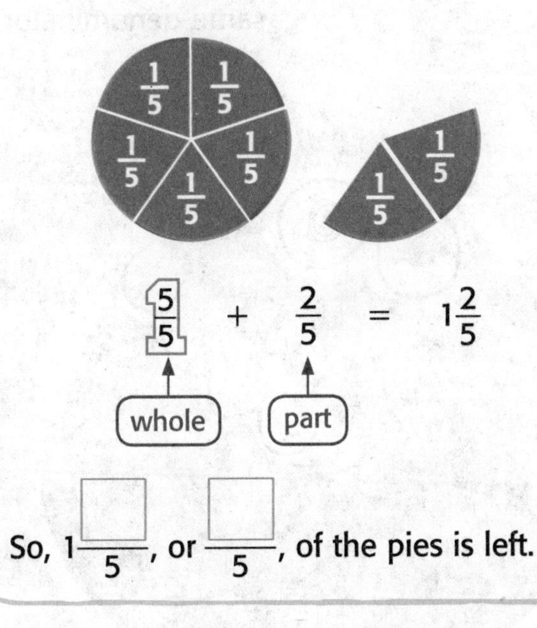

$\frac{5}{5} + \frac{2}{5} = 1\frac{2}{5}$

whole part

Another Way

Count the parts.

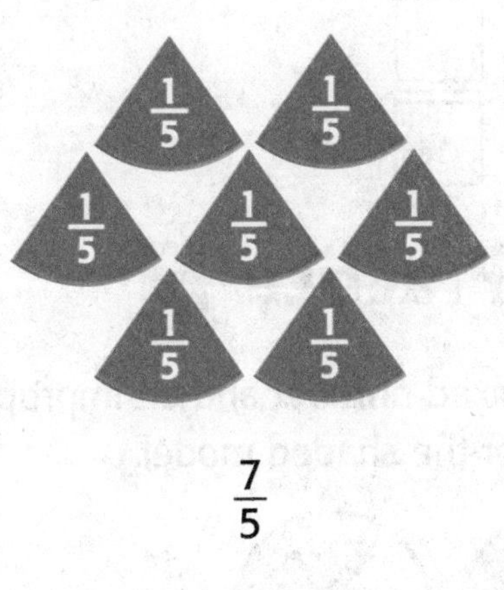

$\frac{7}{5}$

So, $1\frac{\square}{5}$, or $\frac{\square}{5}$, of the pies is left.

You can change a mixed number to an improper fraction.
You can also change an improper fraction to a mixed number.

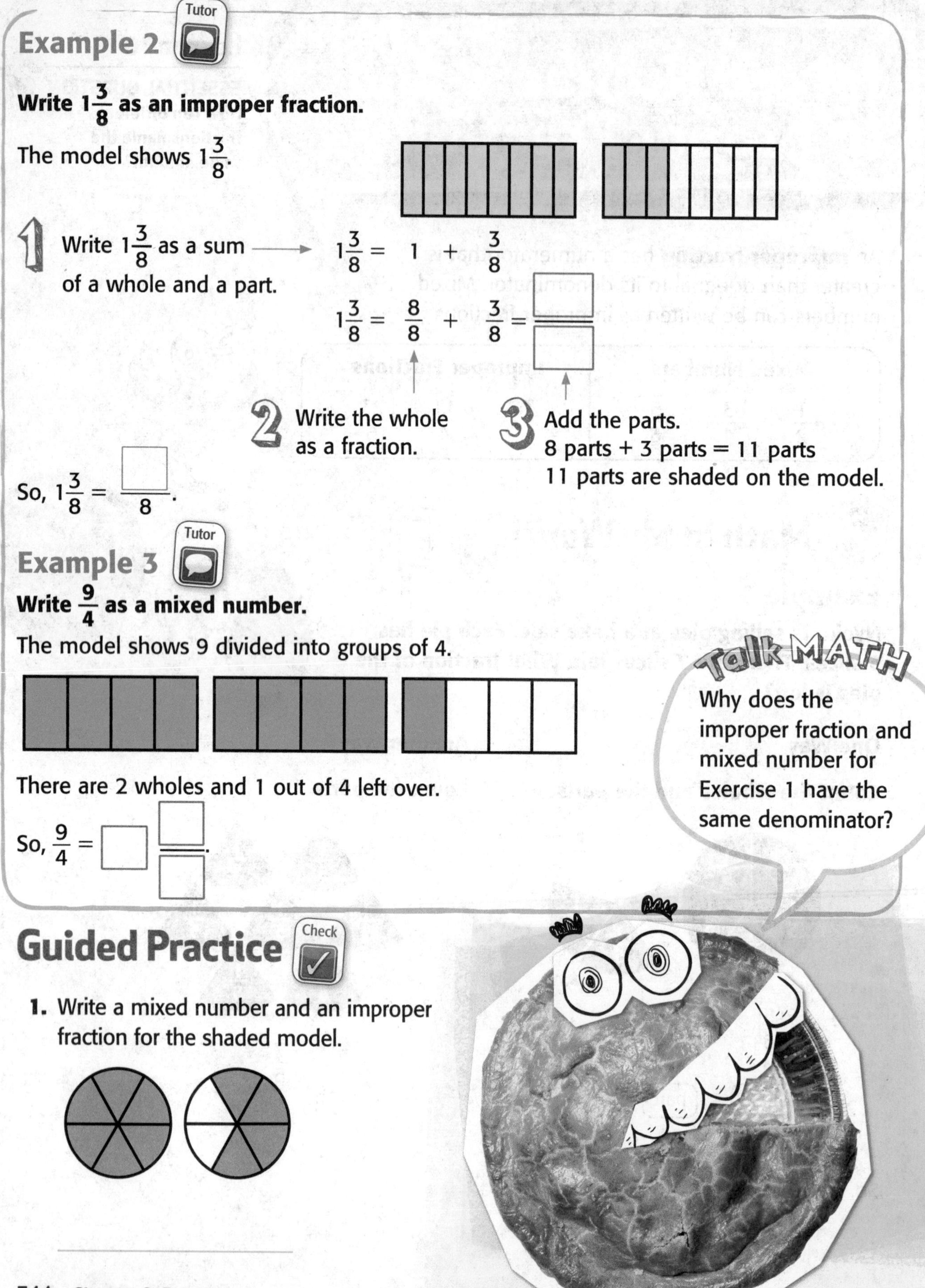

Example 2

Write $1\frac{3}{8}$ as an improper fraction.

The model shows $1\frac{3}{8}$.

1. Write $1\frac{3}{8}$ as a sum of a whole and a part. → $1\frac{3}{8} = 1 + \frac{3}{8}$

$1\frac{3}{8} = \frac{8}{8} + \frac{3}{8} = \frac{\square}{\square}$

2. Write the whole as a fraction.

3. Add the parts.
8 parts + 3 parts = 11 parts
11 parts are shaded on the model.

So, $1\frac{3}{8} = \frac{\square}{8}$.

Example 3

Write $\frac{9}{4}$ as a mixed number.

The model shows 9 divided into groups of 4.

There are 2 wholes and 1 out of 4 left over.

So, $\frac{9}{4} = \square\frac{\square}{\square}$.

Talk MATH

Why does the improper fraction and mixed number for Exercise 1 have the same denominator?

Guided Practice

1. Write a mixed number and an improper fraction for the shaded model.

Name

Independent Practice

Write a mixed number and an improper fraction for each shaded model.

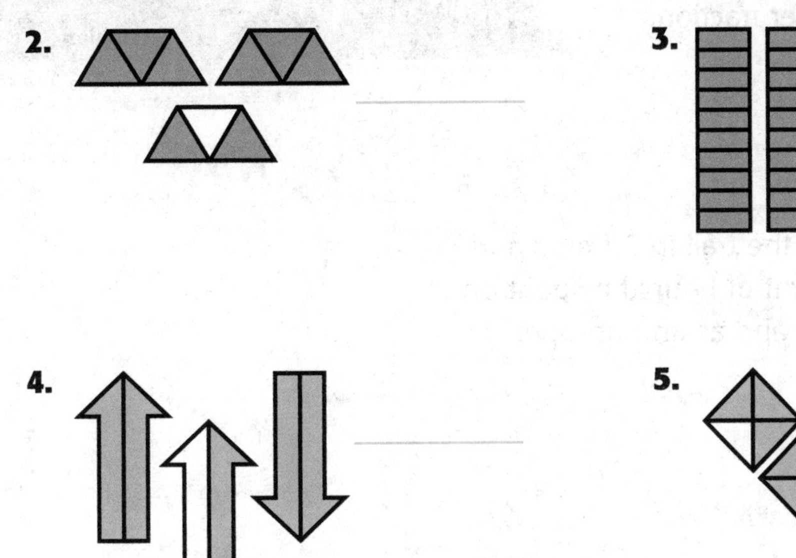

2.

3.

4.

5.

Draw models to write each mixed number as an improper fraction.

6. $1\frac{3}{5} =$ ______

7. $2\frac{3}{4} =$ ______

8. $1\frac{7}{10} =$ ______

Draw models to write each improper fraction as a mixed number.

9. $\frac{11}{8} =$ ______

10. $\frac{9}{6} =$ ______

11. $\frac{7}{3} =$ ______

Problem Solving

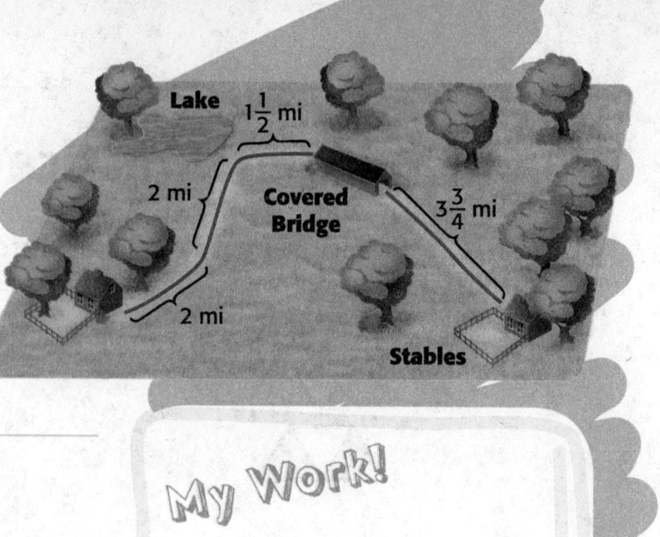

A diagram of a horseback riding tour is shown. There are resting stops along the trail.

12. Write the distance from the Covered Bridge to the Stables as an improper fraction.

My Work!

13. Joaquin reached the end of the trail in 2 hours and 15 minutes. Write the amount of hours he spent on the trail as a mixed number and as an improper fraction.

14. Mathematical PRACTICE 4 **Model Math** Kelly walked 3 miles. Abby walked $\frac{3}{4}$ mile. How far did they walk in all?

HOT Problems

15. Mathematical PRACTICE 2 **Use Number Sense** Name an improper fraction that can be written as a whole number.

16. **Building on the Essential Question** How are improper fractions and mixed numbers alike? How are they different?

Name ....

Lesson 10

Mixed Numbers and Improper Fractions

Homework Helper

eHelp Need help? connectED.mcgraw-hill.com

Kelsey made 2 pitchers of lemonade. Each pitcher holds 6 cups. She poured 4 cups of lemonade from one pitcher. What fraction of the lemonade is left?

One Way **Count the wholes and the parts.**

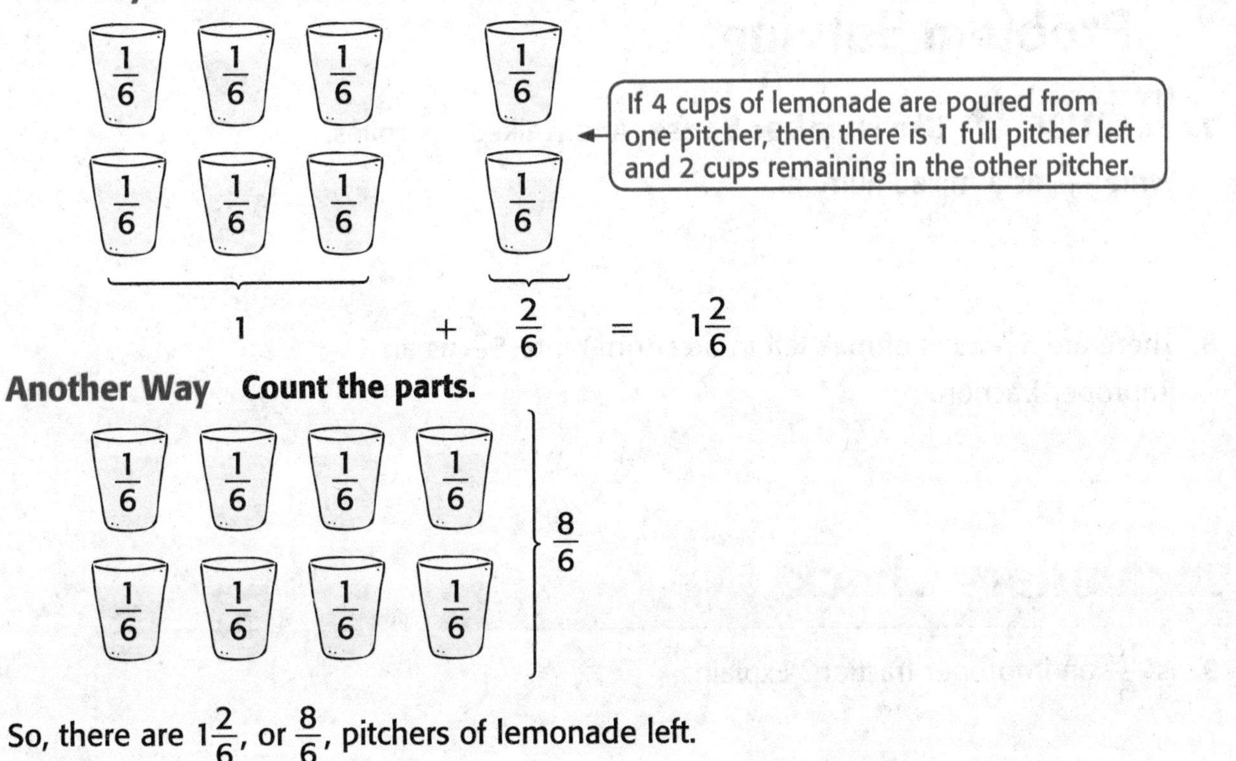

So, there are $1\frac{2}{6}$, or $\frac{8}{6}$, pitchers of lemonade left.

Practice

Write a mixed number and an improper fraction for each shaded model.

1. 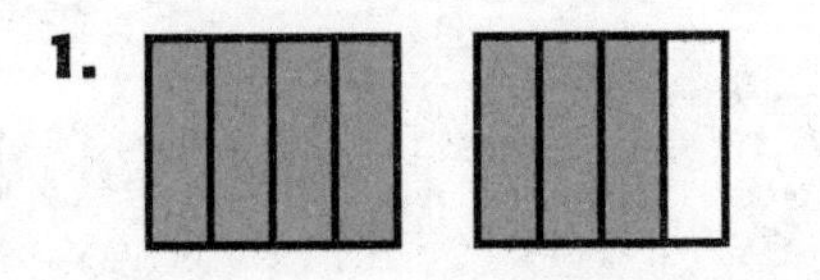

______

2. 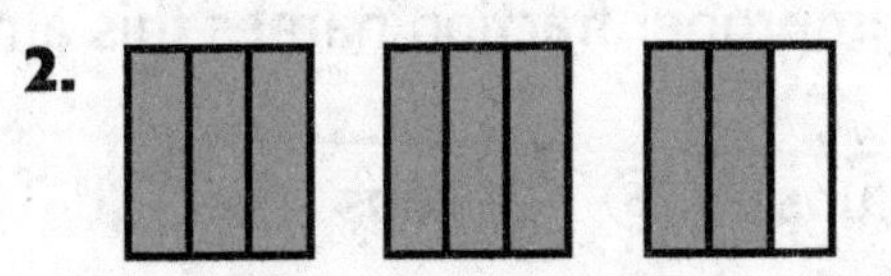

______

Write a mixed number and an improper fraction for each model.

3. **4.**

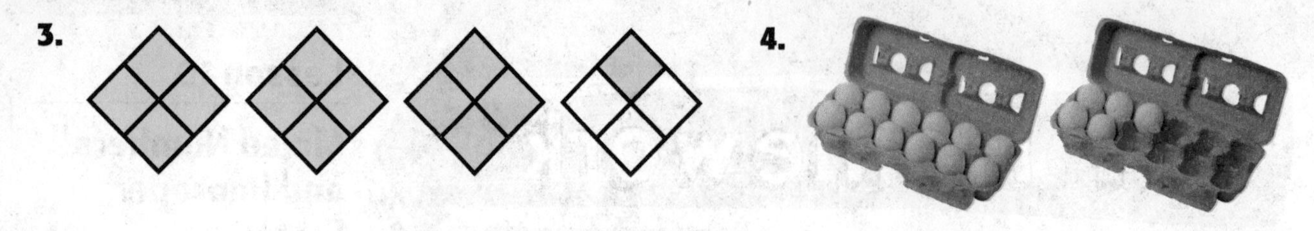

5. Draw a model to write $2\frac{3}{5}$ as an improper fraction.

6. Draw a model to write $\frac{30}{4}$ as a mixed number.

Problem Solving

7. Mathematical PRACTICE 2 **Use Number Sense** Ana walked $\frac{13}{3}$ miles. Write $\frac{13}{3}$ as a mixed number.

8. There are $5\frac{4}{5}$ cups of milk left in a carton. Write $5\frac{4}{5}$ as an improper fraction.

Vocabulary Check

Vocab

9. Is $\frac{10}{3}$ an improper fraction? Explain.

Test Practice

10. Amelia needs $3\frac{2}{3}$ cups of sugar to make cupcakes. Which improper fraction names this amount?

Ⓐ $\frac{5}{3}$ cups
Ⓑ $\frac{8}{3}$ cups
Ⓒ $\frac{11}{3}$ cups
Ⓓ $\frac{18}{3}$ cups

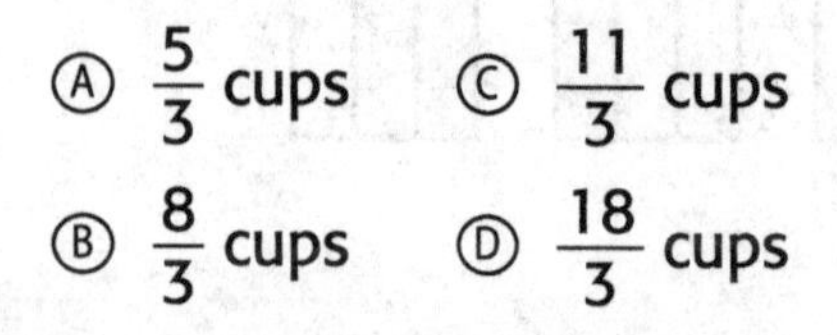

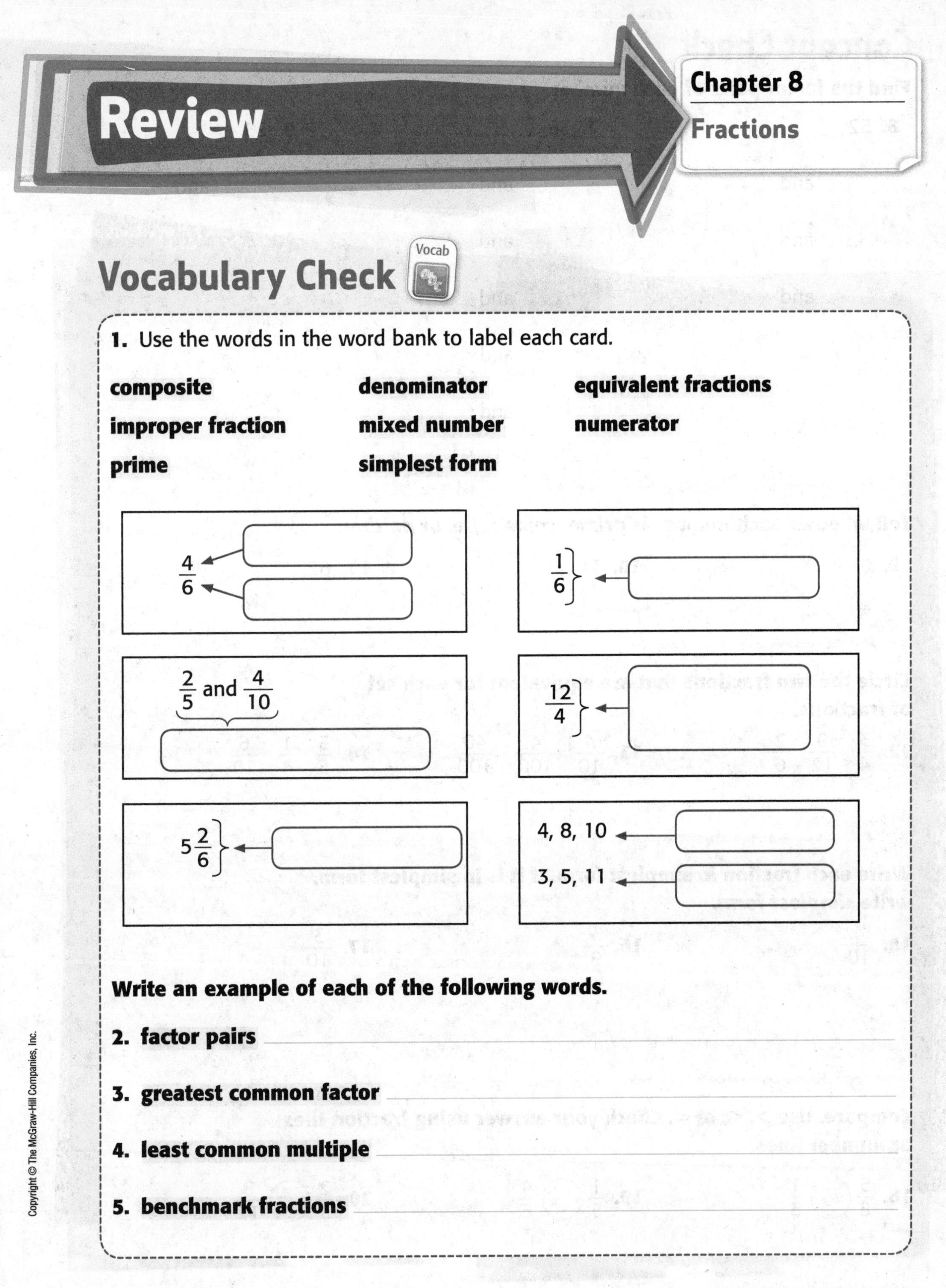

Review

Chapter 8
Fractions

Vocabulary Check

1. Use the words in the word bank to label each card.

composite	**denominator**	**equivalent fractions**
improper fraction	**mixed number**	**numerator**
prime	**simplest form**	

Write an example of each of the following words.

2. factor pairs ____________________

3. greatest common factor ____________________

4. least common multiple ____________________

5. benchmark fractions ____________________

Concept Check

Find the factor pairs of each number.

6. 52

____ and ____

____ and ____

____ and ____

7. 36

____ and ____

____ and ____

____ and ____

____ and ____

____ and ____

8. 23

____ and ____

Tell whether each number is *prime, composite,* or *neither.*

9. 0 ________

10. 31 ________

11. 62 ________

Circle the two fractions that are equivalent for each set of fractions.

12. $\frac{3}{4}$ $\frac{9}{12}$ $\frac{2}{6}$

13. $\frac{4}{10}$ $\frac{4}{100}$ $\frac{40}{100}$

14. $\frac{3}{5}$ $\frac{1}{4}$ $\frac{6}{10}$

Write each fraction in simplest form. If it is in simplest form, write *simplest form.*

15. $\frac{4}{10}$ ______

16. $\frac{3}{9}$ ______

17. $\frac{3}{10}$ ______

Compare. Use >, <, or =. Check your answer using fraction tiles or number lines.

18. $\frac{5}{8}$ ◯ $\frac{1}{3}$

19. $\frac{1}{5}$ ◯ $\frac{4}{6}$

20. $\frac{2}{3}$ ◯ $\frac{8}{12}$

Name

Problem Solving

21. Mrs. Evans has 13 pictures to hang on a wall. Is there any way she can arrange the pictures in rows other than 1×13 or 13×1, so that the same number of pictures is in each row? Tell whether 13 is a composite or prime number. Explain.

22. There are $\frac{2}{8}$ cup of peanuts and $\frac{1}{4}$ cup of walnuts. Is there a greater amount of peanuts or walnuts? Explain.

23. Mia has two whole bananas and one-fifth of another banana. Write a mixed number that represents the amount of bananas she has.

24. Write a real-world problem to compare fractions. Then solve the problem.

My Work!

Test Practice

25. Which equation is true?

Ⓐ $2\frac{2}{3} = 2 + 2 + 3$

Ⓑ $2\frac{2}{3} = 1 + 1 + 2 + 3$

Ⓒ $2\frac{2}{3} = 1 + 1 + \frac{1}{3} + \frac{1}{3}$

Ⓓ $2\frac{2}{3} = 1 + 1 + \frac{1}{3}$

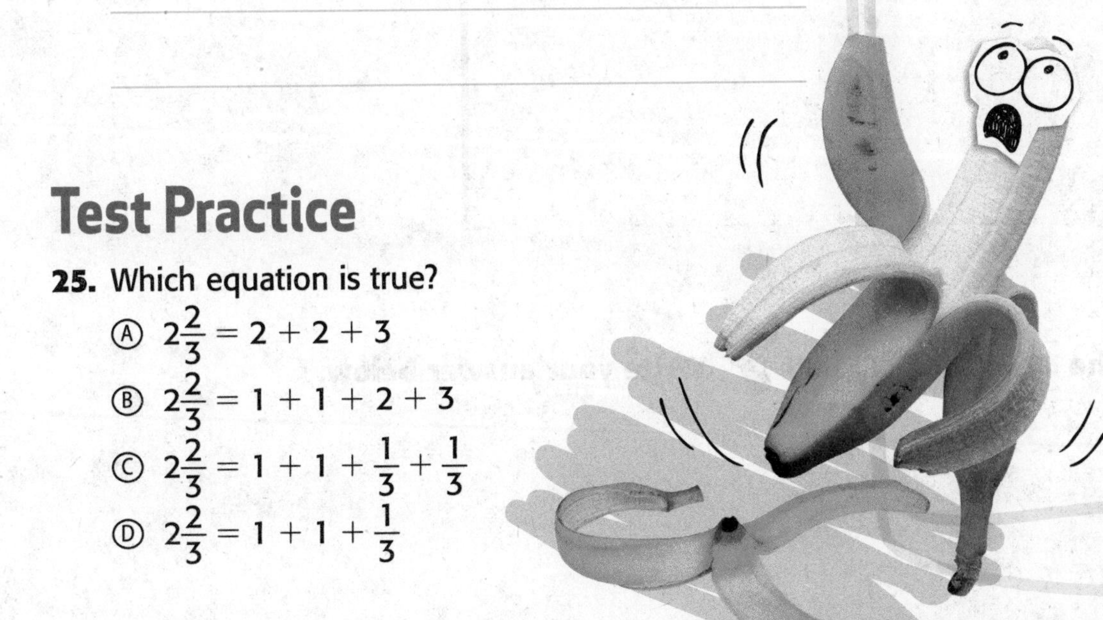

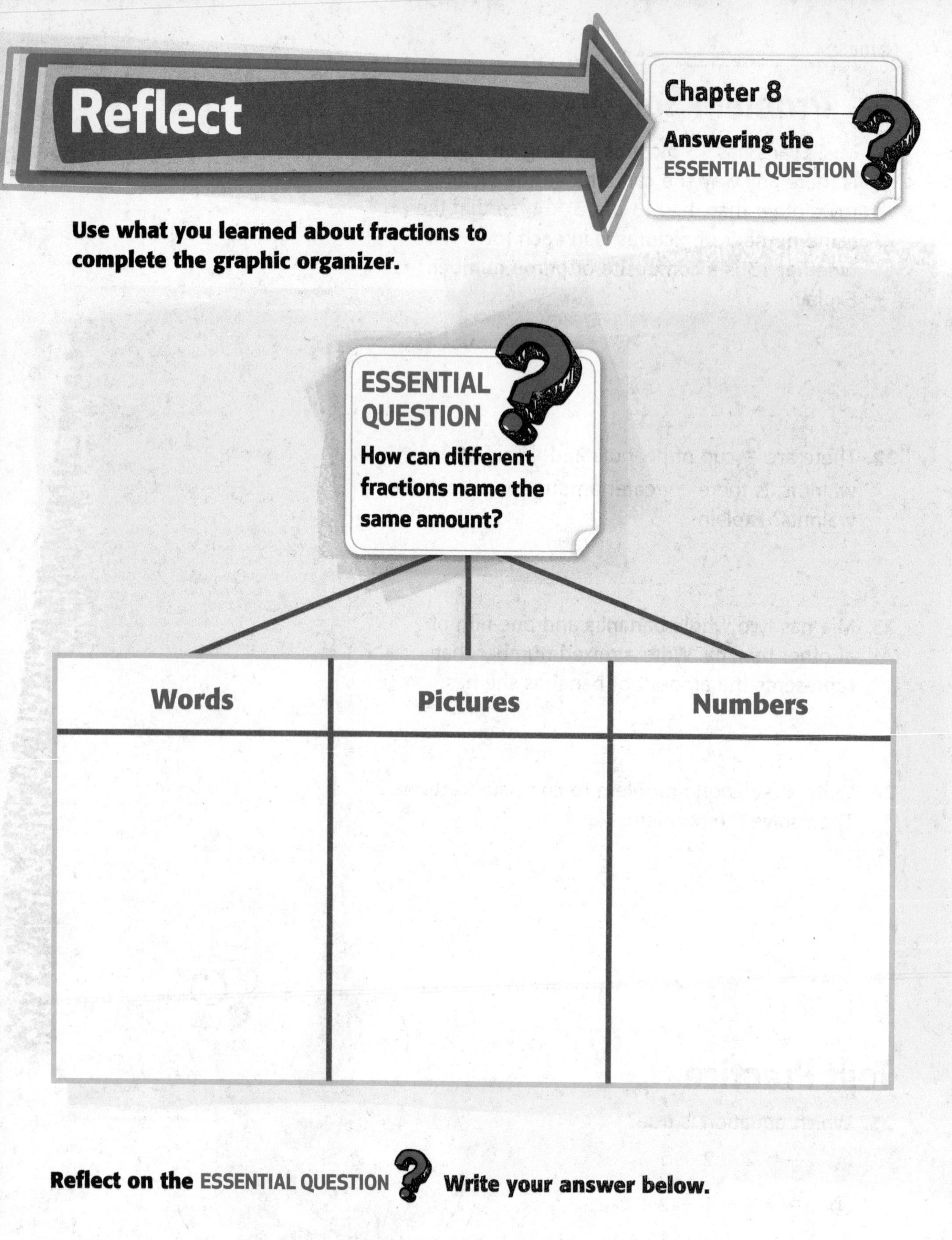

Reflect
Chapter 8
Answering the ESSENTIAL QUESTION
Use what you learned about fractions to complete the graphic organizer.
ESSENTIAL QUESTION
How can different fractions name the same amount?
Words
Pictures
Numbers
Reflect on the ESSENTIAL QUESTION Write your answer below.

Chapter 9

Operations with Fractions

ESSENTIAL QUESTION

How can I use operations to model real-world fractions?

Let's Play Games!

Watch

Watch a video!

MY Common Core State Standards

Number and Operations – Fractions

4.NF.3 Understand a fraction $\frac{a}{b}$ with $a > 1$ as a sum of fractions $\frac{1}{b}$.

4.NF.3a Understand addition and subtraction of fractions as joining and separating parts referring to the same whole.

4.NF.3b Decompose a fraction into a sum of fractions with the same denominator in more than one way, recording each decomposition by an equation. Justify decompositions, e.g., by using a visual fraction model.

4.NF.3c Add and subtract mixed numbers with like denominators, e.g., by replacing each mixed number with an equivalent fraction, and/or by using properties of operations and the relationship between addition and subtraction.

4.NF.3d Solve word problems involving addition and subtraction of fractions referring to the same whole and having like denominators, e.g., by using visual fraction models and equations to represent the problem.

4.NF.4 Apply and extend previous understandings of multiplication to multiply a fraction by a whole number.

4.NF.4a Understand a fraction $\frac{a}{b}$ as a multiple of $\frac{1}{b}$.

4.NF.4b Understand a multiple of $\frac{a}{b}$ as a multiple of $\frac{1}{b}$, and use this understanding to multiply a fraction by a whole number.

4.NF.4c Solve word problems involving multiplication of a fraction by a whole number, e.g., by using visual fraction models and equations to represent the problem.

Looks like I'll be doing a lot in this chapter!

Standards for Mathematical PRACTICE

1. Make sense of problems and persevere in solving them.
2. Reason abstractly and quantitatively.
3. Construct viable arguments and critique the reasoning of others.
4. Model with mathematics.
5. Use appropriate tools strategically.
6. Attend to precision.
7. Look for and make use of structure.
8. Look for and express regularity in repeated reasoning.

= focused on in this chapter

Name

Am I Ready?

Check

← Go online to take the Readiness Quiz

Compare. Use <, >, or =.

1. $\frac{1}{3}$ ◯ $\frac{1}{6}$
2. $\frac{4}{5}$ ◯ $\frac{8}{10}$
3. $\frac{2}{5}$ ◯ $\frac{2}{3}$
4. $\frac{2}{10}$ ◯ $\frac{10}{100}$
5. $\frac{3}{5}$ ◯ $\frac{2}{8}$
6. $\frac{1}{4}$ ◯ $\frac{2}{8}$

Write each fraction in simplest form.

7. $\frac{4}{6}$ ______
8. $\frac{6}{10}$ ______
9. $\frac{3}{12}$ ______
10. $\frac{60}{100}$ ______
11. $\frac{5}{10}$ ______
12. $\frac{4}{12}$ ______

Write an equivalent fraction.

13. $\frac{1}{3}$ ______
14. $\frac{3}{4}$ ______
15. $\frac{7}{10}$ ______
16. $\frac{3}{5}$ ______
17. $\frac{2}{10}$ ______
18. $\frac{90}{100}$ ______

19. Hannah has a garden with basil, rosemary, and parsley. One sixth of her garden has basil. One half of her garden has rosemary. One-third of her garden has parsley. Draw and label a picture of Hannah's garden.

My Drawing!

How Did I Do?

Shade the boxes to show the problems you answered correctly.

1	2	3	4	5	6	7	8	9	10	11	12	13	14	15	16	17	18	19

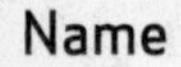

Name

Review Vocabulary

denominator mixed number numerator simplest form

Making Connections
Write or draw an example and a non-example of each review vocabulary.

Fractions

Word	Example	Non-Example
denominator		
mixed number		
numerator		
simplest form		

MY Vocabulary Cards
Vocab
abc
Mathematical PRACTICE
Lesson 9-1
like fractions
1/6
3/6
4/6

Ideas for Use

- Write a tally mark on each card every time you read the word in this chapter or use it in your writing. Challenge yourself to use at least 5 tally marks.
- Ask students to use the blank cards to draw or write phrases or examples that will help them with concepts like *adding like fractions* and *subtracting mixed numbers.*

Fractions that have the same denominator.

Like can mean "of the same form or kind." How does this help you remember the meaning of *like fractions*?

MY Foldable

FOLDABLES® Follow the steps on the back to make your Foldable.

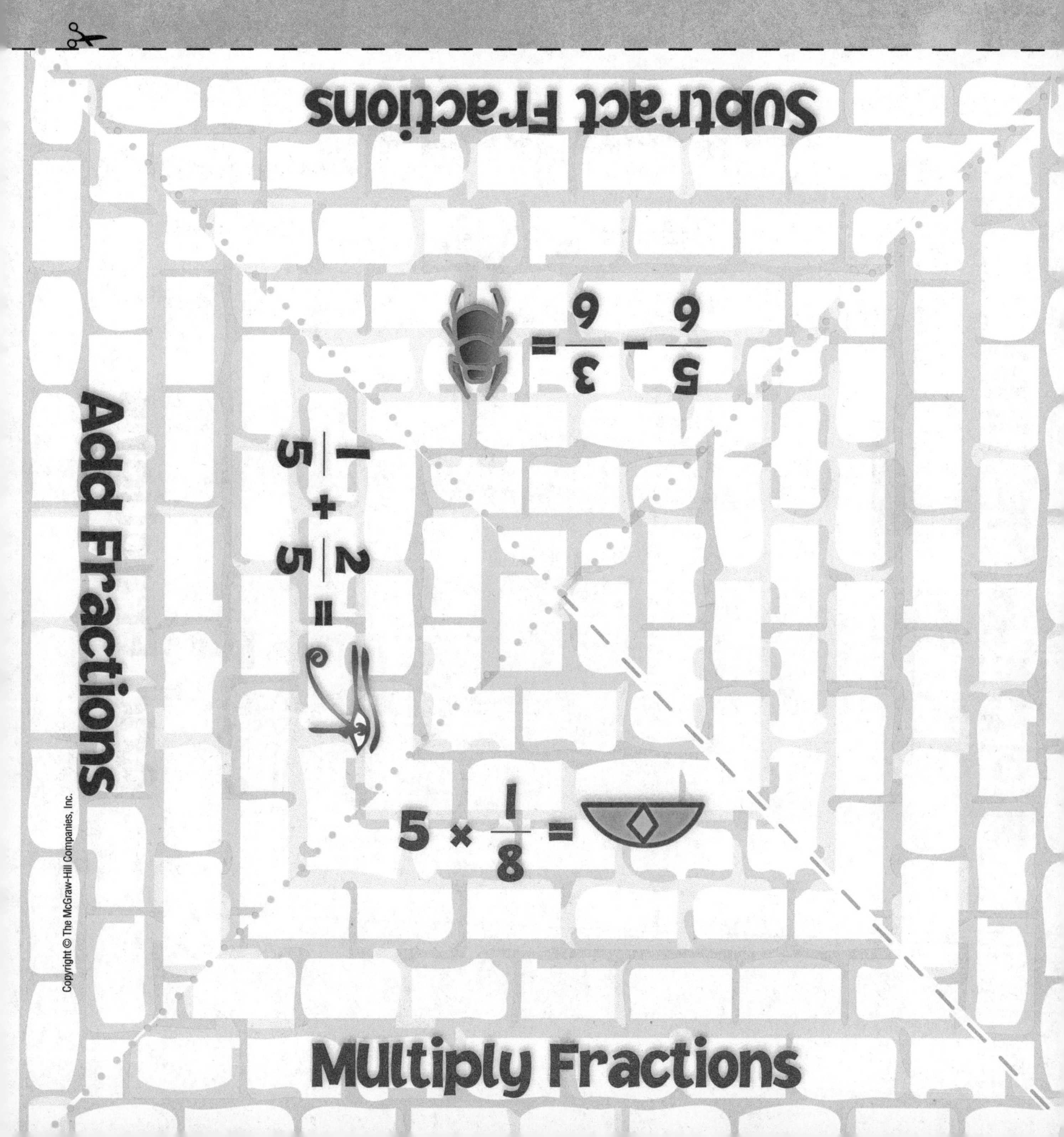

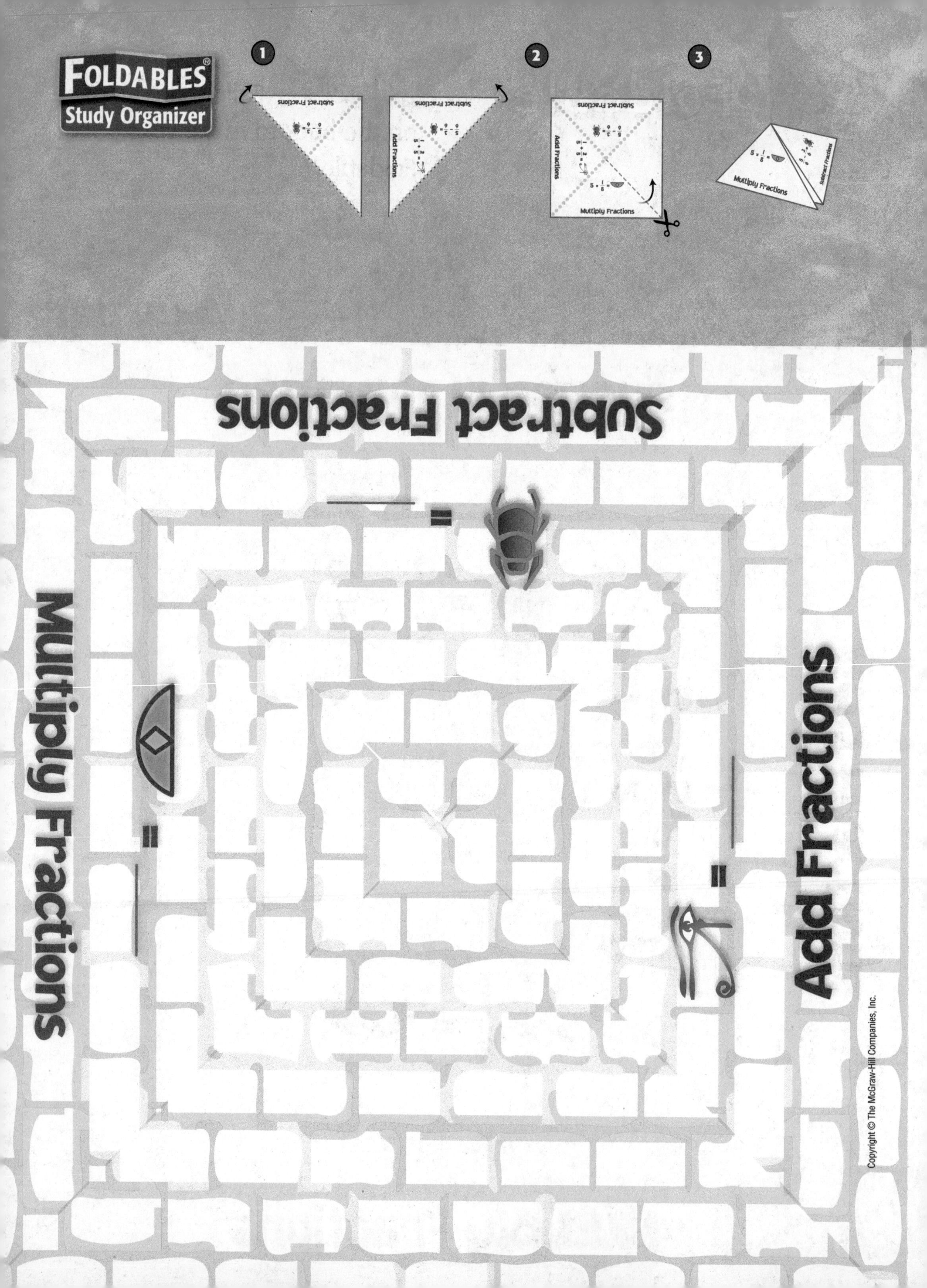
FOLDABLES®
Study Organizer
Subtract Fractions
Multiply Fractions
Add Fractions

Hands On
Use Models to Add Like Fractions

Lesson 1

ESSENTIAL QUESTION
How can I use operations to model real-world fractions?

A unit fraction has a numerator of 1. It is one of the equal parts of the whole. Fractions that have the same denominators are called **like fractions**. You can use models to add like fractions.

Build It

Abigail is playing a board game with her friends. There are 5 game pieces. There is a red piece, a blue piece, a yellow piece, a green piece, and a purple piece. What is the total fraction of the pieces that are yellow, green, or purple?

Find $\frac{1}{5} + \frac{1}{5} + \frac{1}{5}$.

1 **Model three $\frac{1}{5}$-fraction tiles.**

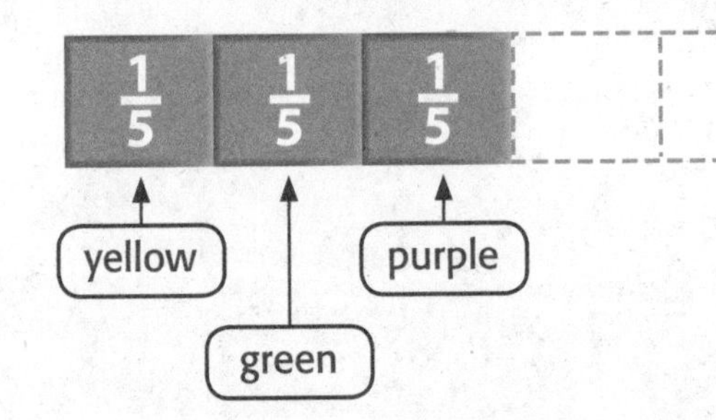

Each game piece represents a unit fraction.

2 **Add the like fractions.**

There are three $\frac{1}{5}$-fraction tiles altogether.

So, $\frac{1}{5} + \frac{1}{5} + \frac{1}{5} = \frac{3}{5}$. $\frac{3}{5}$ is the sum of three unit fractions of $\frac{1}{5}$.

So, the total fraction of game pieces that are yellow, green, or purple is $\frac{\square}{\square}$.

Not all fractions are unit fractions. You can decompose, or break down, a fraction into a sum of unit fractions.

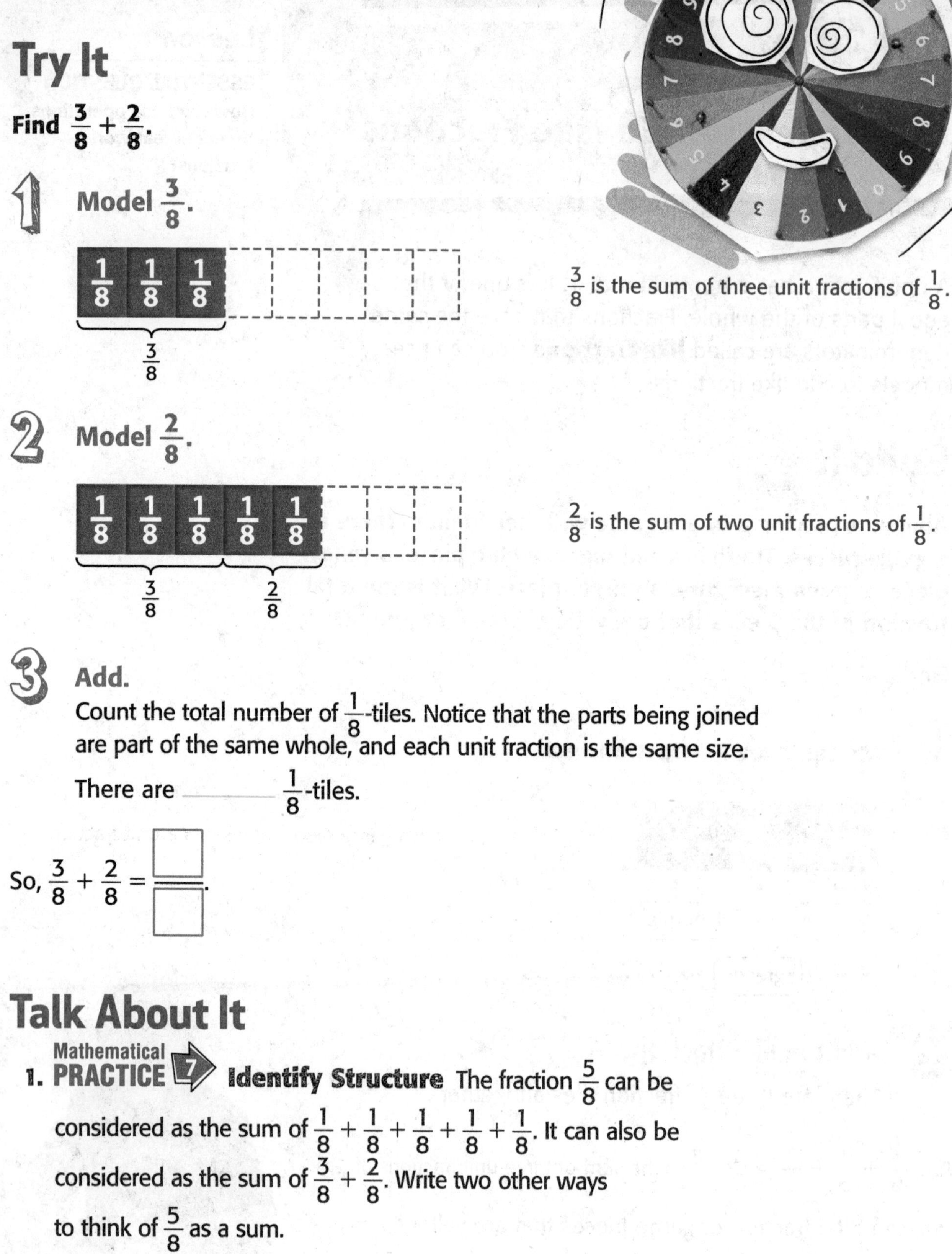

Try It

Find $\frac{3}{8} + \frac{2}{8}$.

1 **Model $\frac{3}{8}$.**

$\frac{1}{8}$	$\frac{1}{8}$	$\frac{1}{8}$					

$\frac{3}{8}$

$\frac{3}{8}$ is the sum of three unit fractions of $\frac{1}{8}$.

2 **Model $\frac{2}{8}$.**

$\frac{1}{8}$	$\frac{1}{8}$	$\frac{1}{8}$	$\frac{1}{8}$	$\frac{1}{8}$			

$\frac{3}{8}$ $\frac{2}{8}$

$\frac{2}{8}$ is the sum of two unit fractions of $\frac{1}{8}$.

3 **Add.**

Count the total number of $\frac{1}{8}$-tiles. Notice that the parts being joined are part of the same whole, and each unit fraction is the same size.

There are ______ $\frac{1}{8}$-tiles.

So, $\frac{3}{8} + \frac{2}{8} = \frac{\square}{\square}$.

Talk About It

1. **Mathematical PRACTICE 7 Identify Structure** The fraction $\frac{5}{8}$ can be considered as the sum of $\frac{1}{8} + \frac{1}{8} + \frac{1}{8} + \frac{1}{8} + \frac{1}{8}$. It can also be considered as the sum of $\frac{3}{8} + \frac{2}{8}$. Write two other ways to think of $\frac{5}{8}$ as a sum.

____

____

Name

CCSS

Practice It

Model the sum using fraction tiles. Draw the model. Then add.

2. $\frac{1}{6} + \frac{4}{6} =$ ______

3. $\frac{4}{8} + \frac{2}{8} =$ ______

Use the table to answer Exercises 4 and 5.

Addends	Sum
$\frac{1}{5} + \frac{1}{5} + \frac{1}{5}$	$\frac{3}{5}$
$\frac{3}{8} + \frac{2}{8}$	$\frac{5}{8}$
$\frac{1}{10} + \frac{1}{10} + \frac{7}{10}$	$\frac{9}{10}$

4. The table shows the sums of several like fractions. Study the pattern in the table.

Write a rule that you can use to add like fractions without using models.

______

______

______

5. Use your rule from Exercise 4 to find $\frac{1}{4} + \frac{1}{4} + \frac{1}{4}$.

______

Algebra **Write each fraction as a sum of unit fractions. Then write an equation to decompose the fraction in a different way.**

6. $\frac{7}{8}$ ______

______

7. $\frac{5}{10}$ ______

______

Apply It

My Work!

8. Mathematical PRACTICE 2 **Use Number Sense** Claire walked $\frac{1}{3}$ mile on Saturday and $\frac{1}{3}$ mile on Sunday. How many miles did she walk in all?

______

9. Miguel cut a cantaloupe into 8 slices. He ate 1 slice and his friends ate 4 slices. What fraction of the cantaloupe did they eat?

______

10. There are 12 children at lunch. Six children are eating apples, four children are eating grapes, and two children are eating celery. What fraction of the children are eating fruit?

______

11. Mathematical PRACTICE 4 **Model Math** Write three different ways to decompose $\frac{9}{10}$ into a sum.

______

______

______

Write About It

12. When adding like fractions, why is the numerator the only part of the fraction that changes?

______

______

______

Name

MY Homework

Lesson 1

Hands On: Use Models to Add Like Fractions

Homework Helper

eHelp Need help? connectED.mcgraw-hill.com

Caroline has 6 cups. She fills one cup with juice for each of her 5 friends. What is the fraction of the total number of cups that Caroline gives to her friends?

Each cup is $\frac{1}{6}$ of the total number of cups.

Find $\frac{1}{6} + \frac{1}{6} + \frac{1}{6} + \frac{1}{6} + \frac{1}{6}$.

1 Use five $\frac{1}{6}$-fraction tiles to model the addition.

$\frac{1}{6}$	$\frac{1}{6}$	$\frac{1}{6}$	$\frac{1}{6}$	$\frac{1}{6}$	

2 Add the like fractions.

Count the total number of $\frac{1}{6}$-fraction tiles. There are 5.

$\frac{1}{6} + \frac{1}{6} + \frac{1}{6} + \frac{1}{6} + \frac{1}{6} = \frac{5}{6}$

So, the fraction of the total number of cups that Caroline gives to her friends is $\frac{5}{6}$.

Practice

Model the sum using fraction tiles. Draw the model. Then add.

1. $\frac{1}{3} + \frac{1}{3} =$ ______

2. $\frac{2}{12} + \frac{6}{12} =$ ______

3. $\frac{3}{10} + \frac{4}{10} =$ ______

4. $\frac{5}{8} + \frac{1}{8} =$ ______

Write an addition sentence for each model. Then find the sum.

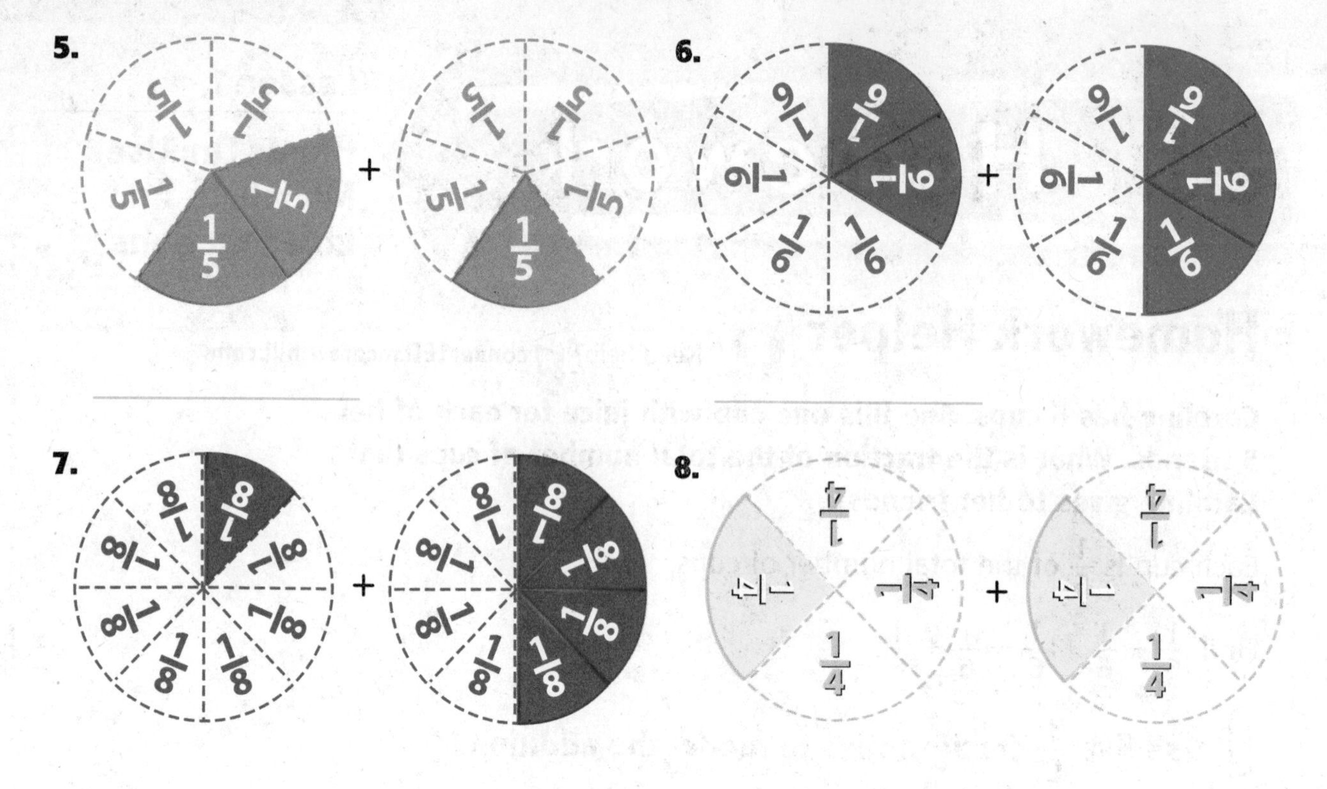

Problem Solving

9. **Mathematical PRACTICE 1 Make Sense of Problems** Nelson got a pack of 12 pencils. He took 3 pencils to put in his desk at school and 2 pencils to put in his locker at school. What is the fraction of total pencils that Nelson took to school?

10. There are 10 chicks on Ginger's farm. She has 2 chicks in one outdoor pen and 5 chicks in another. The rest of the chicks are in the barn. What is the fraction of total chicks in outdoor pens?

Vocabulary Check

Vocab

11. Explain what is the same about two fractions that are like fractions.

Name ....

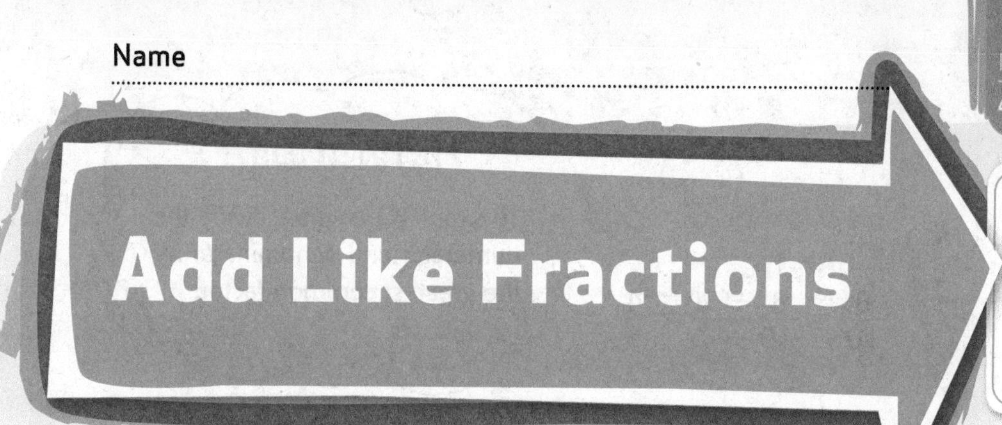

Add Like Fractions

Lesson 2

ESSENTIAL QUESTION
How can I use operations to model real-world fractions?

Think of adding like fractions as joining parts that refer to the same whole. To add like fractions, add the numerators and keep the same denominator.

$\frac{1}{5} + \frac{2}{5} = \frac{3}{5}$

$\frac{3}{10} + \frac{4}{10} = \frac{7}{10}$

Math in My World

Tools Watch Tutor

Example 1

Pablo spent $\frac{2}{6}$ of an hour on a jigsaw puzzle. Conrad spent $\frac{1}{6}$ of an hour on the puzzle. How much time did they spend working on the puzzle in all?

Find $\frac{2}{6} + \frac{1}{6}$.

1 Add the numerators.
Keep the same denominator.

$$\frac{2}{6} + \frac{1}{6} = \frac{2+1}{6}$$
$$= \frac{3}{6}$$

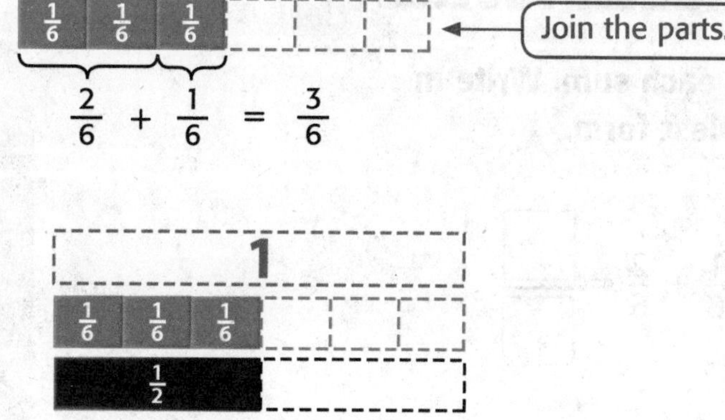

2 Write the sum in simplest form.

$$\frac{3}{6} = \frac{3 \div 3}{6 \div 3} = \frac{1}{2}$$

So, they spent $\frac{\square}{\square}$ hour on the puzzle.

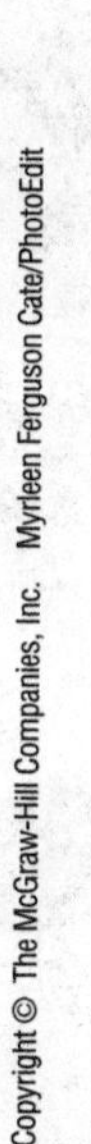

Example 2

Find $\frac{3}{10} + \frac{5}{10}$.

Helpful Hint

To simplify a fraction, divide the numerator and denominator by the greatest common factor.

1 Add.

$$\frac{3}{10} + \frac{5}{10} = \frac{3+5}{10}$$

$$= \frac{8}{10}$$

2 Simplify.

$$\frac{8}{10} = \frac{8 \div 2}{10 \div 2} = \frac{4}{5}$$

So, $\frac{3}{10} + \frac{5}{10} = \frac{\square}{\square}$.

Check Use models to check.

$\frac{3}{10}$ $\frac{5}{10}$

$\frac{1}{10}$	$\frac{1}{10}$	$\frac{1}{10}$	$\frac{1}{10}$	$\frac{1}{10}$	$\frac{1}{10}$	$\frac{1}{10}$	$\frac{1}{10}$		

$\frac{3}{10} + \frac{5}{10} = \frac{8}{10}$

$\frac{1}{5}$	$\frac{1}{5}$	$\frac{1}{5}$	$\frac{1}{5}$	

$\frac{4}{5}$

$\frac{8}{10} = \frac{4}{5}$

Talk MATH

Describe two ways to decompose $\frac{4}{5}$ into a sum.

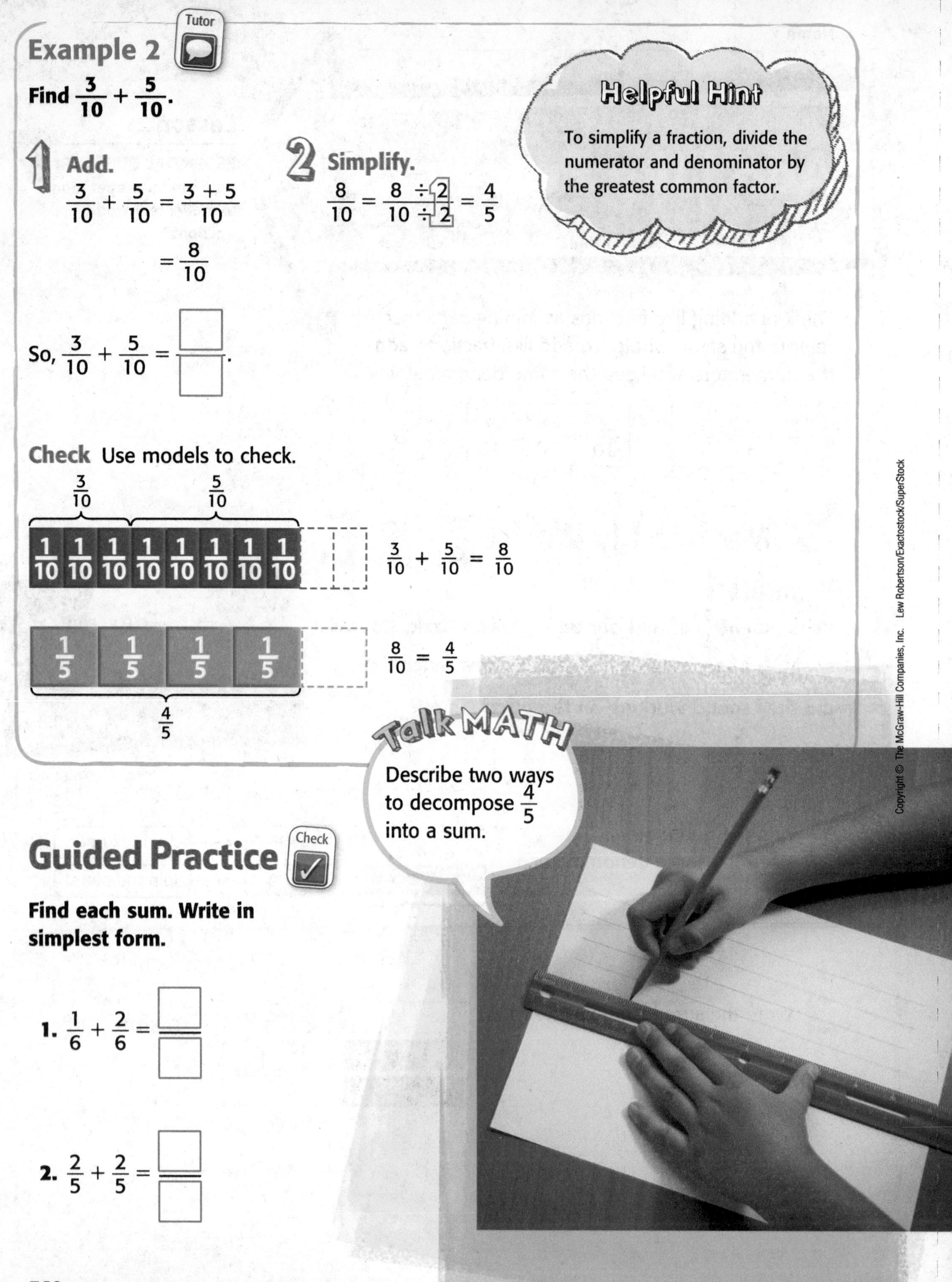

Guided Practice

Find each sum. Write in simplest form.

1. $\frac{1}{6} + \frac{2}{6} = \frac{\square}{\square}$

2. $\frac{2}{5} + \frac{2}{5} = \frac{\square}{\square}$

CCSS

Name ______________________________

Independent Practice

Find each sum. Write in simplest form.

3. $\frac{1}{3} + \frac{1}{3} =$ ______

4. $\frac{2}{6} + \frac{3}{6} =$ ______

5. $\frac{3}{8} + \frac{1}{8} =$ ______

6. $\frac{5}{8} + \frac{2}{8} =$ ______

7. $\frac{3}{10} + \frac{7}{10} =$ ______

8. $\frac{2}{5} + \frac{1}{5} =$ ______

9. $\frac{2}{12} + \frac{5}{12} =$ ______

10. $\frac{1}{4} + \frac{2}{4} =$ ______

11. $\frac{1}{6} + \frac{2}{6} =$ ______

Algebra **Write each fraction as a sum of unit fractions. Then write an equation to decompose the fraction in a different way.**

12. $\frac{4}{5}$ ______________________________

____

13. $\frac{3}{4}$ ______________________________

____

14. $\frac{3}{8}$ ______________________________

____

Real World! Problem Solving

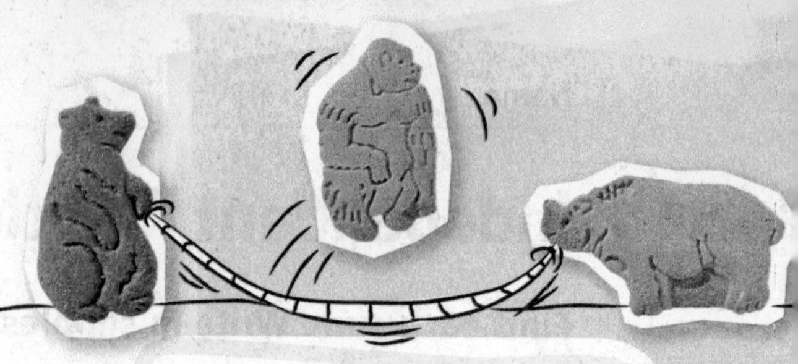

My Work!

15. Marcela ate $\frac{4}{10}$ of a box of crackers last week.

Then she ate $\frac{2}{10}$ of the box of crackers this week.

What fraction of the box of crackers did Marcela eat altogether? Write in simplest form.

16. Greyson's mom used 4 out of 12 eggs to make pancakes. Then she used 3 out of 12 eggs to make cupcakes. What fraction of a dozen eggs did she use in all? *(Hint: 1 dozen =12)*

17. Mathematical PRACTICE 4 **Model Math** Bryce has to complete 100 minutes of writing every ten days. He writes for 10 minutes each day. What fraction of his writing requirement does Bryce complete in five days? Write in simplest form.

HOT Problems

18. Mathematical PRACTICE 2 **Stop and Reflect** Write two fractions whose sum is greater than 1.

19. **Building on the Essential Question** How can I add like fractions?

Name

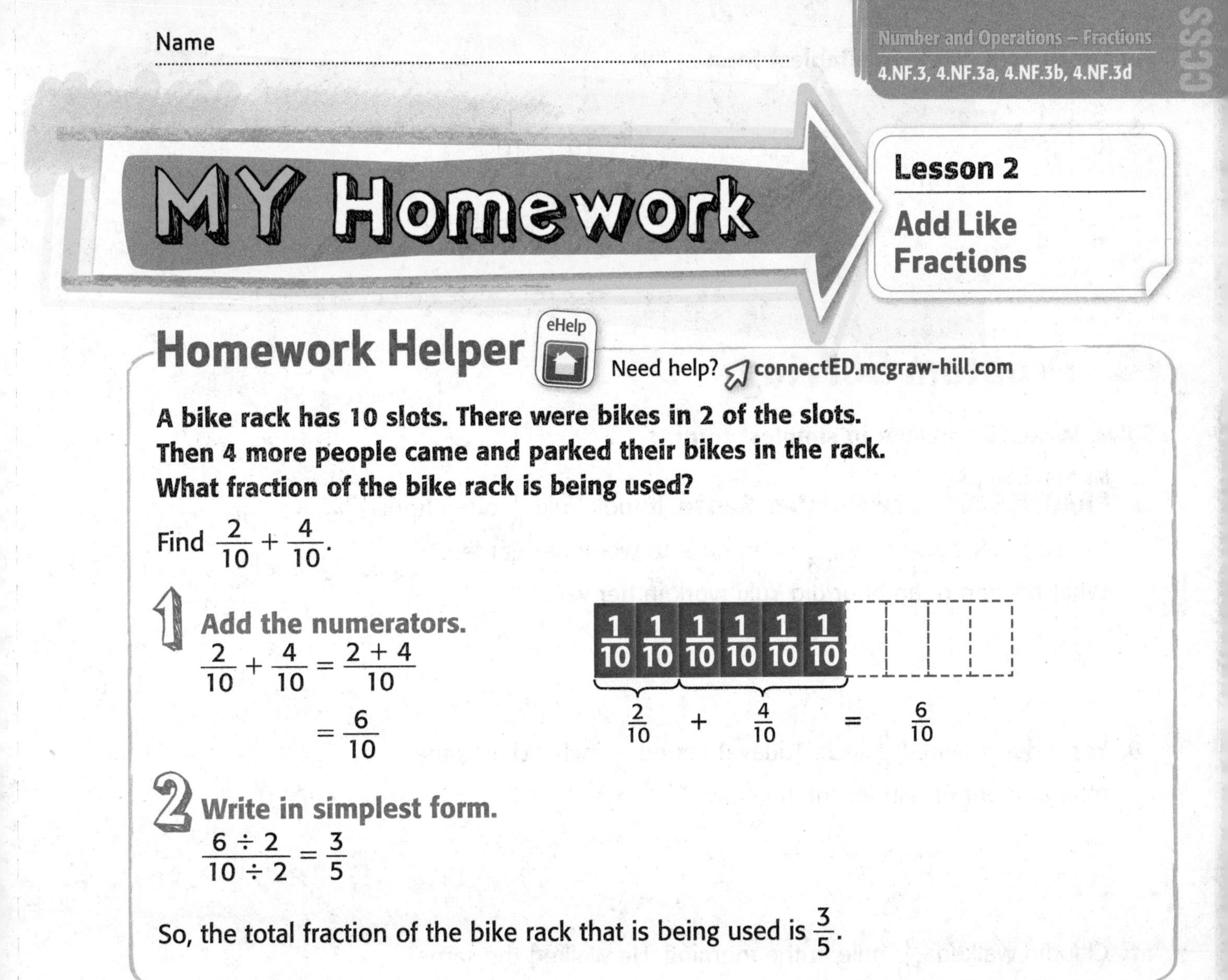

Homework Helper

Need help? connectED.mcgraw-hill.com

A bike rack has 10 slots. There were bikes in 2 of the slots. Then 4 more people came and parked their bikes in the rack. What fraction of the bike rack is being used?

Find $\frac{2}{10} + \frac{4}{10}$.

1 Add the numerators.

$\frac{2}{10} + \frac{4}{10} = \frac{2+4}{10}$

$= \frac{6}{10}$

$\frac{1}{10}$ $\frac{1}{10}$ $\frac{1}{10}$ $\frac{1}{10}$ $\frac{1}{10}$ $\frac{1}{10}$

$\frac{2}{10} + \frac{4}{10} = \frac{6}{10}$

2 Write in simplest form.

$\frac{6 \div 2}{10 \div 2} = \frac{3}{5}$

So, the total fraction of the bike rack that is being used is $\frac{3}{5}$.

Practice

Find each sum. Write in simplest form.

1. $\frac{2}{6} + \frac{3}{6} =$ ________

2. $\frac{3}{8} + \frac{3}{8} =$ ________

3. $\frac{1}{4} + \frac{1}{4} =$ ________

4. $\frac{5}{12} + \frac{3}{12} =$ ________

Find each sum. Write in simplest form.

5. $\frac{3}{5} + \frac{1}{5} =$ ______

6. $\frac{4}{10} + \frac{1}{10} =$ ______

7. $\frac{1}{6} + \frac{3}{6} =$ ______

8. $\frac{50}{100} + \frac{30}{100} =$ ______

Problem Solving

Solve. Write the answer in simplest form.

9. Mathematical PRACTICE 2 **Use Number Sense** It took Yuki $\frac{1}{6}$ of an hour to water her flowers and $\frac{4}{6}$ of an hour to weed her garden. What fraction of an hour did Yuki work in her yard?

______

10. Yesterday it rained $\frac{5}{8}$ inch. Today it rained $\frac{1}{8}$ inch. What is the total amount of rain for the two days?

______

11. Claudio walked $\frac{3}{10}$ mile in the morning. He walked the same distance in the afternoon. How far did Claudio walk altogether?

______

Test Practice

12. Pierre has 12 packages to send. He sent 2 packages on Monday and 2 more packages on Tuesday. What fraction of the packages has Pierre sent so far?

Ⓐ $\frac{1}{4}$

Ⓑ $\frac{1}{3}$

Ⓒ $\frac{8}{12}$

Ⓓ $\frac{2}{3}$

Name

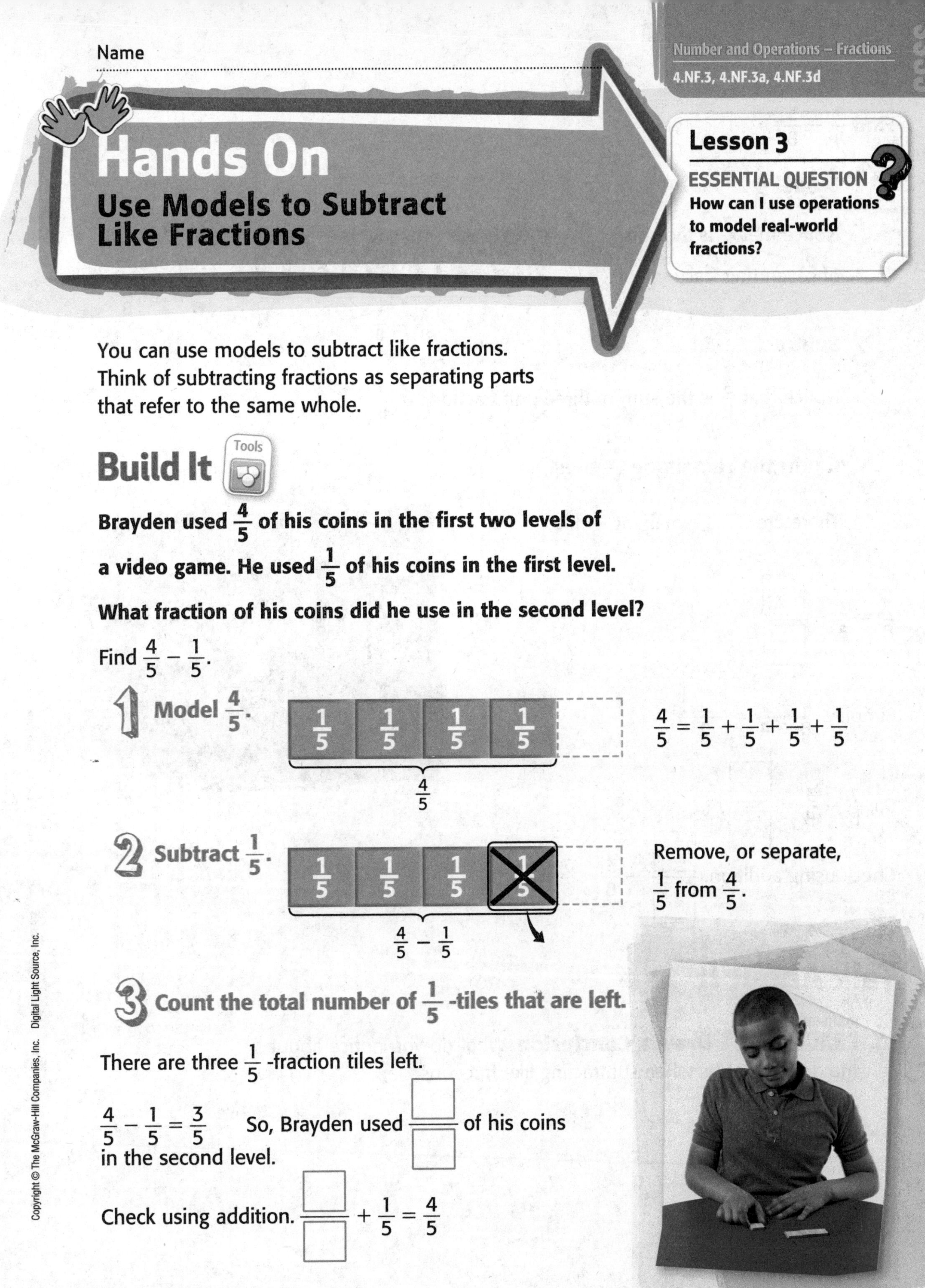

Hands On
Use Models to Subtract Like Fractions

Lesson 3

ESSENTIAL QUESTION
How can I use operations to model real-world fractions?

You can use models to subtract like fractions. Think of subtracting fractions as separating parts that refer to the same whole.

Build It

Tools

Brayden used $\frac{4}{5}$ of his coins in the first two levels of a video game. He used $\frac{1}{5}$ of his coins in the first level. What fraction of his coins did he use in the second level?

Find $\frac{4}{5} - \frac{1}{5}$.

1 Model $\frac{4}{5}$.

$\frac{1}{5}$ $\frac{1}{5}$ $\frac{1}{5}$ $\frac{1}{5}$

$\frac{4}{5}$

$\frac{4}{5} = \frac{1}{5} + \frac{1}{5} + \frac{1}{5} + \frac{1}{5}$

2 Subtract $\frac{1}{5}$.

$\frac{1}{5}$ $\frac{1}{5}$ $\frac{1}{5}$ $\frac{1}{5}$

$\frac{4}{5} - \frac{1}{5}$

Remove, or separate, $\frac{1}{5}$ from $\frac{4}{5}$.

3 Count the total number of $\frac{1}{5}$-tiles that are left.

There are three $\frac{1}{5}$-fraction tiles left.

$\frac{4}{5} - \frac{1}{5} = \frac{3}{5}$ So, Brayden used $\frac{\square}{\square}$ of his coins in the second level.

Check using addition. $\frac{\square}{\square} + \frac{1}{5} = \frac{4}{5}$

Try It

Find $\frac{7}{8} - \frac{3}{8}$.

1 Model $\frac{7}{8}$.

Notice that $\frac{7}{8}$ is the sum of seven unit fractions of $\frac{1}{8}$.

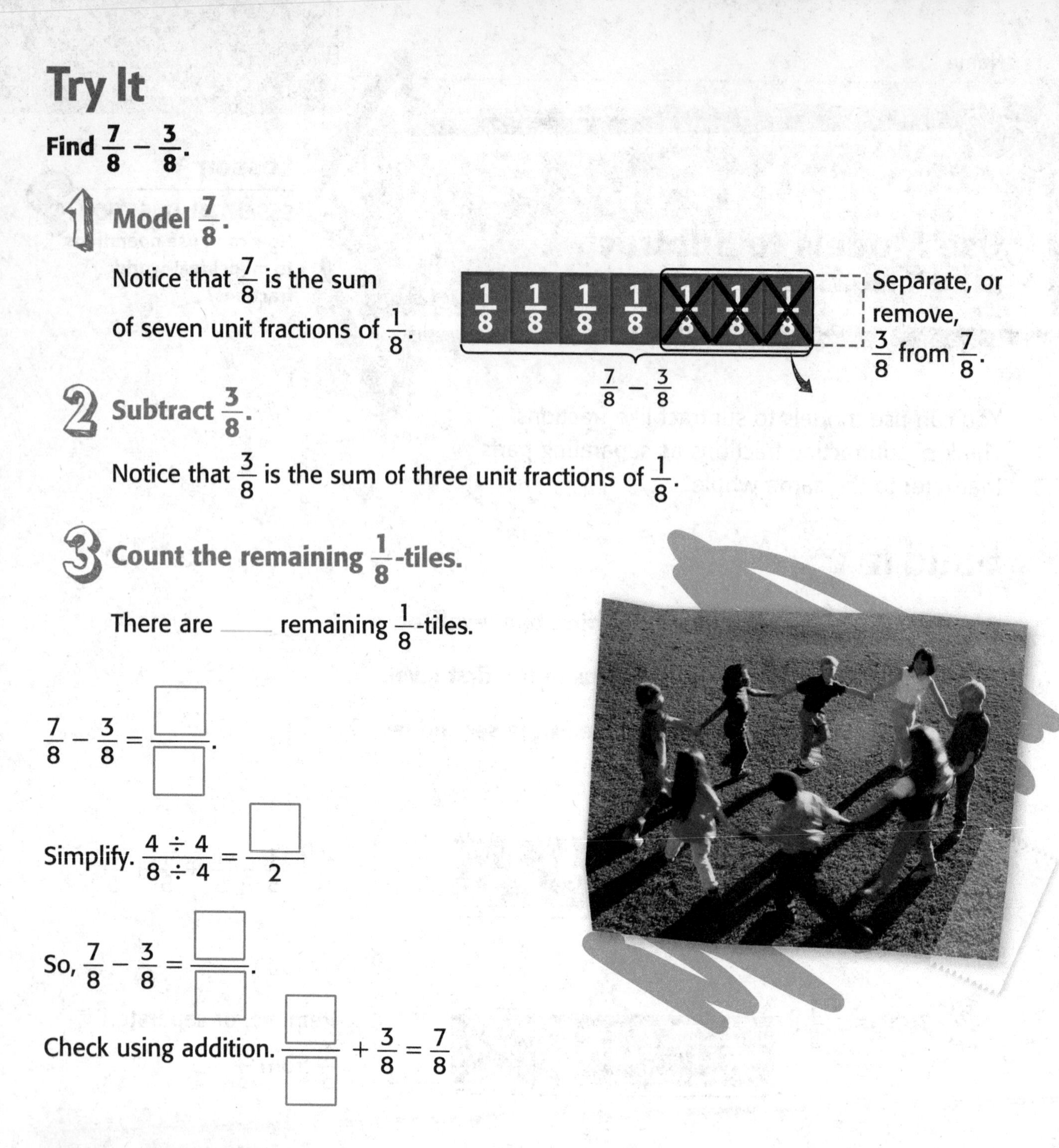

2 Subtract $\frac{3}{8}$.

Notice that $\frac{3}{8}$ is the sum of three unit fractions of $\frac{1}{8}$.

3 Count the remaining $\frac{1}{8}$-tiles.

There are ______ remaining $\frac{1}{8}$-tiles.

$\frac{7}{8} - \frac{3}{8} = \frac{\square}{\square}$.

Simplify. $\frac{4 \div 4}{8 \div 4} = \frac{\square}{2}$

So, $\frac{7}{8} - \frac{3}{8} = \frac{\square}{\square}$.

Check using addition. $\frac{\square}{\square} + \frac{3}{8} = \frac{7}{8}$

Talk About It

1. Mathematical PRACTICE 3 **Draw a Conclusion** What do you notice about the denominators when subtracting like fractions?

____

____

____

Name ....

CCSS

Practice It

Model the difference using fraction tiles. Then subtract.

2. $\frac{4}{6} - \frac{3}{6} =$ ______

3. $\frac{9}{12} - \frac{4}{12} =$ ______

4. $\frac{6}{10} - \frac{3}{10} =$ ______

5. $\frac{4}{5} - \frac{2}{5} =$ ______

Use the table for Exercises 6 and 7.

6. The table shows the difference of several like fractions. Study the pattern in the table. Write a rule that you can use to subtract like fractions without using models.

Like Fractions	Difference
$\frac{4}{5} - \frac{1}{5}$	$\frac{3}{5}$
$\frac{7}{8} - \frac{3}{8}$	$\frac{4}{8}$ or $\frac{1}{2}$
$\frac{9}{10} - \frac{6}{10}$	$\frac{3}{10}$

____

____

____

7. Use your rule from Exercise 6 to find $\frac{11}{12} - \frac{8}{12}$. ______

Algebra Find each unknown. Use fraction tiles.

8. $\frac{2}{4} - \frac{1}{4} = y$

$y =$ ______

9. $\frac{5}{8} - \frac{2}{8} = x$

$x =$ ______

10. $\frac{7}{10} - \frac{3}{10} = t$

$t =$ ______

11. $\frac{5}{6} - \frac{4}{6} = s$

$s =$ ______

Apply It

12. Ann ate 2 slices of a pizza and Teresa ate 3 slices of the pizza. The pizza had 8 slices. What is the difference in the amount of pizza that they ate written as a fraction?

____

13. Mathematical PRACTICE 2 **Use Symbols** Virginia had $\frac{8}{12}$ of the pictures left on her memory card. Then she took off $\frac{7}{12}$ of the pictures. What fraction of the pictures does she have left? Write an equation to solve.

____

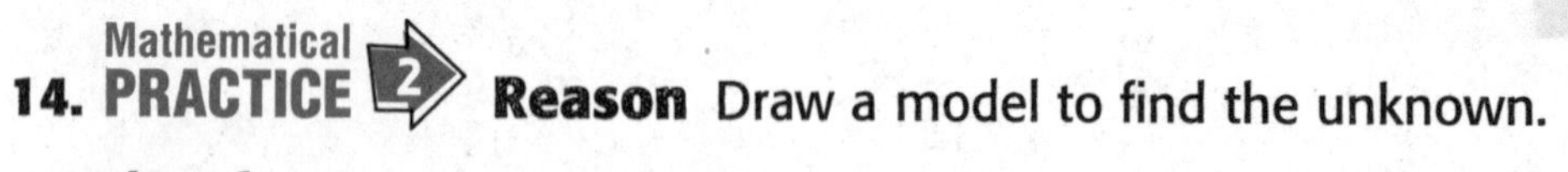

14. Mathematical PRACTICE 2 **Reason** Draw a model to find the unknown.

$\frac{4}{5} - \frac{1}{5} = ?$

____

My Work!

Write About It

15. How is subtracting like fractions similar to adding like fractions?

____

____

MY Homework

Lesson 3
Hands On: Use Models to Subtract Like Fractions

Homework Helper

eHelp Need help? connectED.mcgraw-hill.com

Nicole has 8 charms. She put 7 on her bracelet. At the park, 2 charms fell off the bracelet. What fraction of Nicole's charms are on her bracelet now?

Find $\frac{7}{8} - \frac{2}{8}$.

1 Model $\frac{7}{8}$.

$\frac{7}{8} = \frac{1}{8} + \frac{1}{8} + \frac{1}{8} + \frac{1}{8} + \frac{1}{8} + \frac{1}{8} + \frac{1}{8}$

$\frac{1}{8}$ $\frac{1}{8}$ $\frac{1}{8}$ $\frac{1}{8}$ $\frac{1}{8}$ $\frac{1}{8}$ $\frac{1}{8}$

$\frac{7}{8}$

2 Subtract $\frac{2}{8}$.

Remove, or separate, $\frac{2}{8}$ from $\frac{7}{8}$.

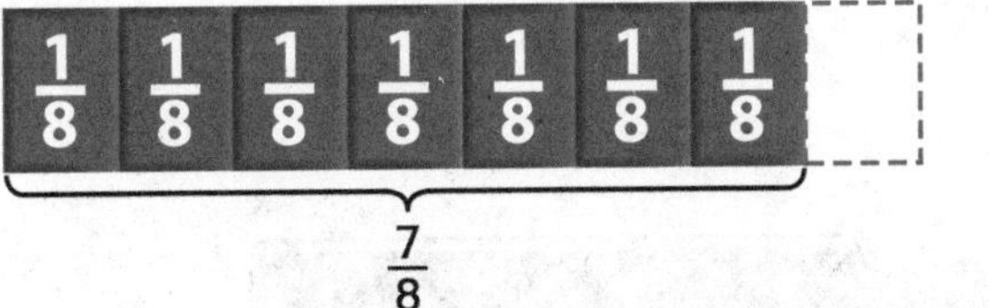

3 Count the total number of $\frac{1}{8}$-tiles that are left.

There are five $\frac{1}{8}$-fraction tiles left.

$\frac{7}{8} - \frac{2}{8} = \frac{5}{8}$

So, Nicole has $\frac{5}{8}$ of her charms left on her bracelet.

Check

$\frac{5}{8} + \frac{2}{8} = \frac{7}{8}$

So, the answer is correct.

Practice

Model the difference using fraction tiles. Then subtract.

1. $\frac{5}{6} - \frac{1}{6} =$ ______

2. $\frac{2}{3} - \frac{1}{3} =$ ______

3. $\frac{9}{10} - \frac{6}{10} =$ ______

4. $\frac{6}{8} - \frac{4}{8} =$ ______

Algebra Write a subtraction equation for each model. Then subtract.

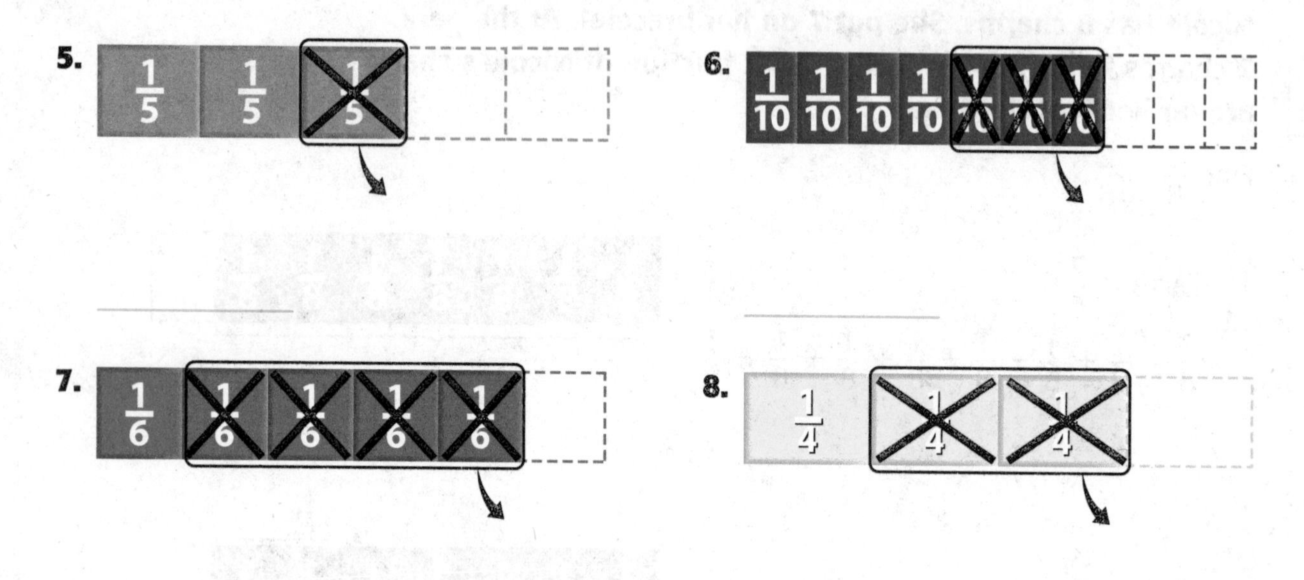

______ ______

______ ______

Problem Solving

9. For Friday night's play performance, $\frac{7}{8}$ of the theater was full. On Saturday night, the theater was only $\frac{5}{8}$ full. How much more of the theater was filled on Friday night than on Saturday night? Write in simplest form.

______

10. **Mathematical PRACTICE 2 Use Number Sense** Liam uses $\frac{3}{4}$ of a cup of butter in his cookie recipe. Gloria uses $\frac{2}{4}$ of a cup of butter in her cookie recipe. How much more butter does Liam use than Gloria?

______

Name

Subtract Like Fractions

Lesson 4

ESSENTIAL QUESTION
How can I use operations to model real-world fractions?

Think of subtracting like fractions as separating parts that refer to the same whole. To subtract like fractions, subtract the numerators and keep the same denominator.

$\frac{4}{5} - \frac{2}{5} = \frac{2}{5}$

$\frac{8}{10} - \frac{5}{10} = \frac{3}{10}$

Math in My World

Real World

Tools Watch Tutor

Example 1

Liliana jogged $\frac{5}{8}$ of a mile on Monday and $\frac{3}{8}$ of a mile on Tuesday. How much farther did she jog on Monday?

Find $\frac{5}{8} - \frac{3}{8}$.

1 Subtract the numerators.
Keep the same denominator.

$$\frac{5}{8} - \frac{3}{8} = \frac{5-3}{8}$$
$$= \frac{2}{8}$$

1 — 1 whole mile

$\frac{1}{8}$ $\frac{1}{8}$ $\frac{1}{8}$ $\frac{1}{8}$ $\frac{1}{8}$ — Separate, or remove, $\frac{3}{8}$.

$\frac{5}{8} - \frac{3}{8} = \frac{2}{8}$

2 Write the difference in simplest form.

$$\frac{2}{8} = \frac{2 \div 2}{8 \div 2} = \frac{1}{4}$$

1

$\frac{1}{8}$ $\frac{1}{8}$

$\frac{1}{4}$

So, Liliana jogged $\frac{\square}{\square}$ mile farther on Monday.

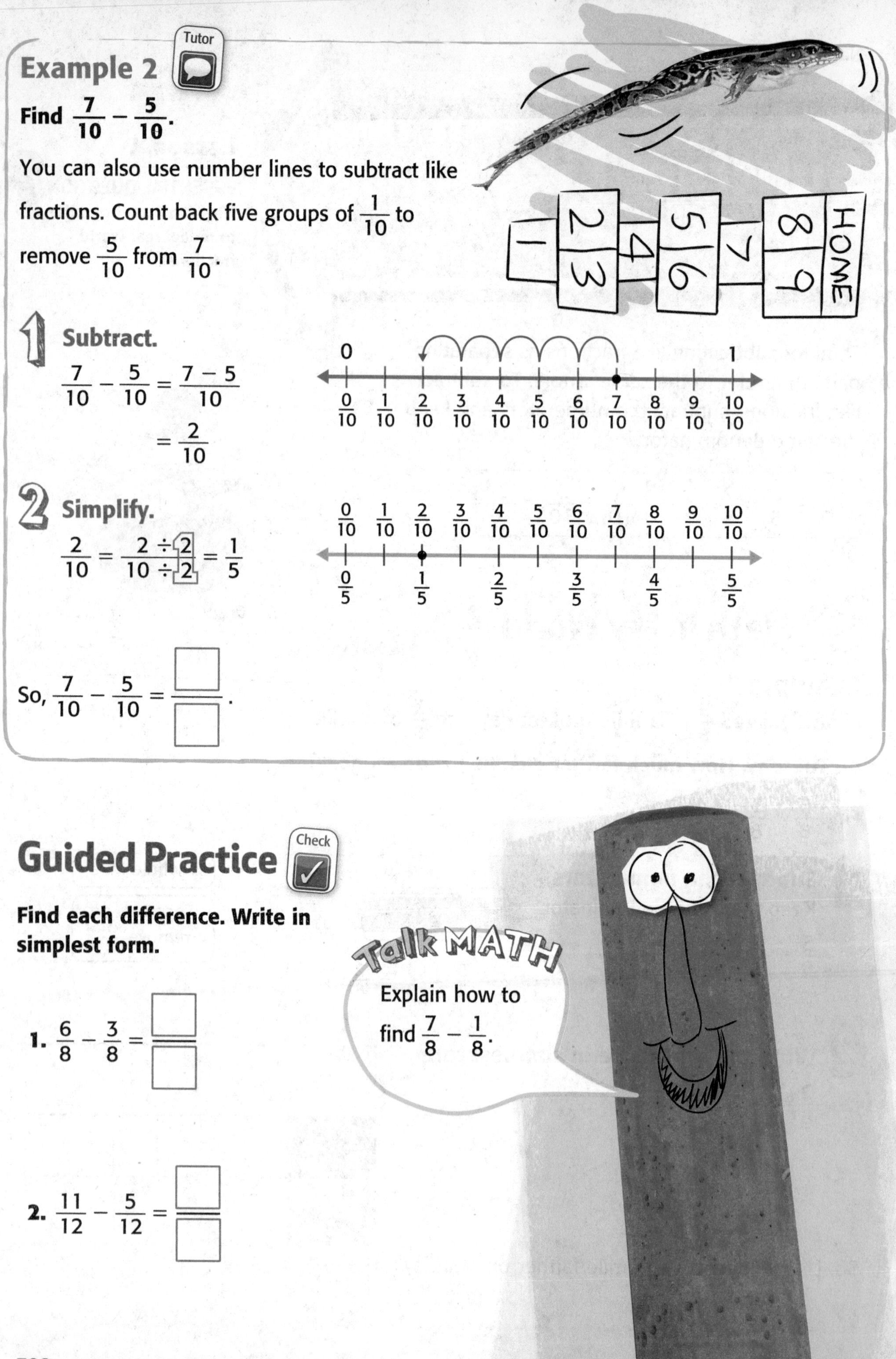

Example 2

Find $\frac{7}{10} - \frac{5}{10}$.

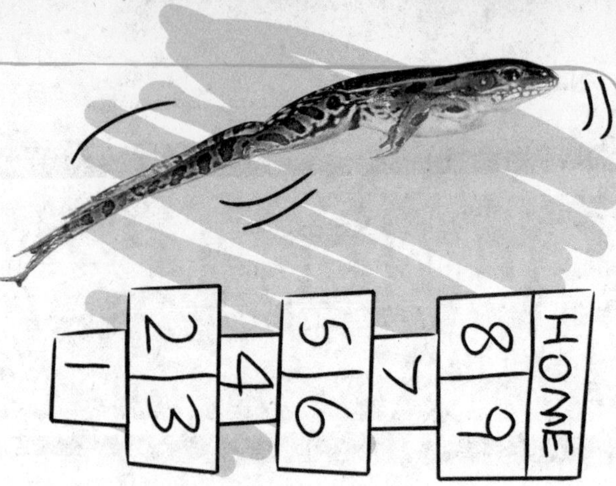

You can also use number lines to subtract like fractions. Count back five groups of $\frac{1}{10}$ to remove $\frac{5}{10}$ from $\frac{7}{10}$.

1 Subtract.

$$\frac{7}{10} - \frac{5}{10} = \frac{7-5}{10}$$
$$= \frac{2}{10}$$

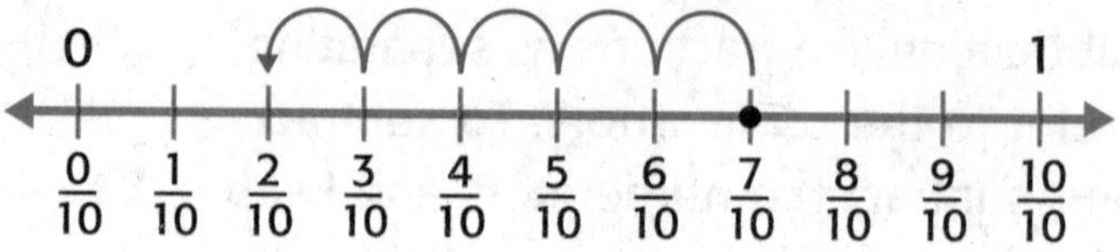

2 Simplify.

$$\frac{2}{10} = \frac{2 \div 2}{10 \div 2} = \frac{1}{5}$$

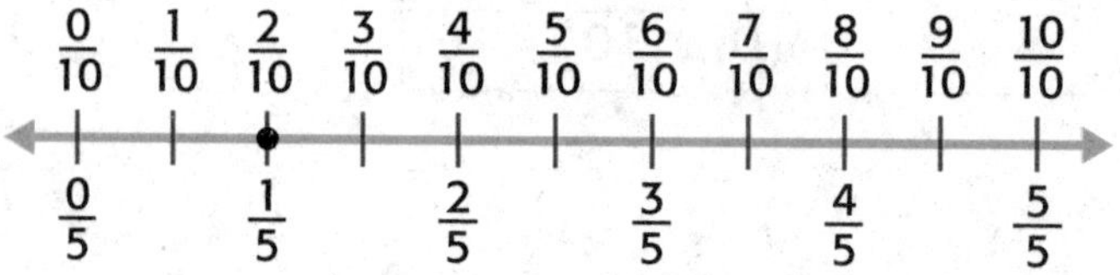

So, $\frac{7}{10} - \frac{5}{10} = \frac{\square}{\square}$.

Guided Practice

Find each difference. Write in simplest form.

1. $\frac{6}{8} - \frac{3}{8} = \frac{\square}{\square}$

2. $\frac{11}{12} - \frac{5}{12} = \frac{\square}{\square}$

Talk MATH

Explain how to find $\frac{7}{8} - \frac{1}{8}$.

Name ....

CCSS

Independent Practice

Find each difference. Write in simplest form.

3. $\frac{7}{8} - \frac{2}{8} =$ ______

4. $\frac{5}{6} - \frac{4}{6} =$ ______

5. $\frac{9}{10} - \frac{6}{10} =$ ______

6. $\frac{2}{3} - \frac{1}{3} =$ ______

7. $\frac{4}{5} - \frac{2}{5} =$ ______

8. $\frac{4}{6} - \frac{2}{6} =$ ______

9. $\frac{9}{12} - \frac{5}{12} =$ ______

10. $\frac{7}{10} - \frac{3}{10} =$ ______

11. $\frac{6}{10} - \frac{1}{10} =$ ______

Algebra Find each unknown. Match each equation to its difference in simplest form.

12. $\frac{4}{5} - \frac{1}{5} = ■$

13. $\frac{6}{12} - \frac{2}{12} = ■$

14. $\frac{7}{10} - \frac{2}{10} = ■$

- $\frac{4}{12}$
- $\frac{3}{5}$
- $\frac{5}{10}$
- $\frac{1}{2}$
- $\frac{1}{3}$
- $\frac{6}{10}$

Problem Solving

15. Kenji's dog ate $\frac{3}{12}$ of the treats in a box of dog treats on Monday. His dog ate $\frac{2}{12}$ of the treats on Tuesday. What fraction more of the box of the treats did his dog eat on Monday than on Tuesday?

16. Mathematical PRACTICE 4 **Model Math** Dan used $\frac{4}{10}$ of a pack of golf balls on Saturday. Then he used $\frac{2}{10}$ of the pack of golf balls on Sunday. What fraction more of the pack of golf balls did Dan use on Saturday than on Sunday? Draw a model to solve.

My Work!

HOT Problems

17. Mathematical PRACTICE 3 **Find the Error** Bella thinks that the fraction $\frac{4}{8}$ in simplest form is $\frac{2}{4}$. Find and correct her mistake.

18. **Building on the Essential Question** When subtracting like fractions, what happens to the numerator and the denominator?

Name ______________________

MY Homework

Lesson 4
Subtract Like Fractions

Homework Helper

eHelp Need help? connectED.mcgraw-hill.com

It takes Margaret $\frac{5}{6}$ of an hour to do her chores. It takes Holly $\frac{3}{6}$ of an hour to do her chores. How much longer does it take Margaret to do her chores than it takes Holly?

Find $\frac{5}{6} - \frac{3}{6}$.

1 Find the difference between the numerators.

$$\frac{5}{6} - \frac{3}{6} = \frac{5-3}{6}$$

$$= \frac{2}{6}$$

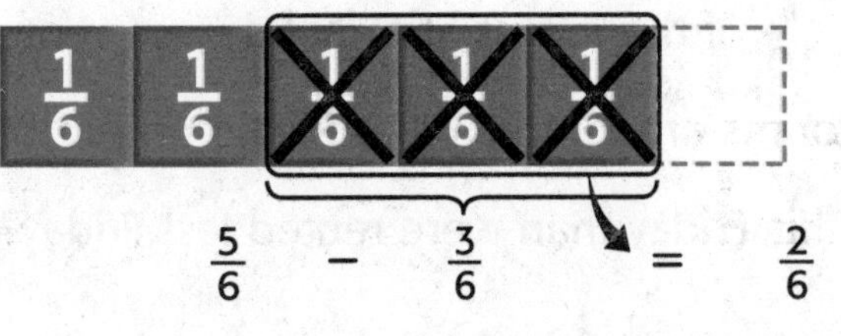

2 Write in simplest form.

$$\frac{2 \div 2}{6 \div 2} = \frac{1}{3}$$

So, it takes Margaret $\frac{1}{3}$ of an hour longer to do her chores.

Practice

Find each difference. Write in simplest form.

1. $\frac{7}{10} - \frac{4}{10} =$ ______

2. $\frac{10}{12} - \frac{3}{12} =$ ______

3. $\frac{4}{5} - \frac{3}{5} =$ ______

4. $\frac{6}{8} - \frac{4}{8} =$ ______

Find each difference. Write in simplest form.

5. $\frac{6}{8} - \frac{2}{8} =$ ______

6. $\frac{4}{10} - \frac{2}{10} =$ ______

7. $\frac{9}{12} - \frac{6}{12} =$ ______

8. $\frac{80}{100} - \frac{20}{100} =$ ______

Real World Problem Solving

Solve. Write the answer in simplest form.

9. A beetle is $\frac{1}{5}$ inch wide and $\frac{2}{5}$ inch long. How much greater is the beetle's length than its width?

______

10. Last Friday, $\frac{7}{10}$ of the rooms at a motel were rented. This Friday, $\frac{9}{10}$ of the rooms are rented. What fraction more of the rooms are rented this Friday than were rented last Friday?

______

11. Mathematical PRACTICE 2 **Use Number Sense** Denise teaches obedience classes. Last session, $\frac{11}{12}$ of the dogs passed her class. This session, $\frac{9}{12}$ of the dogs passed her class. What fraction more of the dogs passed her class last session?

______

Test Practice

12. Find $\frac{4}{8} - \frac{2}{8}$. Write in simplest form.

Ⓐ $\frac{3}{4}$
Ⓑ $\frac{4}{8}$
Ⓒ $\frac{2}{4}$
Ⓓ $\frac{1}{4}$

Check My Progress

Vocabulary Check

1. Circle the pair of **like fractions**.

$\frac{4}{6}$ and $\frac{5}{6}$	$\frac{1}{2}$ and $\frac{3}{4}$	$\frac{1}{3}$ and $\frac{2}{6}$	$\frac{1}{10}$ and $\frac{2}{100}$

Explain how you knew which pair of fractions to circle.

____

____

Concept Check

Algebra Write each fraction as a sum of unit fractions. Then write an equation to decompose the fraction into a different way.

2. $\frac{4}{5}$ ______________________________

____

3. $\frac{3}{8}$ ______________________________

____

Find each sum. Write in simplest form.

4. $\frac{2}{5} + \frac{2}{5} =$ ______ **5.** $\frac{1}{8} + \frac{5}{8} =$ ______ **6.** $\frac{2}{6} + \frac{1}{6} =$ ______

Find each difference. Write in simplest form.

7. $\frac{7}{10} - \frac{4}{10} =$ ______ **8.** $\frac{3}{4} - \frac{1}{4} =$ ______ **9.** $\frac{10}{12} - \frac{6}{12} =$ ______

Problem Solving

10. Elizabeth has $\frac{7}{10}$ of a dollar. If she spends $\frac{4}{10}$ of a dollar, what fraction of a dollar will she have left?

11. A game uses tokens. Two-sixths of the tokens are blue. Three-sixths of the tokens are gold. What fraction of the tokens are either blue or gold?

12. A glass has $\frac{6}{8}$ cup of water. If Jack pours out $\frac{3}{8}$ cup of the water, how much water will be left?

13. A recipe uses $\frac{1}{4}$ cup of apple juice and $\frac{2}{4}$ cup of orange juice. What is the total amount of apple juice and orange juice in the recipe?

My Work!

Test Practice

14. What is the sum of $\frac{2}{10}$ and $\frac{3}{10}$ in simplest form?

Ⓐ $\frac{5}{20}$

Ⓑ $\frac{1}{10}$

Ⓒ $\frac{23}{10}$

Ⓓ $\frac{1}{2}$

Name ________________

Problem-Solving Investigation

STRATEGY: Work Backward

Lesson 5

ESSENTIAL QUESTION
How can I use operations to model real-world fractions?

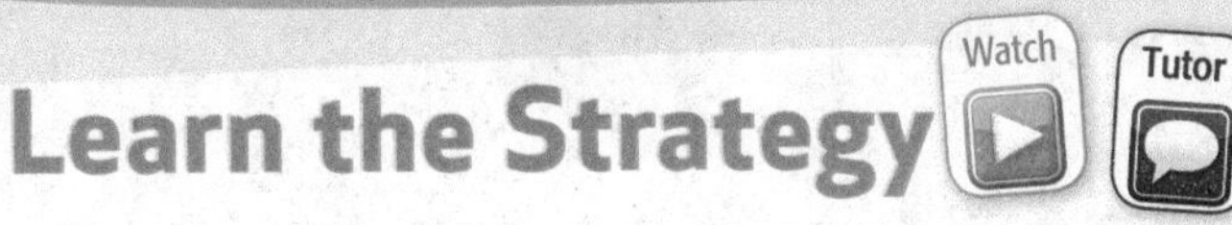

Learn the Strategy

Nathan used some flour for a cake recipe. He used $\frac{1}{4}$ of the bag of flour for a bread recipe. There is $\frac{2}{4}$ of the bag left. What fraction of the bag did Nathan use for the cake?

1 Understand

What facts do you know?

Nathan used some flour for a cake recipe and ________ of the bag for a bread recipe. He has ________ of the bag left.

What do you need to find?

Find the fraction of the bag of flour that was used for the ________.

2 Plan

I will work backward to solve the problem.

3 Solve

$\frac{1}{4}$	$\frac{1}{4}$	$\frac{1}{4}$	$\frac{1}{4}$
bread	flour left		cake

$\frac{1}{4} + \frac{1}{4} + \frac{1}{4} + \frac{1}{4} = 1$ whole bag of flour

So, ________ of the bag of flour was used for the cake.

4 Check

Does your answer make sense? Explain.

____

Practice the Strategy

Lily, Noah, and Kaylee are sharing pizza. Noah ate $\frac{2}{6}$ of the pizza. Kaylee ate $\frac{1}{6}$ of the pizza. There is $\frac{2}{6}$ of the pizza left. What fraction of the pizza did Lily eat?

1 Understand

What facts do you know?

____

____

What do you need to find?

____

2 Plan

____

3 Solve

4 Check

Does your answer make sense? Explain.

____

____

Name

Apply the Strategy

Solve each problem by working backward.

1. Mathematical PRACTICE 1 **Make Sense of Problems** Chloe did some of her homework before dinner She did $\frac{2}{6}$ of her homework after dinner. She has $\frac{1}{6}$ of her homework left. What fraction of her homework did Chloe do before her dinner? Write in simplest form.

2. There were 12 goals scored during the game. Team A scored $\frac{8}{12}$ of the goals. Team B scored 2 goals during the first half of the game. What fraction of the goals did Team B score during the second half of the game? Write in simplest form.

3. Brandi and her mom are at a pet store. The pet store has 12 reptiles. Of the reptiles, $\frac{5}{12}$ are turtles, $\frac{2}{12}$ are snakes, and the rest are lizards. What fraction of the reptiles is lizards?

My Work!

GCSS

Review the Strategies

Use any strategy to solve each problem.

- Work backward.
- Use logical reasoning.
- Look for a pattern.
- Make a model.

4. There are 16 books on a shelf. Four sixteenths of the books are about animals. Two are adventure. The rest are mystery. How many are mystery books?

5. There are 10 pieces of chalk. Two tenths of the chalk is pink. One piece is blue. The rest are white. How many pieces of chalk are white?

6. Giselle played with some friends on Monday. She played with 2 times as many friends on Wednesday. This was 4 more than on Friday. On Friday, she played with 4 friends. How many did she play with on Monday?

7. Mathematical **PRACTICE** 5 **Use Math Tools** Mrs. Vargas is making costumes for a play. She needs 3 buttons for each costume. Complete the table to find how many buttons she will need for 22 costumes.

Costumes	Buttons
8	24
9	27
10	30
12	36
13	39
20	60
21	
22	

My Work!

Name ....

MY Homework

Lesson 5
Problem Solving: Work Backward

Homework Helper

eHelp Need help? connectED.mcgraw-hill.com

Vincent decided to give away his marble collection. He gave some of the marbles to Sam. He gave $\frac{2}{8}$ of the marbles to Jenny, and he gave $\frac{3}{8}$ of the marbles to Molly. What fraction of the marbles did Vincent give to Sam?

1 Understand

What facts do you know?

Vincent gave $\frac{2}{8}$ of his marbles to Jenny and $\frac{3}{8}$ of his marbles to Molly.

What do you need to find?

I need to find the fraction of marbles Vincent gave to Sam.

2 Plan

I will work backward to solve the problem.

3 Solve

Jenny got $\frac{2}{8}$ of the marbles.

Molly got $\frac{3}{8}$ of the marbles.

That leaves another $\frac{3}{8}$.

So, Sam got $\frac{3}{8}$ of the marbles.

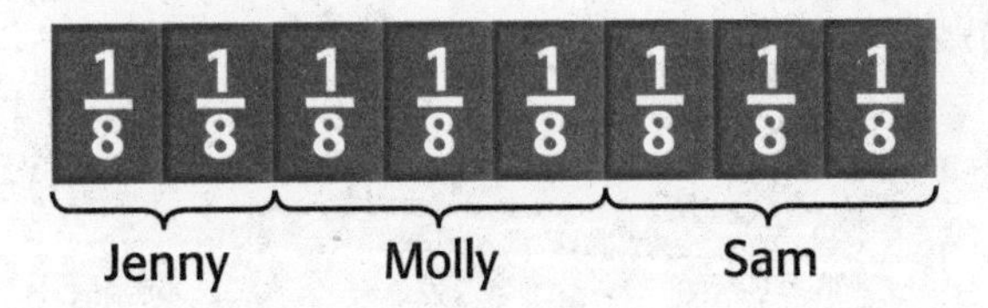

4 Check

Does the answer make sense?

$\frac{2}{8} + \frac{3}{8} + \frac{3}{8} = \frac{8}{8}$ or 1 So, the answer is reasonable.

Real World!

Problem Solving

Solve each problem by working backward.

1. Marla, Jamie, and Sarah have each led their book club's monthly meeting. Jamie has led $\frac{2}{6}$ of the meetings, and Sarah has led $\frac{1}{6}$ of the meetings. What fraction of the meetings has Marla led?

2. Suzanne dropped her penny jar. She found some of the pennies, but some are still missing. She found $\frac{6}{10}$ of the pennies on the rug. She found $\frac{3}{10}$ of the pennies on the couch. What fraction of pennies is still missing?

3. Mathematical **PRACTICE** 2 **Use Number Sense** Noah spent some of his allowance on Monday. He spent $\frac{1}{6}$ of his allowance on Tuesday and $\frac{3}{6}$ of it on Friday. Noah has none of his allowance money left. What fraction of his allowance did he spend on Monday?

My Work!

Name ....

Add Mixed Numbers

Lesson 6

ESSENTIAL QUESTION
How can I use operations to model real-world fractions?

Mixed numbers are numbers with a whole number and a fraction. You can decompose mixed numbers to add them. Use the Associative Property to group the whole numbers and like fractions together.

Math in My World

Tools Watch Tutor

Example 1

Madison made a fruit salad. She used $3\frac{1}{4}$ cups of strawberries and $2\frac{1}{4}$ cups of blueberries. How many cups of berries did Madison use altogether?

Find $3\frac{1}{4} + 2\frac{1}{4}$.

Decompose each mixed number as a sum of whole numbers and unit fractions.

$3\frac{1}{4} + 2\frac{1}{4} = 1 + 1 + 1 + \frac{1}{4} + 1 + 1 + \frac{1}{4}$ Write as a sum of whole numbers and like fractions.

$= (1 + 1 + 1 + 1 + 1) + \left(\frac{1}{4} + \frac{1}{4}\right)$ Associative Property

$=$ _____ $+ \frac{1}{4} + \frac{1}{4}$ Add the wholes. There are 5 wholes.

$= 5 + \frac{2}{4}$ Add the like fractions. $\frac{1}{4} + \frac{1}{4} = \frac{2}{4}$

$= 5\frac{\square}{\square}$ Simplify.

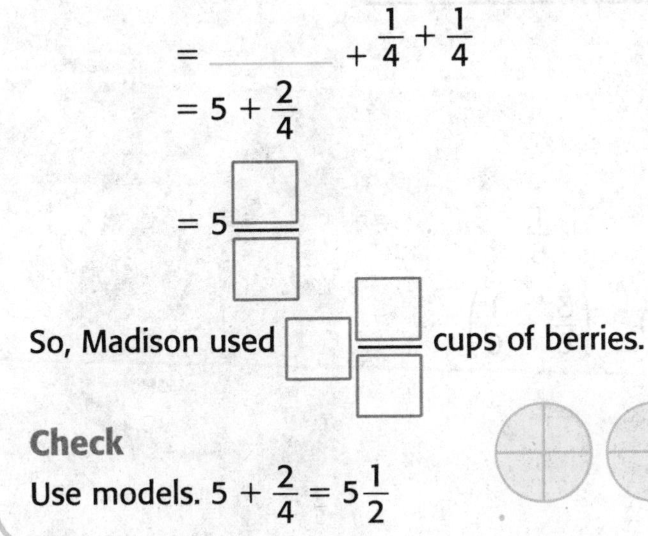

So, Madison used $\square\frac{\square}{\square}$ cups of berries.

Check

Use models. $5 + \frac{2}{4} = 5\frac{1}{2}$

You can also write each mixed number as an equivalent improper fraction.

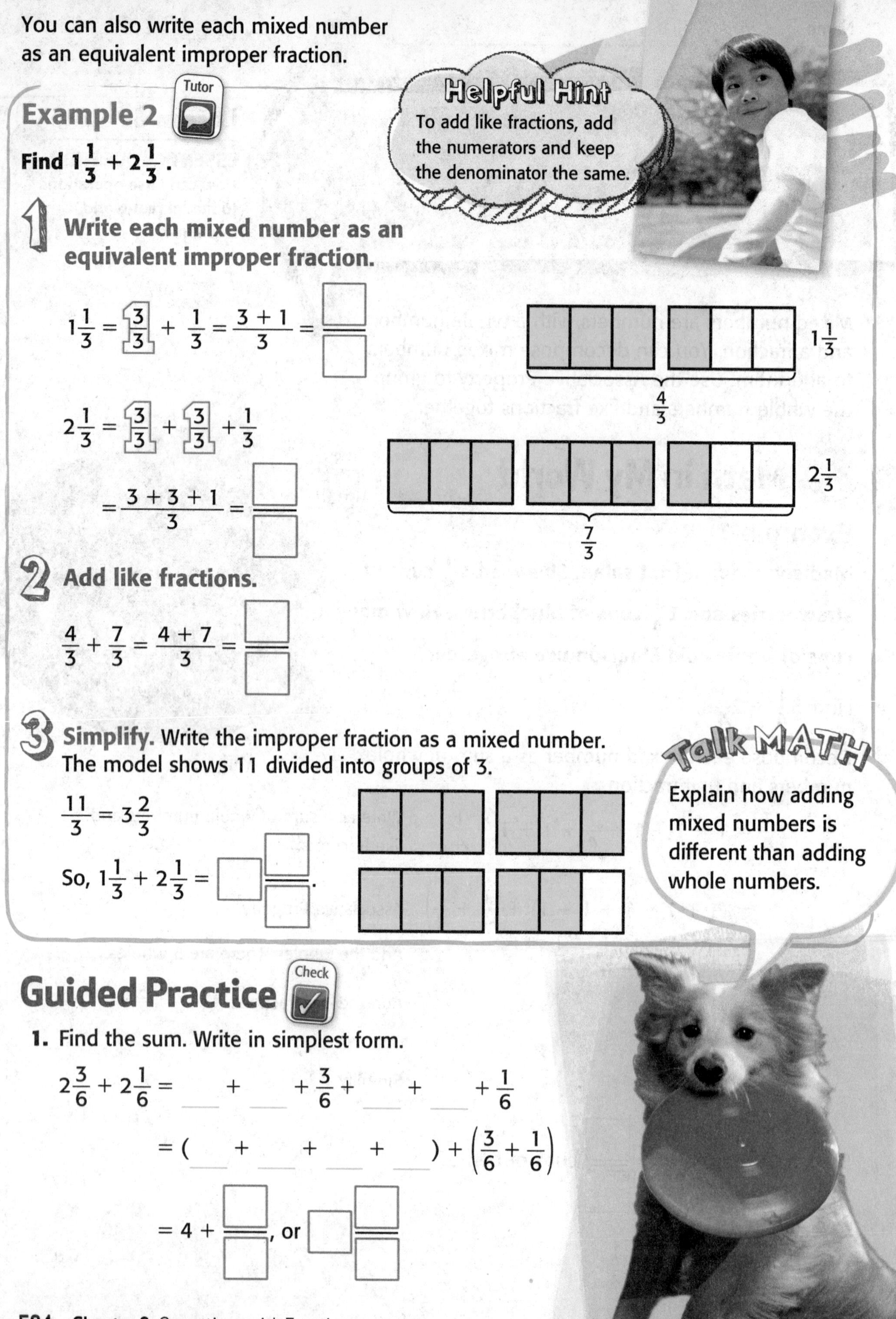

Example 2 (Tutor)

Helpful Hint
To add like fractions, add the numerators and keep the denominator the same.

Find $1\frac{1}{3} + 2\frac{1}{3}$.

1 Write each mixed number as an equivalent improper fraction.

$1\frac{1}{3} = \frac{3}{3} + \frac{1}{3} = \frac{3+1}{3} = \frac{\square}{\square}$

$1\frac{1}{3}$ $\quad \frac{4}{3}$

$2\frac{1}{3} = \frac{3}{3} + \frac{3}{3} + \frac{1}{3}$

$= \frac{3+3+1}{3} = \frac{\square}{\square}$

$2\frac{1}{3}$ $\quad \frac{7}{3}$

2 Add like fractions.

$\frac{4}{3} + \frac{7}{3} = \frac{4+7}{3} = \frac{\square}{\square}$

3 Simplify. Write the improper fraction as a mixed number. The model shows 11 divided into groups of 3.

$\frac{11}{3} = 3\frac{2}{3}$

So, $1\frac{1}{3} + 2\frac{1}{3} = \square\frac{\square}{\square}$.

Talk MATH
Explain how adding mixed numbers is different than adding whole numbers.

Guided Practice (Check)

1. Find the sum. Write in simplest form.

$2\frac{3}{6} + 2\frac{1}{6} = \_\_ + \_\_ + \frac{3}{6} + \_\_ + \_\_ + \frac{1}{6}$

$= (\_\_ + \_\_ + \_\_ + \_\_) + \left(\frac{3}{6} + \frac{1}{6}\right)$

$= 4 + \frac{\square}{\square}$, or $\square\frac{\square}{\square}$

Name ....

CCSS

Independent Practice

Find each sum. Write in simplest form. Use fraction models to check.

2. $2\frac{1}{3} + 1\frac{1}{3} =$ ______

3. $5\frac{1}{8} + 2\frac{3}{8} =$ ______

4. $5\frac{1}{4} + 5\frac{1}{4} =$ ______

5. $4\frac{1}{5} + 4\frac{3}{5} =$ ______

6. $4\frac{3}{8} + 2\frac{4}{8} =$ ______

7. $6\frac{2}{6} + 1\frac{1}{6} =$ ______

8. $3\frac{1}{10} + 1\frac{7}{10} =$ ______

9. $1\frac{2}{12} + 7\frac{2}{12} =$ ______

10. $3\frac{2}{8} + 2\frac{2}{8} =$ ______

Circle the sum that could *not* represent each mixed number.

11. $2\frac{2}{5}$ $\quad 1 + 1 + \frac{2}{5} \quad 1 + 1 + \frac{1}{5} + \frac{1}{5} \quad 2 + \frac{1}{5}$

12. $1\frac{3}{4}$ $\quad 1 + \frac{1}{4} \quad 1 + \frac{1}{4} + \frac{1}{4} + \frac{1}{4} \quad \frac{4}{4} + \frac{2}{4} + \frac{1}{4}$

13. $3\frac{1}{8}$ $\quad 1 + 1 + 1 + \frac{1}{8} \quad \frac{8}{8} + \frac{8}{8} + \frac{1}{8} \quad \frac{8}{8} + \frac{8}{8} + \frac{8}{8} + \frac{1}{8}$

14. $4\frac{1}{2}$ $\quad \frac{2}{2} + \frac{2}{2} + \frac{1}{2} \quad 1 + 1 + 1 + 1 + \frac{1}{2} \quad \frac{2}{2} + \frac{2}{2} + \frac{2}{2} + \frac{2}{2} + \frac{1}{2}$

Problem Solving

15. Brennan ate $2\frac{1}{4}$ apples. Then he ate another $1\frac{2}{4}$ apples the next day. How many apples did Brennan eat altogether? Write in simplest form.

16. The bakery shop purchased $7\frac{2}{10}$ pounds of sugar and $7\frac{5}{10}$ pounds of flour. How much sugar and flour did they purchase in all?

My Work!

17. Mathematical PRACTICE 2 **Use Number Sense** Mrs. Argo has $3\frac{7}{12}$ boxes of pens. She has $4\frac{1}{12}$ boxes of pencils. How many boxes of pens and pencils does she have? Write in simplest form.

HOT Problems

18. Mathematical PRACTICE 4 **Model Math** Write a real-world problem using mixed numbers whose sum is $1\frac{3}{5}$.

19. **Building on the Essential Question** How can a mixed number be written as a sum?

MY Homework

Lesson 6
Add Mixed Numbers

Homework Helper

eHelp Need help? connectED.mcgraw-hill.com

Gavin put $2\frac{1}{3}$ scoops of chili in his bowl. He put $4\frac{1}{3}$ scoops of chili in his dad's bowl. How many scoops of chili do Gavin and his dad have in all?

Find $2\frac{1}{3} + 4\frac{1}{3}$.

1 Change the mixed numbers into improper fractions.

$2\frac{1}{3} = \frac{3}{3} + \frac{3}{3} + \frac{1}{3} = \frac{3+3+1}{3} = \frac{7}{3}$

$4\frac{1}{3} = \frac{3}{3} + \frac{3}{3} + \frac{3}{3} + \frac{3}{3} + \frac{1}{3} = \frac{3+3+3+3+1}{3} = \frac{13}{3}$

2 Add the like fractions.

$\frac{7}{3} + \frac{13}{3} = \frac{20}{3}$

3 Simplify.

$\frac{20}{3} = 6\frac{2}{3}$

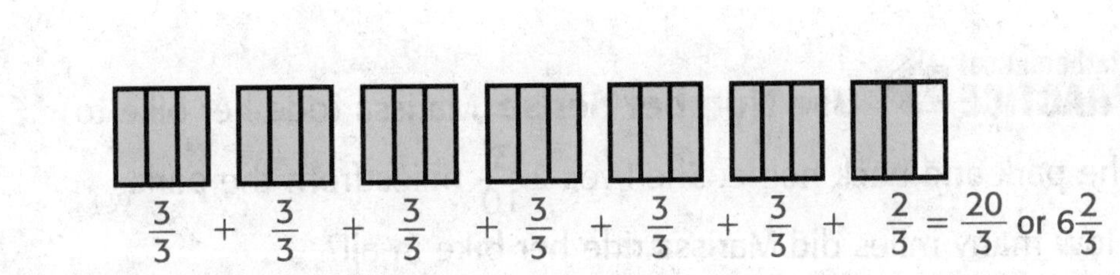

$\frac{3}{3} + \frac{3}{3} + \frac{3}{3} + \frac{3}{3} + \frac{3}{3} + \frac{3}{3} + \frac{2}{3} = \frac{20}{3}$ or $6\frac{2}{3}$

So, Gavin and his dad have $6\frac{2}{3}$ scoops of chili in all.

Practice

Find each sum. Write in simplest form.

1. $4\frac{1}{4} + 2\frac{2}{4} =$ ______

2. $3\frac{3}{6} + 6\frac{1}{6} =$ ______

Find each sum. Write in simplest form.

3. $6\frac{2}{5} + 3\frac{2}{5} =$ ______

4. $4\frac{1}{6} + 1\frac{2}{6} =$ ______

5. $2\frac{1}{4} + 9\frac{1}{4} =$ ______

6. $7\frac{4}{8} + 1\frac{3}{8} =$ ______

7. $5\frac{6}{10} + 8\frac{3}{10} =$ ______

8. $12\frac{5}{10} + 6\frac{1}{10} =$ ______

Problem Solving

Solve. Write the answer in simplest form.

9. James cut $1\frac{1}{4}$ dozen flowers for a bouquet. Gwen added $1\frac{2}{4}$ dozen flowers to the bouquet. How many dozen flowers are there altogether?

______

10. On Monday, Simon's class filled $3\frac{2}{5}$ boxes with books to donate to charity. On Wednesday, the class filled $4\frac{2}{5}$ more boxes with books to donate. How many boxes of books will Simon's class donate in all?

______

11. Mathematical PRACTICE 2 **Use Number Sense** Marissa rode her bike to the park and back home. She lives $2\frac{3}{10}$ miles from the park. How many miles did Marissa ride her bike in all?

______

Test Practice

12. Nate is $10\frac{9}{12}$ years old. How old will he be in $2\frac{1}{12}$ more years?

Ⓐ $13\frac{1}{3}$ years old

Ⓑ $12\frac{5}{6}$ years old

Ⓒ $12\frac{1}{4}$ years old

Ⓓ $12\frac{3}{12}$ years old

Name

Subtract Mixed Numbers

Lesson 7

ESSENTIAL QUESTION
How can I use operations to model real-world fractions?

You can use equivalent fractions to subtract mixed numbers.

Math in My World

Tools Watch Tutor

Example 1

Olivia had $4\frac{3}{4}$ boxes of games. She gave $1\frac{1}{4}$ of the boxes to her brother. How many boxes of games does she have left?

Find $4\frac{3}{4} - 1\frac{1}{4}$.

1 Write each mixed number as an equivalent improper fraction.

$$4\frac{3}{4} = \frac{4}{4} + \frac{4}{4} + \frac{4}{4} + \frac{4}{4} + \frac{3}{4} = \frac{4+4+4+4+3}{4} = \frac{\square}{\square}$$

$$1\frac{1}{4} = \frac{4}{4} + \frac{1}{4} = \frac{4+1}{4} = \frac{\square}{\square}$$

2 Subtract like fractions.

$$\frac{19}{4} - \frac{5}{4} = \frac{19-5}{4} = \frac{\square}{\square}$$

So, Olivia has ________ boxes of games left.

Check Use addition to check.

$$3\frac{2}{4} + 1\frac{1}{4} = 1 + 1 + 1 + \frac{2}{4} + 1 + \frac{1}{4}$$
$$= 4\frac{3}{4}$$

3 Simplify. The models below show 14 divided into groups of 4.

$$\frac{14}{4} = 3\frac{2}{4} = 3\frac{\square}{\square}$$

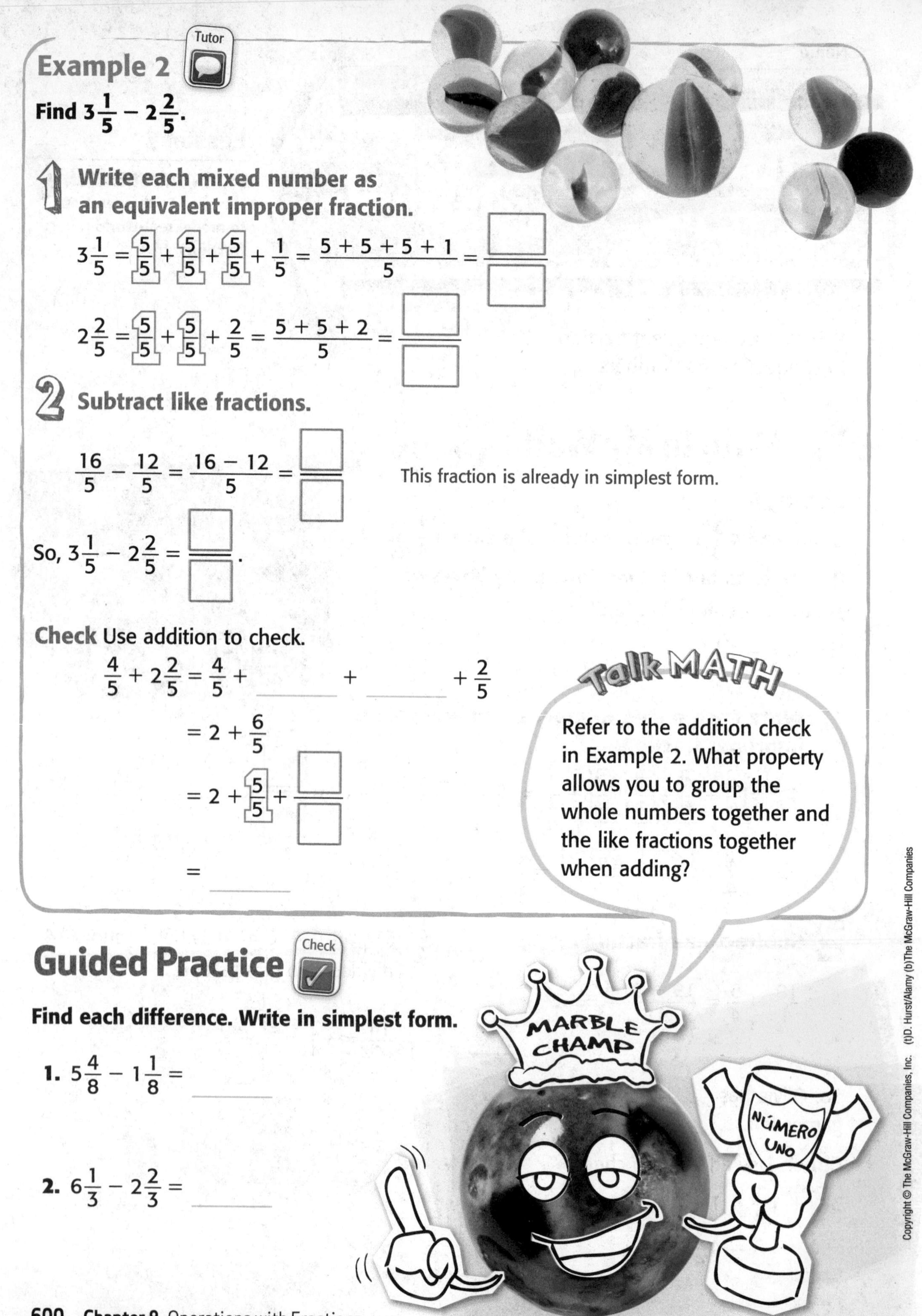

Example 2

Tutor

Find $3\frac{1}{5} - 2\frac{2}{5}$.

1 Write each mixed number as an equivalent improper fraction.

$$3\frac{1}{5} = \frac{5}{5} + \frac{5}{5} + \frac{5}{5} + \frac{1}{5} = \frac{5+5+5+1}{5} = \frac{\square}{\square}$$

$$2\frac{2}{5} = \frac{5}{5} + \frac{5}{5} + \frac{2}{5} = \frac{5+5+2}{5} = \frac{\square}{\square}$$

2 Subtract like fractions.

$$\frac{16}{5} - \frac{12}{5} = \frac{16-12}{5} = \frac{\square}{\square}$$

This fraction is already in simplest form.

So, $3\frac{1}{5} - 2\frac{2}{5} = \frac{\square}{\square}$.

Check Use addition to check.

$$\frac{4}{5} + 2\frac{2}{5} = \frac{4}{5} + \_\_\_\_ + \_\_\_\_ + \frac{2}{5}$$

$$= 2 + \frac{6}{5}$$

$$= 2 + \frac{5}{5} + \frac{\square}{\square}$$

$$= \_\_\_\_$$

Talk MATH

Refer to the addition check in Example 2. What property allows you to group the whole numbers together and the like fractions together when adding?

Guided Practice

Check

Find each difference. Write in simplest form.

1. $5\frac{4}{8} - 1\frac{1}{8} =$ ____

2. $6\frac{1}{3} - 2\frac{2}{3} =$ ____

Name

Independent Practice

Find each difference. Write in simplest form. Use addition to check.

3. $5\frac{4}{5} - 2\frac{2}{5} =$ ______

4. $3\frac{4}{6} - 1\frac{5}{6} =$ ______

5. $8\frac{2}{3} - 3\frac{2}{3} =$ ______

6. $6\frac{1}{4} - 4\frac{2}{4} =$ ______

7. $4\frac{2}{3} - 2\frac{1}{3} =$ ______

8. $6\frac{3}{4} - 3\frac{2}{4} =$ ______

9. $9\frac{5}{6} - 7\frac{2}{6} =$ ______

10. $6\frac{7}{8} - 1\frac{5}{8} =$ ______

11. $4\frac{6}{12} - 3\frac{5}{12} =$ ______

My Work!

12. **Mathematical PRACTICE 7** **Identify Structure** Mrs. Garcia has $5\frac{6}{8}$ boxes of paper. She uses $3\frac{7}{8}$ boxes. How many boxes does she have left? Write in simplest form.

______

13. Erwin had $2\frac{5}{12}$ gallons of lemonade. He spilled $\frac{11}{12}$ gallon of lemonade. How much lemonade does he have left? Write in simplest form.

______

14. **Mathematical PRACTICE 7** **Identify Structure** Abby had $4\frac{7}{10}$ boxes of crackers. She used $1\frac{3}{10}$ boxes of crackers to make a snack for her class. How many boxes of crackers does she have left? Write in simplest form.

______

HOT Problems

15. **Mathematical PRACTICE 1** **Keep Trying** Find the missing mixed number.

$\_\_\_\_ - 2\frac{1}{3} = 3\frac{1}{3}$

16. **Building on the Essential Question** Why should I use addition to check the answer to a subtraction problem?

______

______

______

Name

MY Homework

Lesson 7
Subtract Mixed Numbers

Homework Helper

eHelp Need help? connectED.mcgraw-hill.com

Joanna has $4\frac{1}{6}$ inches of ribbon. She cuts off $1\frac{5}{6}$ inches of the ribbon that are frayed. How long is Joanna's ribbon now?

Find $4\frac{1}{6} - 1\frac{5}{6}$.

1 Write each mixed number as an equivalent improper fraction.

$$4\frac{1}{6} = \frac{6}{6} + \frac{6}{6} + \frac{6}{6} + \frac{6}{6} + \frac{1}{6} = \frac{6+6+6+6+1}{6} = \frac{25}{6}$$

$$1\frac{5}{6} = \frac{6}{6} + \frac{5}{6} = \frac{6+5}{6} = \frac{11}{6}$$

2 Subtract like fractions.

$$\frac{25}{6} - \frac{11}{6} = \frac{14}{6}$$

3 Simplify.

$$\frac{14}{6} = 2\frac{1}{3}$$

$$\frac{6}{6} + \frac{6}{6} + \frac{2}{6} = \frac{14}{6} \text{ or } 2\frac{1}{3}$$

So, Joanna's ribbon is now $2\frac{1}{3}$ inches long.

Practice

Find each difference. Write in simplest form.

1. $7\frac{5}{8} - 4\frac{2}{8} =$ ______

2. $3\frac{1}{4} - 1\frac{3}{4} =$ ______

3. $6\frac{4}{5} - 2\frac{1}{5} =$ ______

4. $8\frac{4}{6} - 2\frac{5}{6} =$ ______

Find each difference. Write in simplest form.

5. $10\frac{2}{6} - 7\frac{3}{6} =$ ______

6. $7\frac{3}{10} - 5\frac{5}{10} =$ ______

7. $15\frac{2}{3} - 8\frac{1}{3} =$ ______

8. $5\frac{2}{5} - 3\frac{4}{5} =$ ______

Problem Solving

Solve. Write the answer in simplest form.

9. Mathematical **PRACTICE** 2 **Use Number Sense** On Saturday Rhonda filled $3\frac{1}{3}$ buckets with shells. On Sunday she filled $4\frac{2}{3}$ buckets with shells. How many more buckets of shells did Rhonda fill on Sunday?

______

10. Cooper worked $6\frac{1}{6}$ hours at the book fair. Amanda worked $4\frac{5}{6}$ hours at the book fair. How much longer did Cooper work?

______

11. Sandy brought $6\frac{1}{4}$ dozen cookies to the bake sale. Alejandro brought $1\frac{3}{4}$ dozen fewer cookies than Sandy brought. How many dozen cookies did Alejandro bring?

______

Test Practice

12. Tara read $4\frac{5}{8}$ pages of the arts section in the newspaper. She read $3\frac{7}{8}$ pages of the sports section. How many more pages of the arts section did Tara read?

Ⓐ $\frac{2}{8}$ pages

Ⓒ $\frac{3}{4}$ pages

Ⓑ $1\frac{2}{8}$ pages

Ⓓ $1\frac{7}{8}$ pages

Check My Progress

Vocabulary Check

Draw lines to match each of the following with its correct description or example.

1. like fractions	• the form a fraction is written in when the numerator and denominator have no common factor other than 1
2. mixed number	• fractions with the same denominator
3. simplest form	• $5\frac{3}{4}$

Concept Check

Find each sum. Write in simplest form.

4. $1\frac{5}{10} + \frac{3}{10} =$ ______

5. $8\frac{8}{12} + 1\frac{1}{12} =$ ______

6. $5\frac{1}{4} + 3\frac{1}{4} =$ ______

7. $7\frac{20}{100} + 2\frac{40}{100} =$ ______

Find each difference. Write in simplest form.

8. $5\frac{7}{8} - 3\frac{2}{8} =$ ______

9. $7\frac{2}{3} - 1\frac{1}{3} =$ ______

10. $9\frac{11}{12} - 4\frac{1}{12} =$ ______

11. $3\frac{60}{100} - 1\frac{20}{100} =$ ______

Problem Solving

Solve. Write in simplest form.

12. Isabella has $2\frac{1}{4}$ oranges. Colleen has $3\frac{1}{4}$ oranges. How many oranges do they have altogether?

13. Liam had $5\frac{7}{8}$ cups of flour. He used $2\frac{3}{8}$ cups of flour to make bread. How much flour does Liam have left?

14. Audrey ran $2\frac{1}{6}$ miles yesterday. Today, she ran $1\frac{3}{6}$ miles. How many miles did she run altogether?

My Work!

Test Practice

15. Isaac's paper airplane flew $5\frac{2}{12}$ feet. Lillian's paper airplane flew $5\frac{5}{12}$ feet. How much farther did Lillian's paper airplane fly?

Ⓐ $\frac{1}{4}$ feet

Ⓑ $\frac{3}{5}$ feet

Ⓒ $5\frac{1}{12}$ feet

Ⓓ $5\frac{7}{12}$ feet

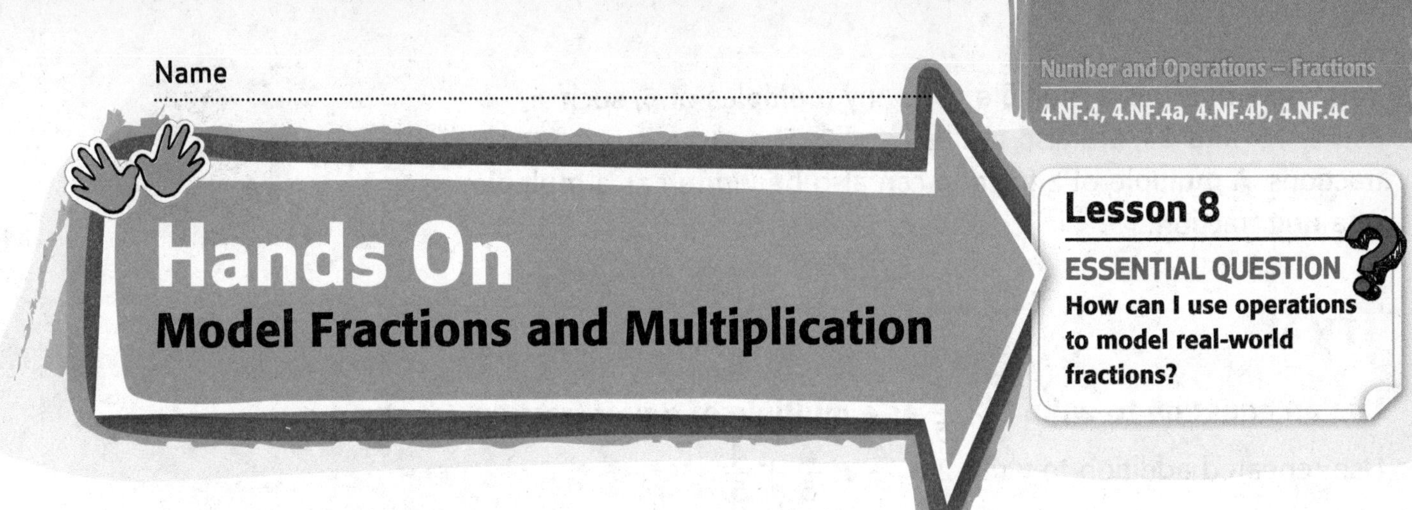

Hands On
Model Fractions and Multiplication

Lesson 8
ESSENTIAL QUESTION
How can I use operations to model real-world fractions?

You have learned to write a fraction as a sum of unit fractions. For example, $\frac{4}{5} = \frac{1}{5} + \frac{1}{5} + \frac{1}{5} + \frac{1}{5}$.

You can also write a fraction as a multiple of a unit fraction.

Build It

Tools

Use an equation to write $\frac{4}{5}$ as a multiple of a unit fraction.

One Way **Use fraction tiles.**

Model $\frac{4}{5}$ using fraction tiles. Draw your result below.

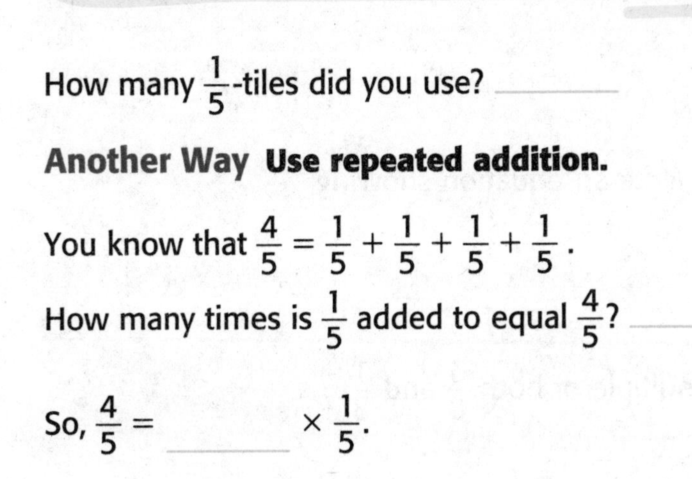

How many $\frac{1}{5}$-tiles did you use? ______

Another Way **Use repeated addition.**

You know that $\frac{4}{5} = \frac{1}{5} + \frac{1}{5} + \frac{1}{5} + \frac{1}{5}$.

How many times is $\frac{1}{5}$ added to equal $\frac{4}{5}$? ______

So, $\frac{4}{5} =$ ______ $\times \frac{1}{5}$.

You know that 6 is a multiple of 2. Any multiples of 6, such as 12, 18, and 24, are also multiples of 2. The same is true for fractions. A multiple of a fraction can also be written as a multiple of a unit fraction.

Try It

Use an equation to write $2 \times \frac{4}{5}$ as a multiple of a unit fraction.

Use repeated addition to write $2 \times \frac{4}{5}$ as $\frac{4}{5} + \frac{4}{5}$.

$\frac{4}{5} + \frac{4}{5} = \frac{8}{5}$ Add like fractions.

Model $\frac{8}{5}$ using fraction tiles. Draw your result below.

My Drawing!

How many $\frac{1}{5}$-tiles did you use? ______

So, $\frac{8}{5}$ is a multiple of $\frac{4}{5}$. It is also a multiple of $\frac{1}{5}$.

$\frac{8}{5} =$ ______ $\times \frac{1}{5}$

Write an equation showing that $\frac{8}{5}$ is a multiple of the unit fraction $\frac{1}{5}$.

$\frac{8}{5} =$ ______ $\times \frac{1}{5}$

So, $2 \times \frac{4}{5} =$ ______ $\times \frac{1}{5}$.

Talk About It

1. Mathematical PRACTICE 7 **Identify Structure** Write an equation showing how $\frac{3}{8}$ is a multiple of $\frac{1}{8}$.

______

2. Write equations showing how $\frac{6}{8}$ is a multiple of both $\frac{3}{8}$ and $\frac{1}{8}$.

______

Name ______________________

CCSS

Practice It

Algebra Use an equation to write each fraction or product as a multiple of a unit fraction.

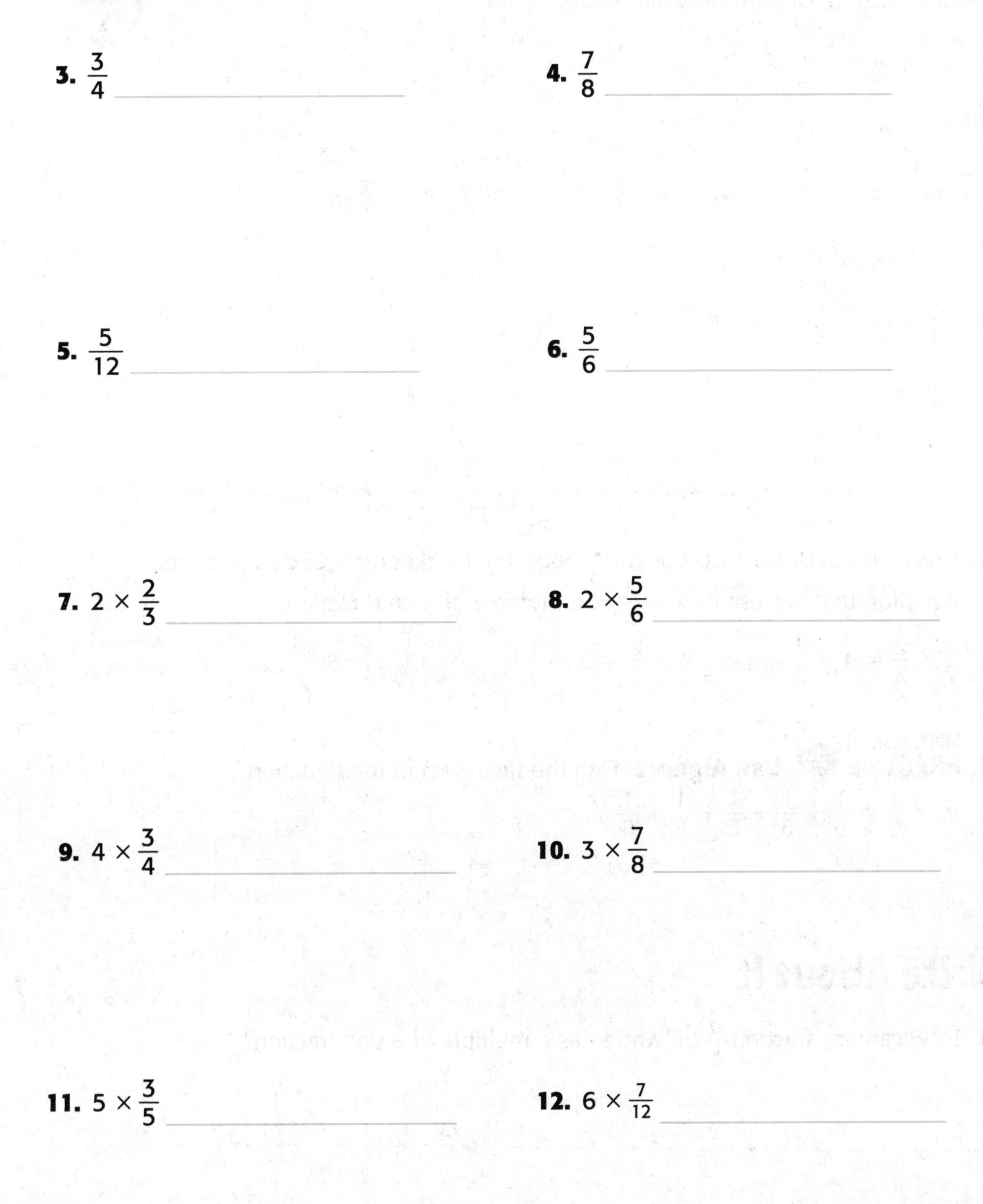

3. $\frac{3}{4}$ ______________________

4. $\frac{7}{8}$ ______________________

5. $\frac{5}{12}$ ______________________

6. $\frac{5}{6}$ ______________________

7. $2 \times \frac{2}{3}$ ______________________

8. $2 \times \frac{5}{6}$ ______________________

9. $4 \times \frac{3}{4}$ ______________________

10. $3 \times \frac{7}{8}$ ______________________

11. $5 \times \frac{3}{5}$ ______________________

12. $6 \times \frac{7}{12}$ ______________________

Apply It

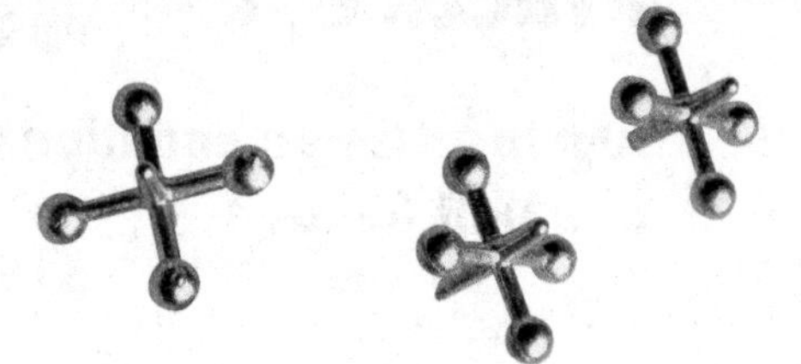

13. Mathematical PRACTICE 4 **Model Math** Use fraction tiles and repeated addition to write $3 \times \frac{3}{4}$ as a multiple of a unit fraction. Draw your result below.

My Drawing!

14. Gracie and Jackson each bought $\frac{2}{3}$ pound of blackberries. Circle the correct equation that represents $2 \times \frac{2}{3}$ as a multiple of a unit fraction.

$2 \times \frac{2}{3} = 4 \times \frac{1}{3}$ $\qquad$ $2 \times \frac{2}{3} = 2 \times \frac{1}{3}$

15. Mathematical PRACTICE 2 **Use Algebra** Find the unknown in the equation $m \times \frac{1}{6} = \frac{1}{6} + \frac{1}{6} + \frac{1}{6} + \frac{1}{6} + \frac{1}{6}$.

Write About It

16. How can any fraction $\frac{a}{b}$ be written as a multiple of a unit fraction?

Name ____________________

MY Homework

Lesson 8
Hands On: Model Fractions and Multiplication

Homework Helper

eHelp Need help? connectED.mcgraw-hill.com

Write $\frac{3}{6}$ as a multiple of a unit fraction.

Model $\frac{3}{6}$ using fraction tiles.

$\frac{1}{6}$	$\frac{1}{6}$	$\frac{1}{6}$			

There are 3 of the $\frac{1}{6}$-tiles.

So, $\frac{3}{6} = 3 \times \frac{1}{6}$.

Helpful Hint

$\frac{3}{6}$ is a multiple of $\frac{1}{6}$ because the product of 3 and $\frac{1}{6}$ is $\frac{3}{6}$.

Use an equation to write $4 \times \frac{2}{3}$ as a multiple of a unit fraction.

Use repeated addition to write $4 \times \frac{2}{3}$ as $\frac{2}{3} + \frac{2}{3} + \frac{2}{3} + \frac{2}{3}$.

$\frac{2}{3} + \frac{2}{3} + \frac{2}{3} + \frac{2}{3} = \frac{8}{3}$

So, $\frac{8}{3}$ is a multiple of $\frac{2}{3}$. It is also a multiple of $\frac{1}{3}$.

$\frac{8}{3} = 8 \times \frac{1}{3}$

So, $4 \times \frac{2}{3} = 8 \times \frac{1}{3}$.

Practice

Algebra Use an equation to write each fraction as a multiple of a unit fraction.

1.

$\frac{1}{6}$	$\frac{1}{6}$	$\frac{1}{6}$	$\frac{1}{6}$	$\frac{1}{6}$	

$\frac{5}{6}$ ____________________

2. 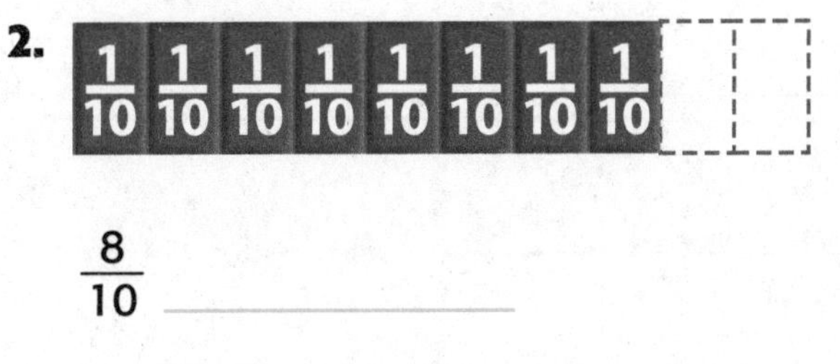

$\frac{8}{10}$ ____________________

Algebra Use an equation to write each fraction or product as a multiple of a unit fraction.

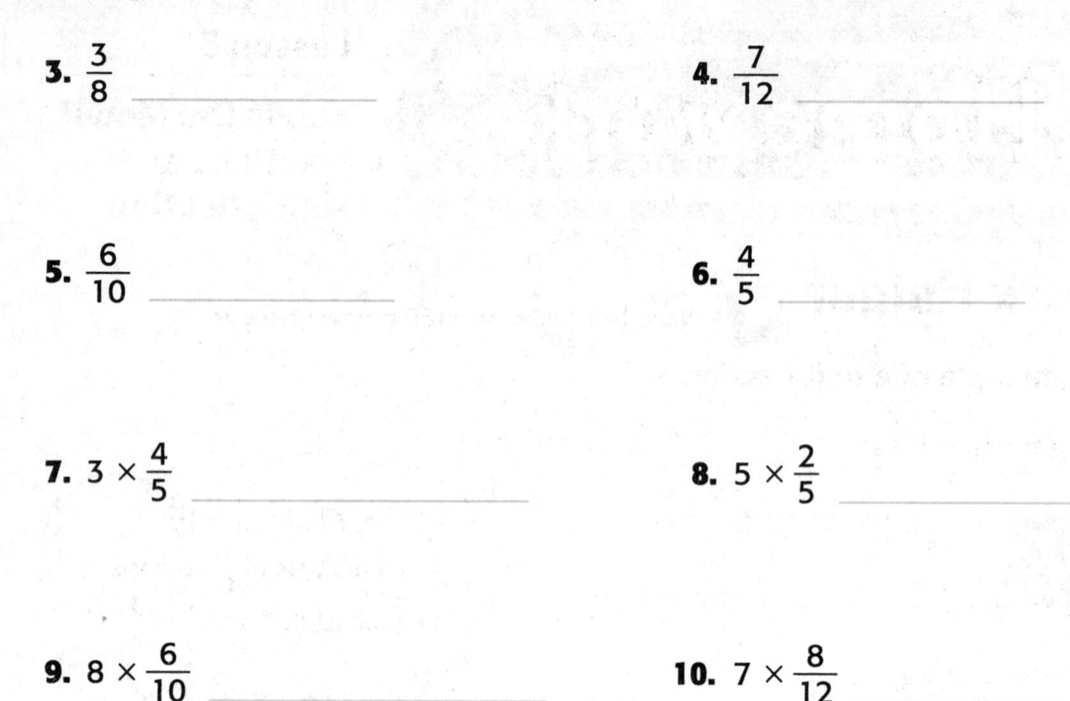

3. $\frac{3}{8}$ ____________

4. $\frac{7}{12}$ ____________

5. $\frac{6}{10}$ ____________

6. $\frac{4}{5}$ ____________

7. $3 \times \frac{4}{5}$ ____________

8. $5 \times \frac{2}{5}$ ____________

9. $8 \times \frac{6}{10}$ ____________

10. $7 \times \frac{8}{12}$ ____________

Problem Solving

11. Mathematical PRACTICE 4 **Model Math** Marcia has one cup of tea each day for 7 days. She puts $\frac{2}{3}$ tablespoons of honey in each cup of tea. Write an equation that represents $7 \times \frac{2}{3}$ as a multiple of a unit fraction.

____

12. Sam buys 4 tropical fish. Each fish is $\frac{5}{8}$ of an inch long. Write an equation that represents $4 \times \frac{5}{8}$ as a multiple of a unit fraction.

____

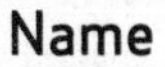

Name ....

Multiply Fractions by Whole Numbers

Lesson 9

ESSENTIAL QUESTION
How can I use operations to model real-world fractions?

You can use models and equations to multiply a fraction by a whole number.

Math in My World

Tools Watch Tutor

Example 1

Each card on a trivia game has 6 questions. Each question represents $\frac{1}{6}$ of the questions on the card. Caleb correctly answered 4 of the questions. What fraction of the questions on a card did he answer correctly?

Find $4 \times \frac{1}{6}$.

One Way

Use repeated addition.

Use repeated addition to write an equation.

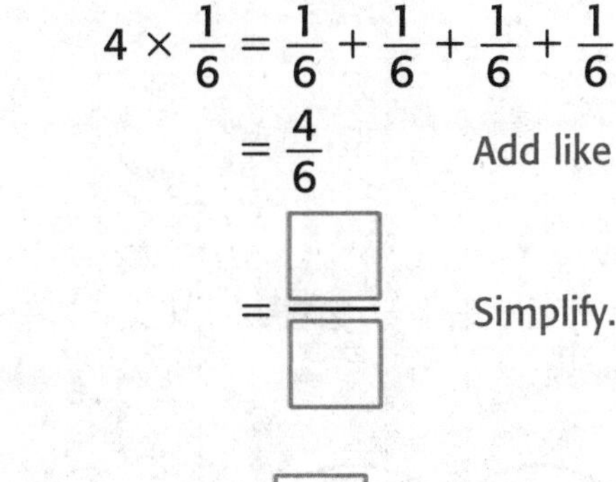

$4 \times \frac{1}{6} = \frac{1}{6} + \frac{1}{6} + \frac{1}{6} + \frac{1}{6}$

$= \frac{4}{6}$ Add like fractions.

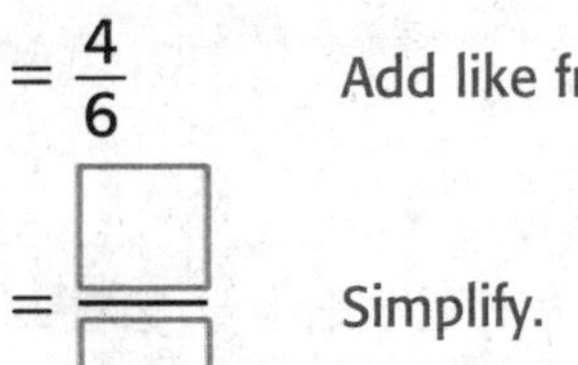

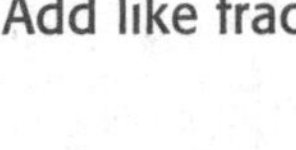

$= \frac{\square}{\square}$ Simplify.

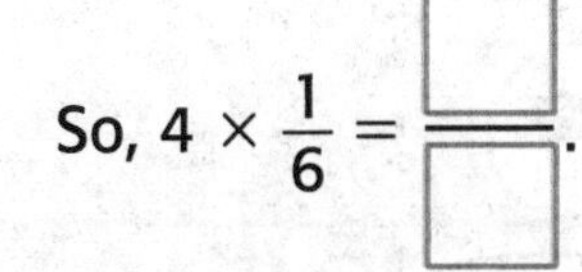

So, $4 \times \frac{1}{6} = \frac{\square}{\square}$.

Another Way

Use models.

The number line shows the first four multiples of $\frac{1}{6}$.

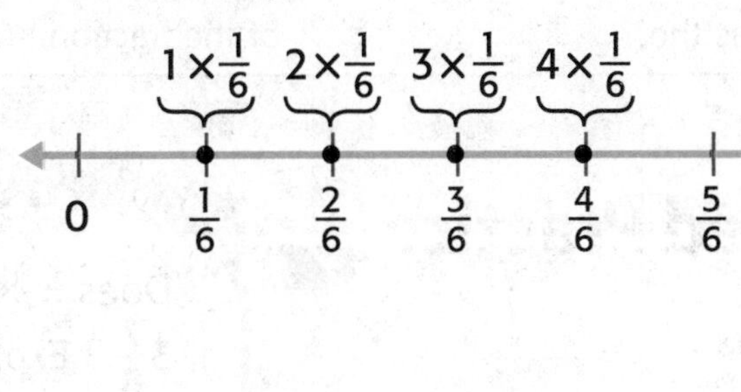

Check Use fraction tiles. $4 \times \frac{1}{6} = \frac{4}{6}$ or $\frac{\square}{\square}$

$\frac{1}{6}$	$\frac{1}{6}$	$\frac{1}{6}$	$\frac{1}{6}$		
$\frac{1}{3}$		$\frac{1}{3}$			

You can use equations and properties to multiply a fraction by a whole number.

Example 2

Tutor

Find $5 \times \frac{3}{10}$. Identify the two whole numbers between which the product lies.

Think of $\frac{3}{10}$ as a multiple of $\frac{1}{10}$.

$5 \times \frac{3}{10} = 5 \times \left(3 \times \frac{1}{10}\right)$	$\frac{3}{10} = 3 \times \frac{1}{10}$
$= (5 \times 3) \times \frac{1}{10}$	Associative Property
$= 15 \times \frac{1}{10}$	Multiply. $5 \times 3 = 15$
$= \frac{15}{10}$	15 groups of $\frac{1}{10} = \frac{15}{10}$
$= 1\frac{5}{10}$	15 divided into groups of $10 = 1\frac{5}{10}$
$= 1\frac{1}{2}$	Simplify. $\frac{5}{10} = \frac{1}{2}$

So, $5 \times \frac{3}{10} =$ ________.

The product lies between the whole numbers 1 and 2.

Look at the product before it was simplified.

$5 \times \frac{3}{10} = \frac{15}{10}$

The numerator of the product is the same as the product of the whole number and the ________ of the fraction. $5 \times 3 = 15$

The denominator of the product is the same as the ________ of the fraction.

Guided Practice

Check

Talk MATH

Does $3 \times \frac{7}{8} = 3\frac{7}{8}$? Explain.

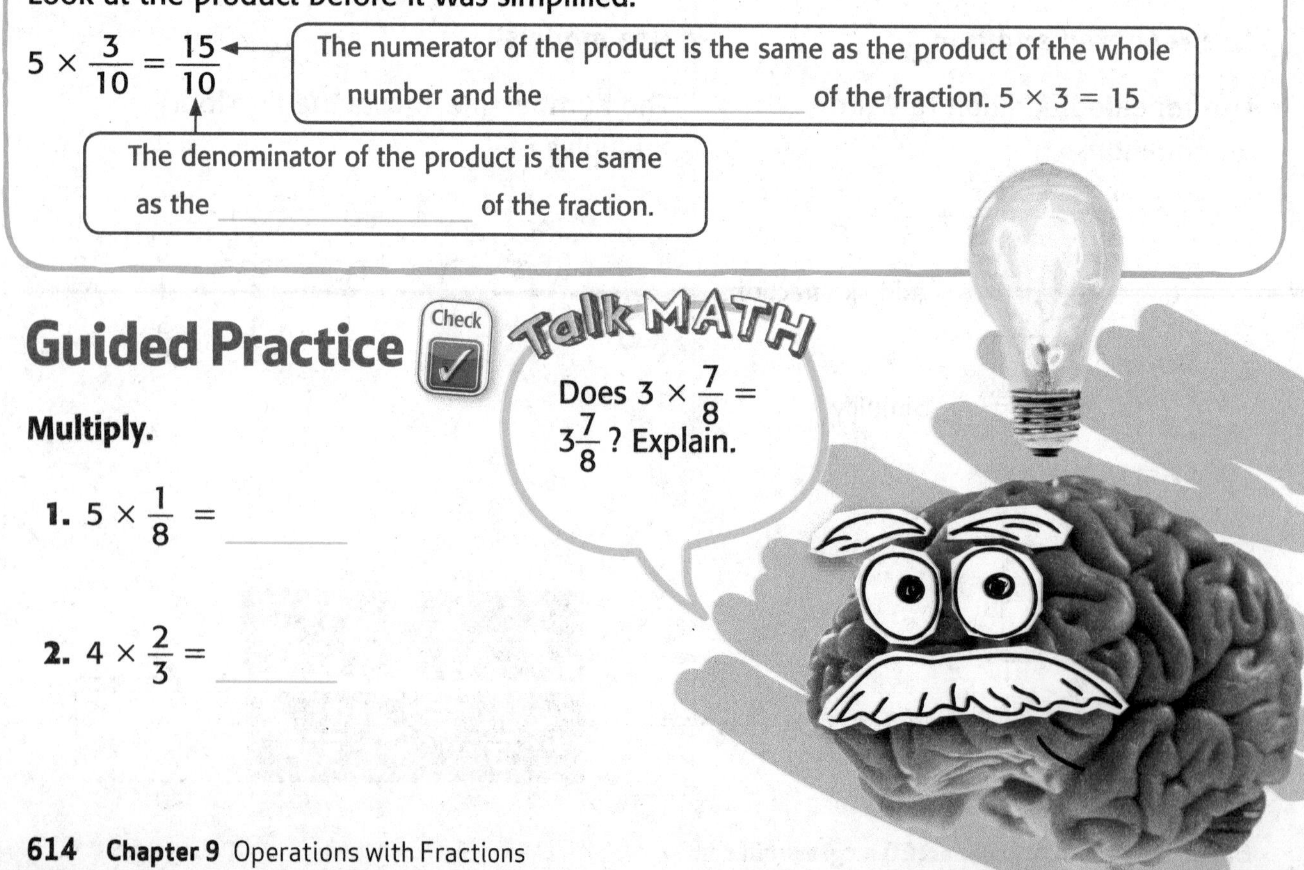

Multiply.

1. $5 \times \frac{1}{8} =$ ________

2. $4 \times \frac{2}{3} =$ ________

Name

Independent Practice

Multiply. Write in simplest form. Identify the two whole numbers between which the product lies.

3. $8 \times \frac{1}{5} =$

4. $25 \times \frac{1}{10} =$

5. $4 \times \frac{3}{4} =$

6. $5 \times \frac{6}{8} =$

7. $11 \times \frac{2}{8} =$

8. $14 \times \frac{2}{12} =$

9. $5 \times \frac{2}{3} =$

10. $2 \times \frac{9}{10} =$

Problem Solving

11. Mario and 3 friends each had $\frac{3}{4}$ foot of rope. They needed to have 5 feet of rope altogether. Do they have enough? Explain.

12. Mathematical PRACTICE 6 **Explain to a Friend** Mrs. Raymond gave each of her students $\frac{1}{12}$ of a box of crackers. She has 30 students in her class. How many boxes of crackers will she need? Explain to a friend.

HOT Problems

13. Mathematical PRACTICE 4 **Model Math** Draw a model of a multiplication problem where the product is an improper fraction.

My Drawing!

14. **Building on the Essential Question** How can changing improper fractions to mixed numbers help me determine between which two whole numbers a fraction lies?

Name ....

MY Homework

Lesson 9
Multiply Fractions by Whole Numbers

Homework Helper

eHelp

Need help? connectED.mcgraw-hill.com

Ms. Randall reads $\frac{1}{10}$ of a book to her class each day. What fraction of the book has Ms. Randall read to the class after 5 days? Find $5 \times \frac{1}{10}$.

Use repeated addition to write an equation.

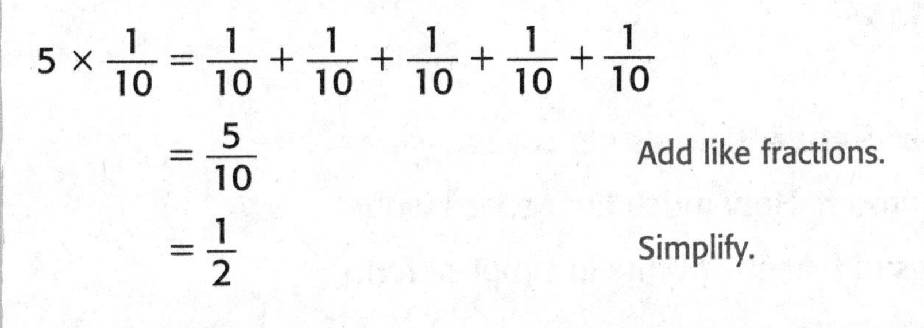

$$5 \times \frac{1}{10} = \frac{1}{10} + \frac{1}{10} + \frac{1}{10} + \frac{1}{10} + \frac{1}{10}$$

$= \frac{5}{10}$ Add like fractions.

$= \frac{1}{2}$ Simplify.

So, Ms. Randall has read $\frac{1}{2}$ of the book after 5 days.

Practice

Multiply.

1. $3 \times \frac{2}{5} =$ ______

2. $7 \times \frac{3}{4} =$ ______

3. $5 \times \frac{5}{6} =$ ______

4. $2 \times \frac{8}{10} =$ ______

5. $8 \times \frac{3}{10} =$ ______

6. $6 \times \frac{5}{8} =$ ______

Find each product. Identify the two whole numbers between which the product lies.

7. $5 \times \frac{7}{10} =$ ______

The product lies between ______ and ______.

8. $7 \times \frac{8}{10} =$ ______

The product lies between ______ and ______.

9. $3 \times \frac{3}{4} =$ ______

The product lies between ______ and ______.

10. $6 \times \frac{4}{5} =$ ______

The product lies between ______ and ______.

Problem Solving

11. Mathematical PRACTICE 2 **Use Number Sense** Calvin's rug covers $\frac{1}{8}$ of the floor space in his bedroom. How much floor space would be covered if Calvin had 4 rugs of that size? Write in simplest form.

______

12. Amy uses $\frac{2}{3}$ of a yard of fabric for each pillow she makes. How many yards of fabric will she need in order to make 8 pillows? Write in simplest form.

______

Test Practice

13. Sheila eats $\frac{3}{4}$ of a bag of baby carrots each week. How many bags of baby carrots does she eat in 6 weeks? Write in simplest form.

Ⓐ $4\frac{1}{2}$ bags

Ⓑ 3 bags

Ⓒ $2\frac{1}{4}$ bags

Ⓓ $1\frac{1}{2}$ bags

Review

Chapter 9

Operations with Fractions

Vocabulary Check

Vocab

1. Use the numbers on the number cubes to create each type of fraction listed below. You can use the numbers more than once.

like fractions

mixed number

simplest form

Concept Check

Check

Find each sum. Write in simplest form.

2. $\frac{1}{6} + \frac{4}{6} =$ ______

3. $3\frac{3}{5} + 2\frac{1}{5} =$ ______

4. $\frac{2}{8} + \frac{4}{8} =$ ______

Find each difference. Write in simplest form.

5. $\frac{3}{4} - \frac{2}{4} =$ ______

6. $5\frac{2}{3} - 2\frac{1}{3} =$ ______

7. $\frac{8}{10} - \frac{3}{10} =$ ______

Algebra Use an equation to write each fraction or product as a multiple of a unit fraction.

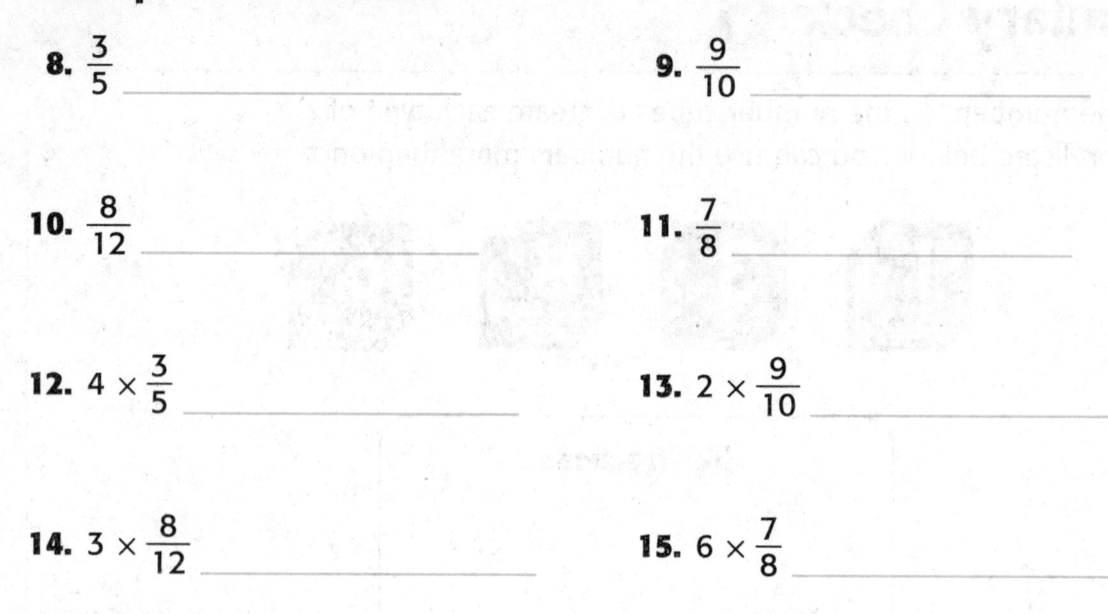

8. $\frac{3}{5}$ ______

9. $\frac{9}{10}$ ______

10. $\frac{8}{12}$ ______

11. $\frac{7}{8}$ ______

12. $4 \times \frac{3}{5}$ ______

13. $2 \times \frac{9}{10}$ ______

14. $3 \times \frac{8}{12}$ ______

15. $6 \times \frac{7}{8}$ ______

Multiply. Write in simplest form. Identify the two whole numbers between which the product lies.

16. $6 \times \frac{1}{4} =$ ______

17. $24 \times \frac{1}{5} =$ ______

18. $7 \times \frac{3}{5} =$ ______

19. $3 \times \frac{4}{6} =$ ______

Name

Problem Solving

20. Kylie and her two sisters each have $\frac{1}{4}$ cup of chopped pineapple. How much pineapple do they have in all?

21. Joshua and his four friends each have $\frac{1}{2}$ package of crackers. How many packages of crackers do they have in all? Explain.

22. Write a real-world problem to add like fractions. Then solve the problem.

My Work!

Test Practice

23. There are 35 students in Miss Klempa's class. Suppose each student has $\frac{1}{10}$ of a box of pencils. How many boxes of pencils do they have in all?

Ⓐ 3 boxes

Ⓑ $3\frac{1}{2}$ boxes

Ⓒ 4 boxes

Ⓓ $4\frac{1}{2}$ boxes

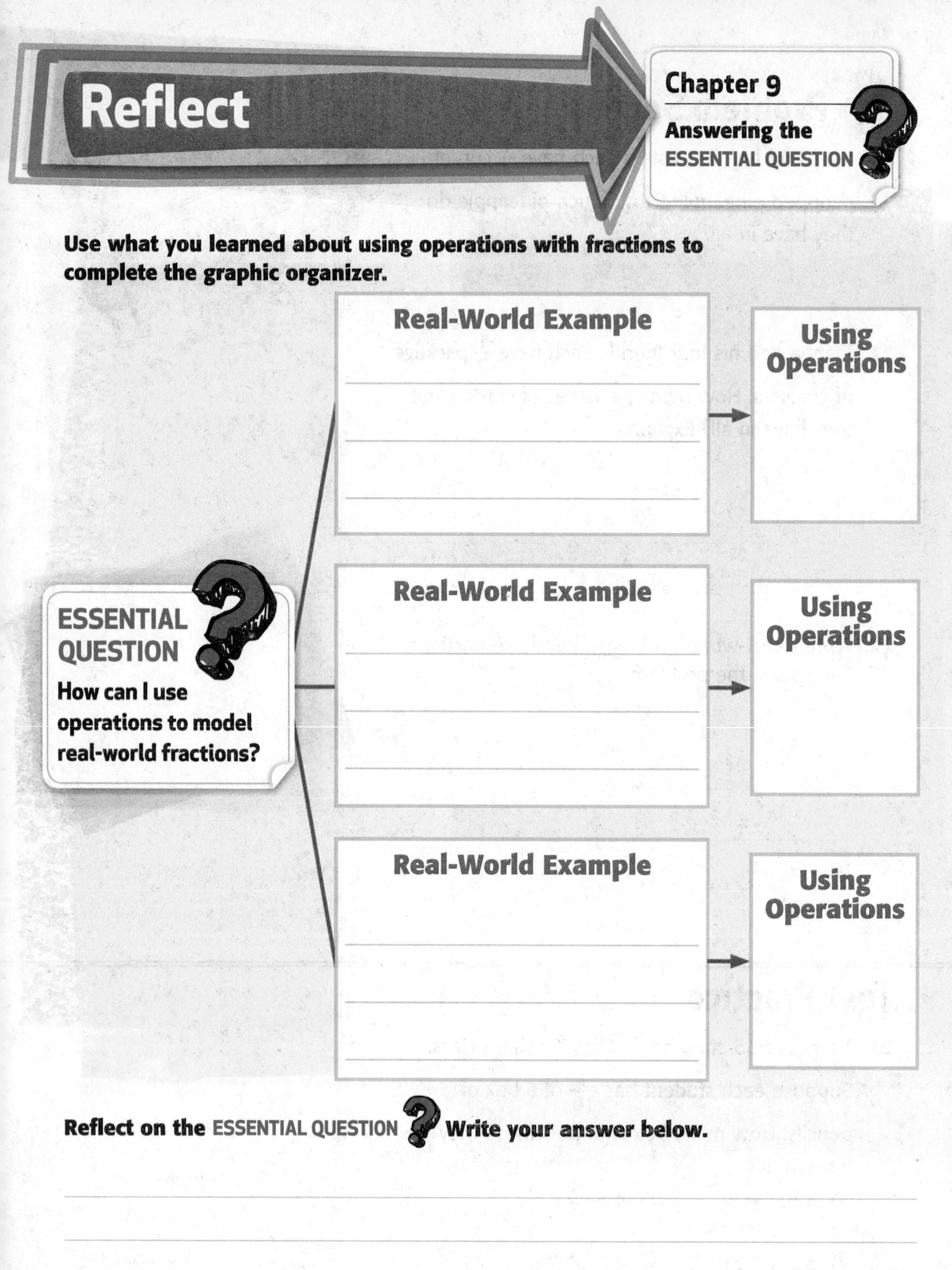

Use what you learned about using operations with fractions to complete the graphic organizer.

Reflect on the ESSENTIAL QUESTION **Write your answer below.**

Chapter

10 Fractions and Decimals

ESSENTIAL QUESTION

How are fractions and decimals related?

Away We Go!

Watch a video!

Watch

MY Common Core State Standards

Number and Operations – Fractions

CCSS

4.NF.5 Express a fraction with denominator 10 as an equivalent fraction with denominator 100, and use this technique to add two fractions with respective denominators 10 and 100.

4.NF.6 Use decimal notation for fractions with denominators 10 or 100.

4.NF.7 Compare two decimals to hundredths by reasoning about their size. Recognize that comparisons are valid only when the two decimals refer to the same whole. Record the results of comparisons with the symbols >, =, or <, and justify the conclusions, e.g., by using a visual model.

OK, this'll be good to know!

Standards for Mathematical PRACTICE

1. Make sense of problems and persevere in solving them.
2. Reason abstractly and quantitatively.
3. Construct viable arguments and critique the reasoning of others.
4. Model with mathematics.
5. Use appropriate tools strategically.
6. Attend to precision.
7. Look for and make use of structure.
8. Look for and express regularity in repeated reasoning.

= focused on in this chapter

Name

← Go online to take the Readiness Quiz

Write a fraction to describe the part that is green.

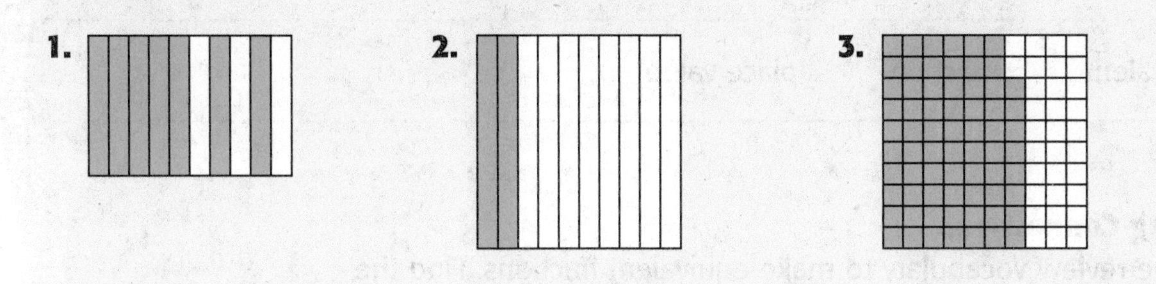

Write each as a fraction.

4. four tenths

5. eight tenths

6. twenty hundredths

7. On Tuesday, seven tenths of an inch of rain fell. Write the amount of rain as a fraction.

Algebra Find each unknown.

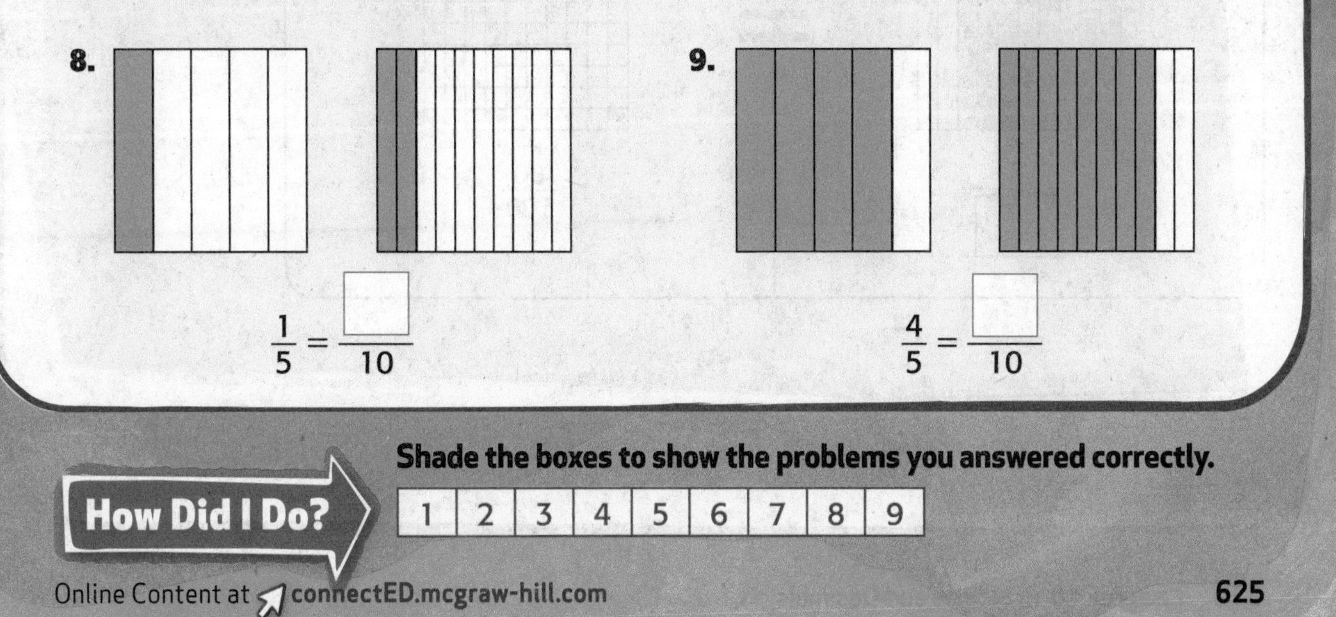

$\frac{1}{5} = \frac{\square}{10}$

$\frac{4}{5} = \frac{\square}{10}$

Shade the boxes to show the problems you answered correctly.

How Did I Do?

1	2	3	4	5	6	7	8	9

Name

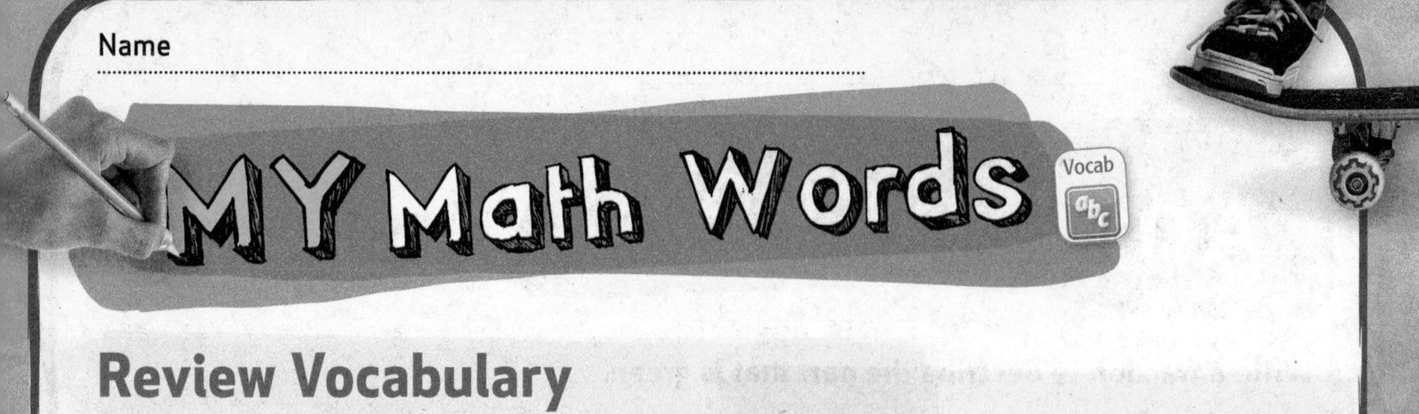

Review Vocabulary

equivalent fraction place value

Making Connections

Use the review vocabulary to make equivalent fractions. Find the unknown in the fraction or shade to complete the area model.

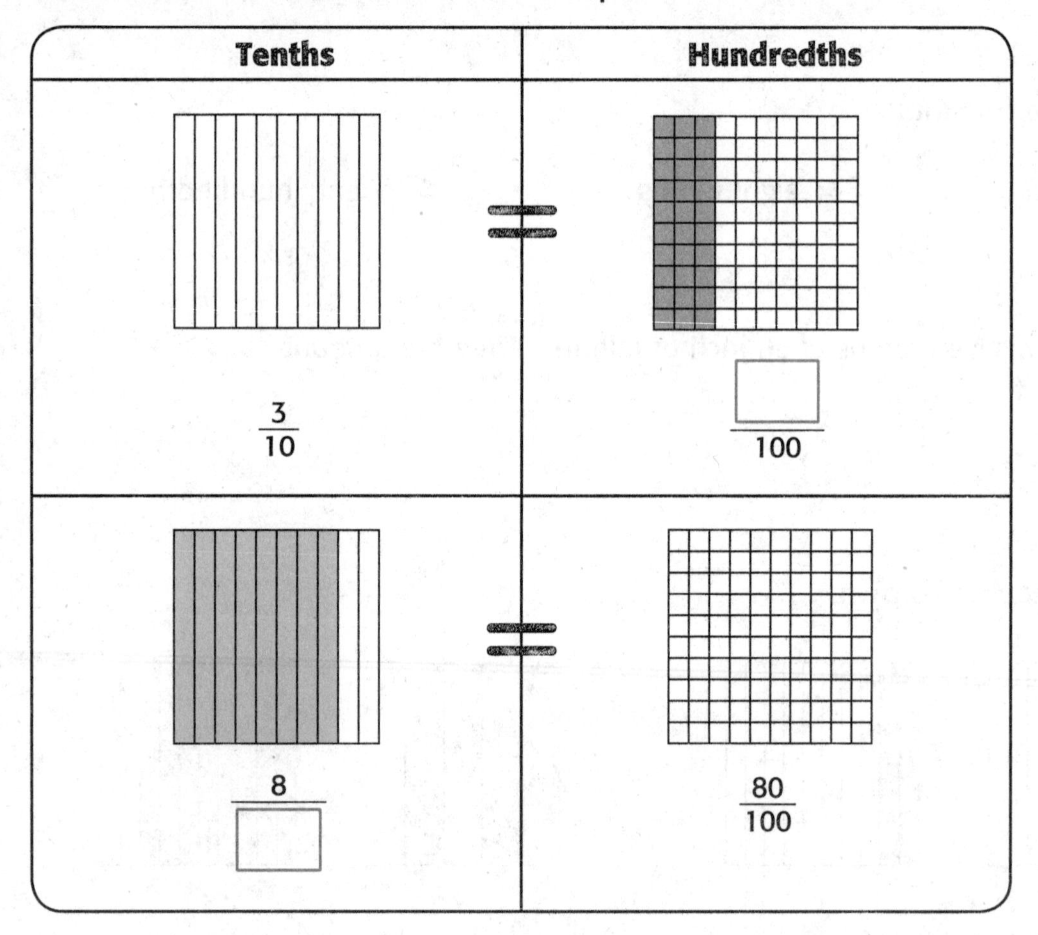

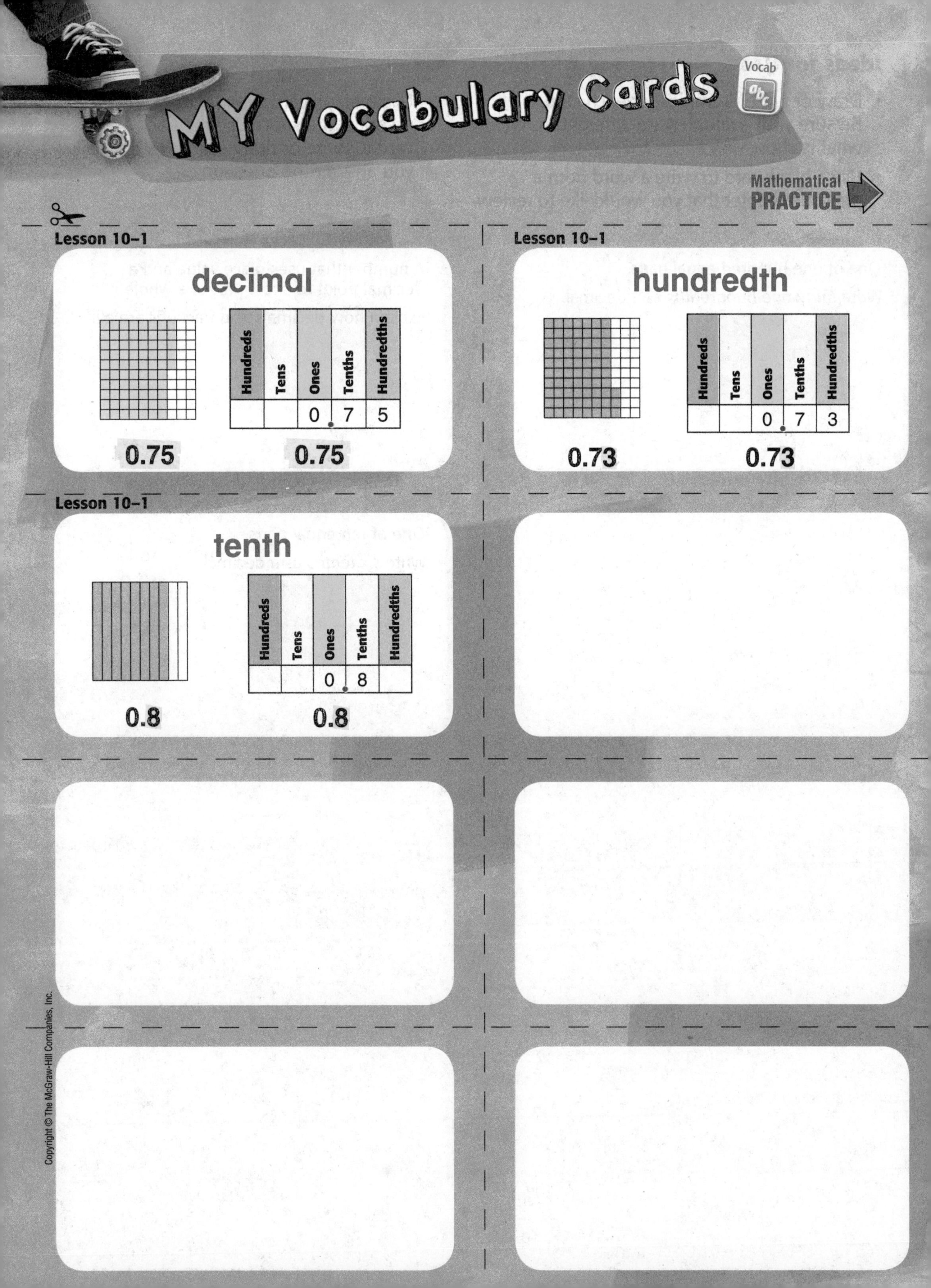

MY Vocabulary Cards

Vocab

Mathematical PRACTICE

Lesson 10-1

decimal

Hundreds	Tens	Ones	Tenths	Hundredths
		0 .	7	5

0.75 0.75

Lesson 10-1

hundredth

Hundreds	Tens	Ones	Tenths	Hundredths
		0 .	7	3

0.73 0.73

Lesson 10-1

tenth

Hundreds	Tens	Ones	Tenths	Hundredths
		0 .	8	

0.8 0.8

Ideas for Use

- Draw or write examples for each card. Be sure your examples are different from what is shown on each card.
- Use a blank card to write a word from a previous chapter that you would like to review.
- Use a blank card to write this chapter's essential question. Use the back of the card to write or draw examples that help you answer the question.

One of one hundred equal parts.

Write *thirty-one hundredths* as a decimal.

A number that uses place value and a decimal point to show part of a whole.

Explain how decimals and fractions are alike.

One of ten equal parts.

Write *six tenths* as a decimal.

MY Foldable

Follow the steps on the back to make your Foldable.

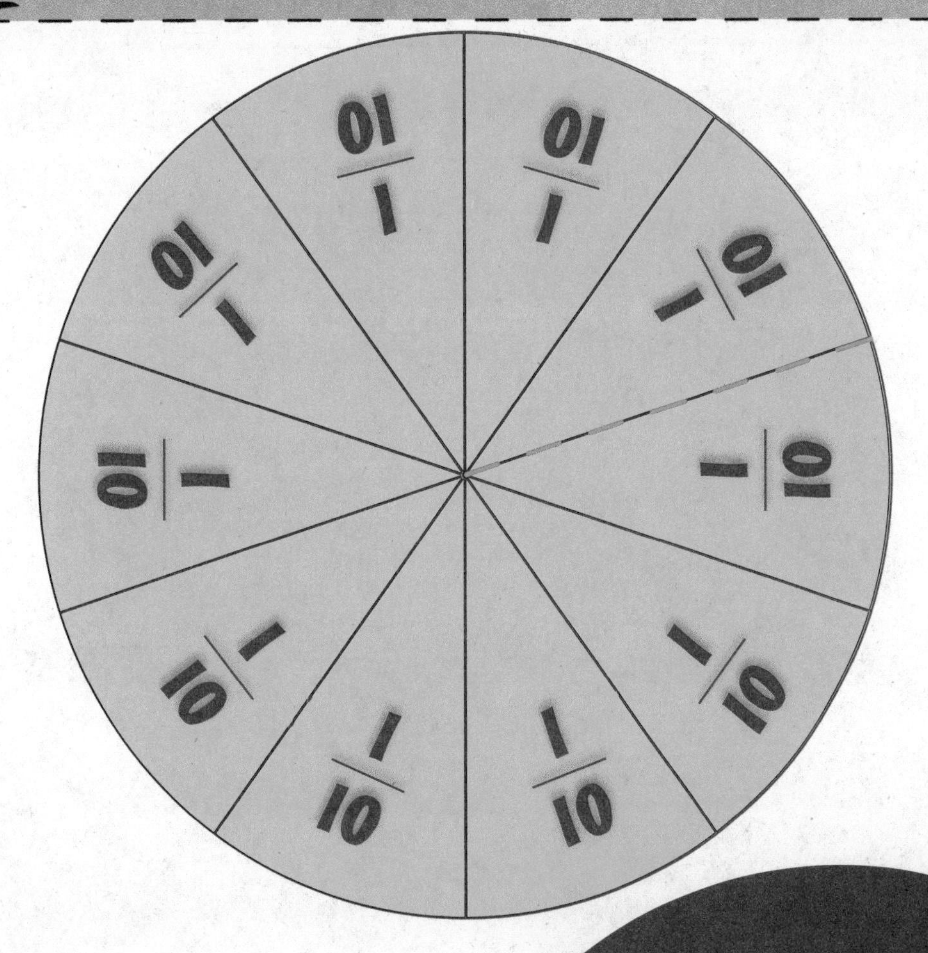

1

2

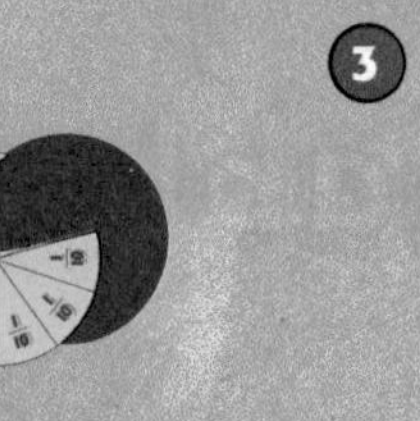

3

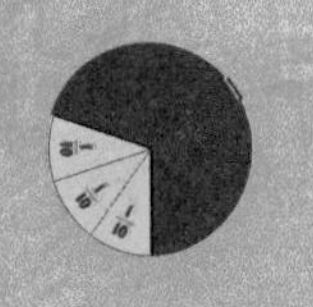

Name

Hands On
Place Value Through Tenths and Hundredths

Lesson 1

ESSENTIAL QUESTION
How are fractions and decimals related?

A **decimal** is a number that uses place value and a decimal point to show part of a whole.

Build It

Tools

Kendra has 1 dollar, 3 dimes, and 5 pennies. Model this amount using a place-value chart.

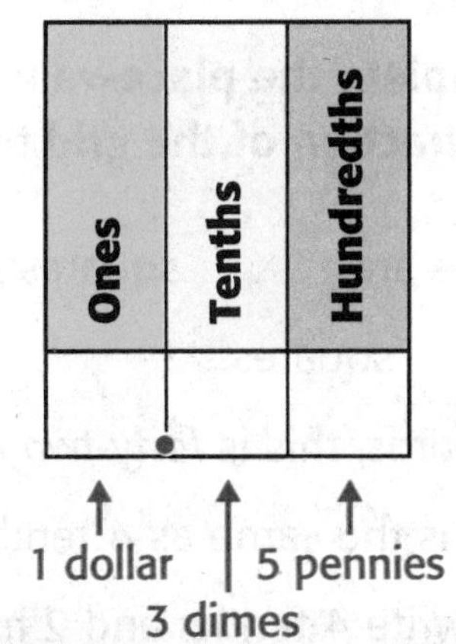

You know that one dollar is equal to ______ cents.

One dollar is one whole. So, write 1 in the ones place.

One dime is one of 10 equal parts of a dollar. One of 10 equal parts is one **tenth**.

So, one dime is *one tenth*, or $\frac{1}{10}$ of a dollar.

Kendra has ______ dimes. Write 3 in the tenths place.

One penny is one of 100 equal parts of a dollar. One of 100 equal parts is one **hundredth**.

So, one penny is *one hundredth*, or $\frac{1}{100}$ of a dollar.

Kendra has ______ pennies. Write 5 in the hundredths place.

The decimal that represents the amount of money Kendra has, in dollars, is $1.35.

You can use models to represent decimals.

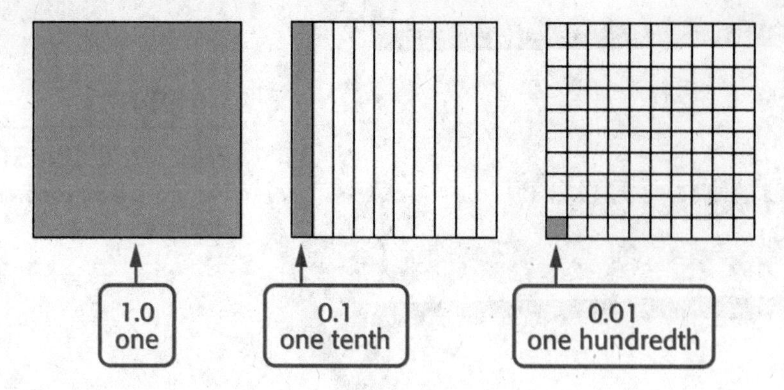

Try It

Complete the place-value chart that represents the fraction of the grid that is shaded at the right.

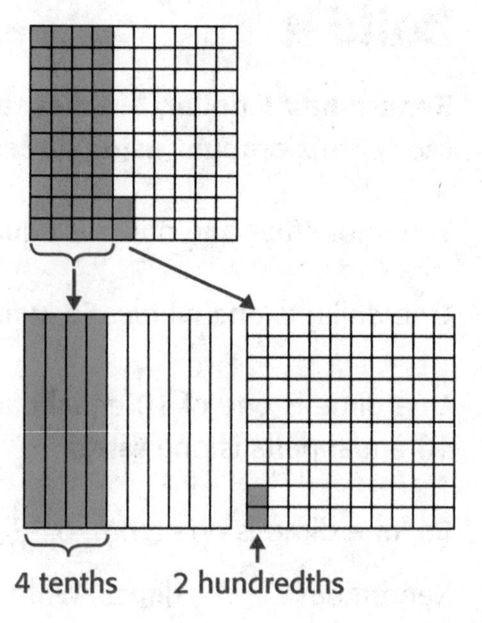

There are ______ squares shaded out of a total of ______ squares.

In words, this is *forty-two hundredths.*

This is the same as 4 tenths and 2 hundredths.

So, write 4 tenths and 2 hundredths in the place-value chart.

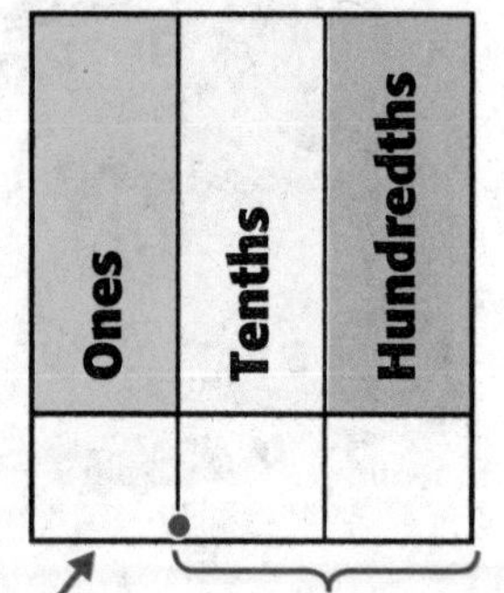

There are 0 whole grids shaded.

4 tenths and 2 hundredths is 42 hundredths.

Talk About It

1. Mathematical PRACTICE 2 **Use Number Sense** Paulo has 6 dimes. Marc has 6 pennies. How many times greater is the value of 6 dimes than 6 pennies? Explain.

______

______

Name

Practice It

Complete the place-value chart that represents each set of dollar bills and coins.

2. 0 dollars, 4 dimes, 8 pennies

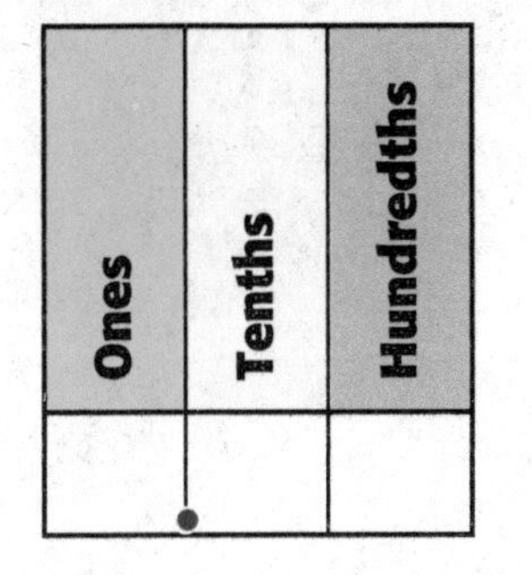

3. 2 dollars, 9 dimes, 1 penny

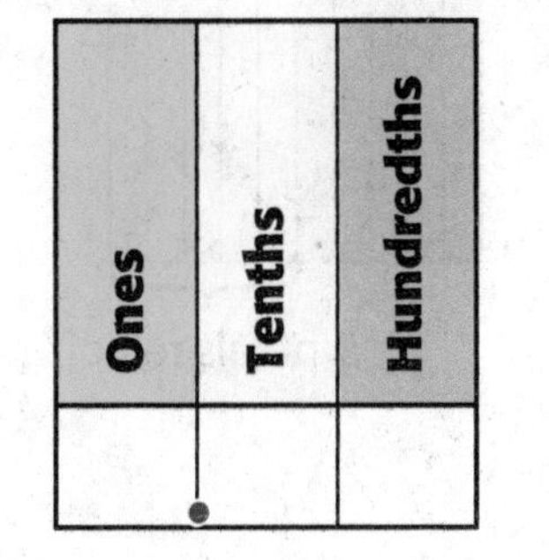

Write the decimal represented by each model.

4.

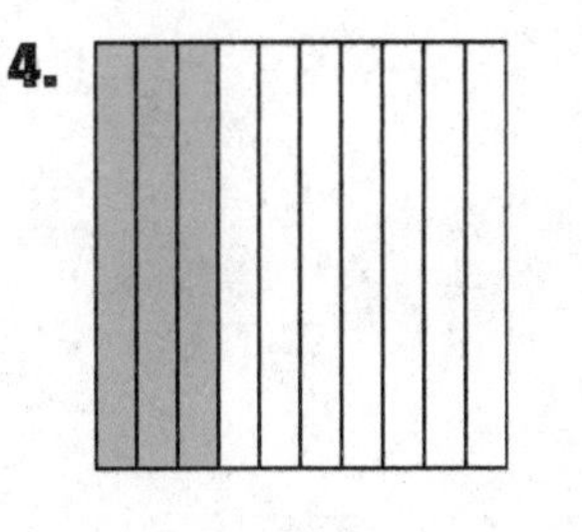

5.

Write each decimal.

6. *two tenths*

7. *twenty-five hundredths*

8. *forty-nine hundredths*

9. *seventy-three hundredths*

Shade each model to represent each decimal.

10. 0.7

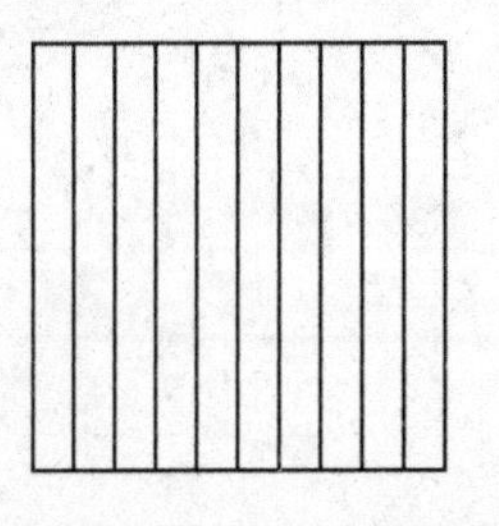

11. 0.51

12. Circle the greater decimal in Exercises 10 and 11.

Apply It

Rain, rain go away!

My Work!

13. A flower has ten petals. During a storm, five of the petals fall off. Write a decimal to show what part of the petals is left.

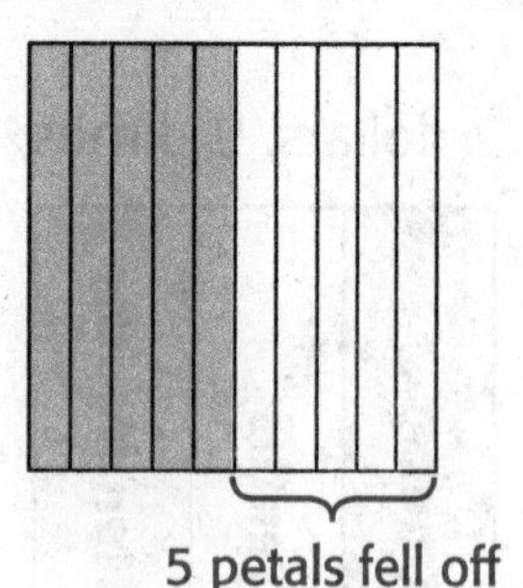

____

14. A movie theater has 100 seats. Seventy-eight of the seats are full. What decimal shows the part of the seats that are full?

78 seats are full

____

15. Mathematical PRACTICE 4 **Model Math** Write a real-world problem that results in an answer of 0.58.

____

____

Write About It

16. How can I relate decimals to a place-value chart?

____

____

Name

MY Homework

Lesson 1

Hands On: Place Value Through Tenths and Hundredths

Homework Helper

Need help? connectED.mcgraw-hill.com

Complete the place-value chart that represents the fraction of the grid that is shaded.

There are 46 squares shaded out of a total of 100 squares.

In words, this is *forty-six hundredths*.

This is the same as 4 tenths and 6 hundredths.

Write 4 tenths and 6 hundredths in the place-value chart below.

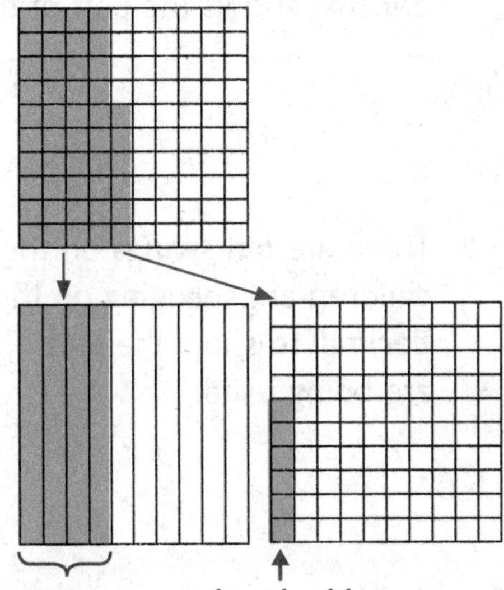

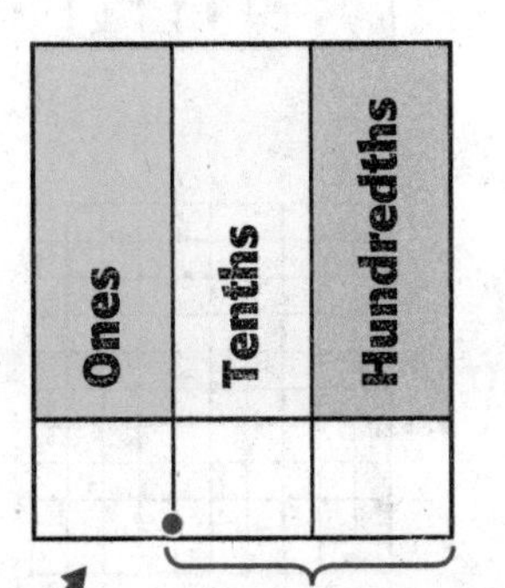

There are 0 whole grids shaded.

4 tenths and 6 hundredths is 46 hundredths.

Practice

Write the decimal represented by each model.

1. 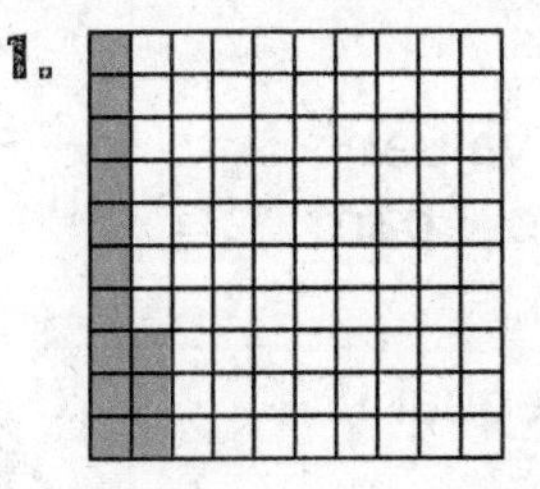

______

2.

______

Problem Solving

3. Kristy has 100 buttons in her button collection. She has 24 red buttons. Write a decimal to show the part of her collection of buttons that is red.

____

4. Harry and Dario went to the county fair. There were ten rides, but they only had time to ride six. What decimal shows the part of the rides that they rode?

____

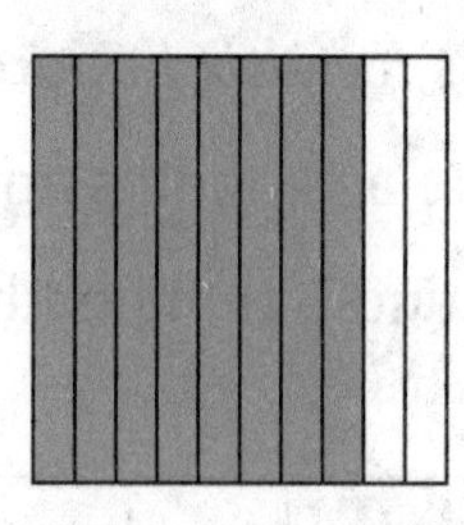

5. There are ten swings on the playground. Eight children are swinging on the swings. Write a decimal to show the part of the swings that are being used.

____

6. There are 100 students in the fourth grade. Five of these students are absent today. Write a decimal to show the part of the students that are absent.

____

Vocabulary Check

Vocab

Draw a line to match each word to its definition.

7. decimal
8. tenth
9. hundredth

- one part of ten equal parts
- a number that uses place value and a decimal point to show part of a whole
- one part of one hundred equal parts

Name

Tenths

Lesson 2

ESSENTIAL QUESTION
How are fractions and decimals related?

Math in My World

Tools Watch Tutor

Example 1

Lorena and Rebecca went to 10 horseback riding lessons. Three of the lessons were outside the barn. What decimal represents the part of the horseback riding lessons that were spent outside?

There were 10 horseback riding lessons. Three of the 10 lessons were spent outside. Three out of ten equal parts means *three tenths*.

1 Use a model.
Three tenths of the model is shaded.

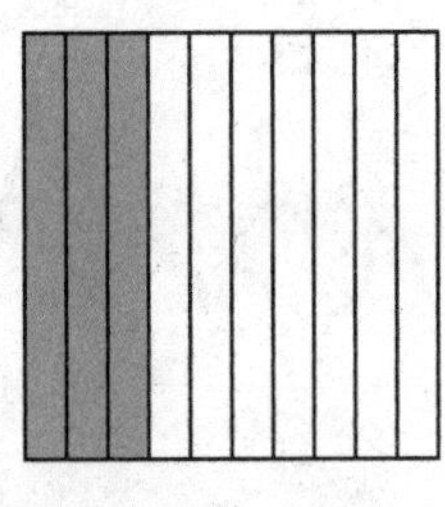

2 Use a place-value chart.
Write 3 tenths in the place-value chart.

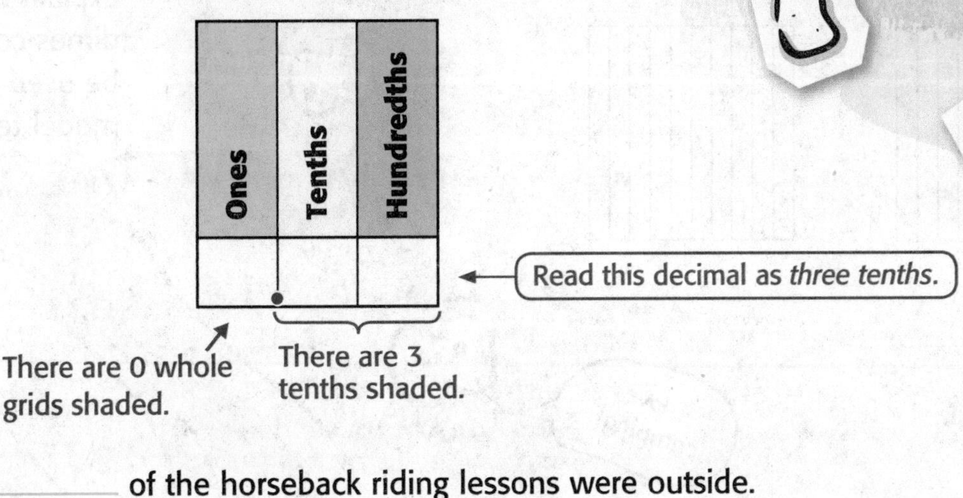

So, ______ of the horseback riding lessons were outside.

Example 2

A campground has 10 campsites. Eight of the campsites have campers. The rest are empty. What part of the campground is empty? Write as a decimal. Use a number line.

If ________ out of ten campsites have campers, then ________ out of ten campsites do *not* have campers.

The number line is divided from 0 to 1 in 10 equal parts.

2 parts out of 10 equal parts

0 0.1 0.2 0.3 0.4 0.5 0.6 0.7 0.8 0.9 1

Two out of ten equal parts is *two tenths*, or 0.2.

So, ________ of the campground is empty.

Example 3

Use words to describe 0.7.

The 7 is in the tenths place. So, 0.7 is *seven tenths*. *Seven tenths* means seven out of ten equal parts.

So, 0.7 can be described as ________ out of ________.

Guided Practice

Shade the model. Then write the decimal.

1. four out of ten

2. eight tenths

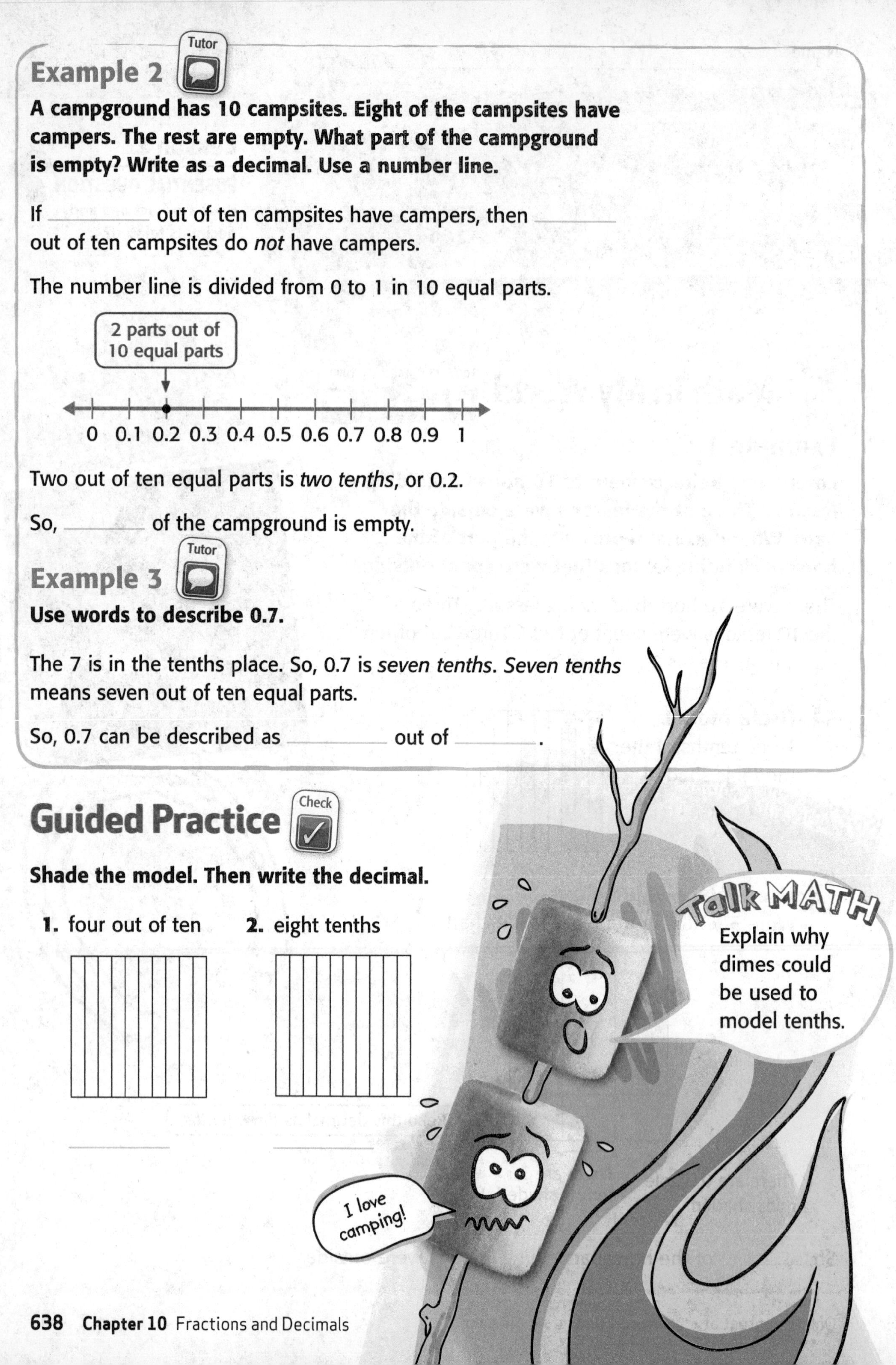

Name ....

CCSS

Independent Practice

Shade the model. Then write the decimal.

3. one out of ten

______

4. five out of ten

______

5. two tenths

______

6. seven tenths

______

Write each of the following as a decimal. Then graph each point on the same number line.

7. six out of ten equal parts

______

8. nine out of ten equal parts

______

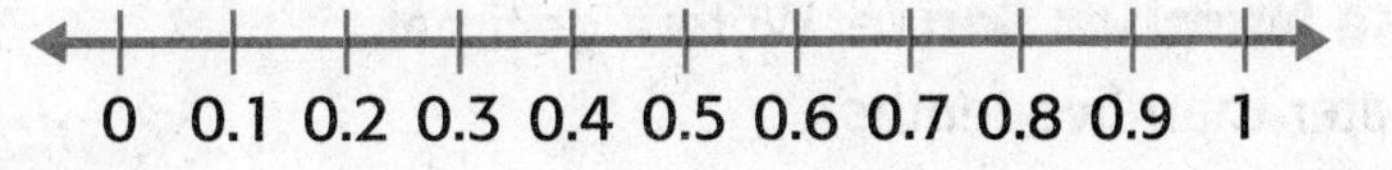

Use words to describe each decimal.

9. 0.2

______

10. 0.3

______

11. 0.8

______

12. 0.9

______

Problem Solving

Squeak!

13. Mathematical PRACTICE 1 **Make Sense of Problems** Cody wanted to play fetch with his dog. His dog had ten chew toys. Cody took six toys outside for his dog. Write a decimal to show what part of the toys is still inside.

______

14. Emilio had a math quiz today. He answered nine of the ten questions correctly. Write a decimal to show what part of the quiz Emilio answered correctly.

______

My Work!

15. Mathematical PRACTICE 4 **Model Math** Jasmine had ten pennies in her pocket. Some fell through a hole in her pocket. Below are the pennies she still has in her pocket. Write a decimal to show what part of the pennies she lost.

______

HOT Problems

16. Mathematical PRACTICE 2 **Use Number Sense** Write a decimal whose value is greater than five tenths.

______

17. **Building on the Essential Question** How can I use decimal grids to model tenths?

______

______

______

Lesson 2
Tenths

Homework Helper

eHelp Need help? connectED.mcgraw-hill.com

Model and write *six out of ten* as a decimal.
Six tenths of the model is shaded.

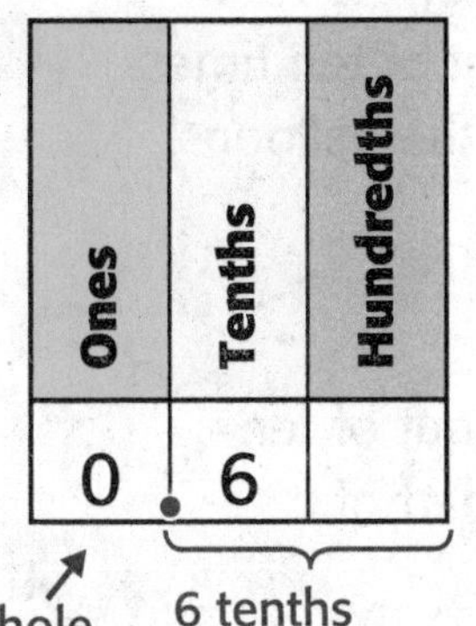

There are 0 whole grids shaded.

6 tenths are shaded.

So, six out of ten is 0.6.

Practice

Model and write each decimal.

1. two out of ten

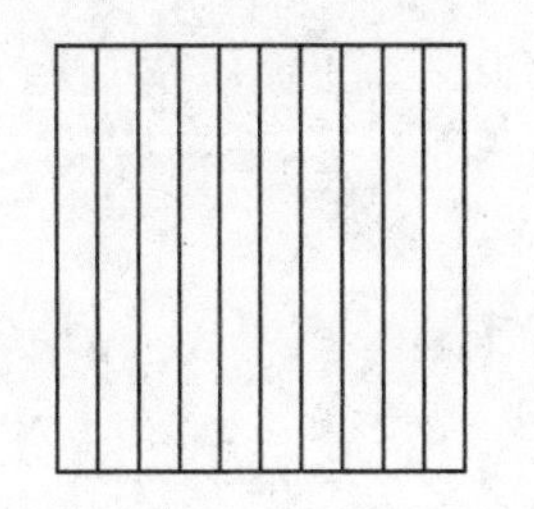

____

2. four out of ten

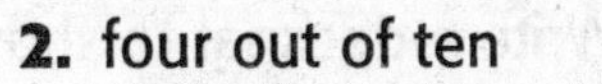

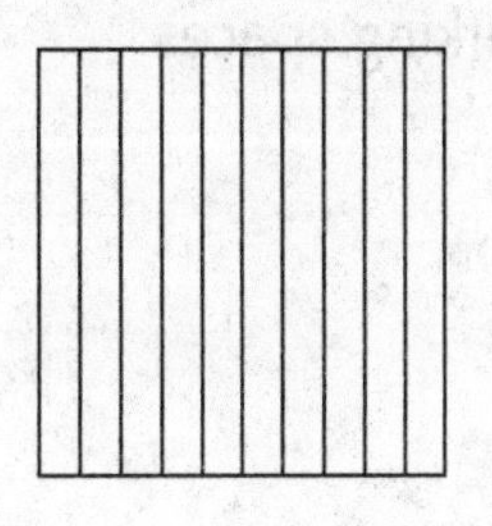

____

Use words to describe each decimal.

3. 0.3

____

4. 0.8

____

Write each of the following as a decimal. Then graph each point on the same number line.

5. one out of ten equal parts

____

6. seven out of ten equal parts

____

0 0.1 0.2 0.3 0.4 0.5 0.6 0.7 0.8 0.9 1

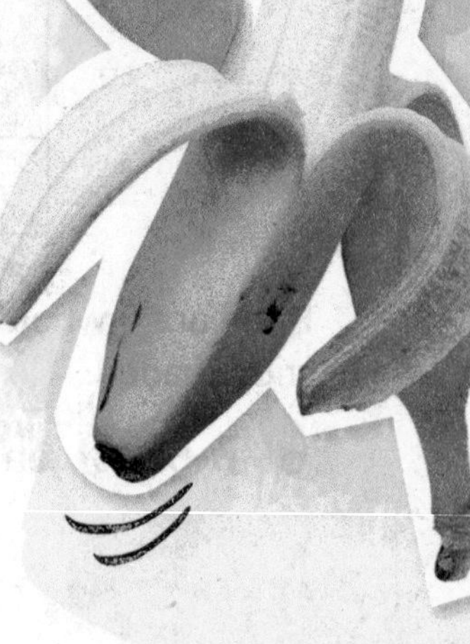

Problem Solving

7. Traci is playing a game at the circus. She has ten chances to throw a dart at a balloon. She hits the balloon 8 out of ten times. Write a decimal to show how many times Traci hit the balloons.

____

8. Jackie took a science quiz. She only answered one out of ten problems incorrectly. Write a decimal to show the part of the quiz that Jackie answered correctly.

____

9. Mathematical PRACTICE 4 **Model Math** Carter's mom bought ten fruits at the grocery store. Seven of the fruits are bananas. Write a decimal to show the part of the fruits that are bananas.

____

10. A parking lot has ten parking spaces. There are five cars parked in the parking lot. Write a decimal to show the part of the parking lot that has empty parking spaces.

____

Test Practice

11. Which decimal represents the part of the model that is shaded?

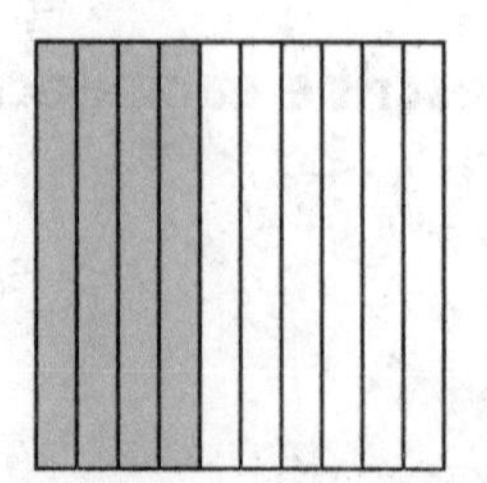

Ⓐ 0.04 Ⓒ 0.4

Ⓑ 0.1 Ⓓ 4.0

Name

Hundredths

Lesson 3

ESSENTIAL QUESTION
How are fractions and decimals related?

Math in My World

Tools Watch Tutor

The pieces are all coming together!

Example 1

Drew has a 100-piece puzzle. There are four corner pieces in the puzzle. What decimal represents the part of the puzzle that is corner pieces?

Four of the 100 puzzle pieces are corner pieces.

1 Use a model.

Shade 4 of the 100 parts.

2 Use a place-value chart.

Write 4 hundredths in the place-value chart.

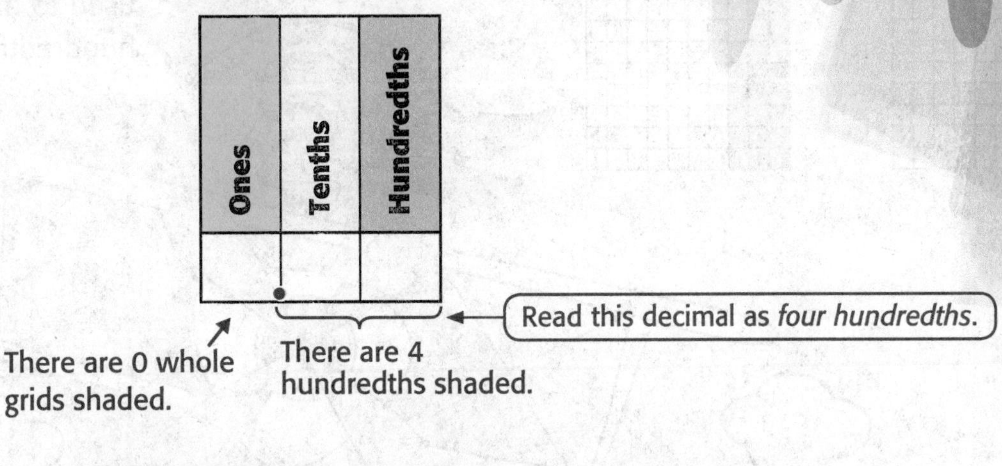

So, ________ of the puzzle pieces are corner pieces.

Tutor

Example 2

It rained 13 out of the past 100 days. What part of the days did it rain? Write as a decimal.

1 Use a model.

The model shows 13 parts of 100 that are shaded.

The model shows ________ out of 100.

Thirteen out of 100 equal parts is *thirteen hundredths*, or ________.

2 Use a place-value chart.

Write 13 hundredths in the place-value chart.

Ones	Tenths	Hundredths
.		

Read this decimal as *thirteen hundredths.*

There are 0 whole grids shaded.

13 hundredths is 1 tenth and 3 hundredths.

So, it rained ________ of the days.

Guided Practice

Check

Write each decimal.

1. ________

2. ________

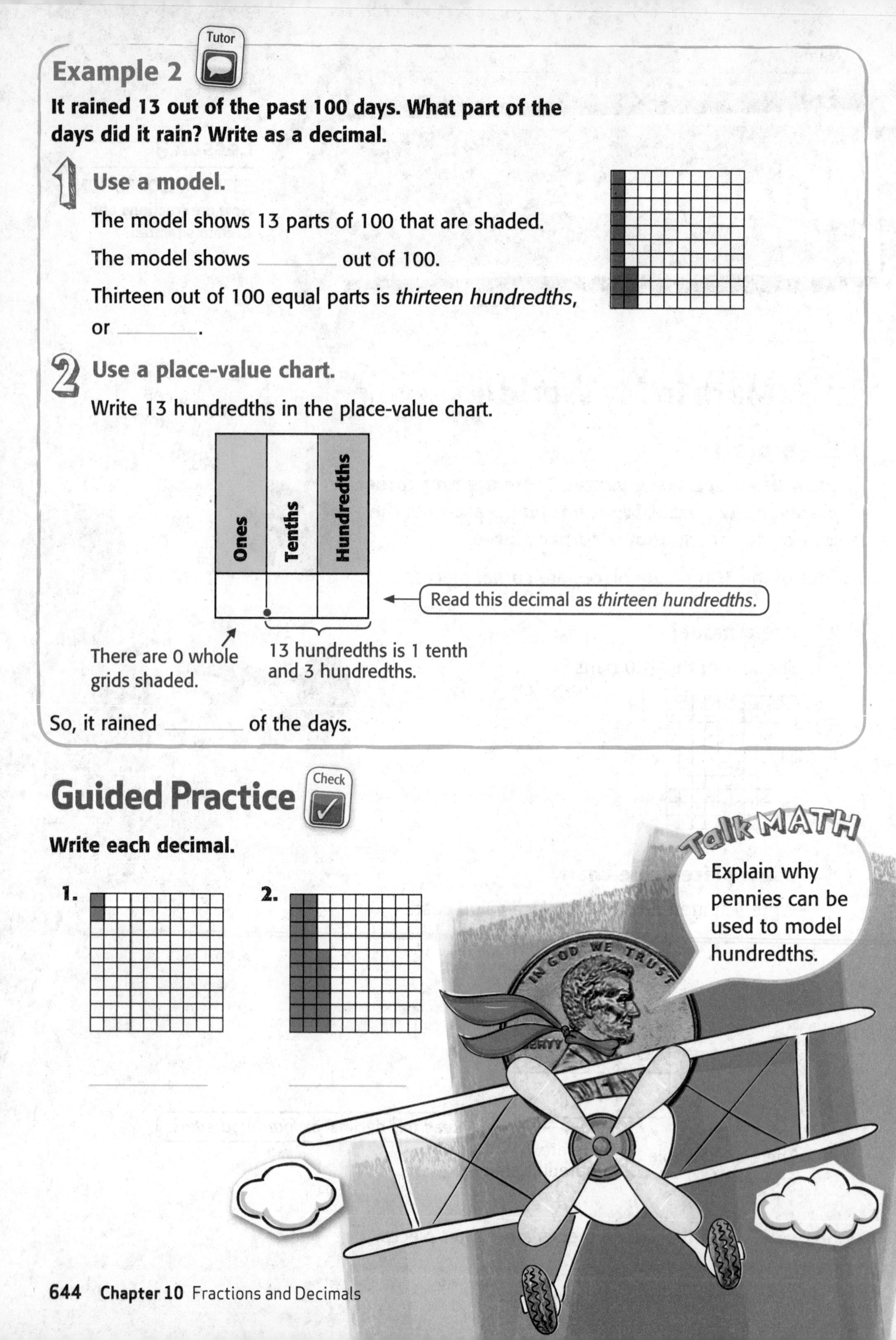

Name

Independent Practice

Write each decimal.

3.

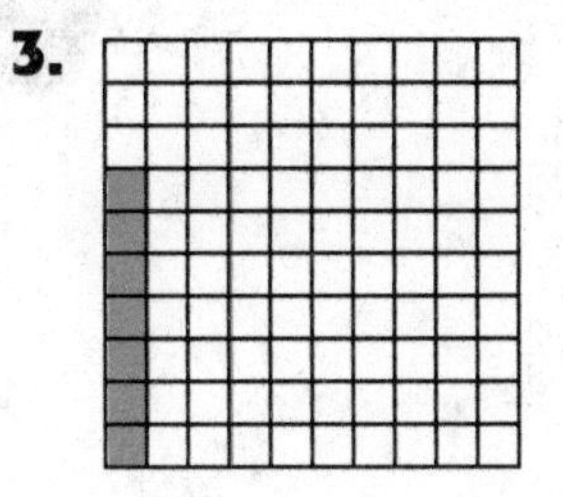

4.

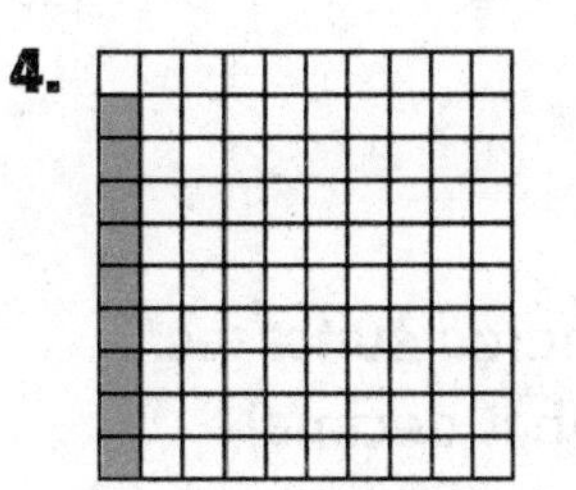

5.

6.

7.

8.

Write a decimal for each part of a dollar shown.

9.

10.

11.

12.

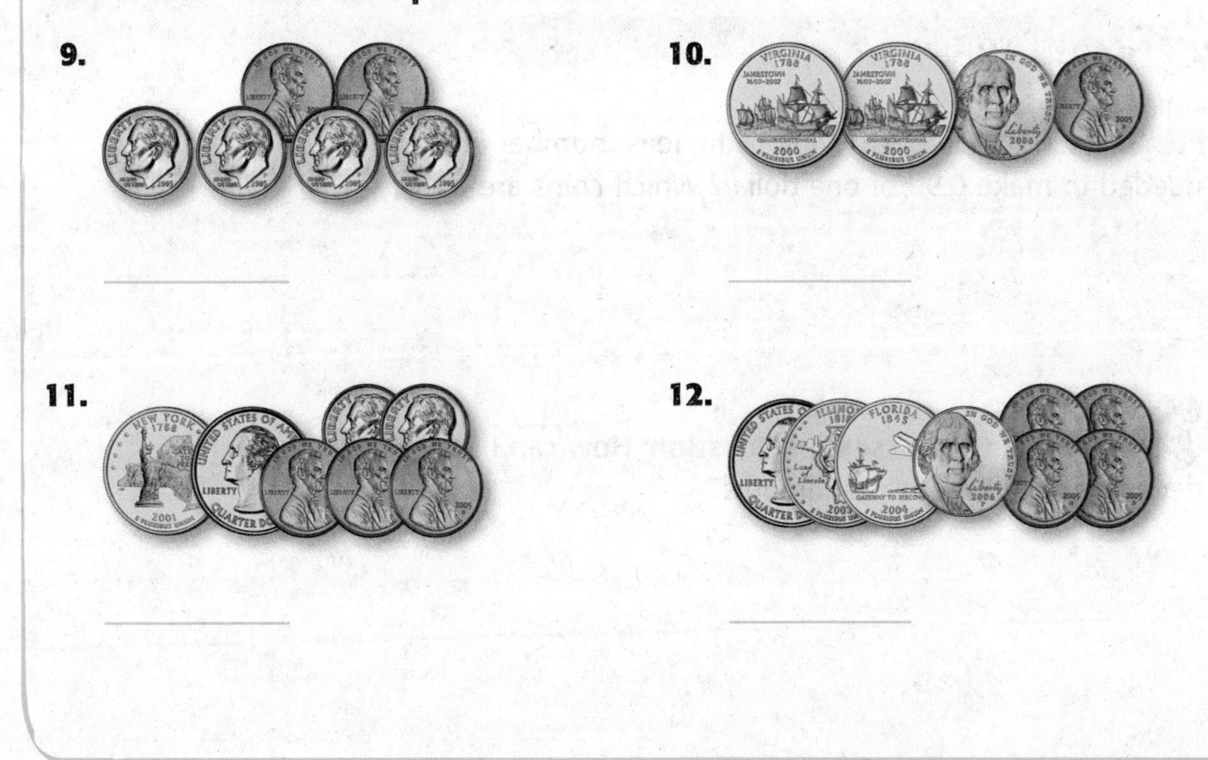

Problem Solving

13. Emma earned eighty-nine out of 100 points on a math test. What decimal shows the part of the points that Emma earned?

______

14. A department store had one hundred suitcases. Six of the suitcases were sold. What decimal represents the part of the suitcases that were left?

______

15. Bena received the coins shown as change after buying pencils at the school store. What part of a dollar are the coins?

______

HOT Problems

16. Mathematical PRACTICE 2 **Reason** What is the least number of coins needed to make 0.90 of one dollar? Which coins are used?

______

______

17. **Building on the Essential Question** How can I use decimal grids to model hundredths?

______

______

______

Name ....

MY Homework

Lesson 3
Hundredths

Homework Helper

eHelp Need help? connectED.mcgraw-hill.com

On a video game, Sandra scored 72 points out of a possible 100 points. What part of the possible points did she score? Write as a decimal.

1 Use a model.

The model shows 72 shaded parts out of 100. This is *seventy-two hundredths*, or 0.72.

2 Use a place-value chart.

Write 72 hundredths in the place-value chart.

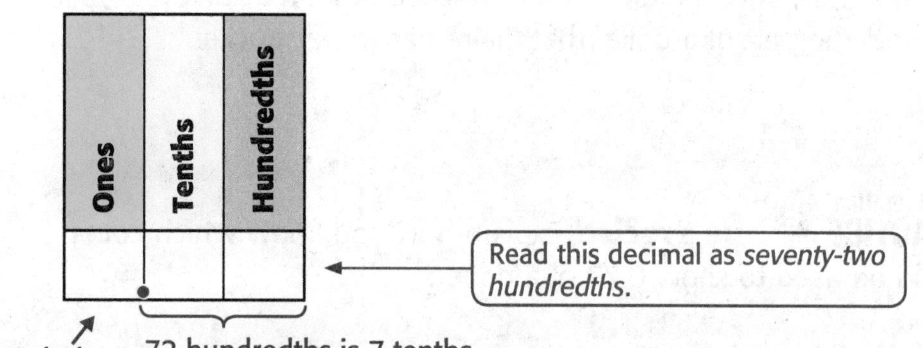

So, Sandra scored 0.72 of the possible points..

Practice

Write a decimal for each part of a dollar shown.

1. ______

2. ______

Problem Solving

3. Dexter bought an apple. He paid using the coins shown below.

Write a decimal to show the part of a dollar that he paid.

______

4. Daniel had a social studies quiz. He answered eighty-seven out of 100 exercises correctly. Write a decimal to show the part of the quiz that Daniel answered correctly.

______

5. Mathematical PRACTICE 4 **Model Math** Claire has three nickels, two quarters, and one dime in her pocket. Write a decimal to show the part of a dollar that Claire has in her pocket.

______

6. Mathematical PRACTICE 6 **Be Precise** Explain how you know which coins could be used to show 0.77 of a dollar.

______

______

______

Test Practice

7. Which decimal shows fifty-seven out of one hundred?

Ⓐ 0.57 Ⓒ 5.70

Ⓑ 0.75 Ⓓ 57.0

Check My Progress

Vocabulary Check

Write *true* or *false*.

1. A **decimal** is a number that uses place value and a decimal point to show part of a whole. ________

2. In the number 0.36, the 3 is in the **tenths** place. ________

3. In the number 0.58, the 5 is in the **hundredths** place. ________

Concept Check

Write each decimal.

4. four tenths ____________

5. thirty-two hundredths ____________

Shade the model. Then write the decimal.

6. three out of ten

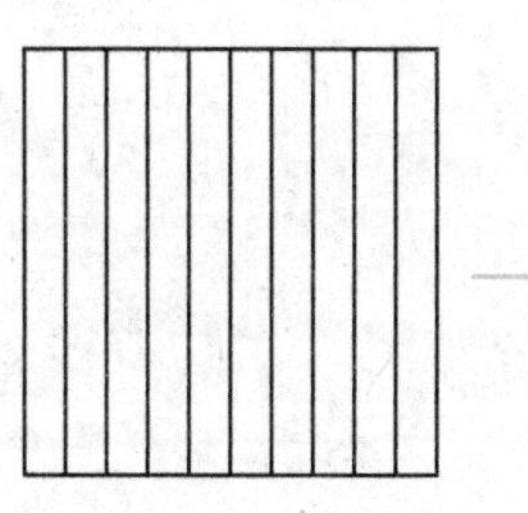

____

7. eight out of ten

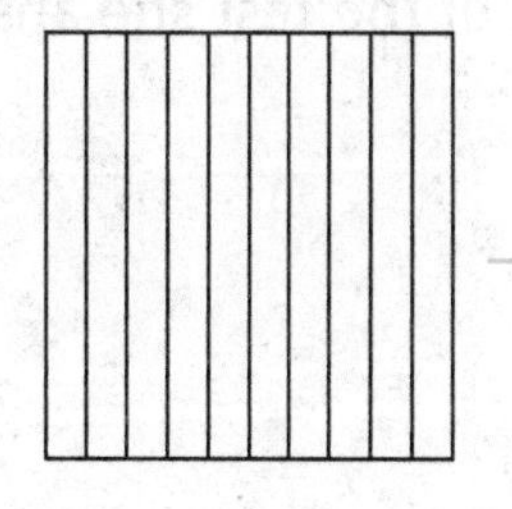

____

Write each decimal.

8.

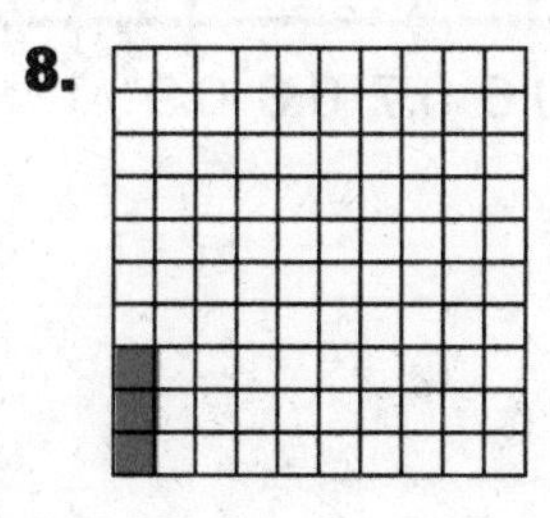

____

9.

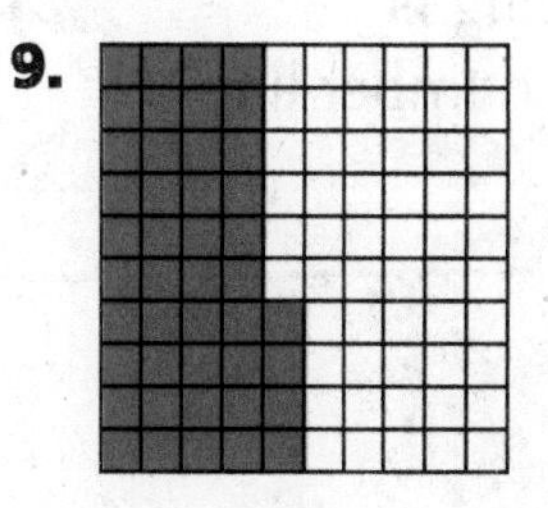

____

Problem Solving

10. The distance between the grocery store and the post office is thirty-four hundredths of a mile. Write this distance as a decimal.

11. There are ten fish in an aquarium. Four of them are blue. Write a decimal to show the part of the fish that are blue.

12. William found the coins shown in his pocket. What part of a dollar are the coins? Write a decimal to show your answer.

13. There are ten questions on a test. Sophie answered 8 of them correctly. Write a decimal to show what part of the test she answered correctly.

My Work!

Test Practice

14. Which of the following is represented by the number line?

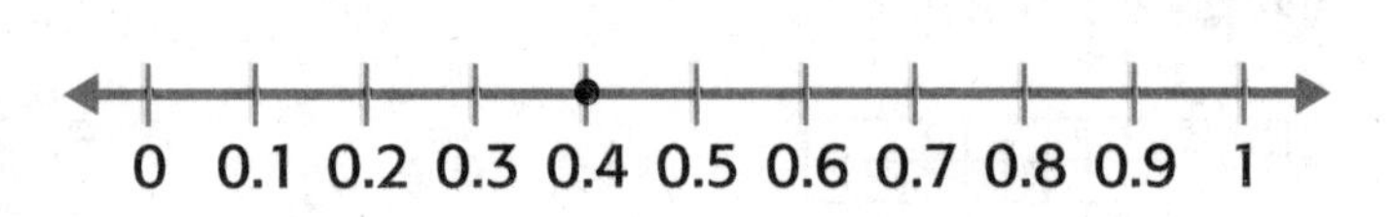

Ⓐ 4.0

Ⓑ four tenths

Ⓒ 4 out of 5

Ⓓ four hundredths

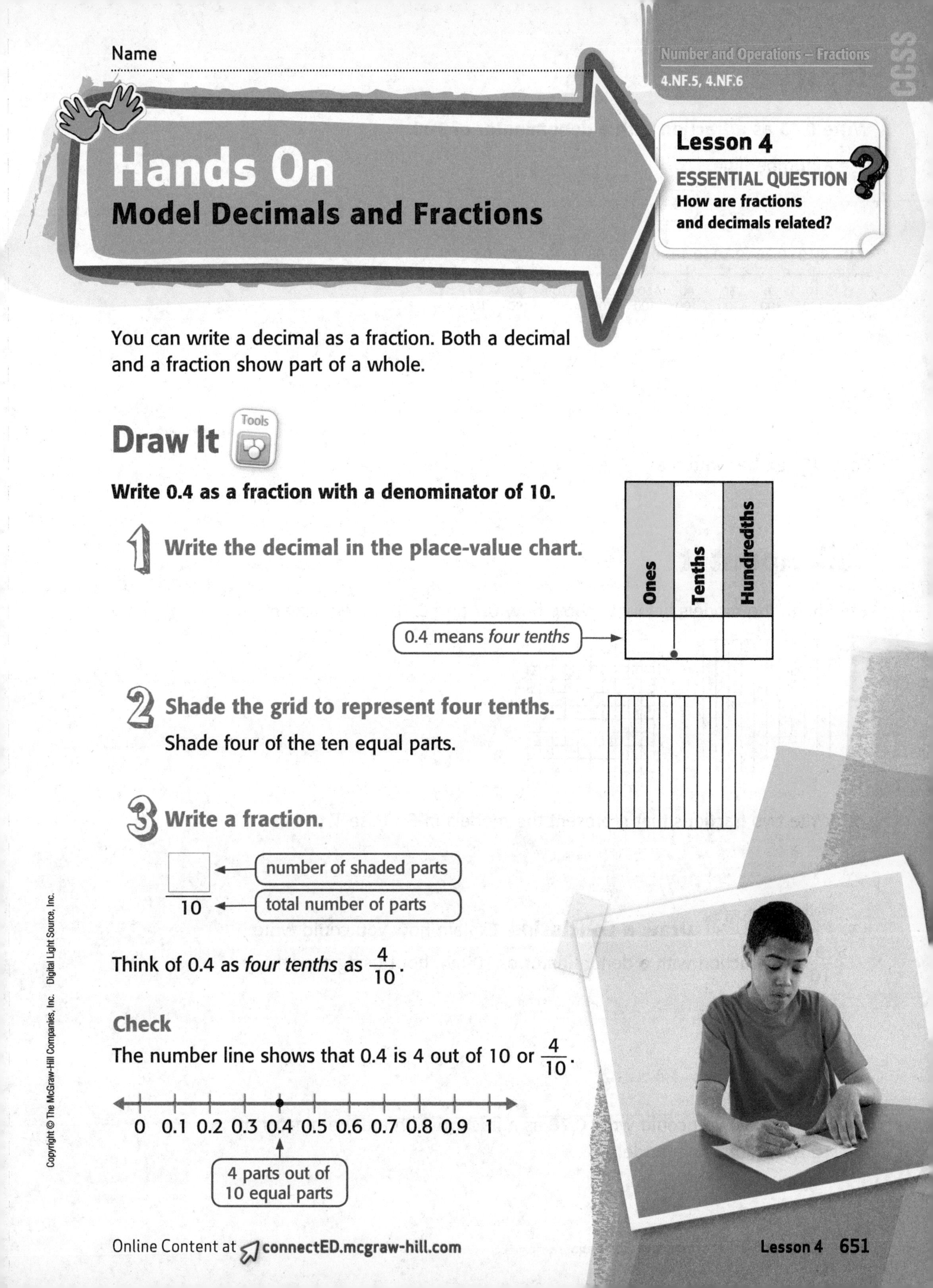

Name

Hands On
Model Decimals and Fractions

Lesson 4

ESSENTIAL QUESTION
How are fractions and decimals related?

You can write a decimal as a fraction. Both a decimal and a fraction show part of a whole.

Draw It

Tools

Write 0.4 as a fraction with a denominator of 10.

1 Write the decimal in the place-value chart.

Ones	Tenths	Hundredths

0.4 means *four tenths*

2 Shade the grid to represent four tenths.

Shade four of the ten equal parts.

3 Write a fraction.

$\frac{\square}{10}$ ← number of shaded parts
← total number of parts

Think of 0.4 as *four tenths* as $\frac{4}{10}$.

Check

The number line shows that 0.4 is 4 out of 10 or $\frac{4}{10}$.

Try It

Write 0.45 as a fraction with a denominator of 100.

Use a number line.

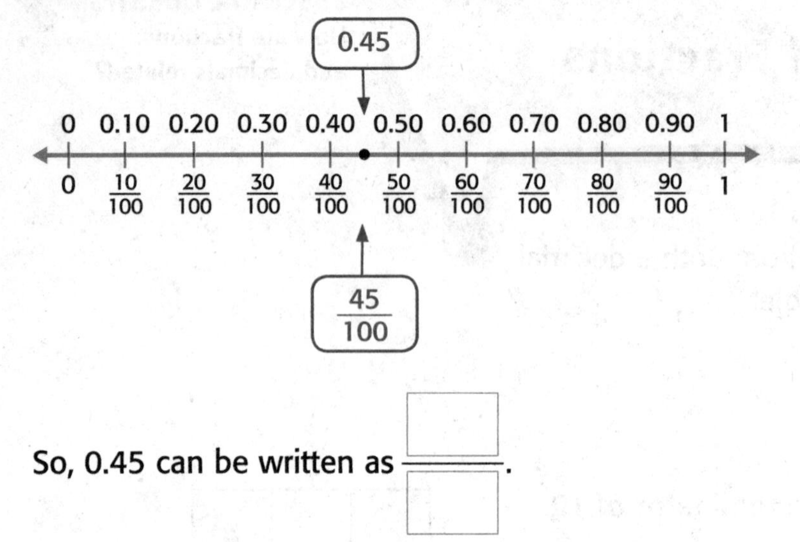

So, 0.45 can be written as $\frac{\square}{\square}$.

Talk About It

1. Shade the models below to show how 0.7 and 0.70 are equivalent.

2. Write two fractions that represent the models in Exercise 1.

____

3. Mathematical PRACTICE 3 **Draw a Conclusion** Explain how you could write $\frac{9}{10}$ as a fraction with a denominator of 100 without using models.

____

____

4. Explain how you could write 0.78 as a fraction with a denominator of 100 without using models.

____

Name ....

Practice It

Write each decimal as a fraction with a denominator of 10. Shade the grid.

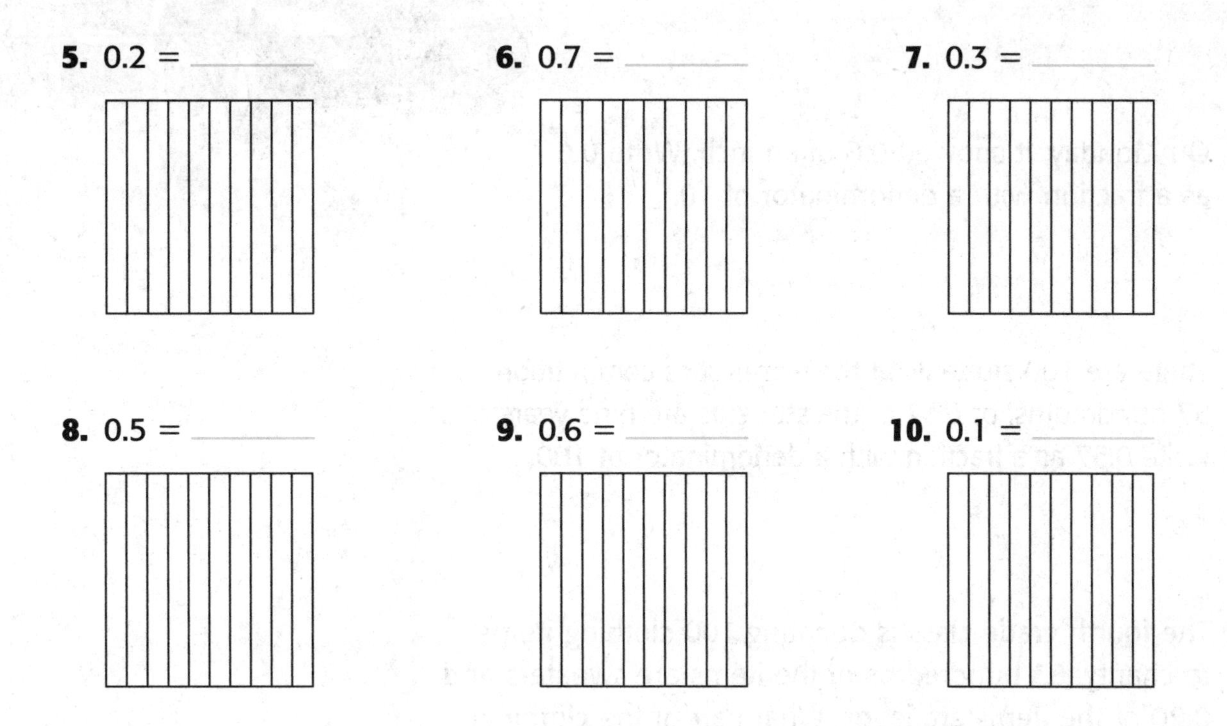

5. 0.2 = ______

6. 0.7 = ______

7. 0.3 = ______

8. 0.5 = ______

9. 0.6 = ______

10. 0.1 = ______

Write each decimal as a fraction with a denominator of 100. Shade the grid.

11. 0.39 = ______

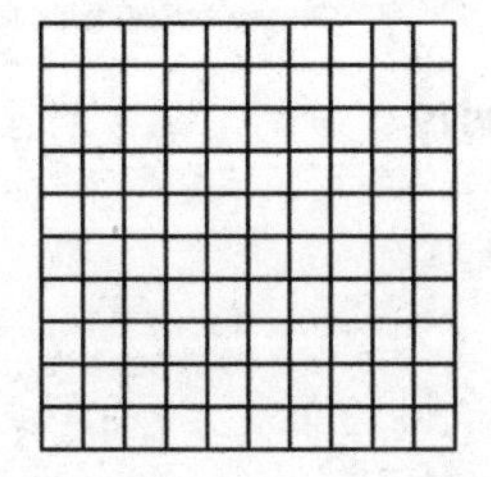

12. 0.71 = ______

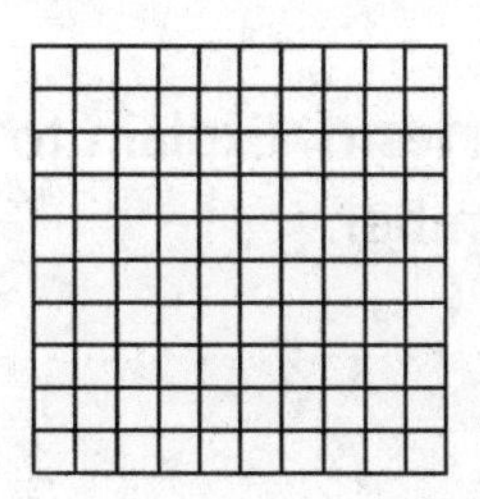

13. 0.12 = ______

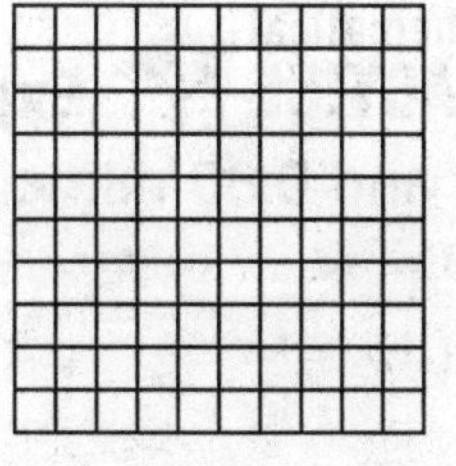

14. 0.09 = ______

15. 0.23 = ______

16. 0.02 = ______

Apply It

17. There are 10 cars on the race track. If 0.3 of the cars are red, what fraction of the cars are not red? Write as a fraction with a denominator of 10.

My Work!

______

18. On Monday, it snowed 0.6 of an inch. Write 0.6 as a fraction with a denominator of 10.

______

19. There are 100 students at the gymnastics competition. 57 hundredths, or 0.57 of the students are nine years old. Write 0.57 as a fraction with a denominator of 100.

______

20. The fourth grade class is donating 100 clothing items to charity. 63 hundredths of the items are sweaters and 0.20 of the items are jeans. What part of the clothing items are neither sweaters nor jeans? Write the number as a fraction with a denominator of 100.

______

21. Mathematical PRACTICE 6 **Explain to a Friend** Explain to a classmate why 0.8 and 0.80 name the same number.

______

______

______

Write About It

22. How can I use models to relate fractions and decimals?

______

______

Name ....

MY Homework

Lesson 4

Hands On: Model Decimals and Fractions

Homework Helper

eHelp Need help? connectED.mcgraw-hill.com

Write 0.7 as a fraction with a denominator of 10.

1 Write the decimal in the place-value chart.

Ones	Tenths	Hundredths
0.	7	

0.7 means *seven tenths*

2 Use a grid to represent seven tenths.

The model shows seven of the ten equal parts are shaded.

3 Write a fraction.

$\frac{7}{10}$ ← 7: number of shaded parts; 10: total number of parts

Think of *seven tenths* as $\frac{7}{10}$.

So, $0.7 = \frac{7}{10}$.

Check

The number line shows that 0.7 is 7 out of 10 or $\frac{7}{10}$.

0 0.1 0.2 0.3 0.4 0.5 0.6 0.7 0.8 0.9 1

7 parts out of 10 equal parts

Practice

Write each decimal as a fraction with a denominator of 10. Shade the grid.

1. 0.1 = ______

2. 0.3 = ______

3. 0.9 = ______

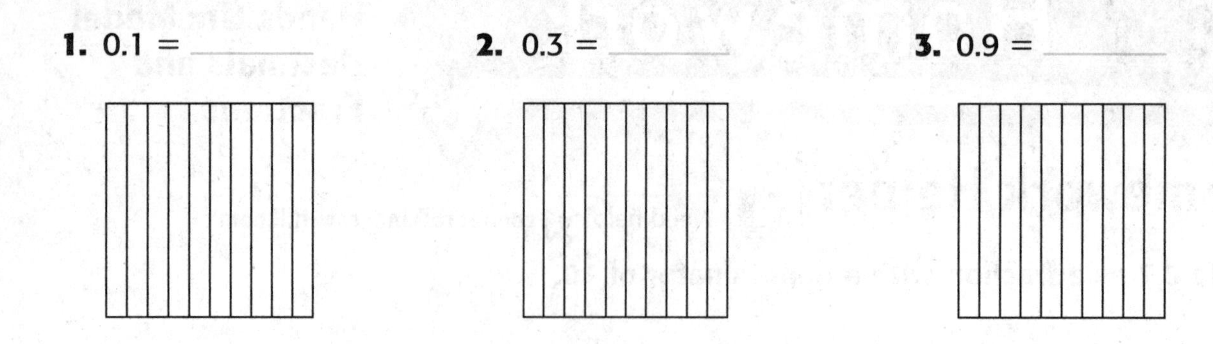

Write each decimal as a fraction with a denominator of 100. Shade the grid.

4. 0.17 = ______

5. 0.24 = ______

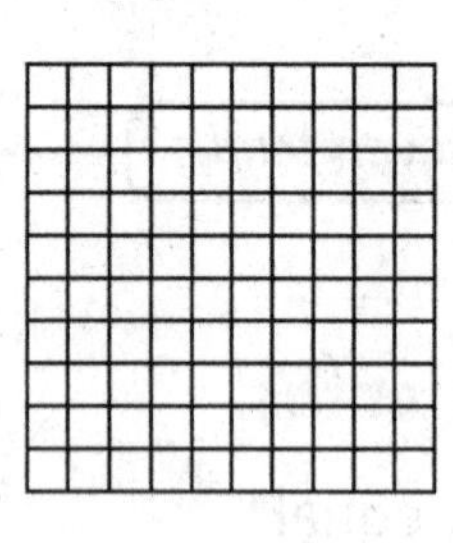

6. 0.88 = ______

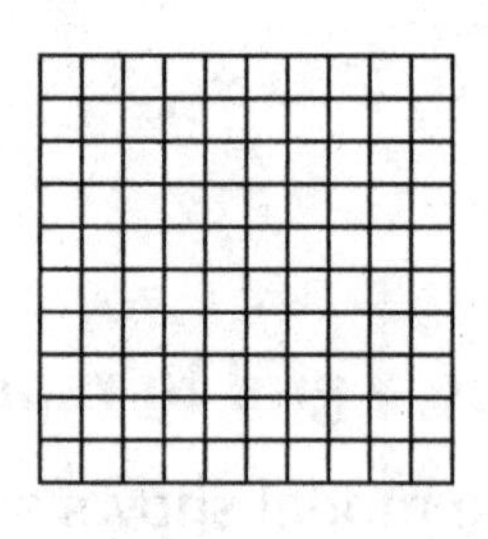

Problem Solving

Mathematical PRACTICE 4 Model Math For Exercises 7–9, refer to the table that shows the part of pets that Josiah's family has for each type of pet.

Type of Pet	Part
Cat	0.2
Dog	0.1
Rabbit	0.2
Fish	0.4
Reptile	0.1

7. Write the part of pets that are cats as a fraction with a denominator of 10.

______

8. Write the part of pets that are fish as a fraction with a denominator of 10.

______

9. Write the part of pets that are fish as a fraction with a denominator of 100.

______

Name

Decimals and Fractions

Lesson 5

ESSENTIAL QUESTION
How are fractions and decimals related?

Decimals and fractions can show equivalent amounts. You can write equivalent fractions and you can also write fractions as decimals.

Math in My World

Example 1

Kara travels $\frac{7}{10}$ mile from her bus stop to the next stop. Write $\frac{7}{10}$ as a fraction with a denominator of 100. Then write the fraction as a decimal.

Use a model to show $\frac{7}{10}$.

1 Write $\frac{7}{10}$ as a fraction with a denominator of 100.

Shade seven tenths of the tenths grid.

On the hundredths grid, shade squares so that the same fraction of the two grids is shaded.

How many squares did you shade on the hundredths grid? ______

The decimal models show that $\frac{7}{10}$ is equivalent to $\frac{70}{100}$.

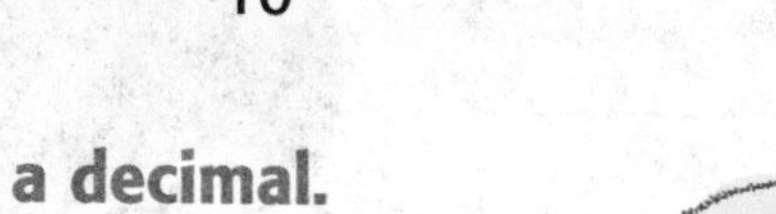

2 Write the fraction as a decimal.

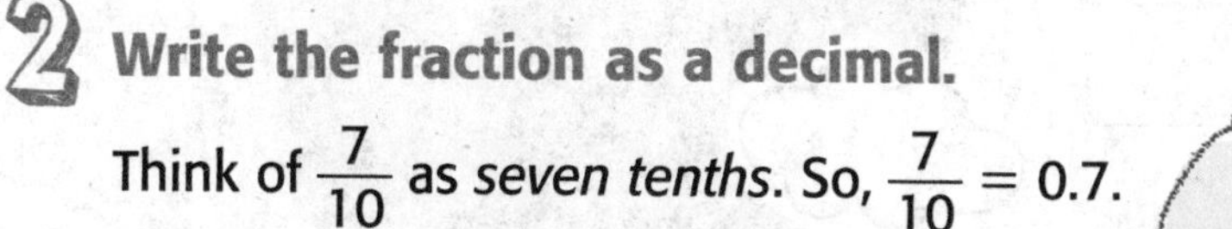

Think of $\frac{7}{10}$ as *seven tenths*. So, $\frac{7}{10} = 0.7$.

Helpful Hint
Notice that seventy hundredths, or 0.70, is the same as seven tenths, or 0.7.

So, $\frac{7}{10}$ can be written as $\frac{70}{100}$ and 0.7.

Example 2 Tutor

Write $\frac{9}{10}$ as an equivalent fraction with a denominator of 100 and as a decimal.

Use a model to show $\frac{9}{10}$.

1 Write $\frac{9}{10}$ as a fraction with a denominator of 100.

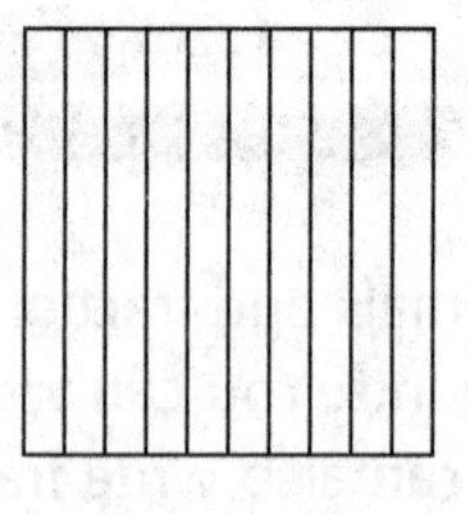

Shade nine tenths of the tenths grid.

On the hundredths grid, shade squares so that the same fraction on the two grids is shaded.

How many squares did you shade on the hundredths grid? ______

The decimal models show that $\frac{9}{10}$ is equivalent to $\frac{90}{100}$.

2 Write the fraction as a decimal.

Think of $\frac{9}{10}$ as *nine tenths*. So, $\frac{9}{10} = 0.9$.

So, $\frac{9}{10}$ can be written as $\frac{90}{100}$ and 0.9.

Notice that ninety hundredths, or 0.90, is the same as *nine tenths,* or 0.9.

Talk MATH

Is one tenth greater than or less than one hundredth? Explain.

Guided Practice Check

Write each fraction as an equivalent fraction with a denominator of 100. Then write the fraction as a decimal.

1. $\frac{1}{10}$

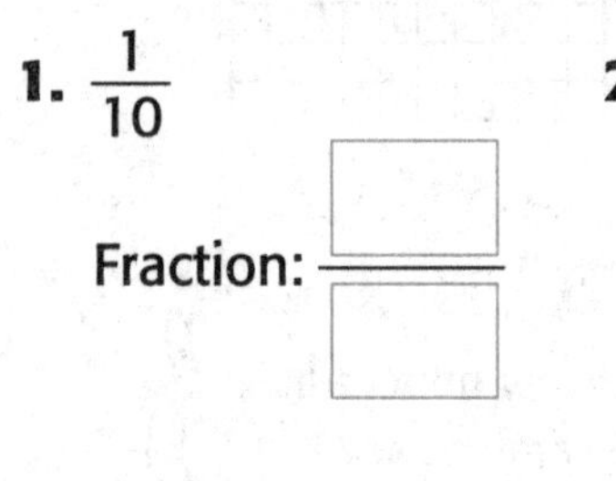

Fraction: ______

Decimal: ______

2. $\frac{5}{10}$

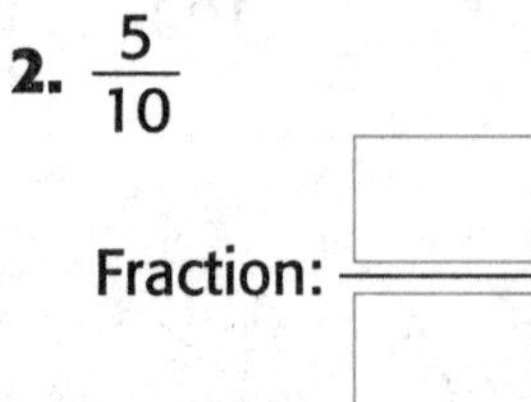

Fraction: ______

Decimal: ______

Name ____________________

Independent Practice

Write each fraction as an equivalent fraction with a denominator of 100. Shade the grids to show that the fractions are equivalent.

3. $\frac{1}{10} =$ ________

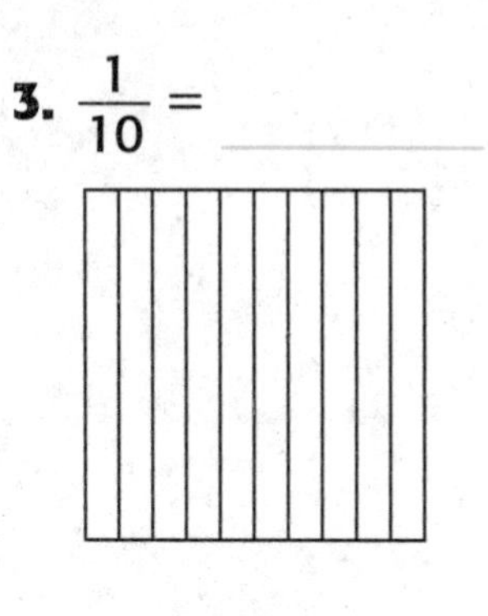

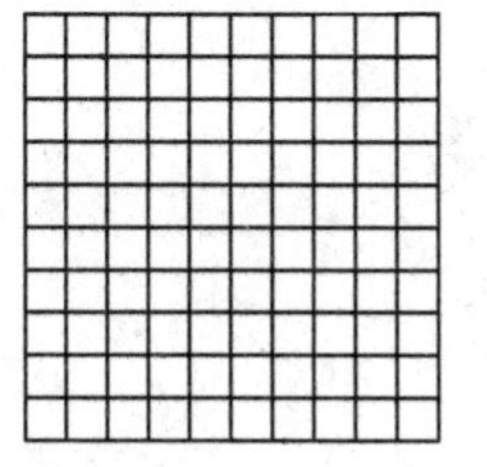

4. $\frac{8}{10} =$ ________

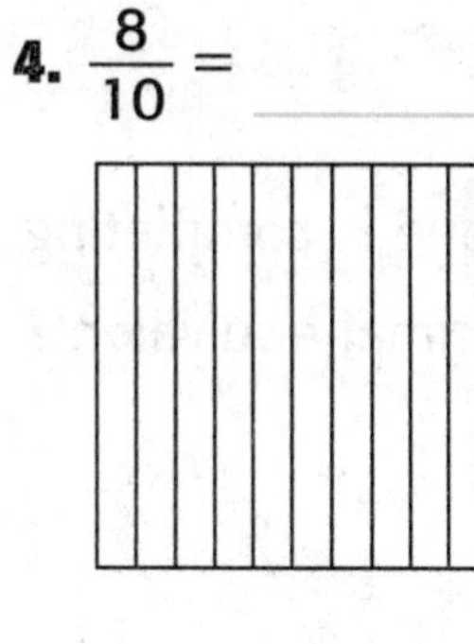

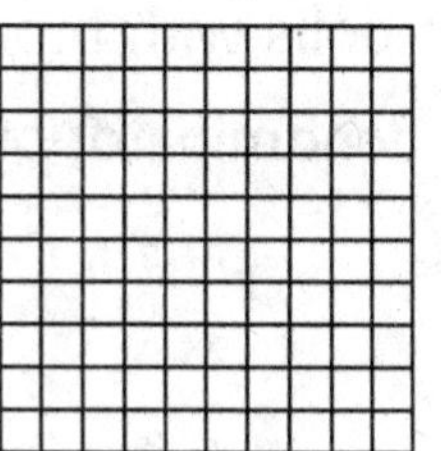

5. $\frac{5}{10} =$ ________

6. Write the fraction in Exercise 3 as a decimal. ________

7. Write the fraction in Exercise 4 as a decimal. ________

8. Write the fraction in Exercise 5 as a decimal. ________

Circle the equivalent ways to represent each number.

9. $\frac{6}{10}$ six tenths   six hundredths $\frac{6}{100}$ 0.6   0.06

10. $\frac{30}{100}$ $\frac{3}{10}$ thirty hundredths $\frac{3}{100}$ 0.03   0.3

11. 0.2   two tenths   twenty hundredths $\frac{2}{100}$ 0.20   0.02

12. 0.7 $\frac{70}{100}$ seven hundredths $\frac{7}{100}$ $\frac{7}{10}$ 0.70

Problem Solving

My Work!

13. Wesley learned that $\frac{3}{10}$ of the students in his class are left handed. Write $\frac{3}{10}$ as a fraction with a denominator of 100 and as a decimal.

14. Mathematical PRACTICE 2 **Use Algebra** Tyrone is completing ■ $= \frac{5}{100}$. What is the unknown decimal equivalent?

15. Of Marissa's doll collection, $\frac{4}{10}$ are dolls with brown hair. Write $\frac{4}{10}$ as a fraction with a denominator of 100. Then write $\frac{4}{10}$ as a decimal.

HOT Problems

16. Mathematical PRACTICE 6 **Be Precise** Write a summary statement about decimals that are equivalent to fractions with denominators of 10 or 100.

17. **Building on the Essential Question** How can I identify the shaded part of a hundredths grid as a fraction and a decimal?

MY Homework

Lesson 5
Decimals and Fractions

Homework Helper

eHelp Need help? connectED.mcgraw-hill.com

Write $\frac{5}{10}$ as an equivalent fraction with a denominator of 100 and as a decimal.

Use a model to show $\frac{5}{10}$.

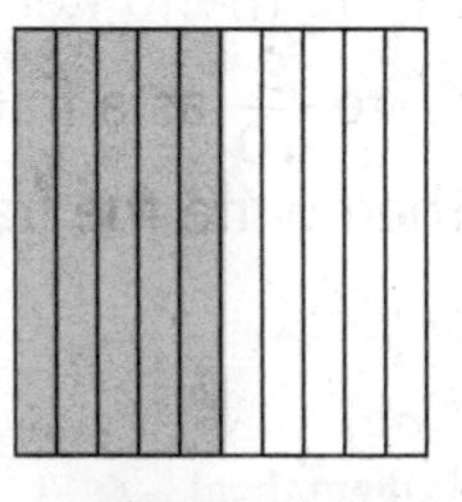

1 Write $\frac{5}{10}$ as a fraction with a denominator of 100.

Five tenths of the tenths grid are shaded.

On the hundredths grid, squares were shaded so that the same fraction on the two grids is shaded.

Fifty squares on the hundredths grid are shaded.

The decimal models show that $\frac{5}{10}$ is equivalent to $\frac{50}{100}$.

2 Write the fraction as a decimal.

Think of $\frac{5}{10}$ as *five tenths*. So, $\frac{5}{10} = 0.5$.

So, $\frac{5}{10}$ can be written as $\frac{50}{100}$ and 0.5.

Practice

1. Write $\frac{4}{10}$ as an equivalent fraction with a denominator of 100. Shade the grids to show that the fractions are equivalent. Then write the fraction as a decimal.

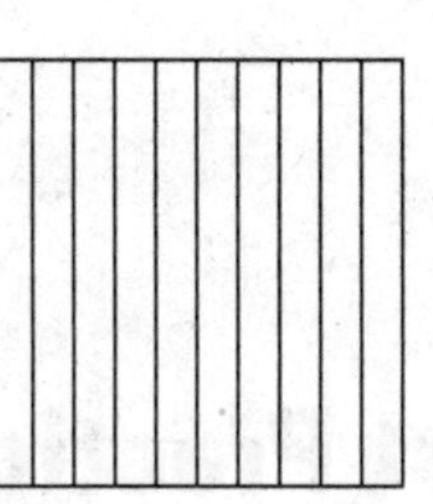

Fraction: $\frac{\square}{\square}$

Decimal: ______

Problem Solving

2. Of Cadence's paintbrushes, $\frac{2}{10}$ were used to paint a picture. Write $\frac{2}{10}$ as a fraction with a denominator of 100. Then write the fraction as a decimal.

3. Of the trees in a park, $\frac{6}{10}$ have red leaves. Write $\frac{6}{10}$ as a fraction with a denominator of 100. Then write the fraction as a decimal.

4. Of the menu items at a deli, $\frac{9}{10}$ are sandwiches. Write $\frac{9}{10}$ as a fraction with a denominator of 100. Then write the fraction as a decimal.

5. Mathematical PRACTICE 3 **Find the Error** Marjorie wrote the fraction $\frac{7}{10}$ as a fraction with a denominator of 100. She then wrote the fraction as a decimal. Find and correct her mistake. Explain.

$$\frac{7}{10} = \frac{7}{100} = 0.07$$

Test Practice

6. Which fraction is equivalent to $\frac{8}{10}$?

Ⓐ $\frac{8}{100}$

Ⓑ $\frac{80}{100}$

Ⓒ $\frac{8}{1}$

Ⓓ $\frac{80}{10}$

Name ....

Use Place Value and Models to Add

Lesson 6

ESSENTIAL QUESTION
How are fractions and decimals related?

That was a long walk. I'm worn out!

Math in My World

Tools Tutor

Example 1

Denny walked $\frac{3}{10}$ mile to the post office. Then he walked $\frac{5}{100}$ mile to the grocery store. How far did he walk in all? Write the answer as a fraction with a denominator of 100 and as a decimal.

Use a model to show $\frac{3}{10} + \frac{5}{100}$.

1 Write $\frac{3}{10}$ as a fraction with a denominator of 100.

The decimal models show that $\frac{3}{10} = \frac{30}{100}$.

2 Add like fractions.

$$\frac{30}{100} + \frac{5}{100} = \frac{30 + 5}{100}$$

$$= \frac{\square}{100}$$

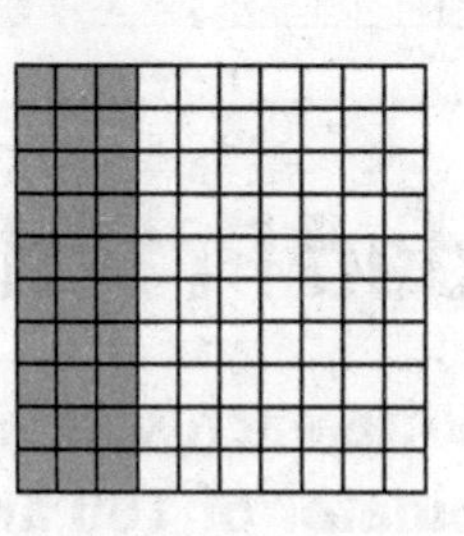

3 Write the sum as a decimal.

Think of $\frac{35}{100}$ as *thirty-five hundredths*. So, $\frac{35}{100} = 0.35$.

So, $\frac{3}{10} + \frac{5}{100} = \frac{35}{100}$, or ______ .

Denny walked $\frac{35}{100}$, or 0.35, mile in all.

Example 2

Find $\frac{4}{10} + \frac{22}{100}$. Write the sum as a fraction with a denominator of 100 and as a decimal.

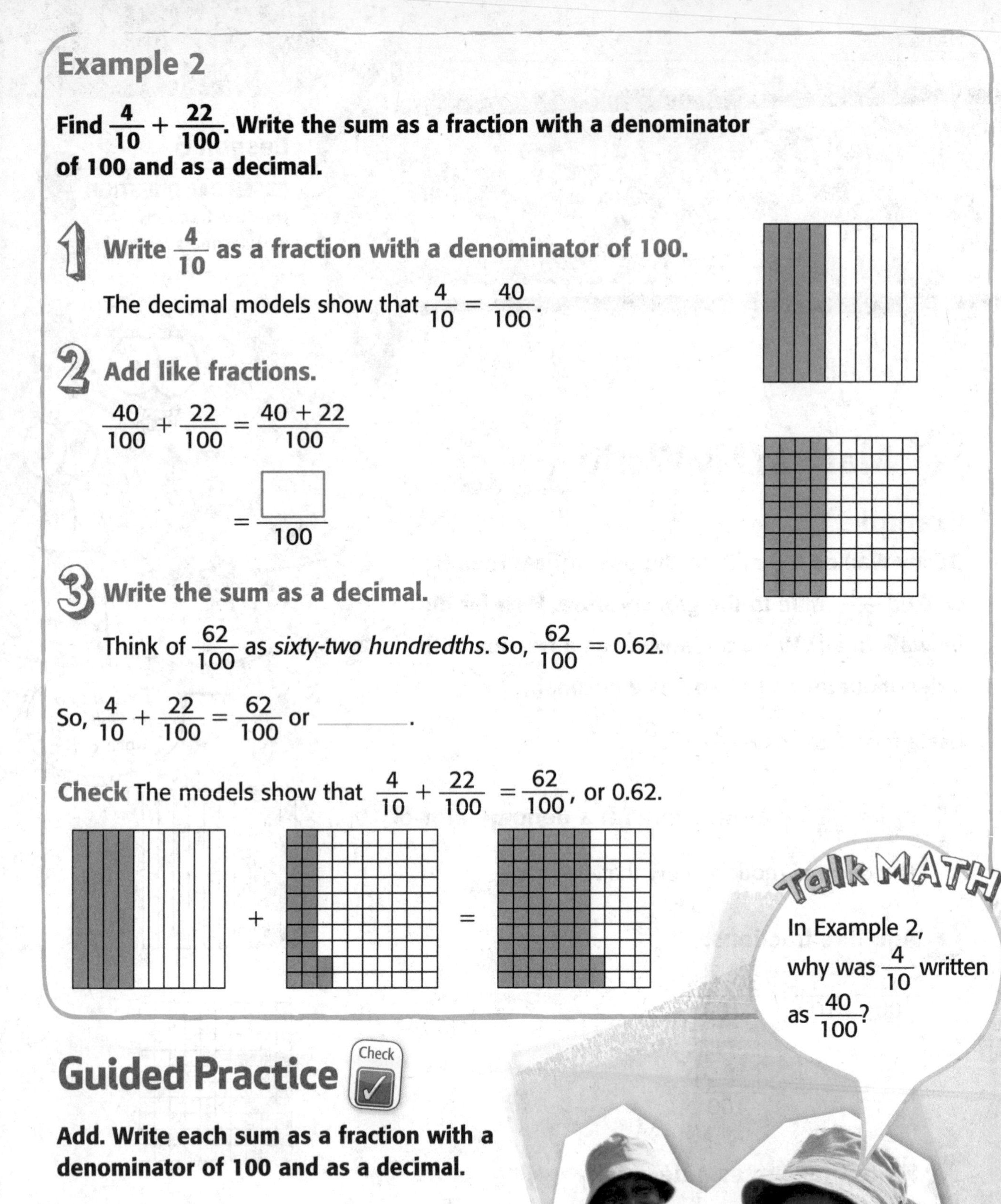

1 Write $\frac{4}{10}$ as a fraction with a denominator of 100.

The decimal models show that $\frac{4}{10} = \frac{40}{100}$.

2 Add like fractions.

$$\frac{40}{100} + \frac{22}{100} = \frac{40 + 22}{100}$$

$$= \frac{\square}{100}$$

3 Write the sum as a decimal.

Think of $\frac{62}{100}$ as *sixty-two hundredths*. So, $\frac{62}{100} = 0.62$.

So, $\frac{4}{10} + \frac{22}{100} = \frac{62}{100}$ or ______.

Check The models show that $\frac{4}{10} + \frac{22}{100} = \frac{62}{100}$, or 0.62.

+ =

Talk MATH

In Example 2, why was $\frac{4}{10}$ written as $\frac{40}{100}$?

Guided Practice

Check

Add. Write each sum as a fraction with a denominator of 100 and as a decimal.

1. $\frac{5}{10} + \frac{1}{100} =$ ______

2. $\frac{7}{10} + \frac{13}{100} =$ ______

Name ______________________

CCSS

Independent Practice

Shade the models to find each sum. Write the sum as a fraction with a denominator of 100.

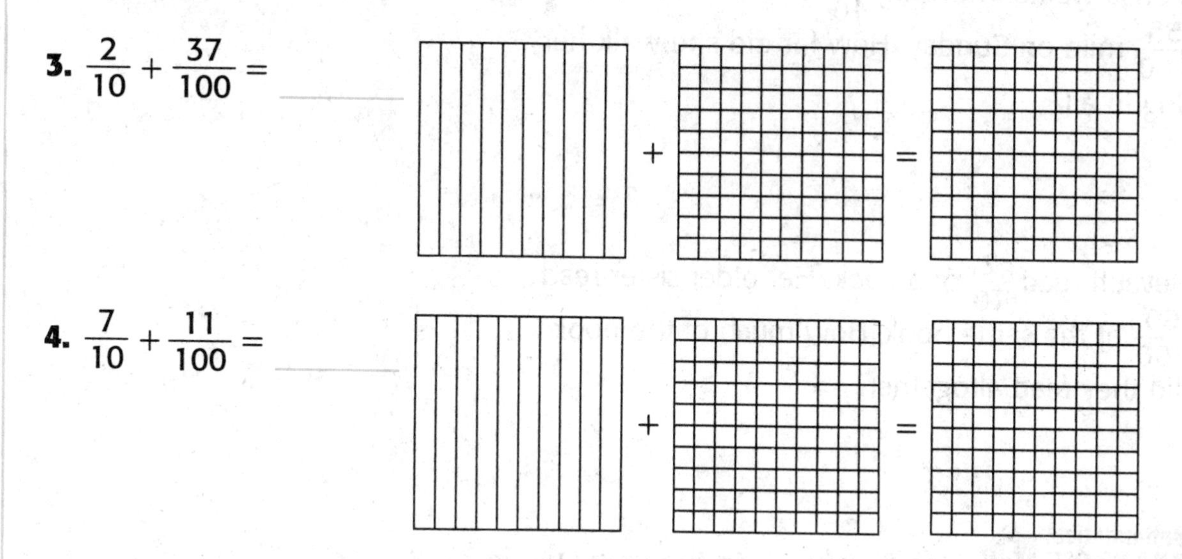

3. $\frac{2}{10} + \frac{37}{100} =$ ______

4. $\frac{7}{10} + \frac{11}{100} =$ ______

Add. Write each sum as a fraction with a denominator of 100 and as a decimal.

5. $\frac{6}{10} + \frac{24}{100} =$ ______

6. $\frac{5}{10} + \frac{21}{100} =$ ______

7. $\frac{3}{10} + \frac{65}{100} =$ ______

8. $\frac{1}{10} + \frac{52}{100} =$ ______

9. $\frac{8}{10} + \frac{17}{100} =$ ______

10. $\frac{2}{10} + \frac{42}{100} =$ ______

11. $\frac{6}{10} + \frac{19}{100} =$ ______

12. $\frac{3}{10} + \frac{35}{100} =$ ______

Problem Solving

For Exercises 13 and 14, write each answer as a fraction with a denominator of 100 and as a decimal.

13. Marisa walked her dog $\frac{1}{10}$ mile on Saturday and $\frac{55}{100}$ mile on Sunday. How far did she walk her dog in all?

14. Nevaeh read $\frac{2}{10}$ of a book. Her older sister read $\frac{60}{100}$ of the same book. How much of the book did they read altogether?

15. Mathematical PRACTICE 2 **Use Algebra** Find the unknown in the number sentence $\frac{3}{10} + \frac{■}{100} = \frac{41}{100}$.

HOT Problems

16. Mathematical PRACTICE 6 **Explain to a Friend** Refer to Exercise 15. Explain to a friend or classmate how you found the unknown.

17. **Building on the Essential Question** How does place value help when adding $\frac{1}{10}$ and $\frac{1}{100}$?

MY Homework

Lesson 6
Use Place Value and Models to Add

Homework Helper

eHelp Need help? connectED.mcgraw-hill.com

Julian sold $\frac{6}{10}$ of the school play tickets on Monday. He sold $\frac{23}{100}$ of the tickets on Tuesday. What part of the tickets did Julian sell in all? Write the answer as a fraction with a denominator of 100 and as a decimal.

Use a model to show $\frac{6}{10} + \frac{23}{100}$.

1 Write $\frac{6}{10}$ as a fraction with a denominator of 100.

The decimal models show that $\frac{6}{10} = \frac{60}{100}$.

2 Add like fractions.

$$\frac{60}{100} + \frac{23}{100} = \frac{60 + 23}{100}$$

$$= \frac{83}{100}$$

3 Write the sum as a decimal.

Think of $\frac{83}{100}$ as *eighty-three hundredths*. So, $\frac{83}{100} = 0.83$.

So, Julian sold $\frac{83}{100}$ or 0.83 of the tickets.

Check

The models show that $\frac{6}{10} + \frac{23}{100} = \frac{83}{100}$.

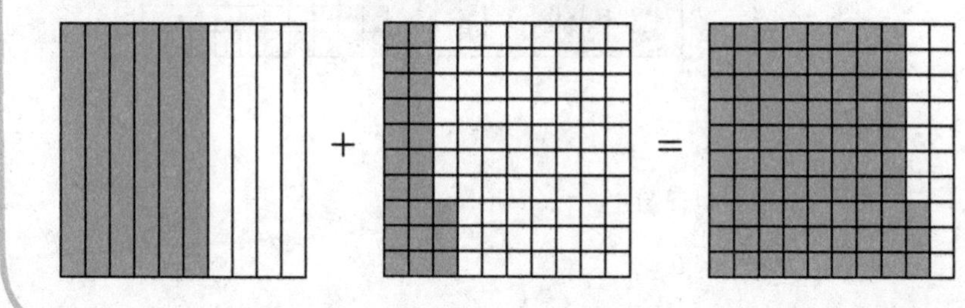

Practice

Add. Write each sum as a fraction with a denominator of 100 and as a decimal.

1. $\frac{2}{10} + \frac{33}{100} =$ __________

2. $\frac{6}{10} + \frac{25}{100} =$ __________

3. $\frac{4}{10} + \frac{17}{100} =$ __________

4. $\frac{2}{10} + \frac{22}{100} =$ __________

Problem Solving

Mathematical PRACTICE 2 Use Number Sense Write each answer as a fraction with a denominator of 100 and as a decimal.

5. An insect's body is $\frac{1}{10}$ inch long. The insect's head is $\frac{3}{100}$ inch long. What is the combined length of the insect's body and head?

____

6. Makenna rode her bike $\frac{6}{10}$ mile in the morning and $\frac{23}{100}$ mile in the afternoon. How far did she ride her bike in all?

____

Test Practice

7. Which addition expression describes the model at the right?

Ⓐ $\frac{70}{10} + \frac{18}{100}$

Ⓑ $\frac{7}{10} + \frac{18}{100}$

Ⓒ $\frac{7}{100} + \frac{18}{100}$

Ⓓ $\frac{7}{10} + \frac{18}{10}$

+

Name ____________________

Compare and Order Decimals

Lesson 7

ESSENTIAL QUESTION
How are fractions and decimals related?

To compare decimals, you can use a number line or place value.

I WIN!

Math in My World

Tools Watch Tutor

Example 1

At recess, Austin and his friends decided to have two races. One was 0.2 mile, and the other was 0.4 mile. Which race was longer?

One Way **Use a number line.**

Graph 0.4 and 0.2.

0 0.1 0.2 0.3 0.4 0.5 0.6 0.7 0.8 0.9 1.0

0.4 is to the right of 0.2.

So, 0.4 > 0.2.

Another Way **Use place value.**

Line up the decimal points. Then compare the digits in each place-value position.

In the tenths place, 4 > 2.

So, 0.4 > 0.2.

Ones	Tenths
0.	2
0.	4

So, the longer race was the one that was ____________.

Notice that you can only compare 0.4 and 0.2 because they refer to the same whole, which is one mile.

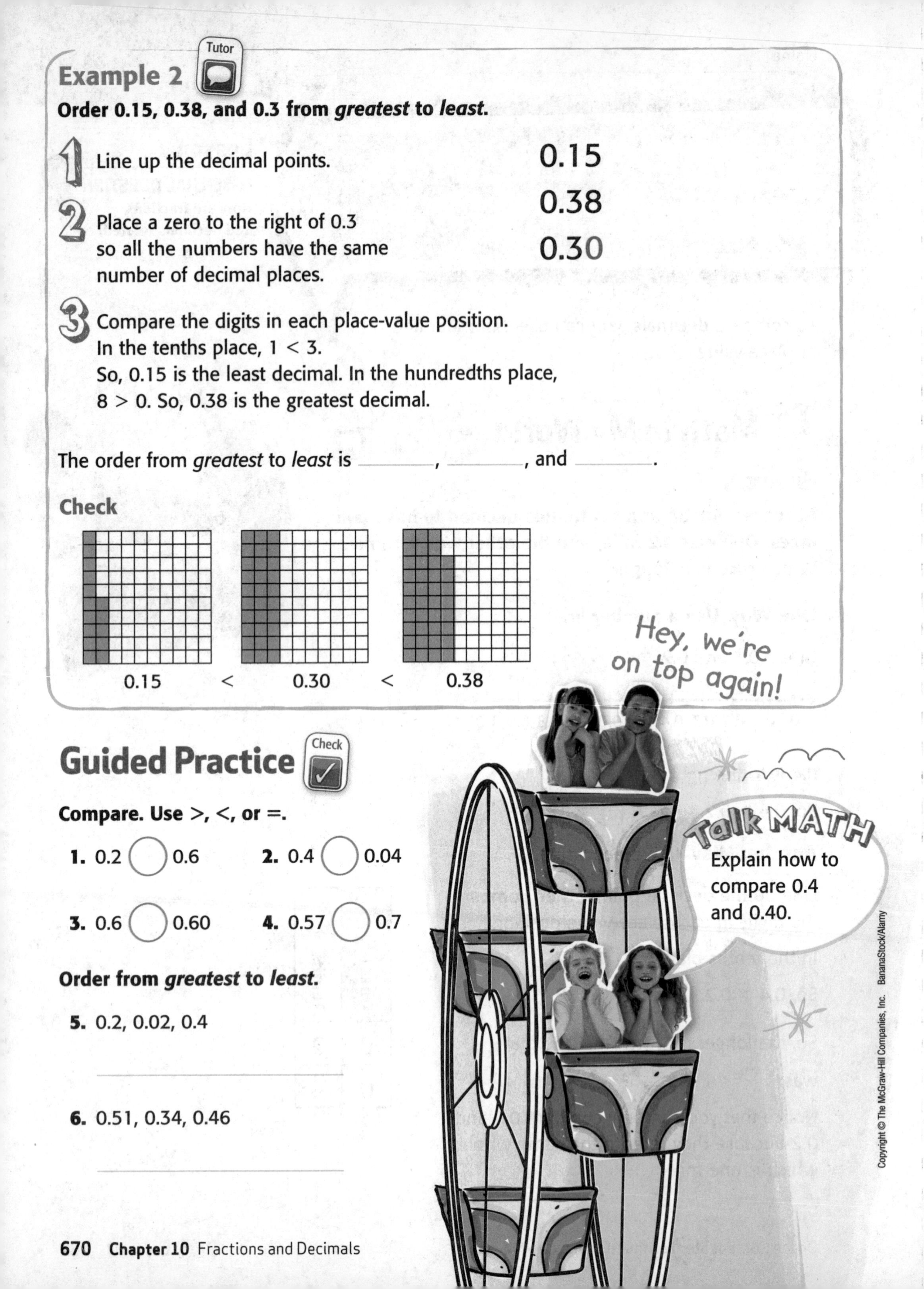

Example 2

Tutor

Order 0.15, 0.38, and 0.3 from *greatest* to *least*.

1. Line up the decimal points.

2. Place a zero to the right of 0.3 so all the numbers have the same number of decimal places.

0.15
0.38
0.30

3. Compare the digits in each place-value position. In the tenths place, 1 < 3. So, 0.15 is the least decimal. In the hundredths place, 8 > 0. So, 0.38 is the greatest decimal.

The order from *greatest* to *least* is ______, ______, and ______.

Check

0.15 < 0.30 < 0.38

Guided Practice

Check

Compare. Use >, <, or =.

1. 0.2 ◯ 0.6 **2.** 0.4 ◯ 0.04

3. 0.6 ◯ 0.60 **4.** 0.57 ◯ 0.7

Order from *greatest* to *least*.

5. 0.2, 0.02, 0.4

______

6. 0.51, 0.34, 0.46

______

Talk MATH

Explain how to compare 0.4 and 0.40.

Name ....

Independent Practice

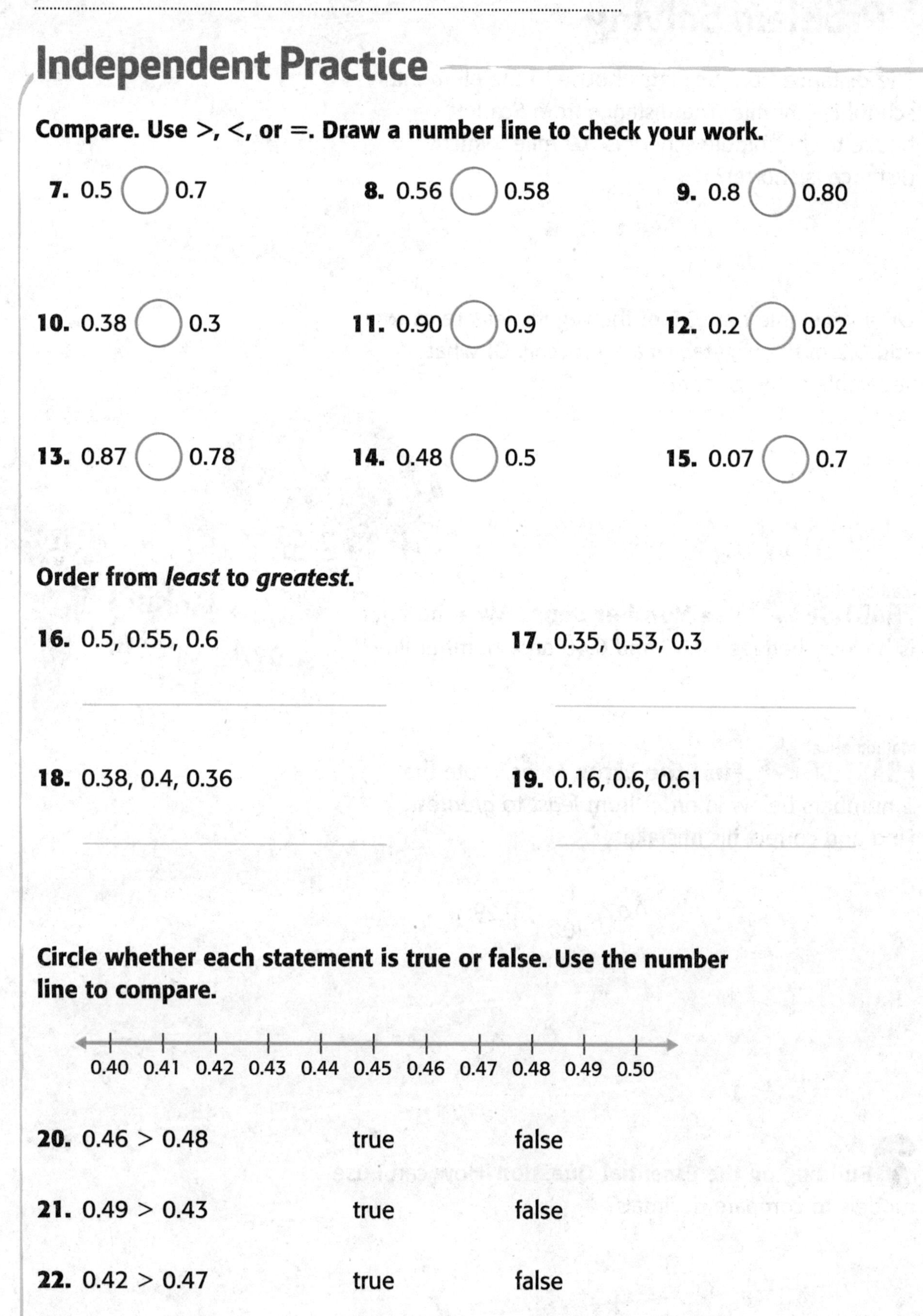

Compare. Use >, <, or =. Draw a number line to check your work.

7. 0.5 ◯ 0.7 **8.** 0.56 ◯ 0.58 **9.** 0.8 ◯ 0.80

10. 0.38 ◯ 0.3 **11.** 0.90 ◯ 0.9 **12.** 0.2 ◯ 0.02

13. 0.87 ◯ 0.78 **14.** 0.48 ◯ 0.5 **15.** 0.07 ◯ 0.7

Order from *least* to *greatest*.

16. 0.5, 0.55, 0.6 ______________

17. 0.35, 0.53, 0.3 ______________

18. 0.38, 0.4, 0.36 ______________

19. 0.16, 0.6, 0.61 ______________

Circle whether each statement is true or false. Use the number line to compare.

0.40 0.41 0.42 0.43 0.44 0.45 0.46 0.47 0.48 0.49 0.50

20. $0.46 > 0.48$ true false

21. $0.49 > 0.43$ true false

22. $0.42 > 0.47$ true false

23. $0.50 < 0.45$ true false

Problem Solving

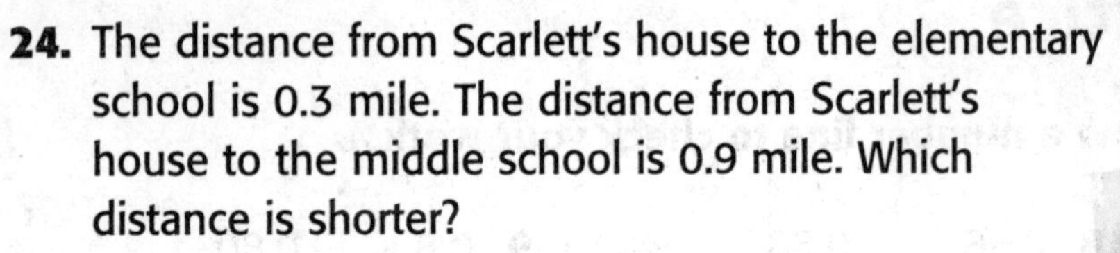

24. The distance from Scarlett's house to the elementary school is 0.3 mile. The distance from Scarlett's house to the middle school is 0.9 mile. Which distance is shorter?

____

25. On a vegetable tray, 0.5 of the vegetables are carrots and 0.2 of the vegetables are broccoli. Of what vegetable is there more?

____

HOT Problems

26. Mathematical PRACTICE 2 **Use Number Sense** What number is halfway between 0.36 and 0.48 on a number line?

____

27. Mathematical PRACTICE 3 **Find the Error** Miles wrote the 3 numbers below in order from *least* to *greatest*. Find and correct his mistake.

$$0.27, \frac{19}{100}, 0.29$$

____

____

28. **Building on the Essential Question** How can I use models to compare decimals?

____

____

____

Name ....

MY Homework

Lesson 7
Compare and Order Decimals

Homework Helper

eHelp Need help? connectED.mcgraw-hill.com

Compare 0.83 and 0.85. Use >, <, or =.

One Way **Use a number line.**

Graph 0.83 and 0.85.

0.80 0.81 0.82 0.83 0.84 0.85 0.86 0.87 0.88 0.89 0.90

0.83 is to the left of 0.85.

So, $0.83 < 0.85$.

Another Way **Use place value.**

Line up the decimal points. Then compare the digits in each place-value position.

The digits in the ones places are the same.

The digits in the tenths places are the same.

In the hundredths place, $3 < 5$.

So, $0.83 < 0.85$.

Ones	Tenths	Hundredths
0.	8	3
0.	8	5

Practice

Compare. Use >, <, or =.

1. 0.9 ◯ 0.4

2. 0.32 ◯ 0.37

3. 0.9 ◯ 0.90

Order from *least* to *greatest*.

4. 0.7, 0.27, 0.43

5. 0.4, 0.22, 0.72

Problem Solving

6. Eli lives 0.7 mile from his school. He lives 0.25 mile from his friend's house. Does Eli live closer to his school or his friend's house?

7. Alana spent 0.35 of a dollar on a ring at the arcade and 0.72 of a dollar on a snack. Did Alana spend more on the ring or the snack?

8. 

Identify Structure Tristan has a box of toys. Four out of 100 toys are red, 0.52 of the toys are green, and 0.1 of the toys are blue. Write the decimals in order from *greatest* to *least*.

9. Mathematical PRACTICE 4 **Model Math** Write a real-world problem that compares 0.4 and 0.29. Then solve the problem.

Test Practice

10. Which decimals are ordered from *least* to *greatest*?

    Ⓐ 0.25, 0.9, 0.35

    Ⓑ 0.9, 0.25, 0.35

    Ⓒ 0.25, 0.35, 0.9

    Ⓓ 0.9, 0.35, 0.25

Name

Real World

Problem-Solving Investigation

STRATEGY: Extra or Missing Information

Lesson 8

ESSENTIAL QUESTION
How are fractions and decimals related?

Learn the Strategy

Watch Tutor

On the first day, Gabriella's family traveled $\frac{3}{10}$ of a road trip. On the second day, they traveled $\frac{27}{100}$ of the trip. They traveled 4 days. What part of their trip did they travel in the first two days?

Road Trip!

1 Understand

What facts do you know?

Gabriella's family traveled $\frac{3}{10}$ of a road trip on the first day and $\frac{27}{100}$ of the trip on the second day. They traveled 4 days.

What do you need to find?

the part of their trip that they traveled in the first two days

2 Plan

The fact that they traveled 4 days is extra information. Find $\frac{3}{10} + \frac{27}{100}$.

3 Solve

$\frac{3}{10} + \frac{27}{100} = \frac{30}{100} + \frac{27}{100} = \frac{30 + 27}{100} = \frac{57}{100}$ Write $\frac{3}{10}$ as $\frac{30}{100}$. Then add.

So, Gabriella's family traveled $\frac{\square}{100}$ of their trip in the first two days.

4 Check

Does your answer make sense? Explain.

Practice the Strategy

Charlotte walked $\frac{6}{10}$ mile to school. After school, she walked $\frac{24}{100}$ mile to her friend's house. How much time does it take Charlotte to walk to school and to her friend's house?

1 Understand

What facts do you know?

What do you need to find?

Plan

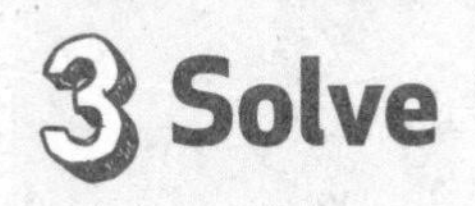

Solve

4 Check

Does your answer make sense? Explain.

Name

Apply the Strategy

Determine if there is extra or missing information to solve each problem. Then solve if possible.

My Work!

1. There are 100 movies at the store. $\frac{30}{100}$ are action movies, $\frac{50}{100}$ are comedies, and $\frac{20}{100}$ are adventure movies. What part of the movies are action or comedies?

2. Mathematical PRACTICE 1 **Keep Trying** In a basketball game, the red team scored $\frac{3}{10}$ of the baskets during the first half and $\frac{4}{10}$ of the baskets during the second half. The blue team had 10 players. How many baskets did the red team score during the first half and second half of the game?

3. Alexia and her family went on vacation. They walked $\frac{1}{10}$ mile to the beach and $\frac{2}{10}$ mile to the souvenir shop. How far did they walk to the beach and to the souvenir shop?

Review the Strategies

Use any strategy to solve each problem.

- Determine extra or missing information.
- Use logical reasoning.
- Look for a pattern.
- Make a model.

4. Trina is making friendship bracelets. One tenth of the bracelets are blue. Some of the bracelets are red and some are purple. How many bracelets are blue and purple?

____

____

5. Mathematical PRACTICE 8 **Repeated Reasoning** Find the next number in the pattern below. Explain how you found the number.

$\frac{15}{100}, \frac{3}{10}, \frac{45}{100}, \frac{6}{10}, \frac{75}{100}, \ldots$

____

____

____

6. The fourth grade classes voted on their favorite flavor of ice cream. Three tenths of the students voted for strawberry, $\frac{21}{100}$ of the students voted for vanilla, and $\frac{4}{10}$ of the students voted for chocolate. How many students voted for vanilla or chocolate?

____

____

7. Harper and his mom are making trail mix for a party. Two tenths of the trail mix is pretzels and $\frac{32}{100}$ of the trail mix is cereal. The party starts at 1:00 P.M. How much of the trail mix is pretzels or cereal?

____

____

My Work!

Name

MY Homework

Lesson 8
Problem Solving: Extra or Missing Information

Homework Helper

eHelp

Need help? connectED.mcgraw-hill.com

Stella walked $\frac{4}{10}$ mile to school. During recess, she played soccer. After school, Stella walked $\frac{13}{100}$ mile to the library. How far did Stella walk before and after school?

1 Understand

What facts do you know?

Stella walked $\frac{4}{10}$ mile to school and $\frac{13}{100}$ mile to the library. During recess, she played soccer.

What do you need to find?

the distance Stella walked before and after school

2 Plan

The fact that Stella played soccer at recess is extra information.

Find $\frac{4}{10} + \frac{13}{100}$.

3 Solve

$\frac{4}{10} + \frac{13}{100} = \frac{40}{100} + \frac{13}{100} = \frac{40 + 13}{100} = \frac{53}{100}$

Write $\frac{4}{10}$ as $\frac{40}{100}$. Then add.

So, Stella walked $\frac{53}{100}$ mile.

4 Check

Does your answer make sense? Explain.

Since $\frac{53}{100} - \frac{40}{100} = \frac{13}{100}$, the answer makes sense.

Problem Solving

Determine if there is extra or missing information to solve each problem. Then solve if possible.

1. Janice bought her mother a bunch of 10 flowers. Two of the flowers are daisies. One half of the remaining flowers are tulips. Write the fraction of the flowers that are daisies.

2. Mathematical PRACTICE 1 **Make a Plan** There are 100 books in the library. There are non-fiction and fiction books. Write the fraction of the books that are fiction.

3. Sean has a collection of coins. One tenth of the coins are from Europe. Thirty-two hundredths are from Asia. The rest are from Africa. Write a decimal to show the total part of the coins that are from Europe or Asia.

4. Kenley has 100 songs on her digital music player. Of the songs, seventeen hundredths are country songs, two tenths are musicals, and four tenths are classical music songs. What part of the songs are either country or musicals? Write as a decimal.

My Work!

Review

Chapter 10

Fractions and Decimals

Vocabulary Check

1. Write three examples of numbers that are **decimals**.

   ____

   Write three examples of numbers that are not **decimals**.

   ____

2. Write three examples of decimals that have a 4 in the **tenths** place.

   ____

3. Write three examples of decimals that have a 5 in the **hundredths** place.

   ____

Concept Check

Write the decimal represented by each model.

4. 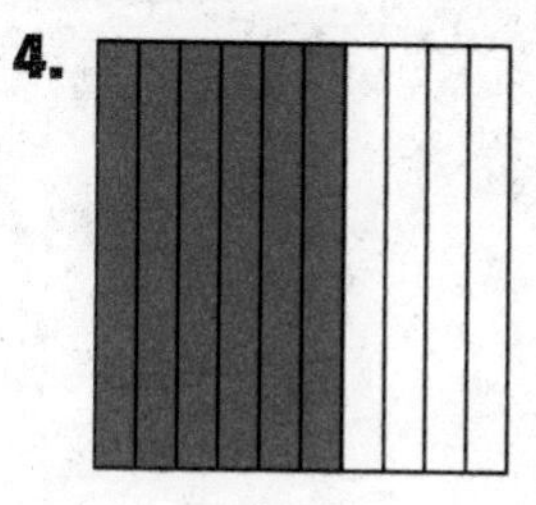

   ____

5. _______________

6. _______________

Shade the model. Then write the decimal.

7. four out of ten

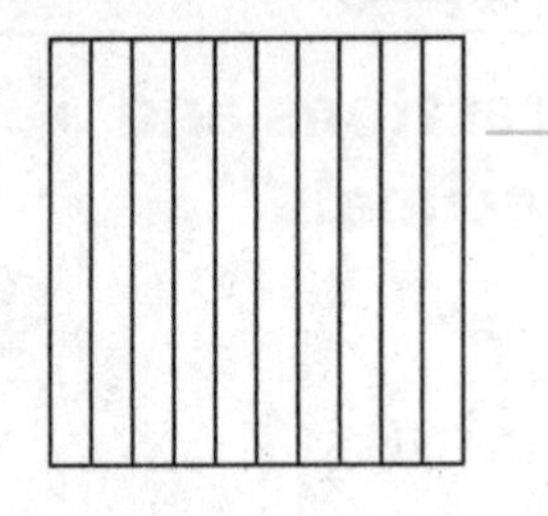

______

8. thirty-nine out of 100

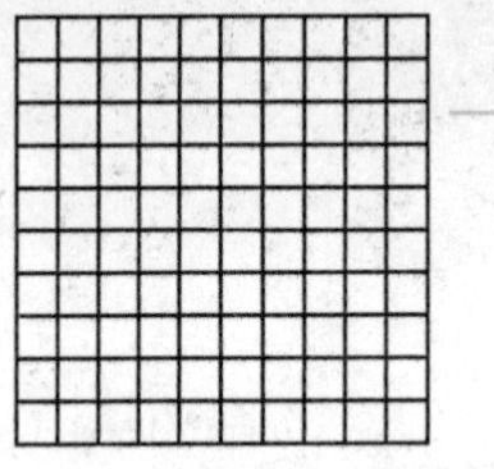

______

9. Write 0.8 as a fraction with a denominator of 10. Shade the grid.

0.8 = ______

10. Write 0.41 as a fraction with a denominator of 100. Shade the grid.

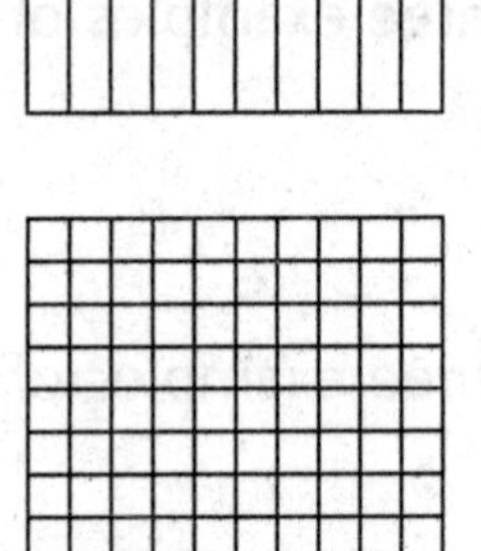

0.41 = ______

Write each fraction as an equivalent fraction with a denominator of 100. Then write the fraction as a decimal.

11. $\frac{9}{10}$ = ______

Decimal: ______

12. $\frac{3}{10}$ = ______

Decimal: ______

Add. Write each sum as a fraction with a denominator of 100 and as a decimal.

13. $\frac{2}{10} + \frac{36}{100} =$ ______

14. $\frac{7}{10} + \frac{13}{100} =$ ______

Order each set of decimals from *greatest* to *least*.

15. 0.3, 0.23, 0.61 ______

16. 0.72, 0.5, 0.69 ______

Name

17. A barn has 10 animals. Three of the animals are horses. Write a decimal to show what part of the barn animals are horses.

18. Allison bought a snack using the coins shown. Write a decimal to show what part of a dollar she spent.

19. John lives 0.67 mile from the recreation center. Bella lives 0.8 mile from the recreation center. Who lives a greater distance from the recreation center? Explain.

20. Joseph ran $\frac{65}{100}$ mile in a race. Write a decimal to show what part of a mile Joseph ran.

Test Practice

21. Chase poured three-tenths liter of lemonade and twenty-nine-hundredths liter of raspberry juice in a pitcher. Write the total amount of liquid Chase poured into the pitcher as a decimal.

Ⓐ 0.32 liter
Ⓑ 0.59 liter
Ⓒ 0.69 liter
Ⓓ 0.95 liter

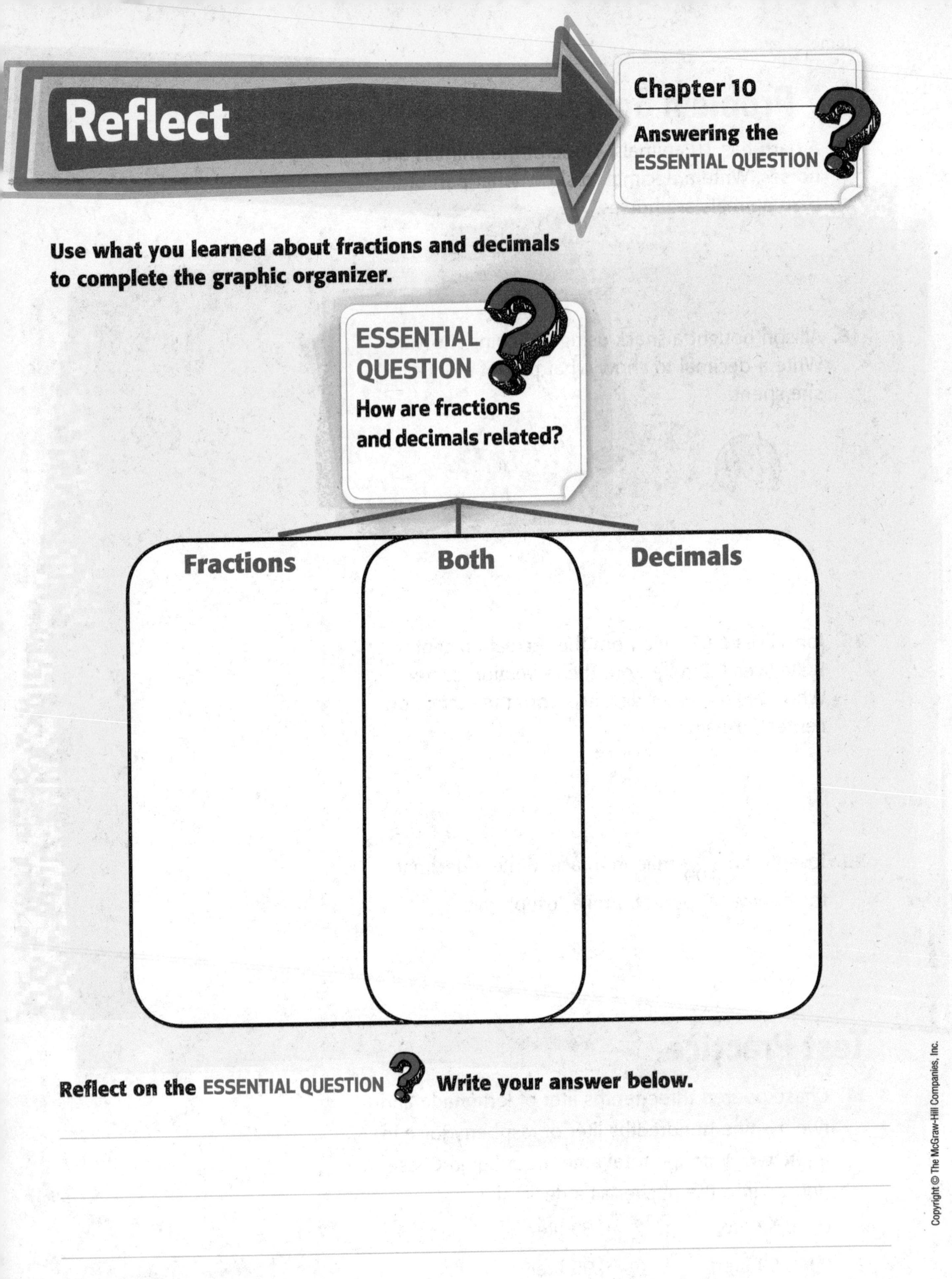

Use what you learned about fractions and decimals to complete the graphic organizer.

Reflect on the ESSENTIAL QUESTION Write your answer below.

ESSENTIAL QUESTION

Why do we convert measurements?

Let's Discover Nature!

Watch

Watch a video!

MY Common Core State Standards

Measurement and Data

CCSS

4.MD.1 Know relative sizes of measurement units within one system of units including km, m, cm; kg, g; lb, oz.; l, ml; hr, min, sec. Within a single system of measurement, express measurements in a larger unit in terms of a smaller unit. Record measurement equivalents in a two column table.

4.MD.2 Use the four operations to solve word problems involving distances, intervals of time, liquid volumes, masses of objects, and money, including problems involving simple fractions or decimals, and problems that require expressing measurements given in a larger unit in terms of a smaller unit. Represent measurement quantities using diagrams such as number line diagrams that feature a measurement scale.

4.MD.4 Make a line plot to display a data set of measurements in fractions of a unit ($\frac{1}{2}, \frac{1}{4}, \frac{1}{8}$). Solve problems involving addition and subtraction of fractions by using information presented in line plots.

I'll be able to get this—no problem!

Standards for Mathematical PRACTICE

1. Make sense of problems and persevere in solving them.
2. Reason abstractly and quantitatively.
3. Construct viable arguments and critique the reasoning of others.
4. Model with mathematics.
5. Use appropriate tools strategically.
6. Attend to precision.
7. Look for and make use of structure.
8. Look for and express regularity in repeated reasoning.

= focused on in this chapter

Name

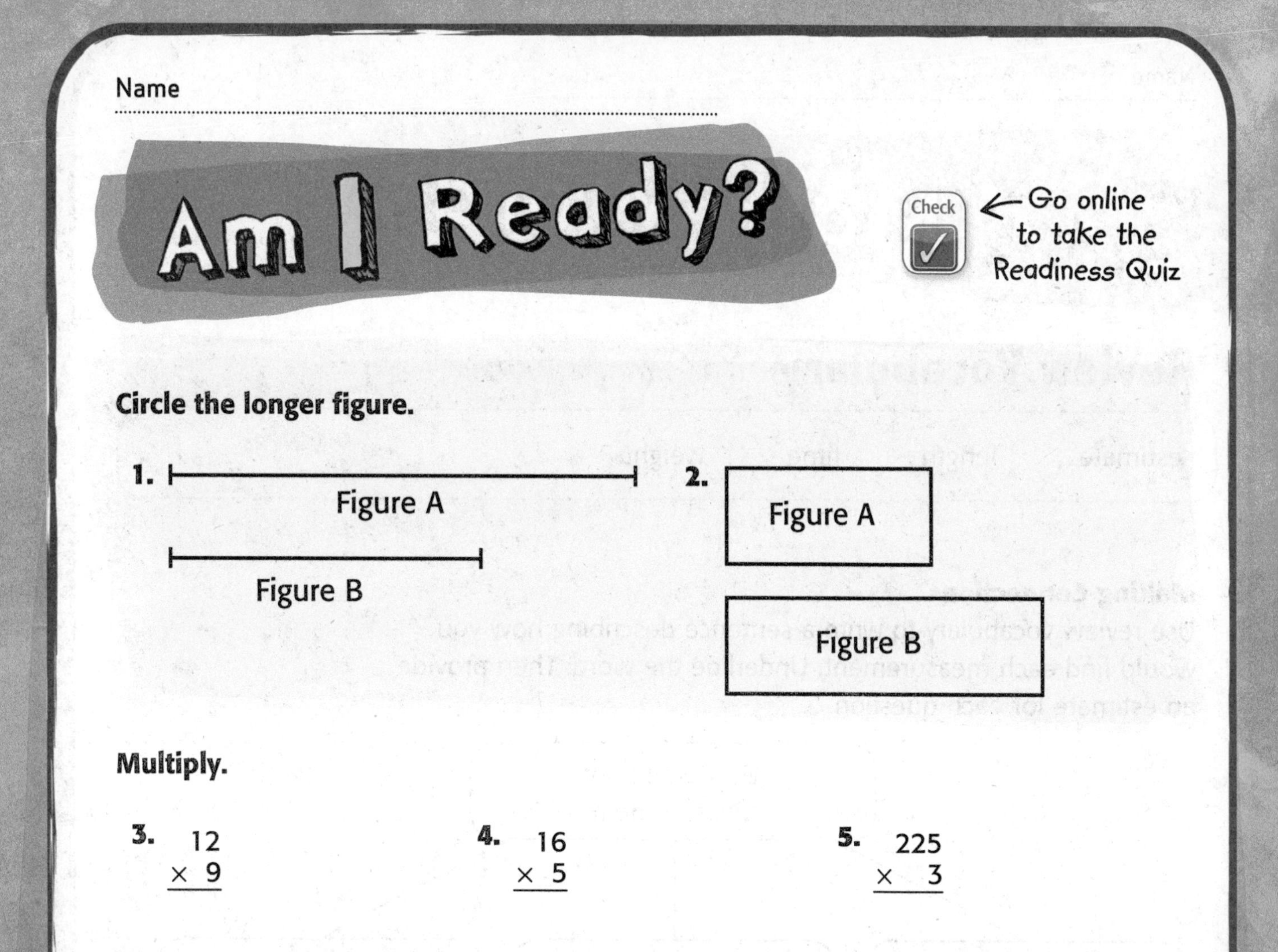

Circle the longer figure.

1. Figure A

Figure B

2. Figure A

Figure B

Multiply.

3. $\begin{array}{r} 12 \\ \times\ 9 \\ \hline \end{array}$

4. $\begin{array}{r} 16 \\ \times\ 5 \\ \hline \end{array}$

5. $\begin{array}{r} 225 \\ \times\ \ 3 \\ \hline \end{array}$

Add. Write in simplest form.

6. $\frac{2}{10} + \frac{5}{10} =$ ______

7. $\frac{1}{4} + \frac{1}{4} =$ ______

8. $\frac{1}{6} + \frac{3}{6} =$ ______

Write the time shown on each clock.

9. **10.** **11.**

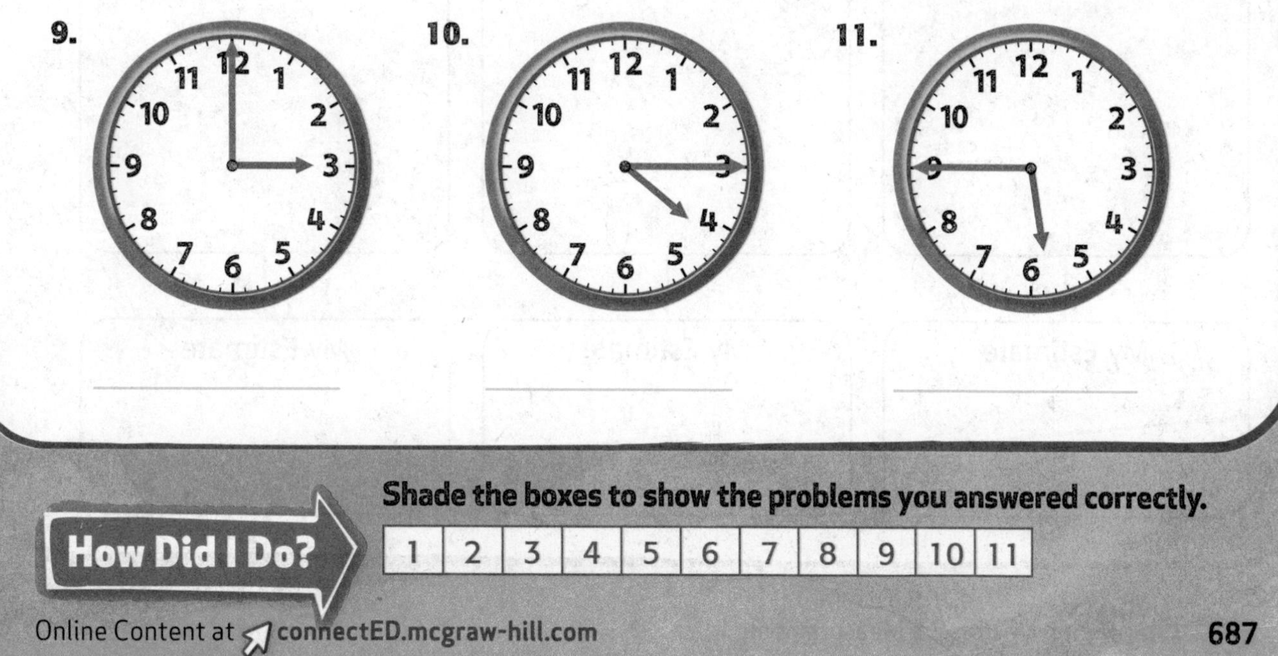

Shade the boxes to show the problems you answered correctly.

How Did I Do?

1	2	3	4	5	6	7	8	9	10	11

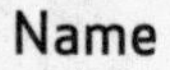

Name

Review Vocabulary

estimate   length   time   weight

Making Connections

Use review vocabulary to write a sentence describing how you would find each measurement. Underline the word. Then provide an estimate for each question.

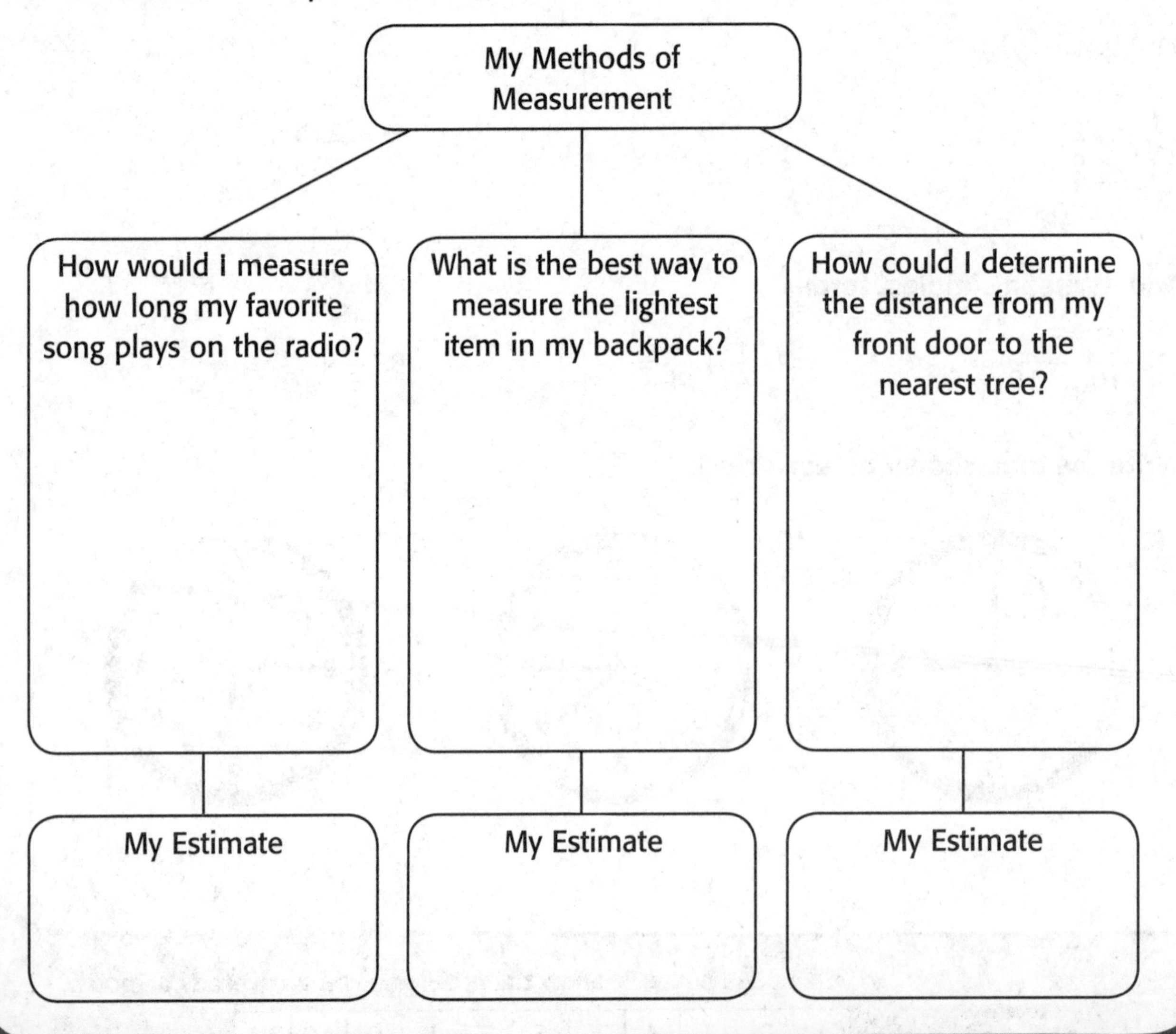

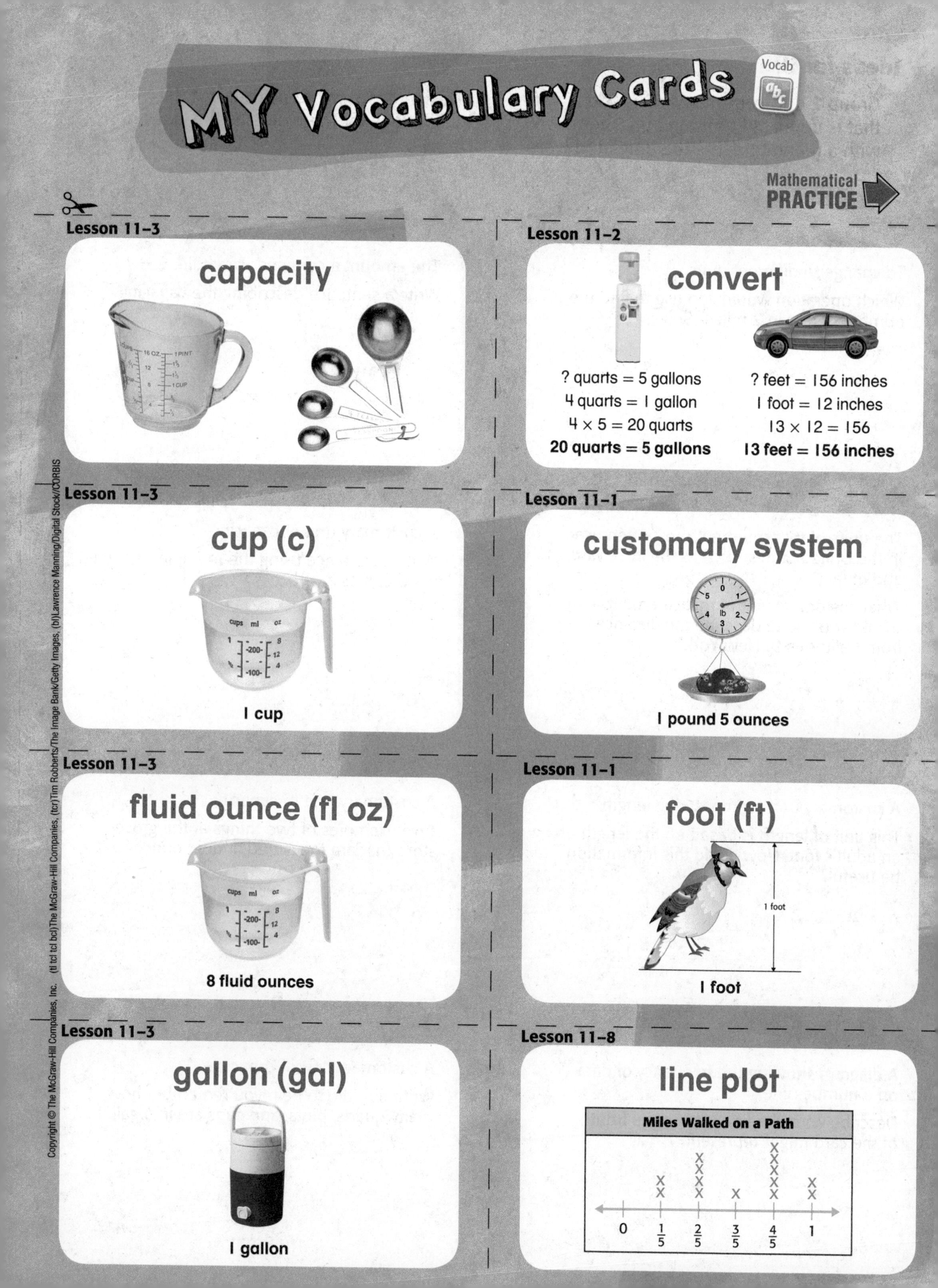
MY Vocabulary Cards
Vocab
abc
Mathematical PRACTICE
Lesson 11-3
capacity
Lesson 11-2
convert
? quarts = 5 gallons
4 quarts = 1 gallon
4 × 5 = 20 quarts
20 quarts = 5 gallons
? feet = 156 inches
1 foot = 12 inches
13 × 12 = 156
13 feet = 156 inches
Lesson 11-3
cup (c)
1 cup
Lesson 11-1
customary system
1 pound 5 ounces
Lesson 11-3
fluid ounce (fl oz)
8 fluid ounces
Lesson 11-1
foot (ft)
1 foot
1 foot
Lesson 11-3
gallon (gal)
1 gallon
Lesson 11-8
line plot
Miles Walked on a Path
0
1/5
2/5
3/5
4/5
1
Copyright © The McGraw-Hill Companies, Inc. (tl tcl tcl bcl)The McGraw-Hill Companies, (tcr)Tim Robberts/The Image Bank/Getty Images, (bl)Lawrence Manning/Digital Stock/CORBIS

Ideas for Use

- Group 2 or 3 common words. Add a word that is unrelated to the group. Then work with a friend to name the unrelated word.
- Design a crossword puzzle. Use the definitions for the words as the clues.

To change to different units.

Which operation would you use to find the number of feet in 2 miles? Solve.

The amount a container can hold.

Write a sentence describing the capacity of a bathtub.

The units of measurement most often used in the United States, such as the inch, yard, and mile.

What customary unit of measurement would you use to describe the distance from California to New York?

A customary unit of capacity.

Write a sentence using the multiple-meaning word *cup* as a verb.

A customary unit for measuring length.

This unit of length is based on the length of an adult's foot. How could this information be useful?

A customary unit of capacity.

Give examples of two things at the grocery store that are measured in fluid ounces.

A diagram showing the frequency of data on a number line.

Describe what the line plot on the front of the card might represent.

A customary unit of capacity.

Write a riddle to help you remember how many quarts, pints, and cups are in a gallon.

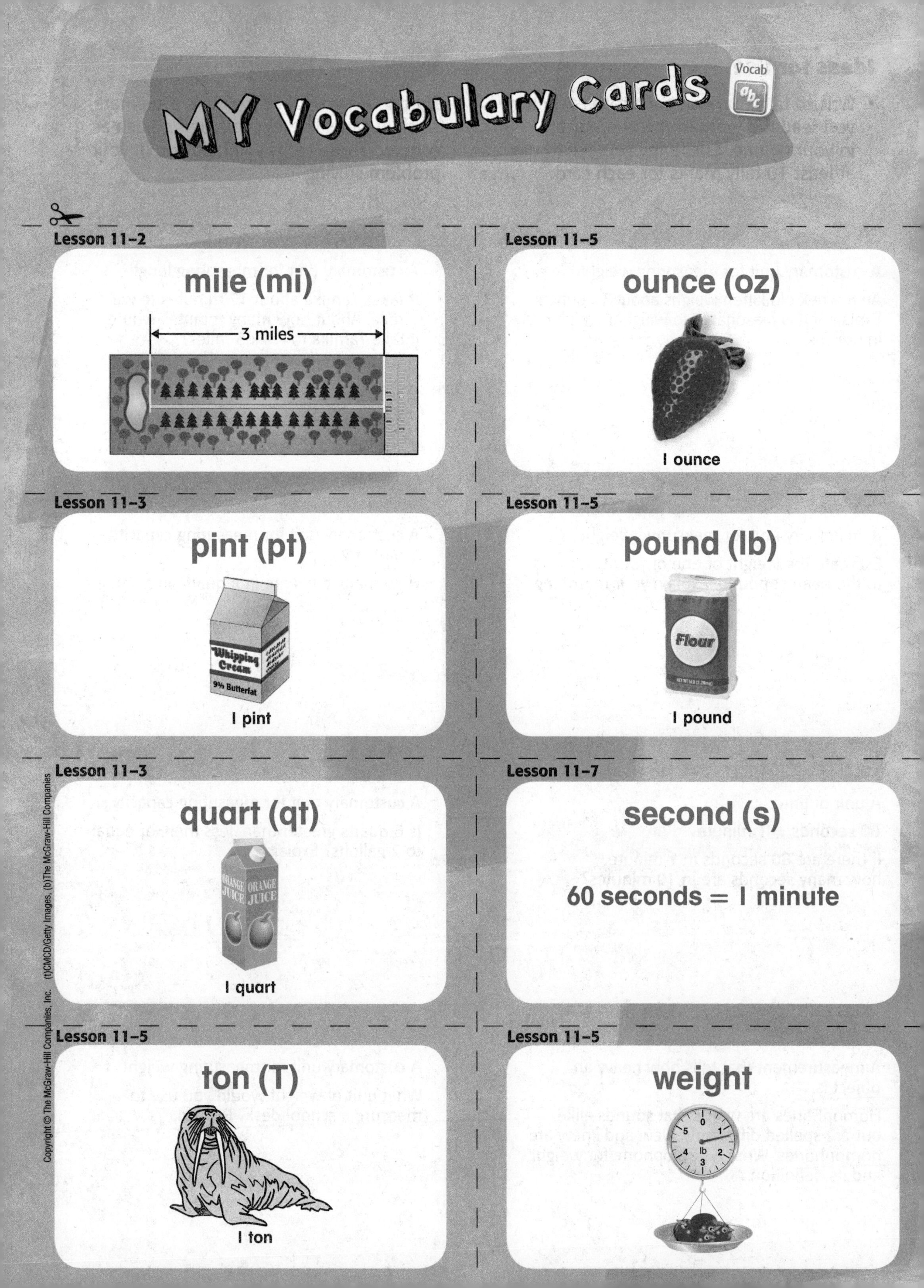
MY Vocabulary Cards
Vocab
abc
Lesson 11-2
mile (mi)
3 miles
Lesson 11-5
ounce (oz)
1 ounce
Lesson 11-3
pint (pt)
Whipping Cream
9% Butterfat
1 pint
Lesson 11-5
pound (lb)
Flour
1 pound
Lesson 11-3
quart (qt)
ORANGE JUICE
ORANGE JUICE
1 quart
Lesson 11-7
second (s)
60 seconds = 1 minute
Lesson 11-5
ton (T)
1 ton
Lesson 11-5
weight
0
1
2
3
4
5
lb

Ideas for Use

- Write a tally mark on each card every time you read the word in this chapter or use it in your writing. Challenge yourself to use at least 10 tally marks for each card.
- During this school year, create a separate stack of cards for key math verbs, such as *convert.* These verbs will help you in your problem solving.

A customary unit for measuring weight.

An 8-week old kitten weighs about 32 ounces. Explain if it is reasonable to weigh an adult cat in ounces.

A customary unit for measuring length.

It takes Tamika about 12 minutes to walk 1 mile. About how many minutes would it take Tamika to walk 3 miles?

A customary unit for measuring weight.

Estimate the weight of one of your textbooks to the nearest pound. Explain your reasoning.

A customary unit for measuring capacity.
1 pint = 2 cups

How many pints are in a quart? In a gallon?

A unit of time.

60 seconds = 1 minute

If there are 60 seconds in 1 minute, how many seconds are in 10 minutes?

A customary unit for measuring capacity.

Is 6 quarts greater than, less than, or equal to 2 gallons? Explain.

A measurement that tells how heavy an object is.

Homophones are words that sounds alike, but are spelled differently. New and knew are homophones. Write a homophone for weight and its definition.

A customary unit for measuring weight.

What unit of weight would you use to measure a school desk? Explain.

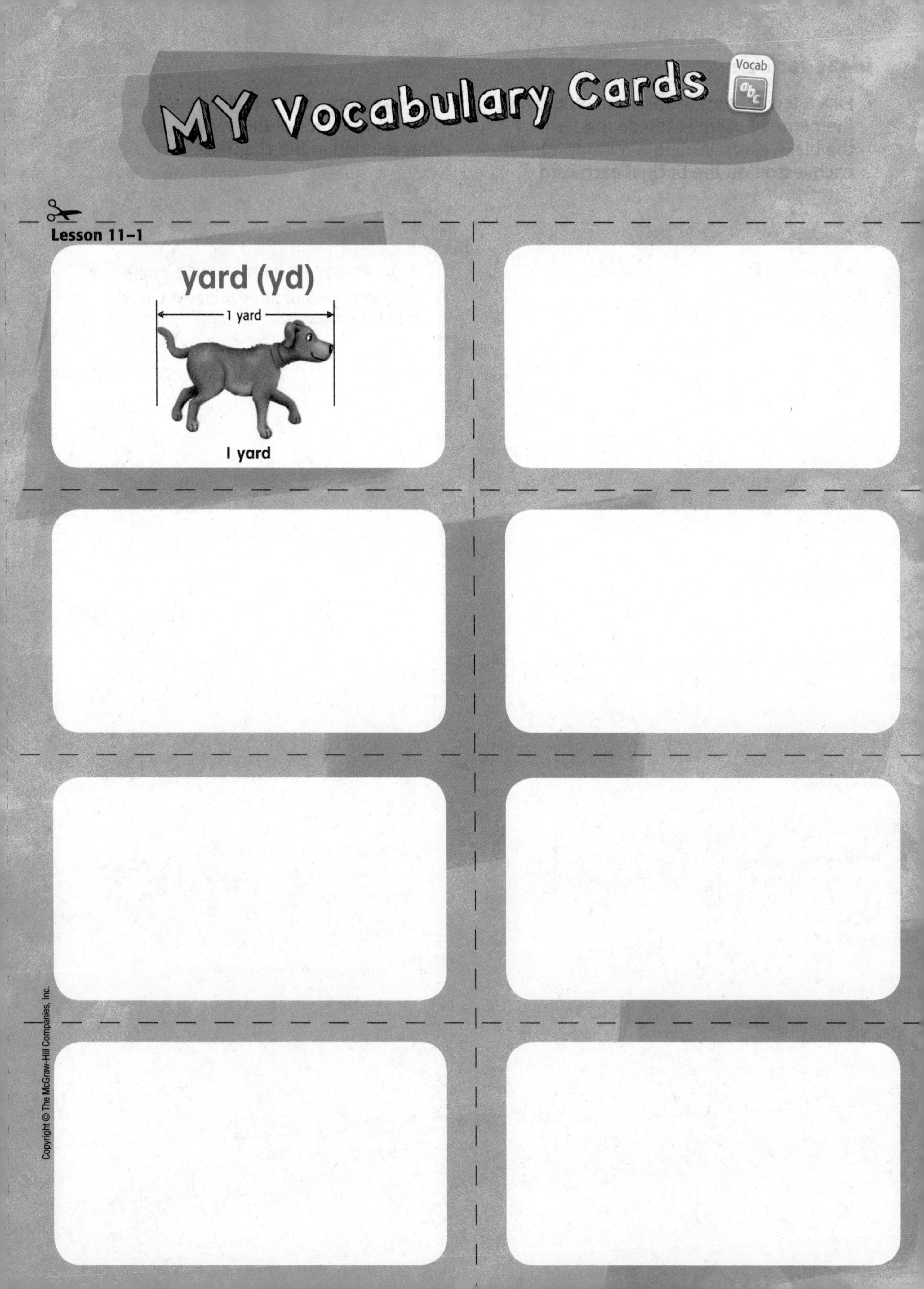

MY Vocabulary Cards

Vocab

Lesson 11–1

yard (yd)

1 yard

Ideas for Use

- Pick a few lessons from the chapter. Write the name of each lesson on the front of the blank cards. Write a few study tips for each lesson on the back of each card.
- Have students use the blank cards to write summaries of the conversions they learned in the chapter.

FOLDABLES® Follow the steps on the back to make your Foldable.

MY Customary Conversions

Weight

Length

Capacity

Capacity

	fluid ounce	cup (c)
fluid ounce (fl oz)	1 fluid ounce	___ fl oz = 1 c
cup (c)	1 c = ___ fl oz	1 cup
pint (pt)	1 pt = 16 fl oz	1 pt = ___ c
quart (qt)	1 qt = 32 fl oz	___ qt = 4 c
gallon (gal)	1 gal = 128 fl oz	1 gal = 16 c

pound (lb)	ton (T)
16 oz = ___ lb	32,000 oz = 1 T
1 pound	___ lb = 1 T
1 T = ___ lb	1 Ton

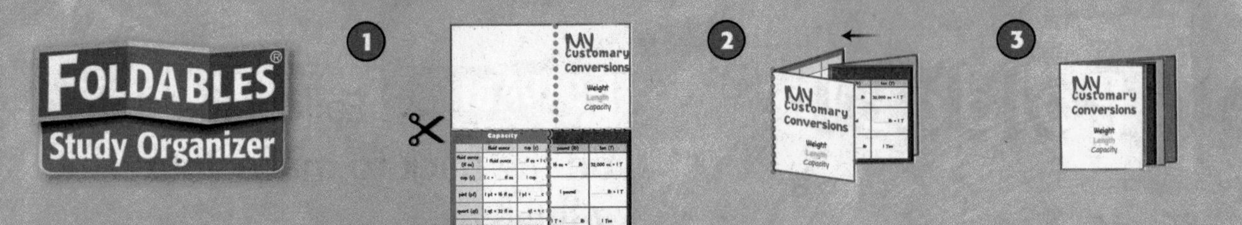

Weight

	ounce (oz)
ounce (oz)	1 ounce
pound (lb)	1 lb = ___ oz
ton (T)	1 T = 32,000 oz

pint (pt)	quart (qt)	gallon (gal)
16 fl oz = 1 pt	32 fl oz = 1 qt	128 fl oz = 1 gal
___ c =1 pt	___ c = 1 qt	___ c = 1 gal
1 pint	___ pt = 1 qt	___ pt = 1 gal
___ qt = 2 pt	1 quart	___ qt = 1 gal
1 gal = 8 pt	___ gal = 4 qt	1 gallon

Length

	inch (in.)	foot (ft)	yard (yd)	mile (mi)
inch (in.)	1 inch	12 in. = ___ ft	___ in. = 1 yd	63,360 in. = 1 mi
foot (ft)	1 ft = ___ in.	1 foot	___ ft = 1 yd	______ ft = 1 mi
yard (yd)	___ yd = 36 in.	___ yd = 3 ft	1 yard	1,760 yd = 1 mi
mile (mi)	1 mi = 63,360 in.	___ mi = 5,280 ft	1 mi = 1,760 yd	1 mile

Name

Customary Units of Length

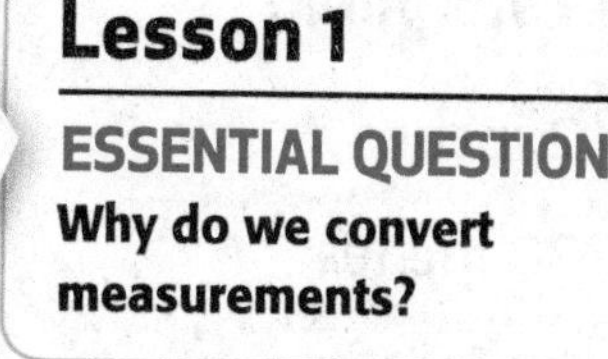

Lesson 1

ESSENTIAL QUESTION
Why do we convert measurements?

Length is the measurement of a line between two points. Inch, foot, and yard are units that are part of the **customary system** of measure for length.

Math in My World

Tools Watch Tutor

Example 1

The actual size of a neon damsel marine fish is shown. How long is this fish to the nearest inch, $\frac{1}{2}$ inch, and $\frac{1}{4}$ inch?

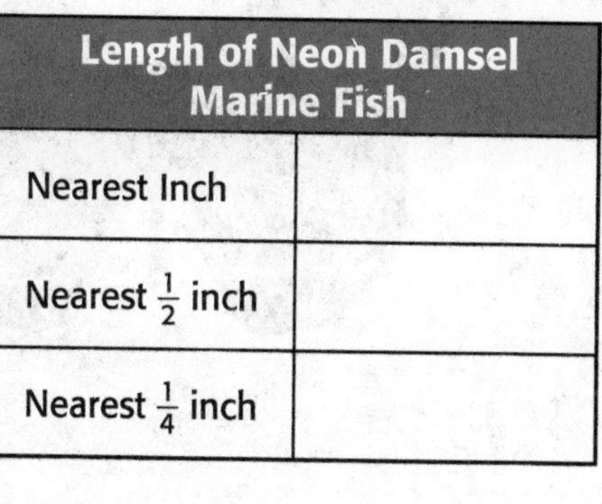

1 Measure to the nearest inch.
Compare the length to what you know about inches. Record the length in the table.

2 Measure to the nearest $\frac{1}{2}$ inch.
Using a ruler, measure the length of the fish to the nearest $\frac{1}{2}$ inch. Record the length in the table.

3 Measure to the nearest $\frac{1}{4}$ inch.
Measure the length of the fish to the nearest $\frac{1}{4}$ inch. Record the length in the table.

Length of Neon Damsel Marine Fish	
Nearest Inch	
Nearest $\frac{1}{2}$ inch	
Nearest $\frac{1}{4}$ inch	

An inch (in.) is about the length of one paper clip.

A **foot (ft)** is about the length of a textbook.

A **yard (yd)** is about the height of a chair.

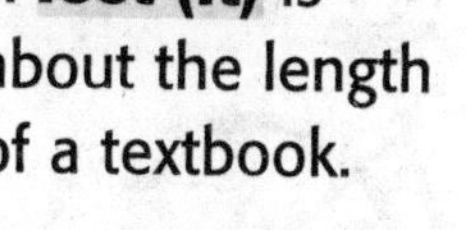
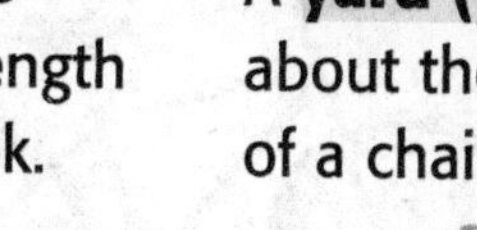
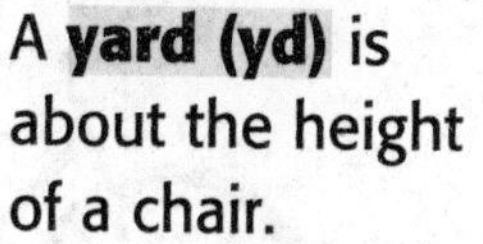

Example 2

Choose the best estimate for the length of the caterpillar.

Ⓐ 2 inches

Ⓑ 20 inches

Ⓒ 20 feet

Ⓓ 2 yards

A caterpillar is small. So, inches are better estimates than feet or yards. Since 20 inches is more than a foot and is too big, the best estimate is ________ .

Guided Practice

Estimate. Then measure each to the nearest inch, $\frac{1}{2}$ inch, and $\frac{1}{4}$ inch.

1. **2.**

________ ________

3. Choose the best estimate for the length.

Ⓐ 12 inches

Ⓑ 4 feet

Ⓒ 12 feet

Ⓓ 4 yards

Talk MATH

Why do you think there is more than one unit of length for measure?

Name

Independent Practice

Estimate. Then measure each to the nearest inch, $\frac{1}{2}$ inch, and $\frac{1}{4}$ inch.

4. **5.** **6.** **7.**

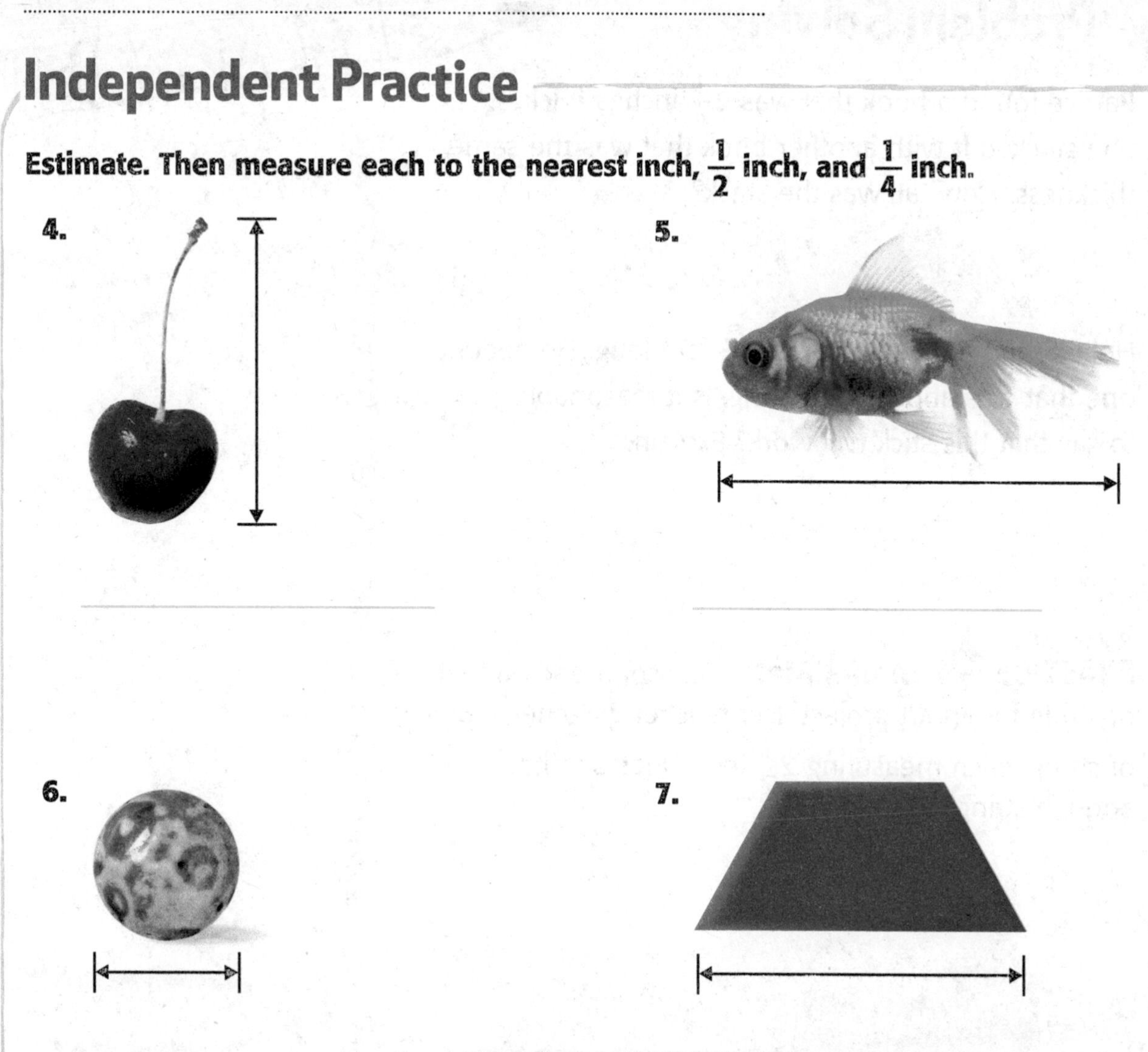

Choose the best estimate for each length.

8. length of a whistle

Ⓐ 2 yards

Ⓑ 2 feet

Ⓒ 12 inches

Ⓓ 2 inches

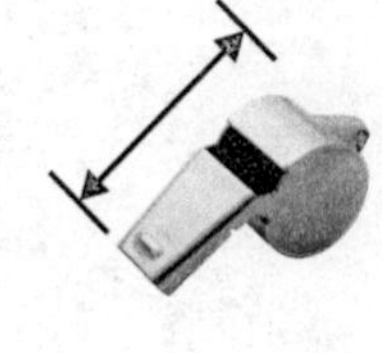

9. length of a chalkboard

Ⓕ 1 foot

Ⓖ 2 feet

Ⓗ 1 yard

Ⓘ 2 yards

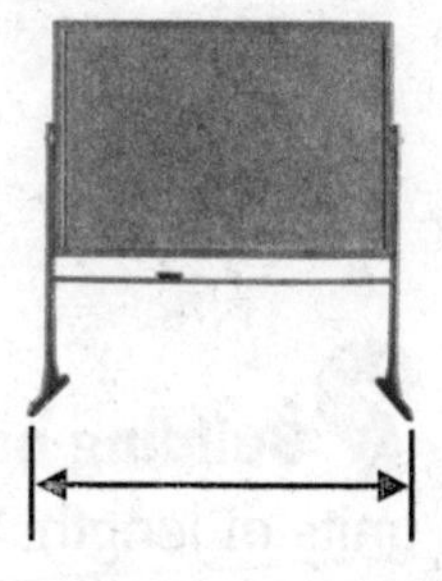

Problem Solving

My Work!

10. Patrice found a book that was $2\frac{1}{4}$ inches thick. She stacked it with another book that was the same thickness. How tall was the stack?

11. Helki found a stick that was $5\frac{3}{4}$ feet long. He needed one that was about 5 feet long. Is it reasonable to say that this stick will work? Explain.

12. Mathematical PRACTICE 4 **Model Math** Addison needs 6 feet of string for an art project. Her teacher gave her 3 pieces of string, each measuring $2\frac{1}{2}$ feet. Does she have enough string? Explain.

HOT Problems

13. Mathematical PRACTICE 5 **Use Math Tools** Find two objects in your classroom that are each longer than two inches and shorter than 4 inches. How did you use estimation in selecting objects?

14. **Building on the Essential Question** Name two customary units of length. Which measurement is more accurate? Explain.

Name ....

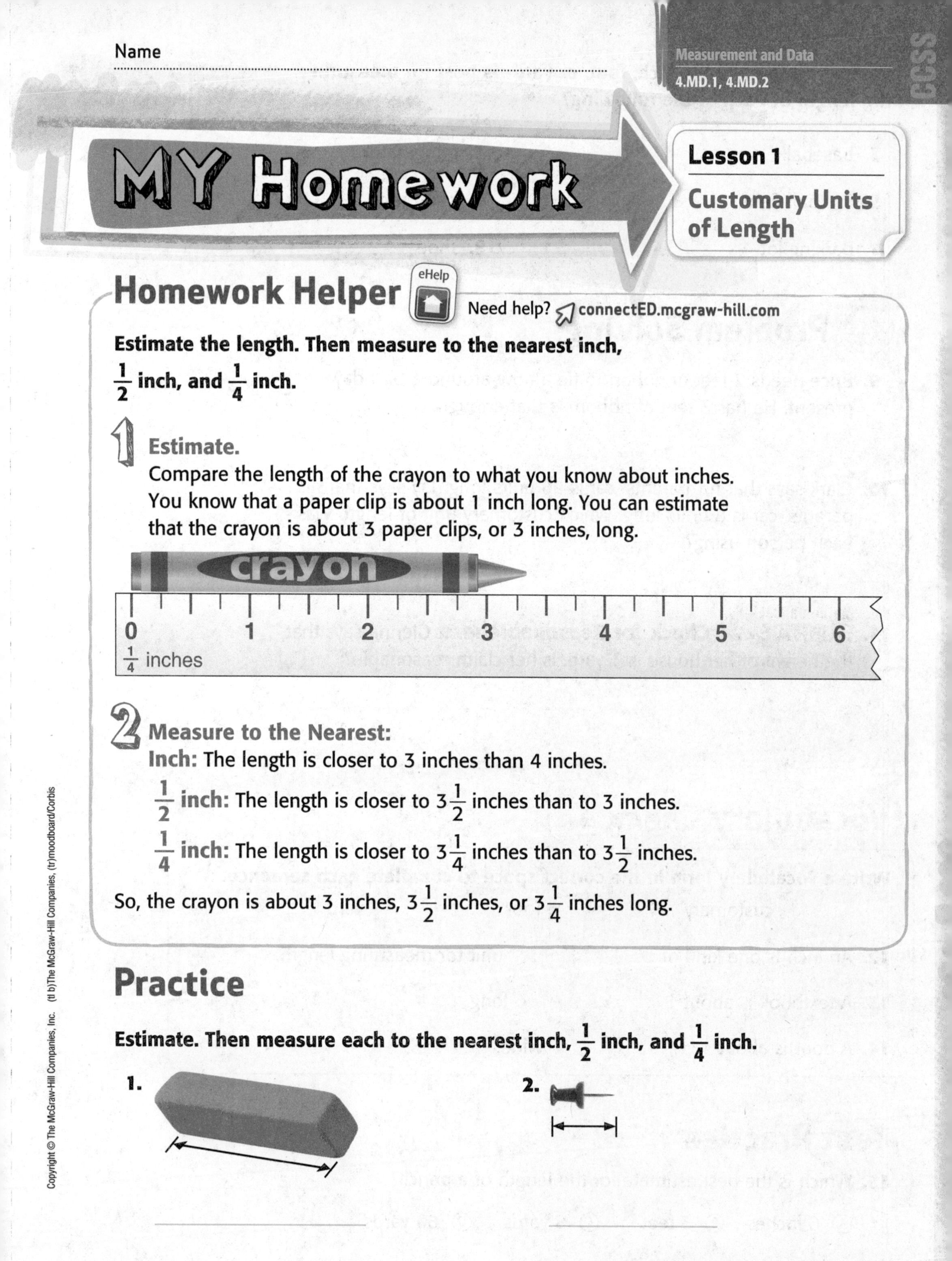

MY Homework

Lesson 1
Customary Units of Length

Homework Helper

eHelp Need help? connectED.mcgraw-hill.com

Estimate the length. Then measure to the nearest inch, $\frac{1}{2}$ inch, and $\frac{1}{4}$ inch.

1 Estimate.
Compare the length of the crayon to what you know about inches. You know that a paper clip is about 1 inch long. You can estimate that the crayon is about 3 paper clips, or 3 inches, long.

2 Measure to the Nearest:
Inch: The length is closer to 3 inches than 4 inches.

$\frac{1}{2}$ inch: The length is closer to $3\frac{1}{2}$ inches than to 3 inches.

$\frac{1}{4}$ inch: The length is closer to $3\frac{1}{4}$ inches than to $3\frac{1}{2}$ inches.

So, the crayon is about 3 inches, $3\frac{1}{2}$ inches, or $3\frac{1}{4}$ inches long.

Practice

Estimate. Then measure each to the nearest inch, $\frac{1}{2}$ inch, and $\frac{1}{4}$ inch.

1. ______________

2. ______________

Which customary unit—inch, foot, or yard—is best for measuring the length of each of the following?

3. baseball ________

4. refrigerator ________

5. football field ________

6. jump rope ________

7. parking lot ________

8. shoe ________

Problem Solving

9. Brice needs 2 feet of ribbon to tie a bow around a birthday present. He has 3 feet of ribbon. Is that enough?

____

10. Clark says that his parents' car is 96 units long. Jay says that his parents' car is 8 units long. Which customary unit of length was each person using?

____

11. Mathematical PRACTICE 1 **Check for Reasonableness** Glenna says that the height of her house is 1 yard. Is her claim reasonable?

____

____

Vocabulary Check

Vocab

Write a vocabulary term in the correct space to complete each sentence.

customary foot yard

12. An inch is one kind of ________ unit for measuring length.

13. A textbook is about 1 ________ long.

14. A door is about 1 ________ wide.

Test Practice

15. Which is the best estimate for the length of a pencil?

Ⓐ 6 inches Ⓑ 6 feet Ⓒ 6 yards Ⓓ 60 yards

Name ....

Convert Customary Units of Length

Lesson 2

ESSENTIAL QUESTION
Why do we convert measurements?

You can multiply to **convert**, or change between, units.

Customary Units of Length
1 foot (ft) = 12 inches (in.)
1 yard (yd) = 3 feet (ft)
1 **mile (mi)** = 5,280 feet (ft)

Math in My World

Watch Tutor

Example 1

Marla's dog, Cory, loves to jump into water. Cory's longest jump is 7 yards. How many feet are in 7 yards?

You know the number of yards and want to find the number of feet. Yards are a larger unit than feet. One yard is 3 times as long as one foot. So, use multiplication.

7 yd = ? ft

Multiply by 3 because 1 yard = 3 feet. $7 \times 3 =$ ______

So, there are ______ feet in 7 yards.

Check

You can check by drawing a picture. You know that 3 feet = 1 yard.

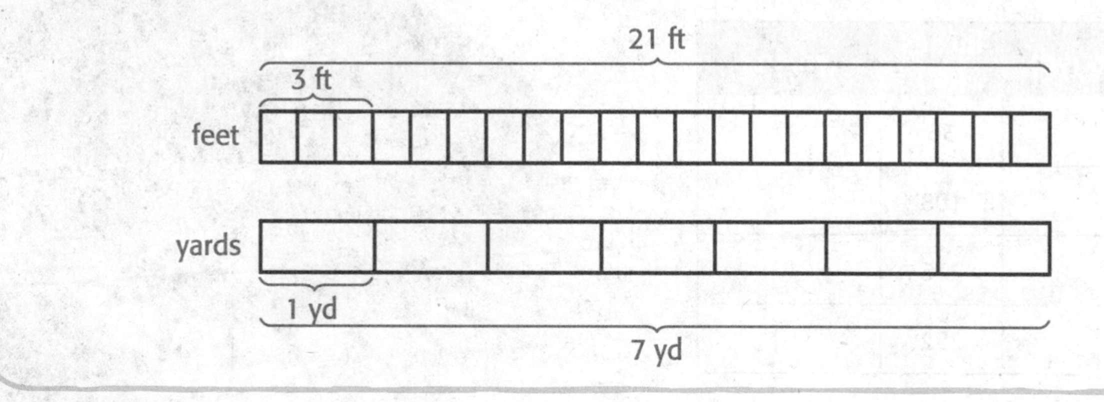

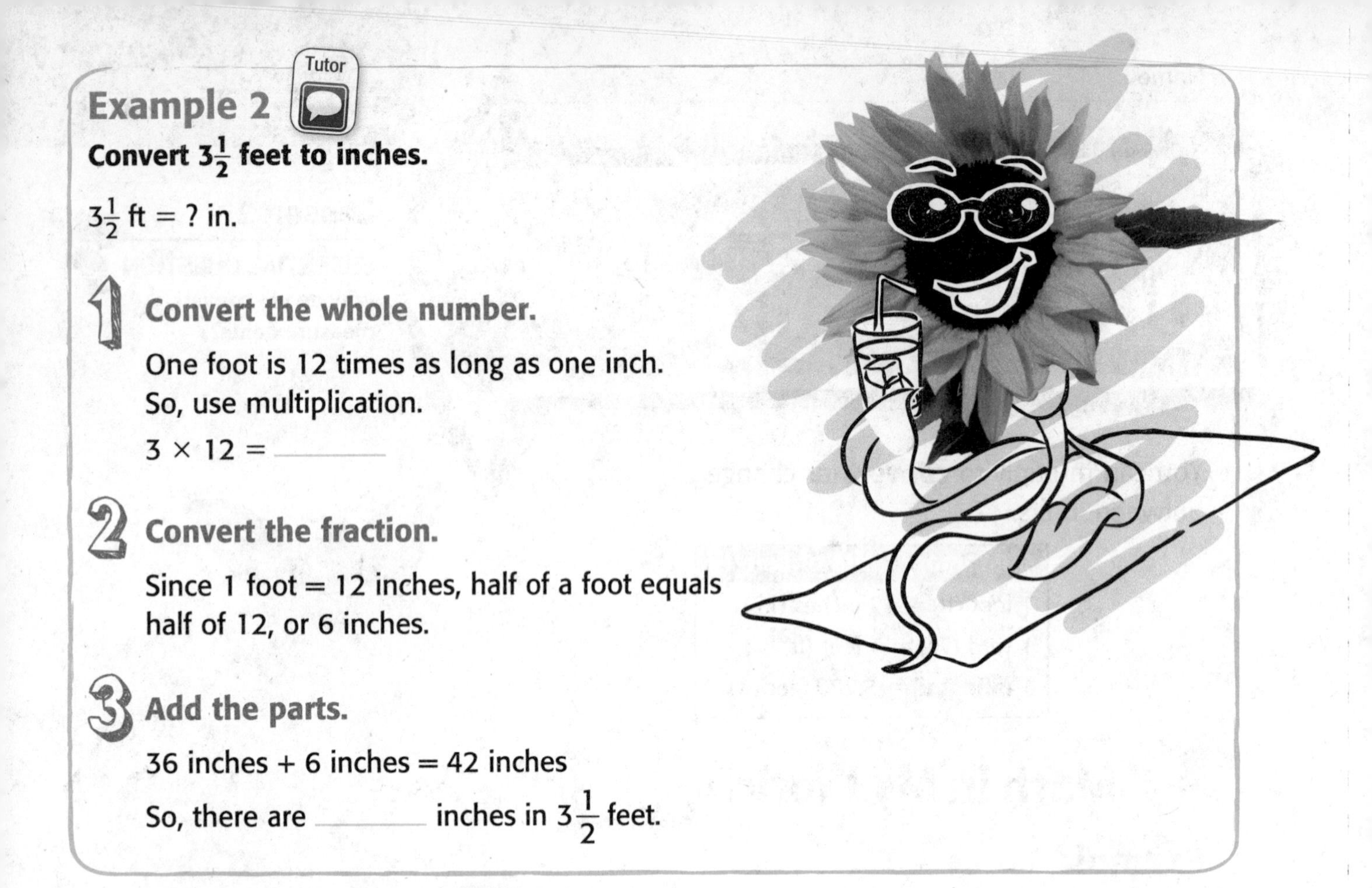

Example 2

Tutor

Convert $3\frac{1}{2}$ feet to inches.

$3\frac{1}{2}$ ft = ? in.

1 Convert the whole number.

One foot is 12 times as long as one inch. So, use multiplication.

$3 \times 12 =$ ______

2 Convert the fraction.

Since 1 foot = 12 inches, half of a foot equals half of 12, or 6 inches.

3 Add the parts.

36 inches + 6 inches = 42 inches

So, there are ______ inches in $3\frac{1}{2}$ feet.

Guided Practice

Check

Complete each conversion table to show measurement equivalents.

1.

inches (in.)	feet (ft)	(in., ft)
	1	(12, 1)
	2	(24, 2)
	3	
	4	

List the number pairs in the last column of the conversion table.

2.

yards (yd)	inches (in.)	(yd, in.)
	36	
	108	
	216	
	252	

Talk MATH

When converting from larger units to smaller units, do you need to multiply or divide?

Name ....

CCSS

Independent Practice

Complete each conversion table.

3.

yards (yd)	feet (ft)	(yd, ft)
2		
3		
4		
5		

4.

yards (yd)	inches (in.)	(yd, in.)
2		
4		
5		
8		

5.

feet (ft)	inches (in.)	(ft, in.)
3		
6		
9		
12		

6.

feet (ft)	miles (mi)	(ft, mi)
	1	
	2	
	3	
	4	

Algebra Find each unknown number.

7. 11 ft = ■ in.

■ = ______

8. 18 yd = ■ ft

■ = ______

9. $2\frac{1}{3}$ yd = ■ ft

■ = ______

10. How many times longer is one yard than one foot? ______

11. How many times longer is one mile than one foot? ______

12. How many times shorter is one inch than one foot? ______

13. How many times shorter is one foot than one yard? ______

Problem Solving

14. Ramiro sits $5\frac{1}{2}$ feet from the bookshelf. Michelle sits 64 inches from the bookshelf. Who sits closer to the bookshelf?

15. The Costa family hiked a trail that was 2 miles in one direction. How many feet was the hike round-trip?

16. Mathematical **PRACTICE 3** **Justify Conclusions** Sumi lives 2 miles from school. Valerie lives 10,542 feet from school. Who lives closer to school? Explain your answer.

HOT Problems

17. Mathematical **PRACTICE 1** **Make Sense of Problems** Darin is 4 feet 10 inches tall. His brother is 68 inches tall. How many inches taller is Darin's brother than Darin?

18. Mathematical **PRACTICE 4** **Model Math** Write a real-world problem involving the conversion of customary lengths. Solve.

19. **Building on the Essential Question** How are yards and feet related?

Name

4.MD.1, 4.MD.2

CCSS

MY Homework

Lesson 2

Convert Customary Units of Length

Homework Helper

Need help? connectED.mcgraw-hill.com

Andrew lives 2 miles from the school. How many feet from the school does Andrew live?

Customary Units of Length
1 foot (ft) = 12 inches (in.)
1 yard (yd) = 3 feet (ft)
1 **mile (mi)** = 5,280 feet (ft)

To convert miles to feet, multiply the number of miles measured by the number of feet in 1 mile. You know that 1 mile = 5,280 feet.

$$\begin{array}{r} \scriptstyle 1 \\ 5{,}280 \\ \times \quad 2 \\ \hline 10{,}560 \end{array} \longrightarrow 2 \text{ miles} = 10{,}560 \text{ feet}$$

So, Andrew lives 10,560 feet from the school.

Conversion table		
miles (mi)	feet (ft)	(mi, ft)
2	10,560	(2, 10,560)
4	21,120	(4, 21,120)
6	31,680	(6, 31,680)
8	42,240	(8, 42,240)

Practice

Convert units to complete each equation.

1. 3 ft = ________ in.

2. $6\frac{1}{2}$ ft = ________ in.

3. 4 yd = ________ in.

4. 760 yd = ________ ft

5. 7 ft = ________ in.

6. 3 yd = ________ in.

7. $2\frac{1}{3}$ yd = ________ ft

8. 5 ft = ________ in.

9. 64 ft = ________ in.

10. ________ in. = 3 yd

11. ________ ft = 7 yd

12. $12\frac{1}{2}$ ft = ________ in.

Draw lines to match the equivalent lengths.

13. 24 ft	• 60 ft
14. 120 in.	• 72 in.
15. 2 yd	• 180 in.
16. 5 yd	• 8 yd
17. 20 yd	• 10 ft

Real World

Problem Solving

18. Mathematical PRACTICE 5 **Use Math Tools** The Millers' house is 25 yards from Mrs. Shapiro's house. How many feet apart are the two houses?

____

19. Kate walks half a mile to the library. How many yards does she walk?

____

20. Pe Ling needs 2 yards of fabric for a craft project. She has 3 feet already. How many inches of fabric does she still need?

____

Vocabulary Check

Vocab

Write a vocabulary term in the correct space to complete each sentence.

convert mile

21. One ______________ equals 5,280 feet.

22. To ______________ from feet to inches, multiply by 12.

Test Practice

23. Matthew is 2 yards tall. How many inches tall is he?

Ⓐ 74 inches Ⓒ 64 inches

Ⓑ 72 inches Ⓓ 60 inches

Name ....

Customary Units of Capacity

Lesson 3

ESSENTIAL QUESTION
Why do we convert measurements?

The amount of liquid a container can hold is its **capacity**. Different containers measure different capacities.

1 fl oz **1 cup** **1 pint** **1 quart** **1 gallon**

Math in My World

Watch Tutor

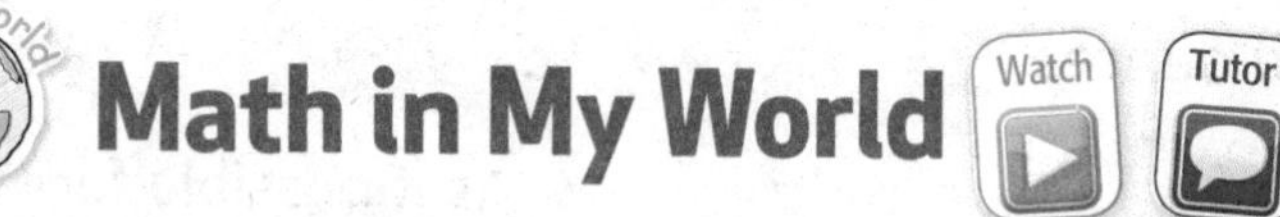

Example 1

Jorge is filling an aquarium. Which container should Jorge use to fill his aquarium most quickly?

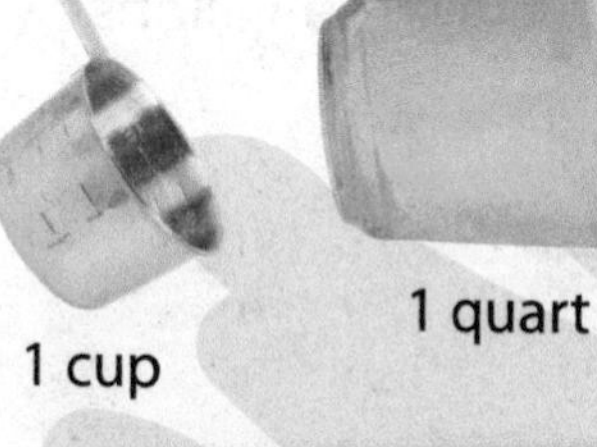

1 cup 1 quart 1 gallon

To fill the aquarium most quickly, Jorge should use the container that will hold the most liquid.

The ______________ is the largest unit. It will fill the aquarium most quickly.

Example 2

Tutor

Nita is pouring salsa into a small bowl. Is the most reasonable estimate for the capacity of the bowl: 8 fluid ounces, 8 cups, 8 quarts, or 8 gallons?

The salsa is a small amount. So, 8 gallons, 8 quarts, and 8 cups are too large.

The most reasonable estimate for the capacity of the bowl is ________.

Guided Practice

Check

Choose the most reasonable estimate for each capacity.

1.

Ⓐ 1 fluid ounce

Ⓑ 1 pint

Ⓒ 1 quart

Ⓓ 100 quarts

2.

Ⓕ 4 fluid ounces

Ⓖ 10 cups

Ⓗ 400 cups

Ⓘ 10 gallons

3.

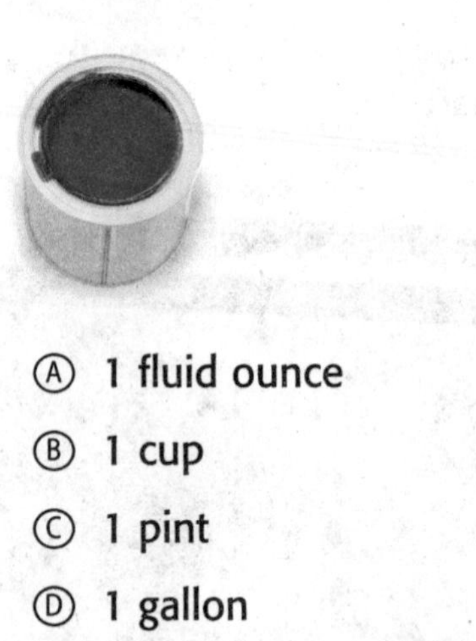

Ⓐ 1 fluid ounce

Ⓑ 1 cup

Ⓒ 1 pint

Ⓓ 1 gallon

Talk MATH

Is it possible for both of us to have a capacity of 1 pint? Explain.

Name ....

Independent Practice

Choose the most reasonable estimate for each capacity.

4. 

Ⓐ 12 fluid ounces
Ⓑ 1 gallon
Ⓒ 12 gallons
Ⓓ 12,000 gallons

5.

Ⓕ 2 fluid ounces
Ⓖ 2 cups
Ⓗ 2 pints
Ⓘ 2 gallons

6. 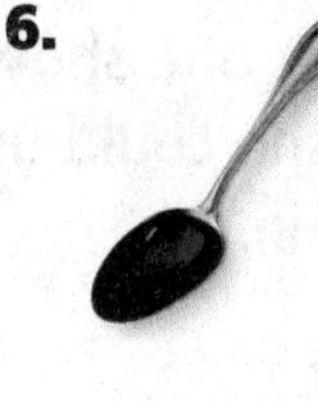

Ⓐ 1 fluid ounce
Ⓑ 1 cup
Ⓒ 1 quart
Ⓓ 1 gallon

7. 

Ⓕ 8 fluid ounces
Ⓖ 8 cups
Ⓗ 8 pints
Ⓘ 8 gallons

8. 

Ⓐ 1 quart
Ⓑ 10 quarts
Ⓒ 100 quarts
Ⓓ 1,000 quarts

9. 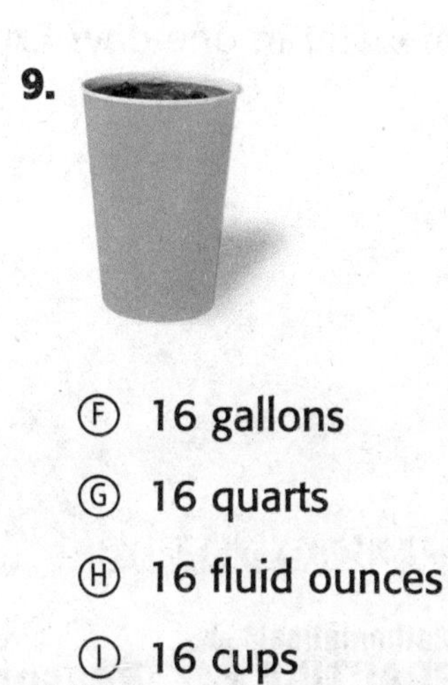

Ⓕ 16 gallons
Ⓖ 16 quarts
Ⓗ 16 fluid ounces
Ⓘ 16 cups

Estimate the capacity of each object.

10. water bottle ____________

11. juice box ____________

12. sink ____________

13. Is it faster to water two large flower pots using a 1-cup pitcher or a 1-quart pitcher? ____________

Some household activities and the amount of water they consume are listed in the table. Use the table for Exercises 14 and 15.

Water Consumption	
Activity	**Water Used (gallons)**
Take shower	15–30
Brush teeth (water running)	1–2
Wash dishes (by hand)	20
Wash dishes (in dishwasher)	9–12
Flush toilet	5–7

My Work!

14. If Callie takes one shower each day, is it reasonable to say that she could use 210 gallons of water in one week? Explain.

15. Mathematical PRACTICE 3 **Draw a Conclusion** Callie brushes her teeth three times each day. She leaves the water running. Is it reasonable to say that she uses 2 cups of water in one day? Explain.

HOT Problems

16. Mathematical PRACTICE 2 **Reason** Name two things in your classroom that can hold more than 1 cup.

17. Mathematical PRACTICE 3 **Find the Error** Alexander estimates the capacity of a small soup can to be about 1 fluid ounce. Find and correct his mistake.

18. **Building on the Essential Question** Why do I measure capacity?

Name ....

MY Homework

Lesson 3
Customary Units of Capacity

Homework Helper

Need help? connectED.mcgraw-hill.com

Ethan is pouring himself a drink of juice. Is it reasonable to say that he will pour about a gallon of juice? Explain.

You know that a large carton of milk is a gallon. That would be way too much for one drink. It is not reasonable to say that Ethan will pour about a gallon of juice.

One cup equals 8 fluid ounces. It would be more reasonable to say that Ethan is pouring one cup, or 8 fluid ounces, of juice.

Practice

Choose the best estimate for each capacity.

1.

Ⓐ 1 fluid ounce
Ⓑ 4 fluid ounces
Ⓒ 1 cup
Ⓓ 4 cups

2.

Ⓕ 6 fluid ounces
Ⓖ 16 fluid ounces
Ⓗ 1 gallon
Ⓘ 6 gallons

3.

Ⓐ 50 gallons
Ⓑ 50 pints
Ⓒ 50 cups
Ⓓ 50 ounces

4.

Ⓕ 8 quarts
Ⓖ 8 pints
Ⓗ 8 cups
Ⓘ 8 fluid ounces

Draw lines to match each object to a reasonable capacity.

5. cereal bowl	• 4 fluid ounces
6. baby food jar	• 2 cups
7. soda can	• 1,000 gallons
8. swimming pool	• 12 fluid ounces
9. large can of paint	• 1 gallon

Problem Solving

10. Zach is filling his dog's bowl with water. Is it reasonable to say that Zach will need about 4 gallons of water?

11. Mathematical PRACTICE 3 **Justify Conclusions** Wes is pouring himself some ketchup to eat with his potato wedges. Is it reasonable to say that Wes will need 2 fluid ounces?

Vocabulary Check

Vocab

12. Explain one relationship between capacity and fluid ounces.

13. Order the following units of capacity from *least* to *greatest:* gallon, pint, cup, quart

Test Practice

14. Henry gargles with mouthwash. Which is a reasonable amount of mouthwash for Henry to use?

Ⓐ 2 fluid ounces

Ⓑ 8 fluid ounces

Ⓒ 12 fluid ounces

Ⓓ 20 fluid ounces

Name ________________

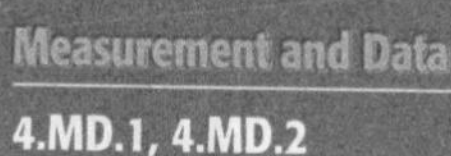

Convert Customary Units of Capacity

Lesson 4

ESSENTIAL QUESTION
Why do we convert measurements?

You can use multiplication to convert units. To change from a larger unit to a smaller unit, multiply.

Customary Units of Capacity	
1 cup (c) = 8 fluid ounces (fl oz)	2 pints (pt) = 1 quart (qt)
2 cups (c) = 1 pint (pt)	4 quarts (qt) = 1 gallon (gal)

Math in My World

Example 1

Marcus has a 2-gallon container of laundry detergent. How many quarts of laundry detergent does he have? How many pints of laundry detergent does he have?

1 Find the number of quarts that are in 2 gallons.

Since quarts are smaller than gallons, multiply. Multiply by 4 because there are 4 quarts in each gallon.

$2 \times 4 =$ ______

So, there are ______ quarts in 2 gallons.

2 Find the number of pints that are in 8 quarts.

Multiply 8 by 2 because there are 2 pints in each quart.

$8 \times 2 =$ ______

So, there are ______ pints in 8 quarts.

Marcus has ______ quarts, or ______ pints, of laundry detergent.

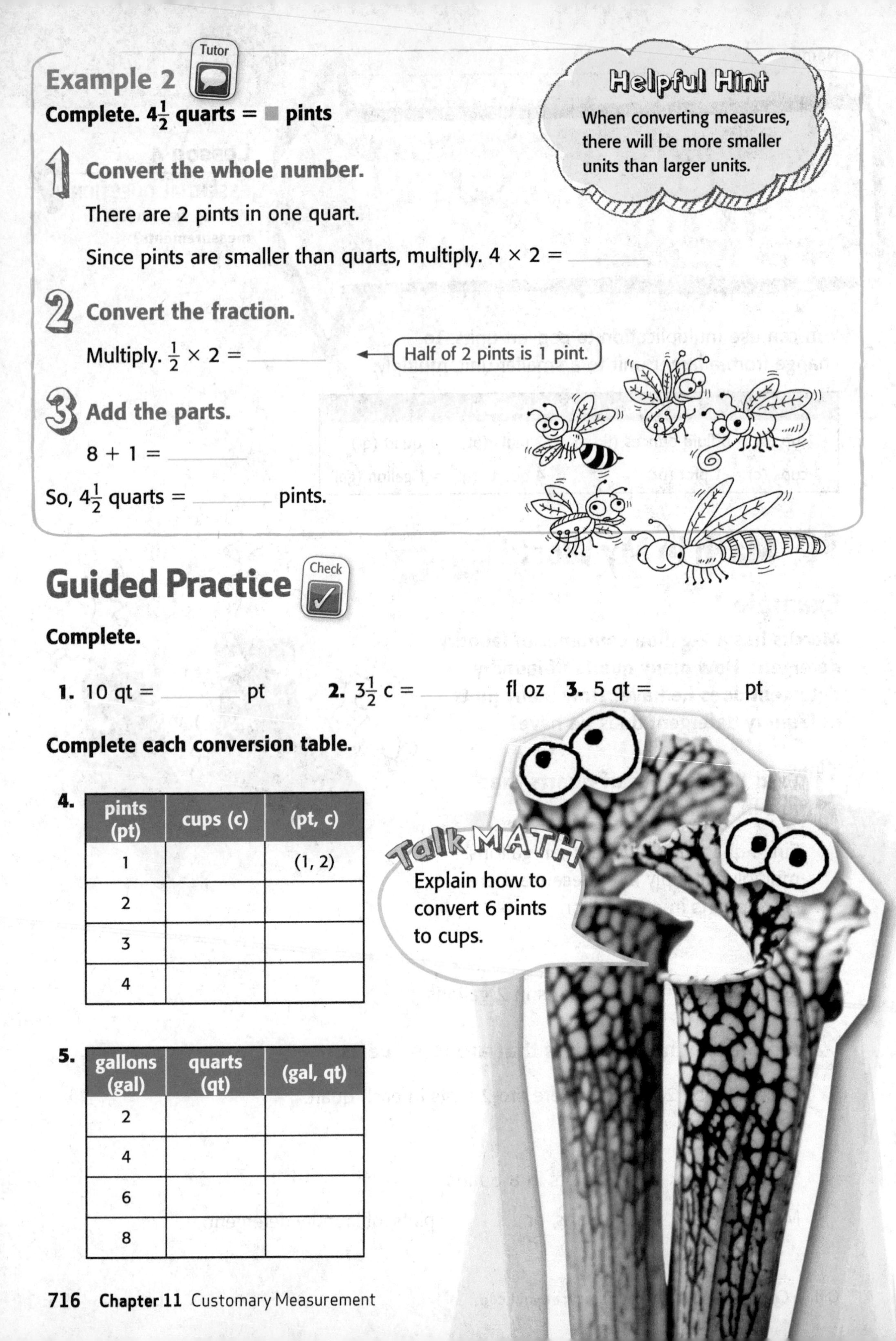

Example 2

Tutor

Complete. $4\frac{1}{2}$ **quarts = ■ pints**

Helpful Hint

When converting measures, there will be more smaller units than larger units.

1 Convert the whole number.

There are 2 pints in one quart.

Since pints are smaller than quarts, multiply. $4 \times 2 =$ ______

2 Convert the fraction.

Multiply. $\frac{1}{2} \times 2 =$ ______ ← Half of 2 pints is 1 pint.

3 Add the parts.

$8 + 1 =$ ______

So, $4\frac{1}{2}$ quarts = ______ pints.

Guided Practice

Check

Complete.

1. 10 qt = ______ pt **2.** $3\frac{1}{2}$ c = ______ fl oz **3.** 5 qt = ______ pt

Complete each conversion table.

4.

pints (pt)	cups (c)	(pt, c)
1		(1, 2)
2		
3		
4		

5.

gallons (gal)	quarts (qt)	(gal, qt)
2		
4		
6		
8		

Talk MATH

Explain how to convert 6 pints to cups.

Name

Independent Practice

Complete each conversion table.

6.

quarts (qt)	pints (pt)	(qt, pt)
1		
2		
3		
4		

7.

pints (pt)	cups (c)	(pt, c)
5		
7		
9		
11		

Algebra Find each unknown number.

8. 8 c = ■ fl oz

■ = ______

9. $6\frac{1}{2}$ gal = ■ qt

■ = ______

10. ■ qt = 5 gal

■ = ______

11. $5\frac{1}{2}$ c = ■ fl oz

■ = ______

12. ■ c = 15 pt

■ = ______

13. 16 c = ■ fl oz

■ = ______

Compare. Use >, <, or =.

14. 4 qt ◯ 10 pt

15. 10 gal ◯ 1,280 fl oz

16. 1 qt ◯ 2 c

17. 1 gal ◯ 16 c

18. 5 qt ◯ 25 c

19. 12 fl oz ◯ 2 c

20. How many times greater is the capacity of one gallon than one quart? ______

21. How many times greater is the capacity of one cup than one fluid ounce? ______

Problem Solving

22. Lucia is making 2 gallons of soup. How many cups of soup is Lucia making?

23. Mathematical PRACTICE 5 **Use Math Tools** Tomas is buying a 2-cup container of liquid dish soap. How many fluid ounces of dish soap is he buying?

24. Danielle is using 2 quarts of water in a recipe. How many cups of water is she using?

25. Karen is buying 4 gallons of orange juice. How many quarts of orange juice is she buying?

HOT Problems

26. Mathematical PRACTICE 3 **Which One Doesn't Belong?** Circle the measurement that does not belong with the other three. Explain your reasoning.

4 pints	2 quarts	8 cups	1 gallon

27. **Building on the Essential Question** How are gallons and fluid ounces related?

Name ....

MY Homework

Lesson 4
Convert Customary Units of Capacity

Homework Helper

eHelp

Need help? connectED.mcgraw-hill.com

Chip needs 1 quart of cream for his ice cream recipe. He has 3 cups of cream. Is that enough for his recipe?

First, convert quarts to pints.

1 quart = 2 pints

Next, convert pints to cups.

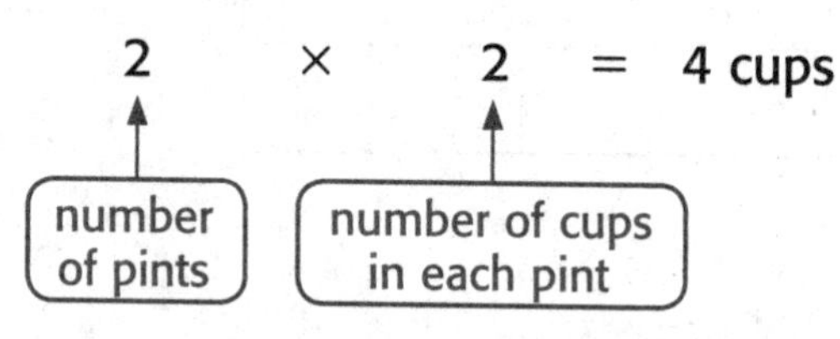

Conversion Table Customary Units of Capacity
1 cup (c) = 8 fluid ounces (fl oz)
2 cups (c) = 1 pint (pt)
2 pints (pt) = 1 quart (qt)
4 quarts (qt) = 1 gallon (gal)

Chip has 3 cups of cream, but he needs 4 cups. He does not have enough cream for the recipe.

Practice

Convert units to complete each equation.

1. 9 pt = ________ c

2. 4 qt = ________ pt

3. 3 c = ________ fl oz

4. $9\frac{1}{2}$ gal = ________ qt

5. $5\frac{1}{2}$ pt = ________ c

6. 6 gal = ________ qt

Complete each conversion table.

7.

cups (c)	fluid ounces (fl oz)	(c, fl oz)
1		
3		
5		
7		

8.

quarts (qt)	pints (pt)	(qt, pt)
1		
2		
3		
4		

9.

gallons (gal)	quarts (qt)	(gal, qt)
4		
5		
6		
7		

10.

gallons (gal)	pints (pt)	(gal, pt)
4		
5		
6		
7		

Problem Solving

11. Mathematical PRACTICE 2 **Reason** Elle can buy 2 quarts of milk for $3 or 1 gallon of milk for $3. Which is the better deal? Explain.

12. Marco needs 2 gallons of paint for his living room. He has 10 quarts of paint. Is that enough? Explain.

13. Ollie drank $\frac{3}{4}$ of a 16-fluid ounce bottle of juice. How many fluid ounces of juice are left in the bottle?

Test Practice

14. Craig drank two 5-cup bottles of water in one day. How many fluid ounces is that?

Ⓐ 10 fluid ounces
Ⓑ 16 fluid ounces
Ⓒ 40 fluid ounces
Ⓓ 80 fluid ounces

Check My Progress

Vocabulary Check

Vocab

Circle the word(s) that completes each sentence.

1. Foot, cup, and gallon are all units in the (**customary system** **capacity**).
2. There are 2 (**quarts** **cups**) in a pint.
3. In order to (**convert** **pint**) yards to feet, multiply the number of yards by 3.
4. Quart and gallon are units of (**capacity** **yards**).
5. There are 5,280 feet in 1 (**yard** **mile**).

Draw lines to match the measurements that are equal.

6. 1 **foot**	• 16 **quarts**
7. 32 **fluid ounces**	• 12 **inches**
8. 4 **gallons**	• 2 **pints**

Concept Check

Check

Complete each conversion table.

9.

yards (yd)	feet (ft)	(yd, ft)
1		
3		
5		
7		

10.

quarts (qt)	cups (c)	(qt, c)
2		
3		
4		
5		

Problem Solving

11. Mateo is making 2 gallons of fruit punch. How many pints of fruit punch is he making?

____

12. Gwenith has 3 gallons of milk. How many quarts of milk does she have?

____

13. Boa constrictors can grow to 13 feet long. How many inches is this?

____

My Work!

Test Practice

14. Which of the following holds about 1 quart of water?

Ⓐ 

Ⓑ

Ⓒ 

Ⓓ

Customary Units of Weight

Lesson 5

ESSENTIAL QUESTION
Why do we convert measurements?

The **weight** of an object is how heavy it is. The customary units of weight are ounce (oz), pound (lb), and ton (T).

1 ounce (oz) **1 pound (lb)** **1 ton (T)**

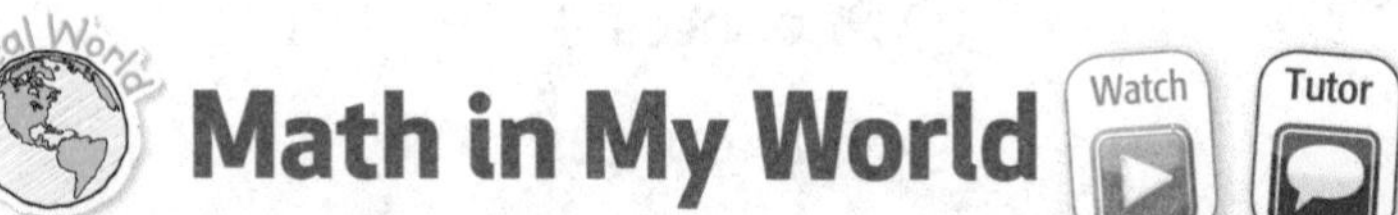

Math in My World

Example 1

Suzie's father bought some sugar for their favorite recipe. Which is a more reasonable unit to use for the weight of a bag of sugar, ounces or pounds?

A small packet of sugar would be weighed in ounces.

A bag of sugar is much larger and would be weighed in pounds.

So, __________ is a reasonable unit to use for the weight of a bag of sugar.

Example 2

Tutor

Which is the most reasonable estimate for the weight of a leaf: 1 ounce, 1 pound, 1 ton, or 10 tons?

Compare the weight of a leaf to the weight of objects that you know.

A leaf weighs less than a pineapple or 1 pound.

Objects that weigh less than 1 pound are weighed in ounces. The only option that contains ounces is 1 ounce.

So, a leaf weighs about ________.

Guided Practice

Check

Choose the best estimate for the weight of each object.

1. paper airplane

Ⓐ 4 ounces

Ⓑ 40 ounces

Ⓒ 4 pounds

Ⓓ 4 tons

2. helicopter

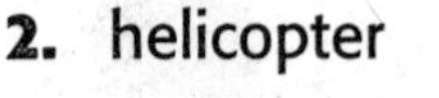

Ⓕ 5 ounces

Ⓖ 500 ounces

Ⓗ 5 tons

Ⓘ 500 tons

3. rabbit

Ⓐ 4 ounces

Ⓑ 4 pounds

Ⓒ 40 pounds

Ⓓ 4 tons

Talk MATH

Does an object that is small always weigh less than an object that is large? Explain.

Name ....

Independent Practice

Choose the best estimate for the weight of each object.

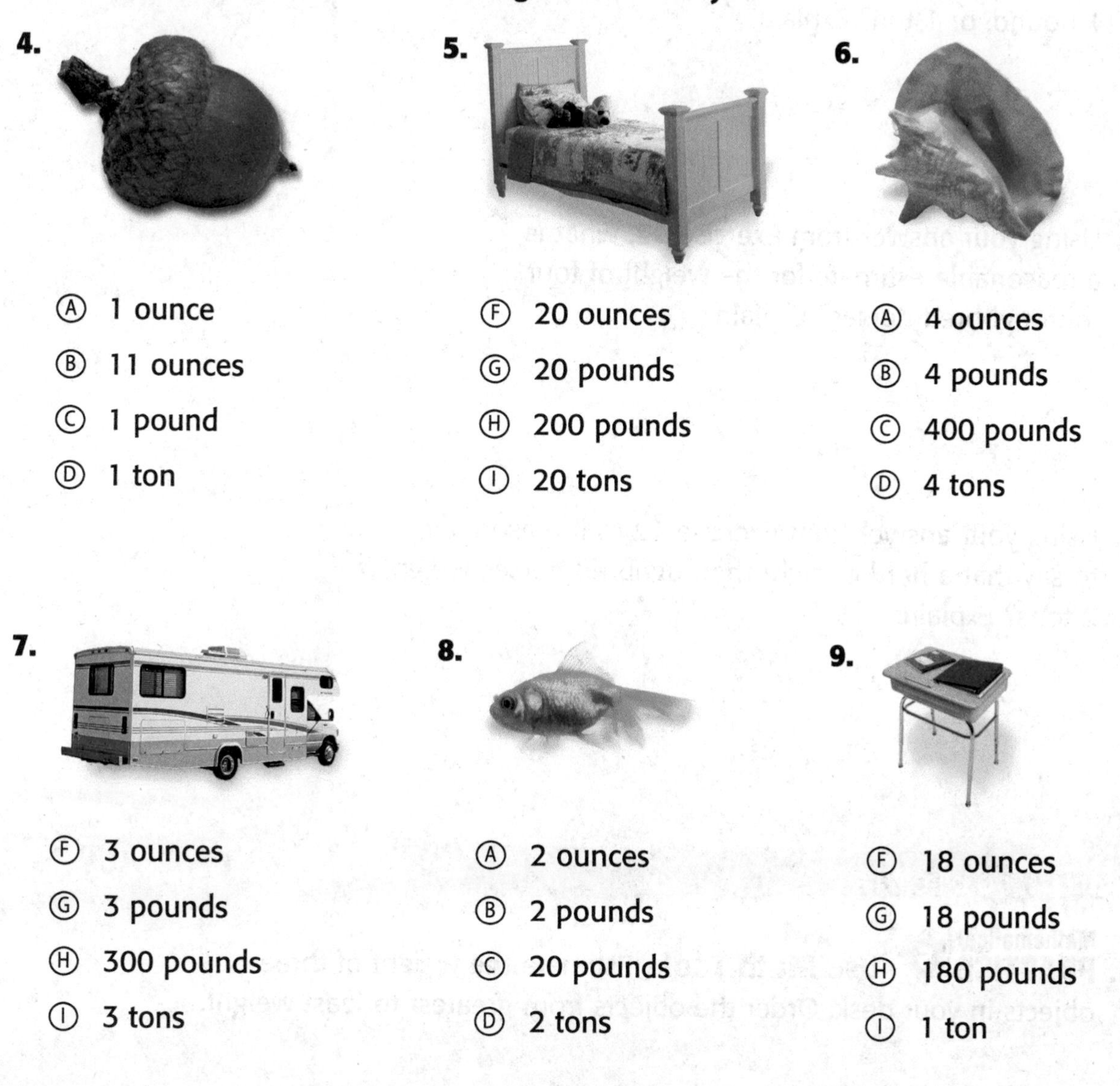

4.

Ⓐ 1 ounce
Ⓑ 11 ounces
Ⓒ 1 pound
Ⓓ 1 ton

5.

Ⓕ 20 ounces
Ⓖ 20 pounds
Ⓗ 200 pounds
Ⓘ 20 tons

6.

Ⓐ 4 ounces
Ⓑ 4 pounds
Ⓒ 400 pounds
Ⓓ 4 tons

7.

Ⓕ 3 ounces
Ⓖ 3 pounds
Ⓗ 300 pounds
Ⓘ 3 tons

8.

Ⓐ 2 ounces
Ⓑ 2 pounds
Ⓒ 20 pounds
Ⓓ 2 tons

9.

Ⓕ 18 ounces
Ⓖ 18 pounds
Ⓗ 180 pounds
Ⓘ 1 ton

10. Is it more reasonable to say that a pair of shoes weighs 1 ounce, 1 pound, or 1 ton?

____

11. Is it more reasonable to say that a pencil weighs 2 ounces, 2 pounds, or 2 tons?

____

Problem Solving

12. Mathematical PRACTICE 6 **Be Precise** Which is most reasonable for the weight of two thoroughbred horses: 1 ounce, 1 pound, or 1 ton? Explain.

My Work!

13. Using your answer from Exercise 12, what is a reasonable estimate for the weight of four thoroughbred horses? Explain.

14. Using your answer from Exercise 12, is it reasonable to say that a herd of eight thoroughbred horses weighs 2 tons? Explain.

HOT Problems

15. Mathematical PRACTICE 5 **Use Math Tools** Estimate the weight of three objects in your desk. Order the objects from greatest to least weight.

16. **Building on the Essential Question** How do you estimate weight?

Name ....

MY Homework

Lesson 5
Customary Units of Weight

Homework Helper

eHelp Need help? connectED.mcgraw-hill.com

Which is the best estimate for the weight of Ellie's little dog: 5 ounces, 5 pounds, or 5 tons?

You know that objects that weigh less than 1 pound are measured in ounces. It is not likely that the dog weighs less than 1 pound. Objects that are quite heavy are weighed in tons. The dog is not extremely heavy.

So, the best estimate for the weight of Ellie's dog is 5 pounds.

Practice

Choose the best estimate for the weight of each object.

1. 

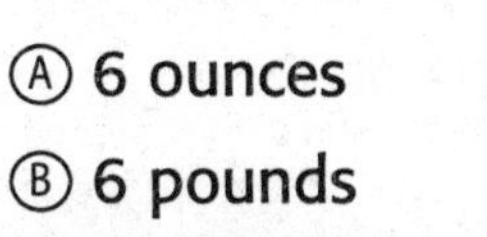

Ⓐ 6 ounces
Ⓑ 6 pounds
Ⓒ 60 pounds
Ⓓ 6 tons

2. 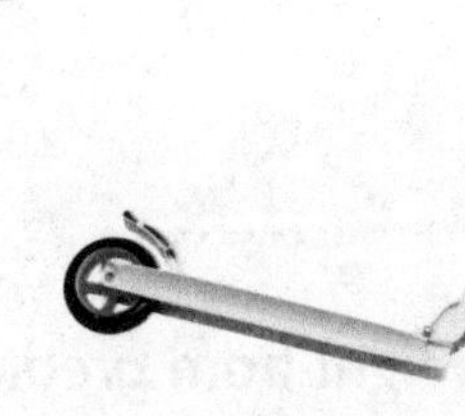

Ⓕ 70 pounds
Ⓖ 7 pounds
Ⓗ 70 ounces
Ⓘ 7 ounces

3. 

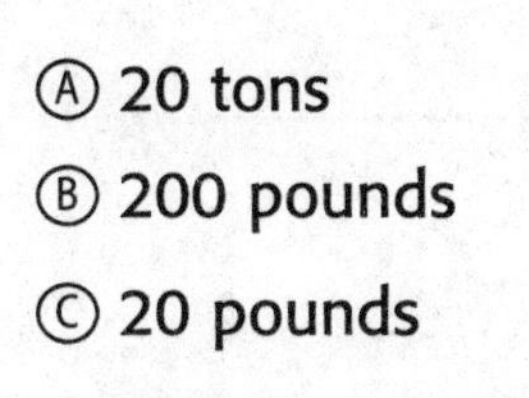

Ⓐ 20 tons
Ⓑ 200 pounds
Ⓒ 20 pounds
Ⓓ 2,000 ounces

4. 

Ⓕ 8 ounces
Ⓖ 80 ounces
Ⓗ 8 pounds
Ⓘ 80 pounds

Draw lines to match each object to its reasonable weight.

5. textbook	• 2 ounces
6. whale	• 40 tons
7. golf ball	• 1 pound
8. car	• 5 pounds
9. can of soup	• 1.5 tons

Real World Problem Solving

10. Mathematical PRACTICE 1 **Check for Reasonableness** Jesse claims that his goldfish weighs about 4 pounds. Is Jesse's claim reasonable? Explain.

____

11. Tasha claims that her skateboard weighs about 3 pounds. Is Tasha's claim reasonable? Explain.

____

Vocabulary Check

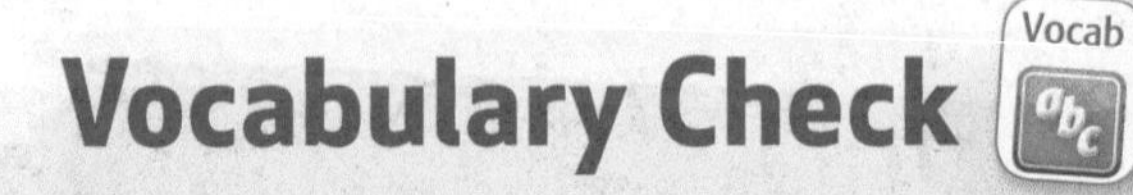

12. Write a definition for the term weight.

____

____

13. Order the units of weight from *greatest* to *least*: pound (lb), ounce (oz), ton (T).

____

Test Practice

14. Which animal's weight is closest to 1 pound?

Ⓐ an ant
Ⓑ a lion
Ⓒ a mosquito
Ⓓ a robin

Name ....

Convert Customary Units of Weight

Lesson 6

ESSENTIAL QUESTION
Why do we convert measurements?

Recall that to convert from a larger unit to a smaller unit, multiply.

1 pound (lb) = 16 ounces (oz)

Multiply by 16.

Math in My World

Example 1

Leigh needs to buy 4 pounds of vegetable burger patties. The table shows the number of ounces in each package of patties. How many packages will Leigh need to buy?

Vegetable Burger Patties	
Packages	**Number of Ounces**
1	32
2	64
3	96
4	128

Convert 4 pounds to ounces.

Because pounds are larger than ounces, multiply.

There are 16 ounces in a pound, so multiply 4 by 16.

$4 \times 16 =$ ______

4 pounds = ______ ounces

So, Leigh will need to buy ______ package(s).

$$\begin{array}{r} \square \\ 16 \\ \times\ 4 \\ \hline \square\square \end{array}$$

A ton is a very heavy unit of measurement. There are 2,000 pounds in 1 ton.

Example 2

Use the table to find how many pounds a Stegosaurus weighed.

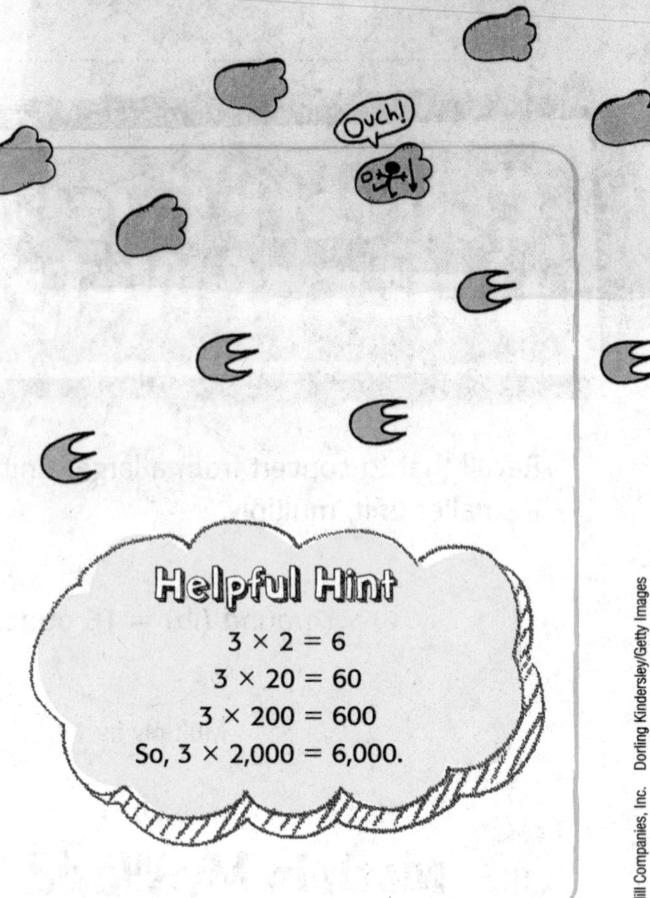

Dinosaur Weights	
Dinosaur	**Weight (tons)**
Allosaurus	2
Megalosaurus	1
Stegosaurus	3
Supersaurus	60
Tyrannosaurus	8

To find the weight of a Stegosaurus in pounds, multiply the number of tons by 2,000.

$3 \times 2{,}000 =$ ______

So, a Stegosaurus weighed ______ pounds.

Guided Practice

Complete each conversion table.

1.

pounds (lb)	ounces (oz)	(lb, oz)
2		(2, 32)
3		
4		
5		

2.

tons (T)	pounds (lb)	(T, lb)
2		
4		
6		
8		

Talk MATH

Explain why you multiply to convert a larger unit of measure to a smaller unit of measure.

Name ____________

CCSS

Independent Practice

Complete each conversion table.

3.

pounds (lb)	ounces (oz)	(lb, oz)
6		
8		
10		
12		

4.

tons (T)	pounds (lb)	(T, lb)
2		
3		
5		
7		

Algebra Find each unknown number.

5. 3 T = ■ lb

■ = ______

6. ■ oz = 2 lb

■ = ______

7. ■ lb = 6 T

■ = ______

8. $4\frac{1}{2}$ lb = ■ oz

■ = ______

9. ■ oz = 10 lb

■ = ______

10. 4 T = ■ lb

■ = ______

11. 5 lb = ■ oz

■ = ______

12. ■ lb = 3 T

■ = ______

13. ■ oz = $11\frac{1}{4}$ lb

■ = ______

14. Circle the table that represents the correct relationship between pounds and ounces.

Pounds	1	2	3	4
Ounces	16	32	48	64

Pounds	1	2	3	4
Ounces	16	24	32	40

Pounds	1	2	3	4
Ounces	8	16	24	32

Pounds	16	32	48	64
Ounces	1	2	3	4

Problem Solving

Baby animals weigh different amounts at birth.

Animal	Birth Weight
Alligator	2 ounces
Giraffe	100–150 pounds
Giant panda	3–5 ounces
Walrus	100–160 pounds

My Work!

15. What is the least weight of a baby giraffe in ounces?

16. What is the greatest weight of a baby walrus in ounces?

17. Mathematical PRACTICE 4 **Model Math** If a baby walrus weighs 100 pounds, how many more ounces does it weigh than a baby alligator?

HOT Problems

18. Mathematical PRACTICE 1 **Plan Your Solution** Tiffany weighed 7 pounds 12 ounces when she was born. Her weight doubled after four months. How much did Tiffany weigh after four months?

19. **Building on the Essential Question** How can I convert measurements of weight?

MY Homework

Lesson 6
Convert Customary Units of Weight

Homework Helper

eHelp Need help? connectED.mcgraw-hill.com

The world's heaviest apple weighed 65 ounces. Did this apple weigh more than 4 pounds?

To answer this question, you need to convert one unit of weight to match the other.

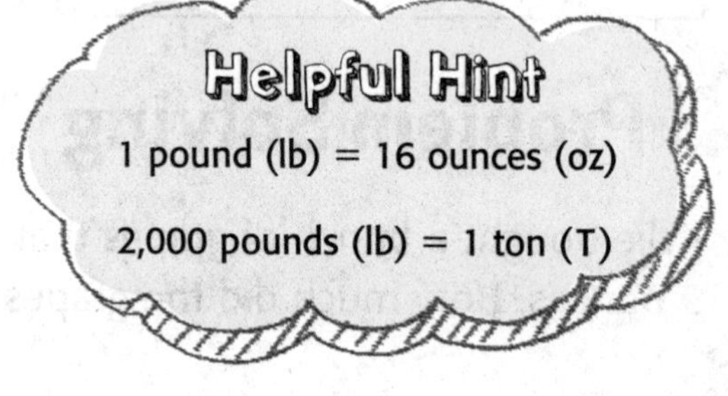

Convert pounds to ounces.

To convert from a larger unit to a smaller unit, you need to multiply.

4 pounds × 16 ounces = 64 ounces

Since 65 > 64, the 65-ounce apple did weigh more than 4 pounds.

Use the conversion table to check.

65 ounces is between 4 and 5 pounds.

pounds (lb)	ounces (oz)	(lb, oz)
4	64	(4, 64)
5	80	(5, 80)
6	96	(6, 96)
7	112	(7, 112)

Practice

Convert units to complete each equation.

1. 3 lb = ________ oz

2. 3 T = ________ lb

3. ________ oz = 10 lb

4. 9 T = ________ lb

Convert units to complete each equation.

5. $\frac{1}{2}$ ton = ________ pounds

6. $\frac{1}{4}$ pound = ________ ounces

Complete each conversion table.

7.

pounds (lb)	ounces (oz)	(lb, oz)
5		
7		
9		
11		

8.

tons (T)	pounds (lb)	(T, lb)
4		
8		
12		
16		

Problem Solving

9. Kylie bought a bunch of grapes that weighed 2 pounds and 9 ounces. How much did the grapes weigh in ounces only?

____

10. Brent's pick-up truck can carry $\frac{1}{5}$ ton. Nina's pick-up truck can carry 500 pounds. Whose pick-up truck can carry more weight? Explain.

____

____

____

11. Mathematical PRACTICE 2 **Use Number Sense** Charlotte's family bought a 2-pound block of cheese. There is $\frac{1}{4}$ of the block of cheese left. How many ounces of cheese has Charlotte's family eaten?

____

Test Practice

12. Which weight is equal to 3 tons?

Ⓐ 2,000 pounds

Ⓑ 3,000 pounds

Ⓒ 5,000 pounds

Ⓓ 6,000 pounds

Name ....

Convert Units of Time

Lesson 7

ESSENTIAL QUESTION
Why do we convert measurements?

The steps that you use to convert units of length, weight, and capacity can be used to convert units of time.

Math in My World

Watch Tutor

Example 1

Cooper watched a butterfly in his garden for 5 minutes. How many seconds did he watch the butterfly?

1 minute = 60 seconds

To convert larger units of time to smaller units of time, multiply.

Helpful Hint
$5 \times 6 = 30$
$5 \times 60 = 300$

$5 \times 60 =$ ______

So, Cooper watched a butterfly for ______ seconds.

Key Concept Units of Time

1 minute (min) = 60 **seconds** (s)
1 hour (h) = 60 min
1 day (d) = 24 h
1 week (wk) = 7 d
1 year (y) = 52 wk = 12 months (mo)

Example 2

Audrey is going to watch a movie at the movie theater. The movie starts at 3:30 P.M. and ends at 5:37 P.M. How long is the movie in minutes?

Find the time interval, or the length of time from the start of the movie to the end of the movie.

The time interval between 3:30 and 5:37 is ______ hours and 7 minutes.

Count the whole hours.
From 3:30 to 5:30 is 2 hours.

Count the remaining minutes.
From 5:30 to 5:37 is 7 minutes.

Since 1 hour = 60 minutes, multiply 60 by 2.

$60 \times 2 =$ ______ $120 + 7 =$ ______

So, the movie is ______ minutes long.

Guided Practice 

Complete each conversion table.

1.

minutes (min)	seconds (s)	(min, s)
1		(1, 60)
2		
3		
4		

2.

hours (h)	days (d)	(h, d)
	1	
	2	
	3	
	4	

3.

years (y)	weeks (wk)	(y, wk)
1		
2		
3		
4		

Talk MATH

What operation would you use to find the number of minutes in 2 hours? Explain.

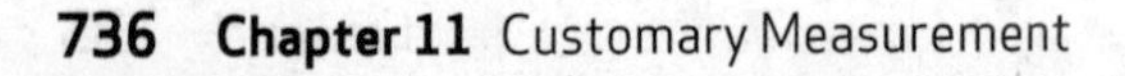

Name

Independent Practice

Complete each conversion table.

4.

days (d)	weeks (wk)	(d, wk)
	2	
	4	
	6	
	8	

5.

hours (h)	minutes (min)	(h, min)
3		
4		
5		
6		

6.

years (y)	months (mo)	(y, mo)
1		
3		
5		
7		

7.

weeks (wk)	years (y)	(wk, y)
	5	
	10	
	15	
	20	

Compare. Use <, >, or =.

8. 5 min ◯ 250 s

9. 36 mo ◯ 4 y

10. 300 s ◯ 1 h

11. How many times greater is one minute than one second? ______

12. How many times greater is one hour than one minute? ______

13. How many times greater is one week than one day? ______

14. How many minutes are in the time interval from 1:22 P.M. to 5:44 P.M.? ______

15. How many minutes are in the time interval from 7:09 A.M. to 10:36 A.M.? ______

Problem Solving

16. Aubrey played on the playground for $2\frac{1}{2}$ hours. How many minutes did she play?

______

17. Alex had to rent the pavilion at the park for a minimum of 3 hours. How many minutes is that?

______

18. Mathematical PRACTICE 5 **Use Math Tools** Justin painted his fence from 8:00 A.M. to 11:47 A.M. For how many minutes did he paint the fence?

______

19. It took Aiden 20 minutes to walk to school. It took Callie 900 seconds to walk to school. Who took less time to walk to school?

______

My Work!

HOT Problems

20. Mathematical PRACTICE 3 **Find the Error** Sadie wrote the following on the board. Find and correct her mistake.

2 years = 24 weeks

______

______

______

21. **Building on the Essential Question** How does multiplication relate to time conversions?

______

______

______

Name ....

MY Homework

Lesson 7
Convert Units of Time

Homework Helper

eHelp

Need help? connectED.mcgraw-hill.com

Chloe is celebrating her 3rd birthday. How many weeks old is she?

To convert a larger unit (years) to a smaller unit (weeks), you need to multiply.

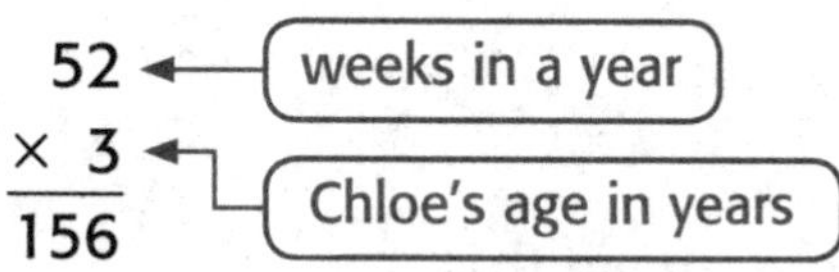

So, Chloe is 156 weeks old.

Conversion Chart Units of Time
1 minute (min) = 60 seconds (s)
1 hour (h) = 60 min
1 day (d) = 24 h
1 week (wk) = 7 d
1 year (y) = 52 wk = 12 months (mo)

Practice

Convert units to complete each equation.

1. 3 minutes = ______ seconds

2. 5 weeks = ______ days

3. ______ months = 5 years

4. ______ minutes = 6 hours

5. 4 days = ______ hours

6. ______ weeks = 8 years

7. $1\frac{1}{2}$ days = ______ hours

8. $3\frac{2}{7}$ weeks = ______ days

9. ______ months = $4\frac{3}{4}$ years

10. ______ minutes = 6 hours and 42 minutes

Complete each conversion table.

11.

weeks (wk)	days (d)	(wk, d)
2		
4		
6		
8		

12.

minutes (min)	hours (h)	(min, h)
	9	
	7	
	5	
	3	

Problem Solving

13. Emma is $9\frac{1}{4}$ years old. How many months old is Emma?

____

14. Mathematical PRACTICE 4 **Model Math** Vincent is watching a movie that lasts 1 hour and 37 minutes. He has watched 52 minutes so far. How many minutes are left in the movie?

____

15. Lexie started her homework at 4:30 P.M. She finished at 5:05 P.M. How many seconds did it take her to finish her homework?

____

Vocabulary Check

Vocab

16. How many seconds are in 1 minute? ______________

Test Practice

17. Cameron's book log shows that he read a total of $4\frac{1}{4}$ hours last month. How many minutes is that?

Ⓐ 240 minutes

Ⓑ 250 minutes

Ⓒ 255 minutes

Ⓓ 270 minutes

Check My Progress

Vocabulary Check

1. Choose a vocabulary word to complete each sentence.

ounce **pound** **second**
ton **weight**

A rhinoceros weighs about 2 ______.

A mouse weighs about 1 ______.

A squirrel weighs about 3 ______.

The ______ of an object describes how heavy it is.

There are 60 ______ in a minute.

Concept Check

Complete each conversion table.

2.

ounces (oz)	pounds (lb)	(oz, lb)
	1	
	3	
	5	
	7	

3.

hours (h)	minutes (min)	(h, min)
2		
4		
6		
8		

Problem Solving

4. Is it more reasonable to say that a pen weighs 2 ounces, 2 pounds, or 2 tons?

5. A hippopotamus eats 100 pounds of food a day. How many days would it take the hippo to eat 1 ton of food?

6. An ostrich egg weighs 64 ounces. Is the weight of the ostrich egg greater than 5 pounds? Explain.

7. Jordan spent 3 hours at a local farm. How many minutes did he spend at the farm?

My Work!

Test Practice

8. Which is the most reasonable estimate for the weight of a bowling ball?

Ⓐ 8 tons

Ⓑ 80 pounds

Ⓒ 8 pounds

Ⓓ 8 ounces

Name ....

Display Measurement Data in a Line Plot

Lesson 8

ESSENTIAL QUESTION
Why do we convert measurements?

You can represent measurement data for fractions of a unit in a **line plot**. The line plot's number line will look like a ruler.

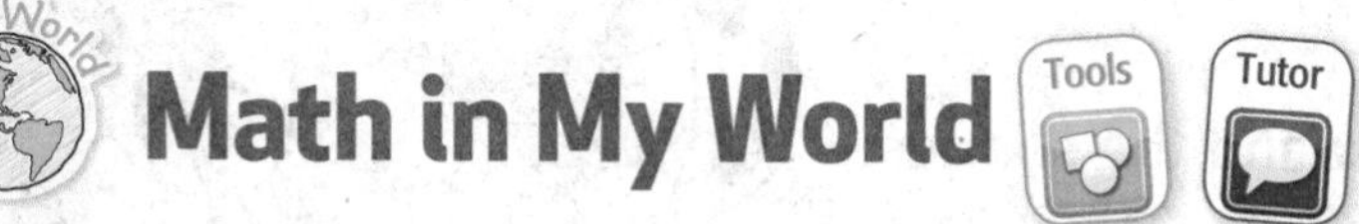

Example 1

The Science room has a collection of bugs. Each bug's length is measured to the nearest eighth of an inch. Make a line plot to represent the data.

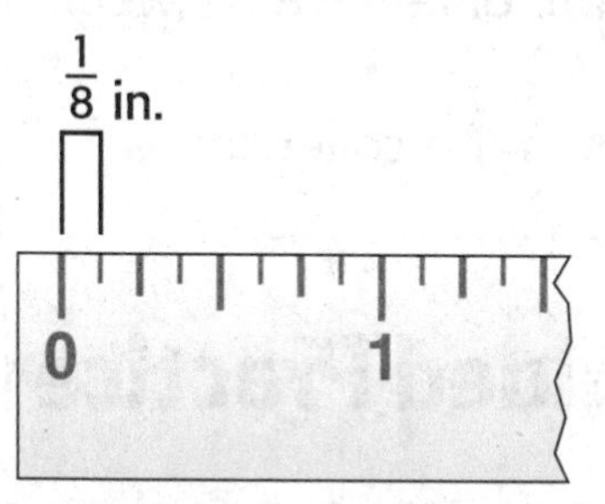

The measurement data are shown in the tally chart.

Bug Length	
$\frac{1}{8}$ in.	
$\frac{2}{8}$ in.	||
$\frac{3}{8}$ in.	|
$\frac{4}{8}$ in.	||
$\frac{5}{8}$ in.	~~||||~~ |
$\frac{6}{8}$ in.	~~||||~~
$\frac{7}{8}$ in.	||||
$\frac{8}{8}$ in.	|

1 First, make a number line to represent the value of each bug's length.

0 $\frac{1}{8}$ $\frac{2}{8}$ $\frac{3}{8}$ $\frac{4}{8}$ $\frac{5}{8}$ $\frac{6}{8}$ $\frac{7}{8}$ $\frac{8}{8}$ or 1

2 Next, place an X above each measurement every time that value occurred.

Bug Length (in.)

0	$\frac{1}{8}$	$\frac{2}{8}$	$\frac{3}{8}$	$\frac{4}{8}$	$\frac{5}{8}$	$\frac{6}{8}$	$\frac{7}{8}$	$\frac{8}{8}$ or 1
		X X	X	X X	X X X X X X	X X X X X	X X X X	X

Example 2

Refer to Example 1. Find the difference in length between the longest and shortest bug.

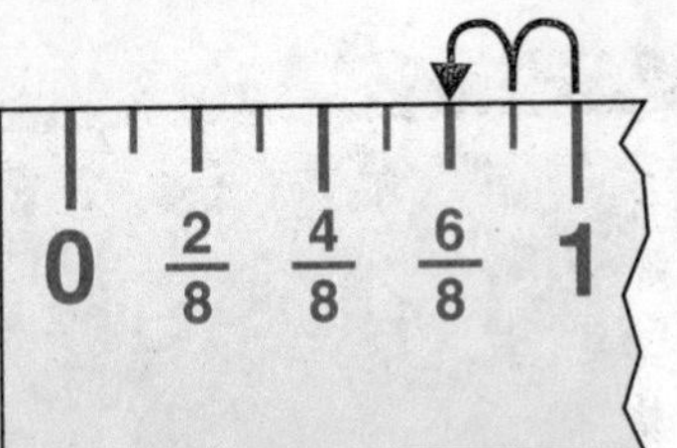

To find the difference between the longest and shortest bugs, subtract the shortest length from the longest length.

Subtract like fractions.

$$\frac{8}{8} - \frac{2}{8} = \frac{8-2}{8} = \frac{6}{8}$$

$$= \frac{3}{4}$$

So, the difference between the longest and the shortest bugs in the collection is ______ of an inch.

Guided Practice

Check

For Exercises 1–2, use the tally chart shown.

Button Sizes	
$\frac{1}{8}$ in.	𝍸 |
$\frac{3}{8}$ in.	|||
$\frac{4}{8}$ in.	|
$\frac{5}{8}$ in.	||
$\frac{7}{8}$ in.	||||

1. The tally chart represents the widths of buttons collected by Bella's mother. Represent this data in a line plot.

Button Sizes (in.)

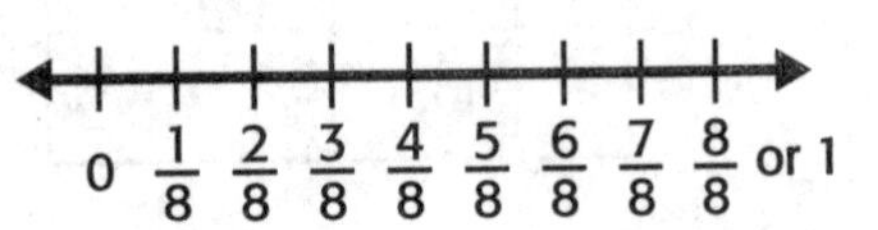

2. Suppose the buttons that were $\frac{3}{8}$-inch wide were laid in a row touching. How far would that row extend?

______

Talk MATH

Describe a real-world situation in which the data in a tally chart and line plot could be helpful.

1 in.

2 in.

Name

Independent Practice

For Exercises 3–6, use the table shown.

3. Mathematical PRACTICE 4 **Model Math** The frequency table represents fractions of an hour Sonja studied each evening over the last 2 weeks. Represent this data in a line plot.

Study Time			
$\frac{1}{4}$ h	$\frac{3}{4}$ h	$\frac{1}{2}$ h	$\frac{1}{4}$ h
$\frac{1}{2}$ h	$\frac{3}{4}$ h	$\frac{1}{4}$ h	1 h
$\frac{1}{2}$ h	$\frac{1}{4}$ h	$\frac{3}{4}$ h	$\frac{1}{2}$ h

Study Time (hr)

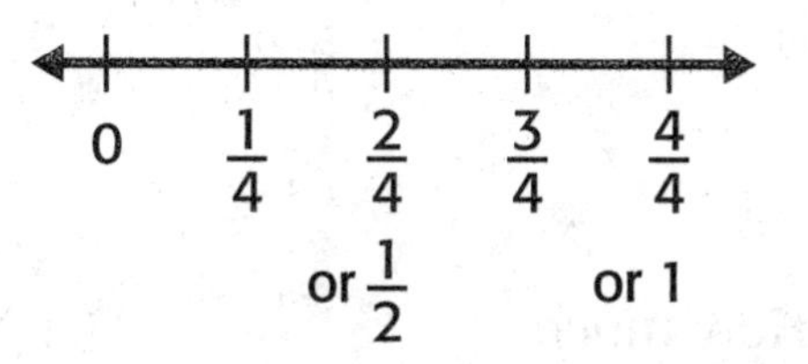

4. Which time interval was least frequent as a study time for Sonja? Explain.

5. What is the difference in the total time spent studying during the $\frac{3}{4}$-hour time intervals and the $\frac{1}{2}$-hour intervals?

6. What is the total time Sonja spent studying over the last two weeks in minutes? What is the equivalent time in hours and minutes?

Problem Solving

Vito Frito Tito Fluffy

For Exercises 7–10, use the table shown.

A Week of Water for Quinn's Pets			
hamster $\frac{1}{8}$ c	dog 1 c	hamster $\frac{2}{8}$ c	rabbit $\frac{1}{8}$ c
rabbit $\frac{2}{8}$ c	hamster $\frac{1}{8}$ c	cat $\frac{4}{8}$ c	dog 1 c
cat $\frac{5}{8}$ c	dog 1 c	rabbit $\frac{1}{8}$ c	dog $\frac{6}{8}$ c

7. Every time one of Quinn's pets needed its water refilled over the past week, he marked the amount of water given in a table. Represent the data in a line plot.

8. What is the difference between the smallest amount of water and the greatest amount of water Quinn gave to his pets over the week? Explain.

9. Mathematical PRACTICE 3 **Draw a Conclusion** How much more water did the cat get than the hamster over the course of the week? Explain.

HOT Problems

10. Mathematical PRACTICE 6 **Explain to a Friend** Explain to a friend the similarity between the number lines on a line plot and the markings on a ruler.

11. **Building on the Essential Question** Line plots can be used to display measurement data. Name another way to display measurement data.

Lesson 8
Display Measurement Data in a Line Plot

Homework Helper

eHelp Need help? connectED.mcgraw-hill.com

At the farmers' market, Jeannie sells beans. Her scale weighs them in fractions of pounds. Jeannie made a tally chart showing how much each handful of beans weighed. How much do the beans weigh altogether?

Weight of Green Beans	
$\frac{1}{4}$ pound	卌 ||
$\frac{1}{2}$ pound	卌 |
$\frac{3}{4}$ pound	||||
1 pound	||

1 Make a line plot to represent the data. First, draw a number line. Then place an X above each measurement for each time that weight occurred.

2 How much do the beans weigh altogether? Multiply to find the total weight for each value.

$7 \times \frac{1}{4}$ lb $= \frac{7}{4}$ lb $= 1\frac{3}{4}$ lb

$6 \times \frac{1}{2}$ lb $= \frac{6}{2}$ lb $= 3$ lb

$4 \times \frac{3}{4}$ lb $= \frac{12}{4}$ lb $= 3$ lb

2×1 lb $= 2$ lb

Then add the weights to find the total.

$1\frac{3}{4}$ lb + 3 lb + 3 lb + 2 lb = $9\frac{3}{4}$ lb

Weight of Green Beans (lb)

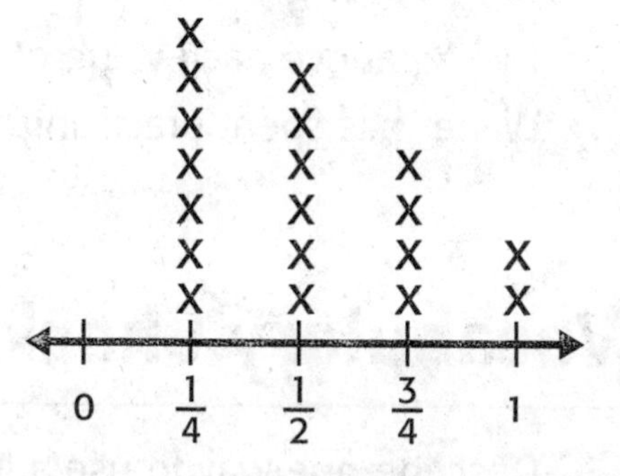

So, altogether the beans weigh $9\frac{3}{4}$ pounds.

Practice

1. Refer to the Homework Helper. What is the difference between the greatest green bean weight and the least green bean weight?

____

For Exercises 2 and 3, use the tally chart shown. The tally chart represents the distance some children were able to ride on a unicycle.

Distance	
$\frac{1}{5}$ mi	𝍸 I
$\frac{2}{5}$ mi	IIII
$\frac{3}{5}$ mi	II
$\frac{4}{5}$ mi	II
1 mi	I

2. Represent this data in a line plot.

3. What is the difference between the greatest distance ridden and the least distance ridden?

Problem Solving

4. **Mathematical PRACTICE 3** **Draw a Conclusion** Walter practices piano for $\frac{1}{4}$ hour, $\frac{1}{2}$ hour, or $\frac{3}{4}$ hour every other day. If a line plot shows two Xs above each value of time, what is the total amount of time Walter has spent practicing?

Vocabulary Check

Vocab

5. Describe one way to use a line plot.

Test Practice

6. Look at the tally chart or line plot from Exercise 2 above. What was the total distance ridden by all of the children?

Ⓐ $5\frac{3}{5}$ hours

Ⓑ $5\frac{4}{5}$ miles

Ⓒ $6\frac{2}{5}$ miles

Ⓓ $6\frac{3}{5}$ miles

Name ....

Solve Measurement Problems

Lesson 9
ESSENTIAL QUESTION
Why do we convert measurements?

Math in My World

Real World · Tutor · Tools

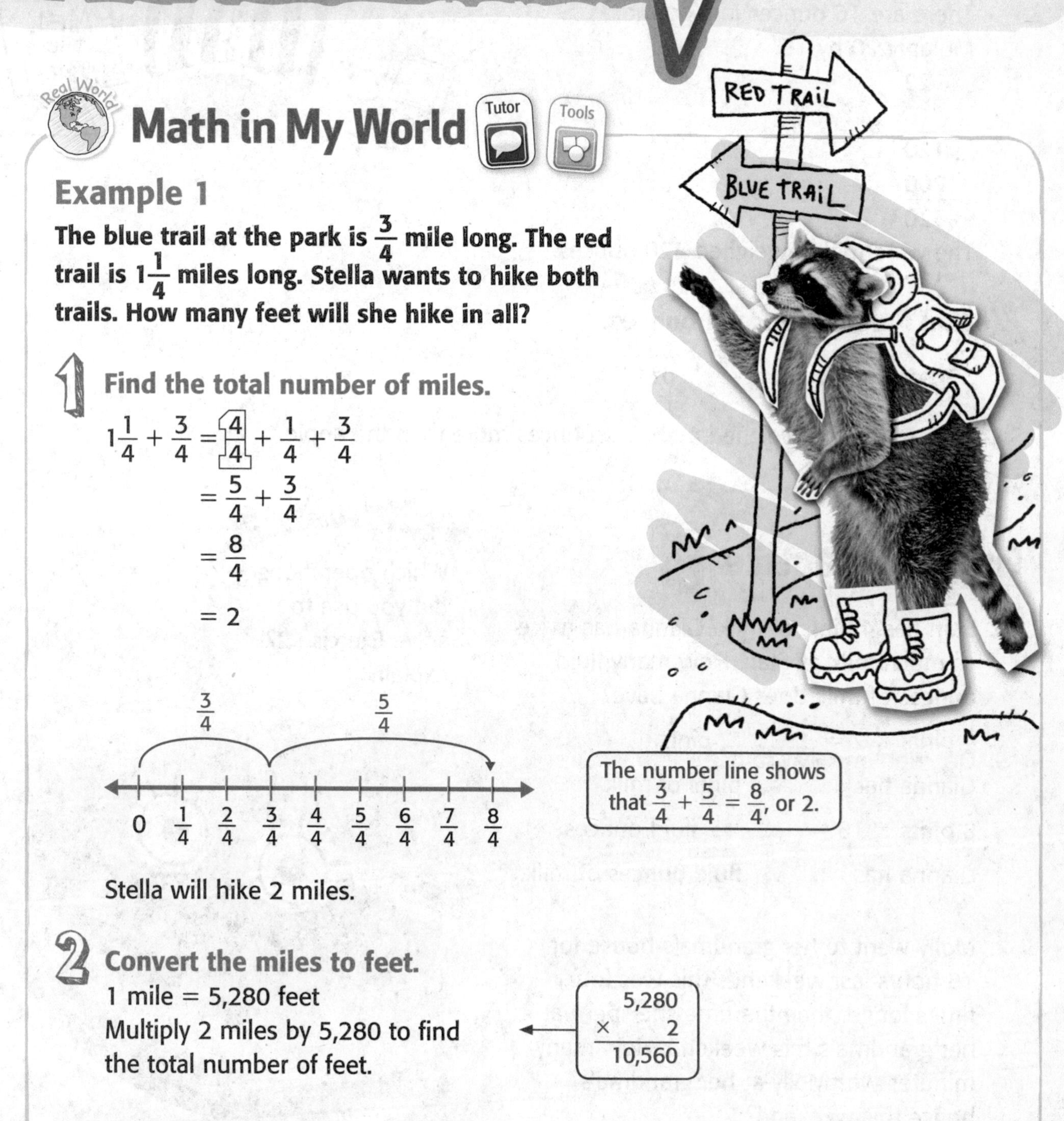

Example 1

The blue trail at the park is $\frac{3}{4}$ mile long. The red trail is $1\frac{1}{4}$ miles long. Stella wants to hike both trails. How many feet will she hike in all?

1 Find the total number of miles.

$$1\frac{1}{4} + \frac{3}{4} = \boxed{\frac{4}{4}} + \frac{1}{4} + \frac{3}{4}$$
$$= \frac{5}{4} + \frac{3}{4}$$
$$= \frac{8}{4}$$
$$= 2$$

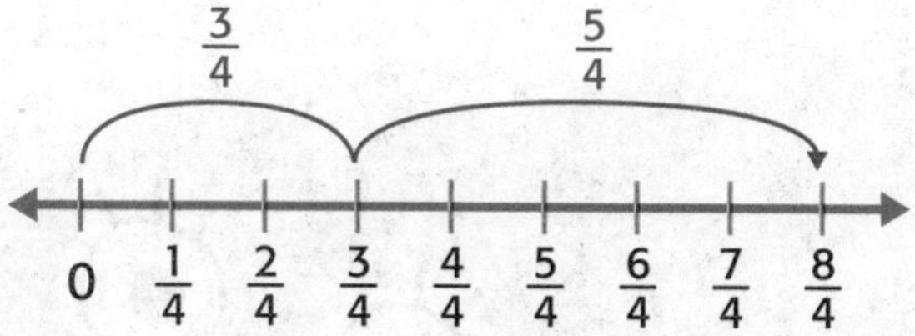

The number line shows that $\frac{3}{4} + \frac{5}{4} = \frac{8}{4}$, or 2.

Stella will hike 2 miles.

2 Convert the miles to feet.

1 mile = 5,280 feet

Multiply 2 miles by 5,280 to find the total number of feet.

$$\begin{array}{r} 5,280 \\ \times \quad 2 \\ \hline 10,560 \end{array}$$

So, Stella will hike ________ feet in all.

Example 2

Dominic weighed an apple and a watermelon. The apple weighed 5 ounces. The watermelon weighed 20 pounds. How many more ounces did the watermelon weigh than the apple?

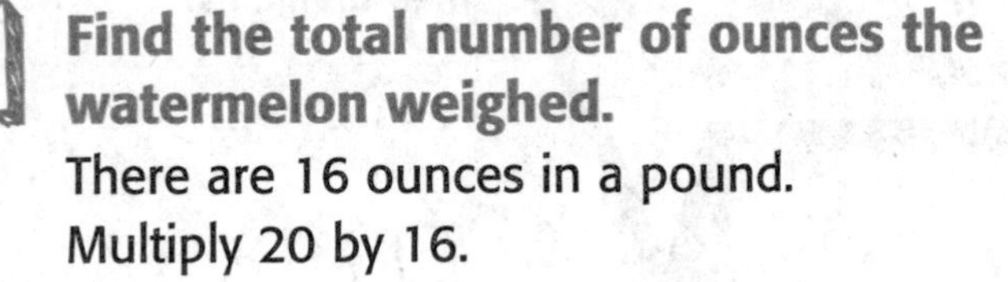

1 Find the total number of ounces the watermelon weighed.

There are 16 ounces in a pound.
Multiply 20 by 16.

$$\begin{array}{r} 20 \\ \times\ 16 \\ \hline 120 \\ +\ 200 \\ \hline 320 \end{array}$$

The watermelon weighed 320 ounces.

2 Find the difference in ounces.

320 oz − 5 oz = ______ oz

So, the watermelon weighed ______ ounces more than the apple.

Guided Practice

Talk MATH

Which operations did you use to solve Exercise 2? Explain.

1. Stan has 4 pints of milk. Gianna has twice as much milk as Stan. How many fluid ounces of milk does Gianna have?

4 pints × 2 = ______ pints

Gianna has ______ pints of milk.

8 pints × 16 = ______ fluid ounces

Gianna has ______ fluid ounces of milk.

2. Molly went to her grandma's house for 16 hours last weekend. This was four times longer than the time she spent at her grandma's this weekend. How many minutes was Molly at her grandma's house this weekend?

______

Name

Independent Practice

3. Jayden read to his little sister for 10 minutes on Saturday and 12 minutes on Sunday. How many seconds did he read to her on Saturday and Sunday? Circle what to do to solve this problem.

- Add 10 and 12 and then multiply by 60.
- Add 10 and 12 and then multiply by 30.
- Add 10 and 12 and then divide by 60.
- Add 10 and 12 and then subtract 60.

Solve.

Jayden read to his little sister for ______________.

4. Mackenzie's dog can jump $2\frac{1}{2}$ feet off the ground. Jordyn's dog can jump $\frac{1}{2}$ foot off the ground. What is the difference in inches between how high Mackenzie's dog can jump and how high Jordyn's dog can jump? Circle what to do to solve this problem.

- Subtract $2\frac{1}{2}$ ft $-$ $\frac{1}{2}$ ft and then multiply by 3.
- Subtract $2\frac{1}{2}$ ft $-$ $\frac{1}{2}$ ft and then divide by 3.
- Subtract $2\frac{1}{2}$ ft $-$ $\frac{1}{2}$ ft and then multiply by 12.
- Add $2\frac{1}{2}$ ft $+$ $\frac{1}{2}$ ft and then multiply by 12.

Use the number line to help solve.

0 $\frac{1}{2}$ $\frac{2}{2}$ $\frac{3}{2}$ $\frac{4}{2}$ $\frac{5}{2}$

The difference in inches is ______________.

Problem Solving

5. Brooklyn bought 1 pound of cucumbers for a salad. She bought twice as much lettuce. How many ounces of lettuce did Brooklyn buy for the salad?

6. Mathematical PRACTICE 5 **Use Math Tools** Sebastian needs 2 gallons of juice for his party. How many cups of juice is this?

My Work!

7. The practice field is 13 yards long. How many feet is this?

8. Jacob needs $\frac{1}{2}$ pint of milk for his recipe. He needs $\frac{1}{2}$ pint of water. How many cups will he have after he combines both ingredients?

HOT Problems

9. Mathematical PRACTICE 2 **Use Number Sense** The moving company charges extra money to move boxes weighing more than 25 pounds. Circle the packages that weigh less than 25 pounds. Put an X on the packages that weigh more than 25 pounds.

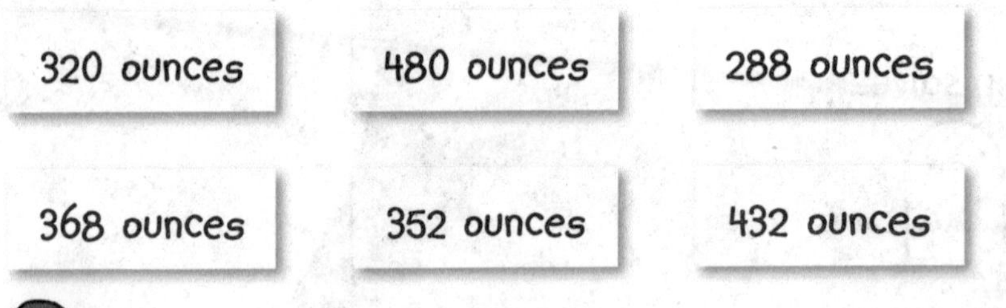

10. **Building on the Essential Question** What information do I need to solve word problems that involve measurement?

MY Homework

Lesson 9

Solve Measurement Problems

Homework Helper

eHelp

Need help? connectED.mcgraw-hill.com

Harlan walked his dog $\frac{1}{6}$ mile on Monday, $\frac{3}{6}$ mile on Wednesday, and $\frac{5}{6}$ mile on Friday. What is the total distance Harlan walked his dog that week?

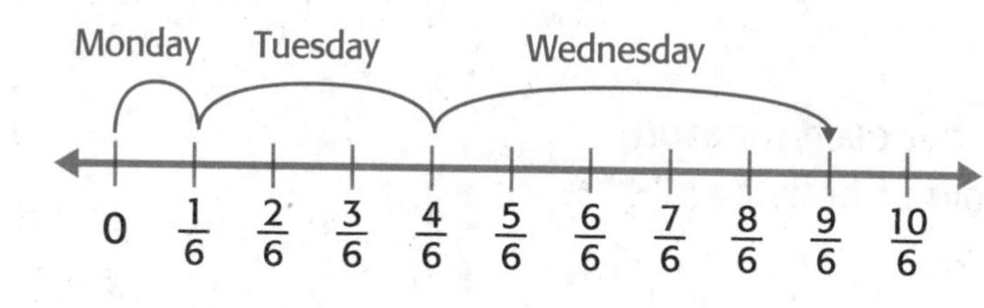

Add the three distances. $\frac{1}{6} + \frac{3}{6} + \frac{5}{6} = \frac{9}{6}$

Simplify. $\frac{9}{6} = 1\frac{3}{6}$ or $1\frac{1}{2}$

So, Harlan walked his dog a total of $1\frac{1}{2}$ miles.

Practice

For Exercises 1–3, use the following information.

Marissa bought $2\frac{1}{2}$ gallons of orange juice to make punch.

1. How many quarts of orange juice did Marissa buy?

____

2. Marissa bought 6 fewer pints of soda than of orange juice. How many cups of soda did she buy?

____

3. Marissa pours $1\frac{1}{2}$ cups of punch in each glass. How many fluid ounces is each serving?

____

Problem Solving

My Work!

For Exercises 4–6, use the following information.

Martin sorted the bolts in his toolbox by size. They measured $\frac{3}{8}$ inch, $\frac{1}{2}$ inch, $\frac{5}{8}$ inch, $\frac{7}{8}$ inch, and $1\frac{1}{4}$ inches.

4. What is the difference in length between the longest bolt and the shortest bolt?

______

5. If you laid one of each size bolt end-to-end, how long would the row of bolts be?

______

6. If you laid 9 bolts end-to-end that each measure $\frac{7}{8}$ inch, how long would the row of bolts be?

______

7. Aurora is moving to a new house. She has lived in her current house for 6 years. How many months is that? How many days?

______

8. Mathematical PRACTICE 5 **Use Math Tools** Nolan bought 26 yards of heavy-duty rope at the hardware store. The rope costs $2 per foot. How much did Nolan pay for the rope?

______

Test Practice

9. Kim and her two brothers each use $1\frac{1}{2}$ cups of milk for breakfast. How many fluid ounces of milk do they use in 4 days?

Ⓐ 144 fl oz
Ⓑ 108 fl oz
Ⓒ 72 fl oz
Ⓓ 36 fl oz

Name

Real World

Problem-Solving Investigation

STRATEGY: Guess, Check, and Revise

Lesson 10

ESSENTIAL QUESTION
Why do we convert measurements?

Learn the Strategy

Three elk are walking in the mountains. Two of the elk weigh the same amount. The other elk weighs 150 pounds more than the other two. If the total weight of the three elk is 1,650 pounds, how much does each elk weigh?

1 Understand

What facts do you know?
Two elk weigh the same amount. A third elk weighs 150 pounds more than the other two.

The total weight of the three elk is ________ pounds.

What do you need to find?
the weight of each elk

2 Plan

I will guess, check, and revise to solve the problem.

3 Solve

First Elk Weight (lb)	Second Elk Weight (lb)	Third Elk Weight (lb)	Total Weight (lb)	Check
400	400	550	400 + 400 + 550 = 1,350	too low
500	500	650	500 + 500 + 650 = 1,650	correct

So, two of the elk weigh ______ pounds. The third elk weighs

______ + 150, or ______ pounds.

4 Check

Does your answer make sense? Explain.

Practice the Strategy

Corrinne is making twice as much fruit punch as lemonade. She is making 12 gallons total. How many gallons will be fruit punch and how many will be lemonade?

1 Understand

What facts do you know?

What do you need to find?

2 Plan

3 Solve

4 Check

Does your answer make sense? Explain.

Name

Apply the Strategy

Guess, check, and revise to solve each problem.

1. Mathematical **PRACTICE** 3 **Draw a Conclusion** Theo lives twice as far from Cassidy as Jarvis. How far do Theo and Jarvis live from Cassidy?

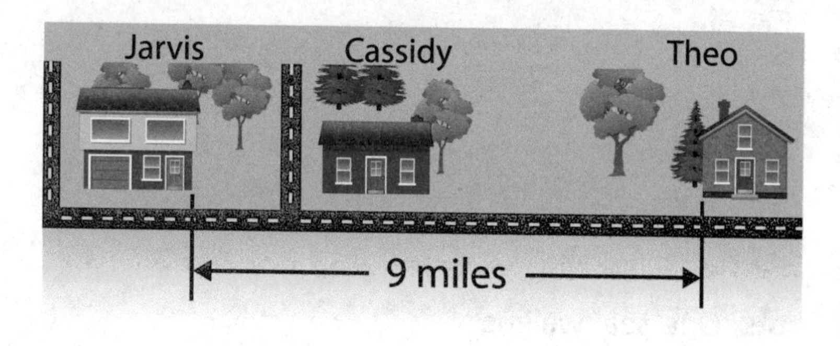

My Work!

2. Jorge and his two brothers each water the plants. Jorge's watering can holds half as much water as his older brother's watering can. His younger brother's watering can holds 8 cups. Altogether, the watering cans hold 2 gallons. How many cups does Jorge's watering can hold?

3. Mathematical **PRACTICE** 1 **Make Sense of Problems** Rhonda filled her aquarium with water. Then, she took out 3 gallons to make room for the gravel and fish. She added 2 quarts back in. Then, she added 4 fish. Each fish was in a separate bag with 1 pint of water. If she has a 20-gallon aquarium, how many pints are in the aquarium now?

Review the Strategies

Use any strategy to solve each problem.

- Guess, check, and revise.
- Find extra or missing information.
- Use logical reasoning.
- Look for a pattern.

4. The distance from Ian's home to the museum is 2,640 yards. Is it reasonable to say that Ian's home is more than 9,000 feet away from the museum? (*Hint*: 3 feet = 1 yard)

My Work!

5. Mathematical **PRACTICE** 5 **Use Math Tools** One seal weighs 22 pounds. Another seal weighs three times as much. How much do the seals weigh altogether?

6. Mathematical **PRACTICE** 4 **Model Math** Mario wants to download 12 songs on his digital music player. He has only 5 minutes to download the songs. If it takes 30 seconds for Mario to download one song, will he have enough time to download all of the songs? Explain.

7. A stunt person jumps from the roof of a 54-foot building. A skydiver jumps from a plane that is 186 times as high. From what height did the skydiver jump?

Name

MY Homework

Lesson 10
Problem Solving: Guess, Check, and Revise

Homework Helper

eHelp

Need help? connectED.mcgraw-hill.com

Cassie is doing sand art. She fills a bottle with 10 inches of sand. She makes two equal layers of yellow sand with a layer of blue sand in between. The layer of blue sand is 1 inch greater than each layer of yellow sand. How many inches of each color of sand does she use?

1 Understand

What facts do you know?

The bottle holds 10 inches of sand. There are two equal layers of yellow sand. There is one layer of blue sand that is 1 inch greater than each yellow layer.

What do you need to find?

the number of inches of each color of sand used

2 Plan

Guess, check, and revise to solve the problem.

3 Solve

Layers of Sand (in.)				
1st Yellow Layer (in.)	2nd Yellow Layer (in.)	Blue Layer (in.)	Total (in.)	Check
4	4	5	$4 + 4 + 5 = 13$	too high
3	3	4	$3 + 3 + 4 = 10$	correct

So, there are two 3-inch layers of yellow sand with a 4-inch layer of blue sand in between the yellow sand layers.

4 Check

Does the answer make sense?

Yes; $3 + 3 + 4 = 10$ and $4 - 3 = 1$.

Problem Solving

Guess, check, and revise to solve each problem.

1. Max was on vacation twice as long as Jared and half as long as Wesley. The boys were on vacation a total of 3 weeks. How many days was each boy on vacation?

2. Anu drinks 2 cups of water each day. Jan drinks twice as much water as Anu. How many fluid ounces does Jan drink?

3. Mathematical PRACTICE 1 **Plan Your Solution** Casey likes to run. She runs an additional $\frac{1}{4}$ mile each day. On the last day, she ran $1\frac{1}{4}$ miles. If she ran $\frac{1}{2}$ mile her first day, for how many days has she been running?

4. There are 4 semi-trailer trucks parked in a line at the rest stop. After the first truck, each truck in the line weighs 2 tons more than the truck before it. The trucks weigh a total of 32 tons. How many pounds does each truck weigh?

My Work!

Review

Chapter 11

Customary Measurement

Vocabulary Check

Draw a line from each word to its definition.

1. **customary system**
2. **convert**
3. **line plot**
4. **capacity**
5. **weight**

- A measurement that tells the amount of liquid a container can hold.
- A measurement that tells how heavy an object is.
- A graph that uses columns of Xs above a number line to show frequency of data.
- To change one unit to another.
- Units of measure most often used in the United States.

Write an example for each measurement.

6. 1 **yard** ______________
7. 1 **quart** ______________
8. 1 **foot** ______________
9. 1 **gallon** ______________
10. 1 **mile** ______________
11. 1 **ounce** ______________
12. 1 **fluid ounce** ______________
13. 1 **pound** ______________
14. 1 **cup** ______________
15. 1 **ton** ______________
16. 1 **pint** ______________
17. 1 **second** ______________

Concept Check

Complete each conversion table.

18.

inches (in.)	yards (yd)	(in., yd)
	1	
	3	
	5	
	7	

19.

cups (c)	gallons (gal)	(c, gal)
	2	
	3	
	4	
	5	

20.

feet (ft)	yards (yd)	(ft, yd)
	2	
	4	
	6	
	8	

21.

minutes (min)	seconds (s)	(min, s)
1		
3		
5		
7		

22. How many times greater is one hour than one minute? ______

23. How many times greater is one gallon than one quart? ______

24. How many times greater is one mile than one foot? ______

25. How many times greater is one pound than one ounce? ______

26. How many minutes have passed between the time shown on the first clock and the time shown on the second clock?

______

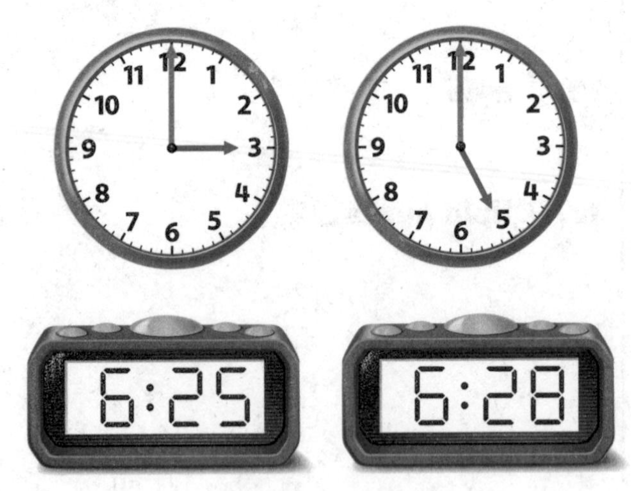

27. How many seconds have passed between the time shown on the first clock and the time shown on the second clock?

______

Name

Problem Solving

For Exercises 28–31, use the table shown. The table represents the lengths of beads that Zoe is using to make a necklace.

Lengths of Beads		
$\frac{1}{6}$ in.	$\frac{4}{6}$ in.	$\frac{5}{6}$ in.
$\frac{2}{6}$ in.	$\frac{1}{6}$ in.	$\frac{5}{6}$ in.
$\frac{5}{6}$ in.	$\frac{1}{6}$ in.	$\frac{4}{6}$ in.

28. Represent this data in a line plot.

29. What is the difference in length between the longest bead and the shortest bead?

____

30. What is the total length of the beads that are $\frac{4}{6}$ inch long?

____

Test Practice

31. Which bead length has enough beads to total exactly $\frac{1}{2}$ inch if the beads of that size are laid end to end?

Ⓐ $\frac{1}{6}$ inch

Ⓑ $\frac{2}{6}$ inch

Ⓒ $\frac{4}{6}$ inch

Ⓓ $\frac{5}{6}$ inch

My Work!

Reflect

Chapter 11

Answering the ESSENTIAL QUESTION

Use what you learned about customary measurements to complete the graphic organizer.

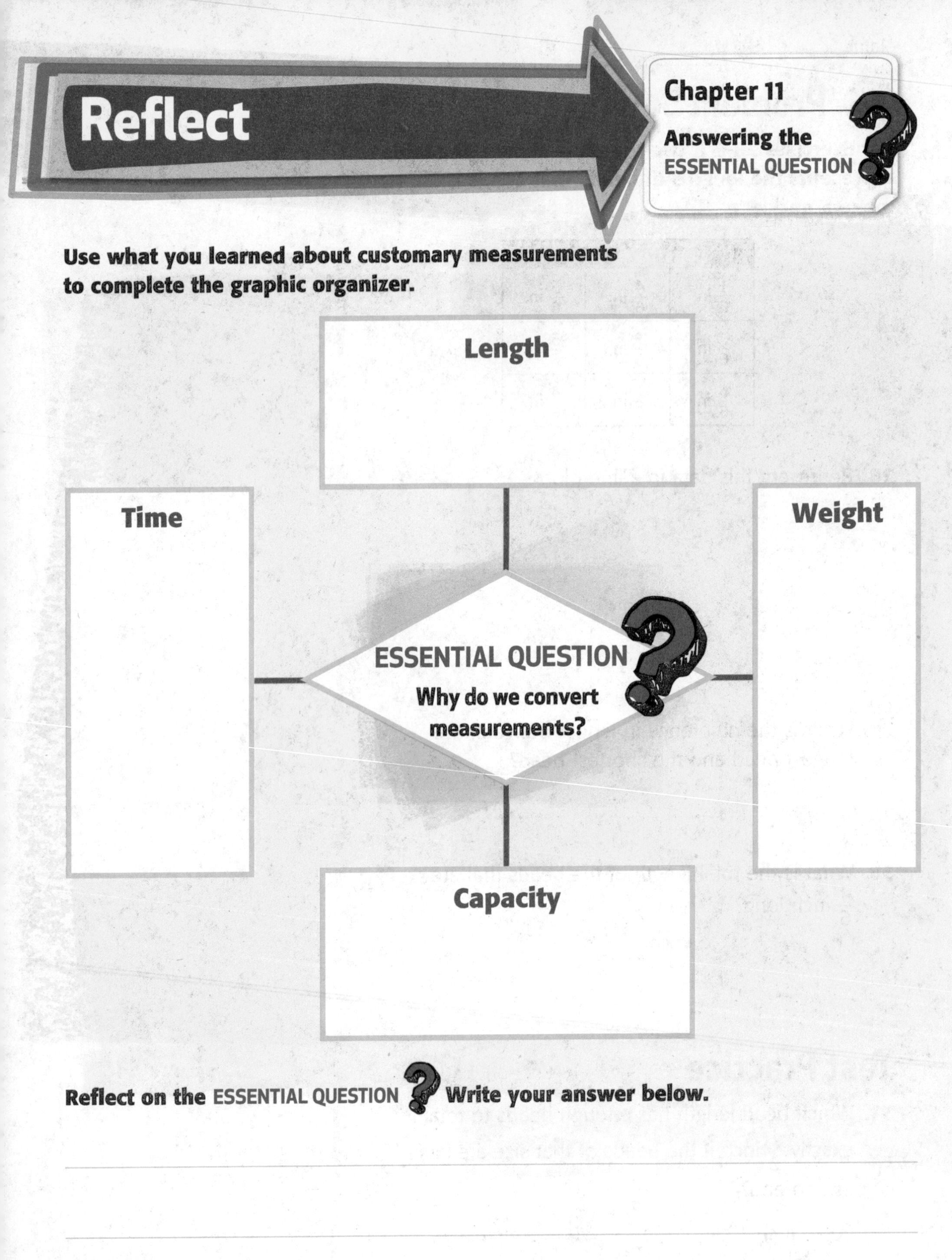

Reflect on the ESSENTIAL QUESTION Write your answer below.

Chapter

12 Metric Measurement

ESSENTIAL QUESTION

How can conversion of measurements help me solve real-world problems?

Watch a video!

Watch

MY Common Core State Standards

Measurement and Data

4.MD.1 Know relative sizes of measurement units within one system of units including km, m, cm; kg, g; lb, oz.; l, ml; hr, min, sec. Within a single system of measurement, express measurements in a larger unit in terms of a smaller unit. Record measurement equivalents in a two-column table.

4.MD.2 Use the four operations to solve word problems involving distances, intervals of time, liquid volumes, masses of objects, and money, including problems involving simple fractions or decimals, and problems that require expressing measurements given in a larger unit in terms of a smaller unit. Represent measurement quantities using diagrams such as number line diagrams that feature a measurement scale.

It looks hard, but I think I can get it!

Standards for Mathematical PRACTICE

1. Make sense of problems and persevere in solving them.
2. Reason abstractly and quantitatively.
3. Construct viable arguments and critique the reasoning of others.
4. Model with mathematics.
5. Use appropriate tools strategically.
6. Attend to precision.
7. Look for and make use of structure.
8. Look for and express regularity in repeated reasoning.

= focused on in this chapter

Name

Am I Ready?

Check ← Go online to take the Readiness Quiz

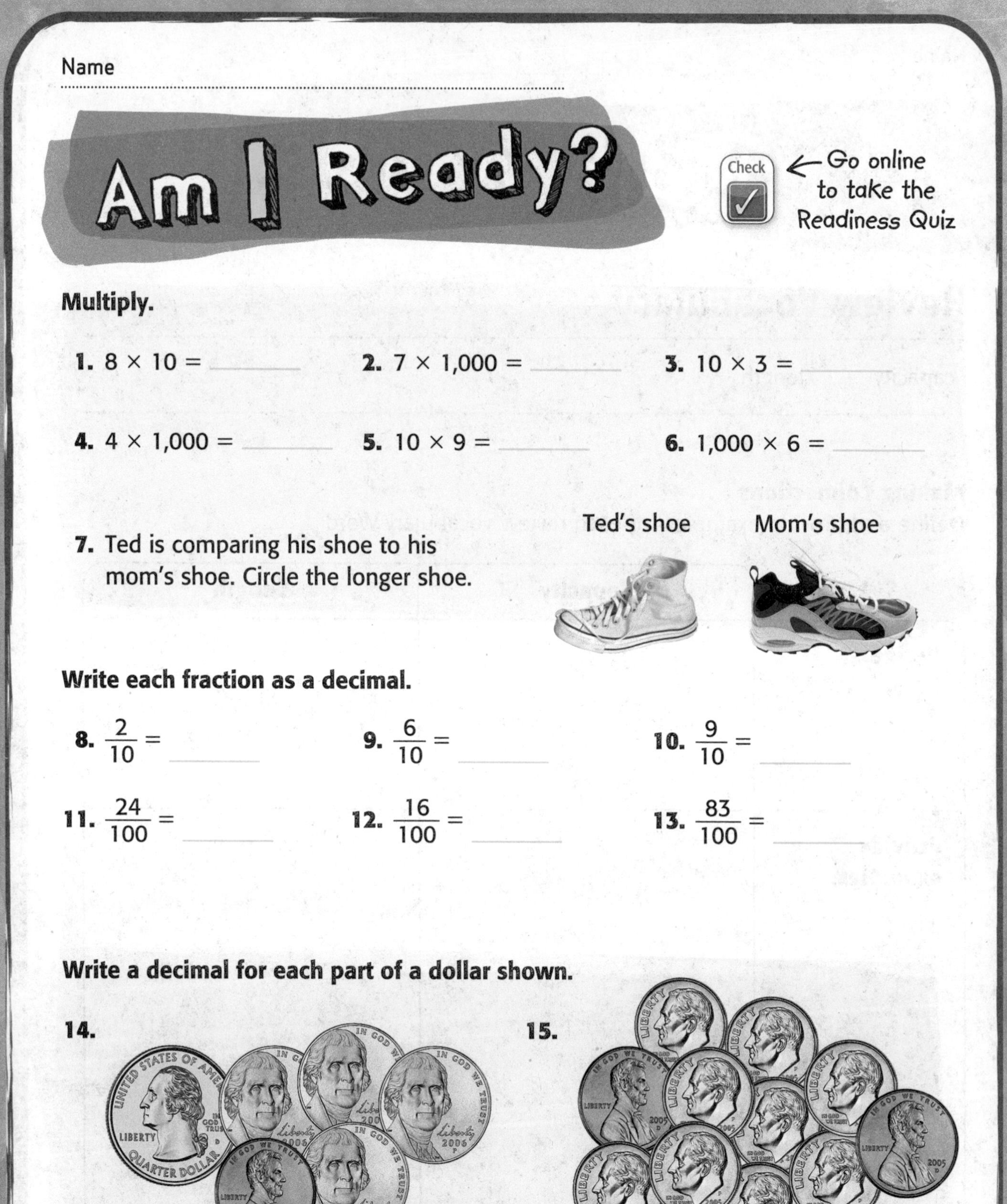

Multiply.

1. $8 \times 10 =$ ______ **2.** $7 \times 1{,}000 =$ ______ **3.** $10 \times 3 =$ ______

4. $4 \times 1{,}000 =$ ______ **5.** $10 \times 9 =$ ______ **6.** $1{,}000 \times 6 =$ ______

7. Ted is comparing his shoe to his mom's shoe. Circle the longer shoe.

Write each fraction as a decimal.

8. $\frac{2}{10} =$ ______ **9.** $\frac{6}{10} =$ ______ **10.** $\frac{9}{10} =$ ______

11. $\frac{24}{100} =$ ______ **12.** $\frac{16}{100} =$ ______ **13.** $\frac{83}{100} =$ ______

Write a decimal for each part of a dollar shown.

14. ______

15. ______

How Did I Do? Shade the boxes to show the problems you answered correctly.

1	2	3	4	5	6	7	8	9	10	11	12	13	14	15

Name

Review Vocabulary

capacity length

Making Connections

Define and provide examples for each review vocabulary word.

	Capacity	Length
Define.		
Provide examples.		

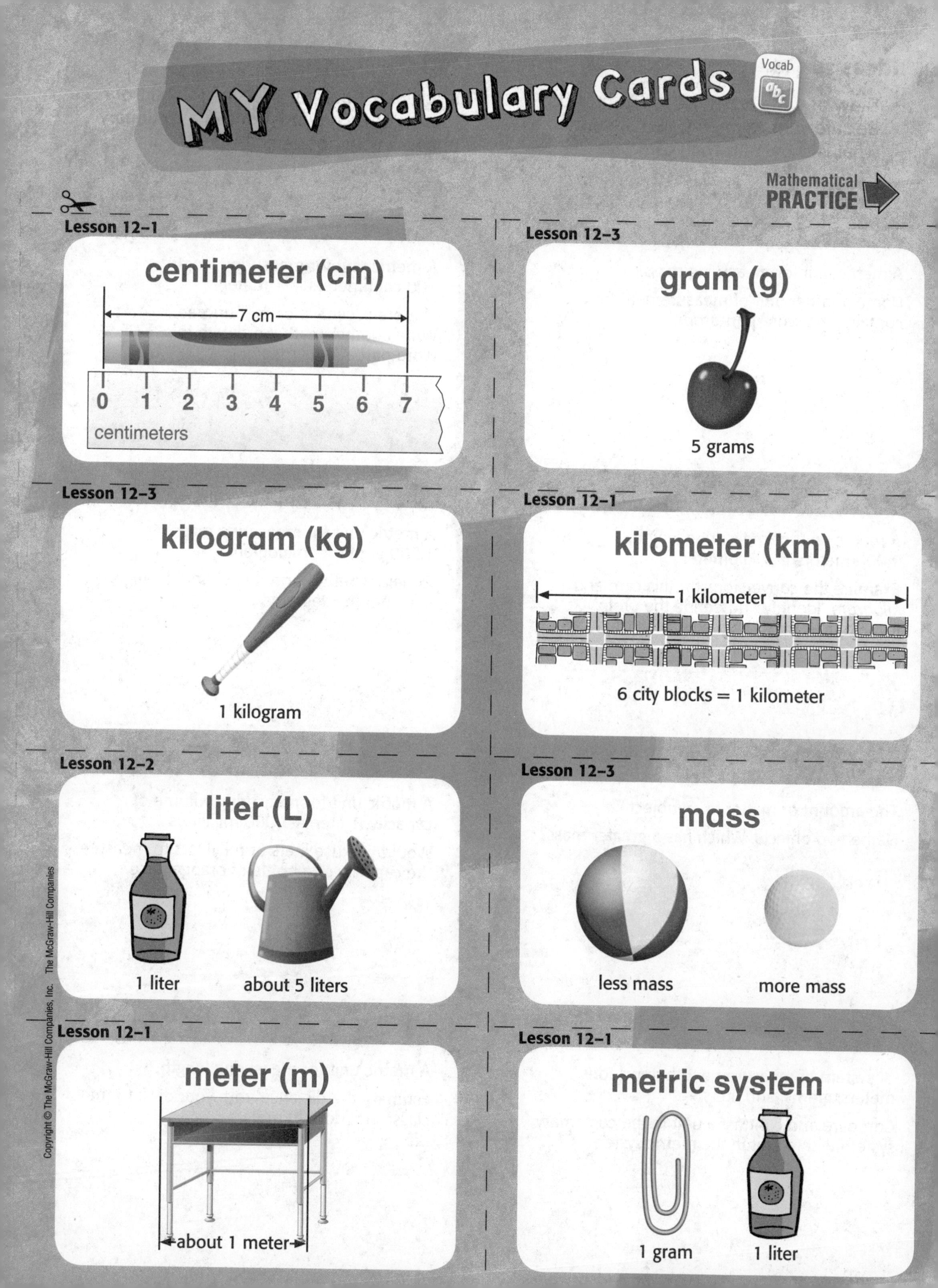
MY Vocabulary Cards
Vocab
abc
Mathematical PRACTICE
Lesson 12-1
centimeter (cm)
7 cm
0 1 2 3 4 5 6 7
centimeters
Lesson 12-3
gram (g)
5 grams
Lesson 12-3
kilogram (kg)
1 kilogram
Lesson 12-1
kilometer (km)
1 kilometer
6 city blocks = 1 kilometer
Lesson 12-2
liter (L)
1 liter
about 5 liters
Lesson 12-3
mass
less mass
more mass
Lesson 12-1
meter (m)
about 1 meter
Lesson 12-1
metric system
1 gram
1 liter
Copyright © The McGraw-Hill Companies, Inc.

Ideas for Use

- Draw or write examples for each card. Be sure your examples are different from what is shown on each card.
- Work with a partner to name the part of speech of each word. Consult a dictionary to check your answers.

A metric unit for measuring mass.

Name another unit of measurement that contains the word part *gram*.

A metric unit for measuring length.
100 centimeters = 1 meter

The prefix *cent-* means "hundred." Write another math word with this word part and its meaning.

A metric unit for measuring length.
1,000 meters = 1 kilometer

Examine the conversions for this card and *kilogram*. Identify and define the prefix. Explain how it can help you recall the words.

A metric unit for measuring mass.
1,000 grams = 1 kilogram

Explain whether or not you would weigh a feather in kilograms.

The amount of matter in an object.

Name two objects. Which has a greater mass?

A metric unit for measuring volume or capacity. 1 liter = 1,000 milliliters

Would you use liters or milliliters to measure the capacity of a bottle of orange juice?

A system of measurement that includes meters, grams, and liters.

Compare and contrast a unit in the customary system with a unit in the metric system.

A metric unit for measuring length.

Estimate the distance from your desk to the classroom door in meters.

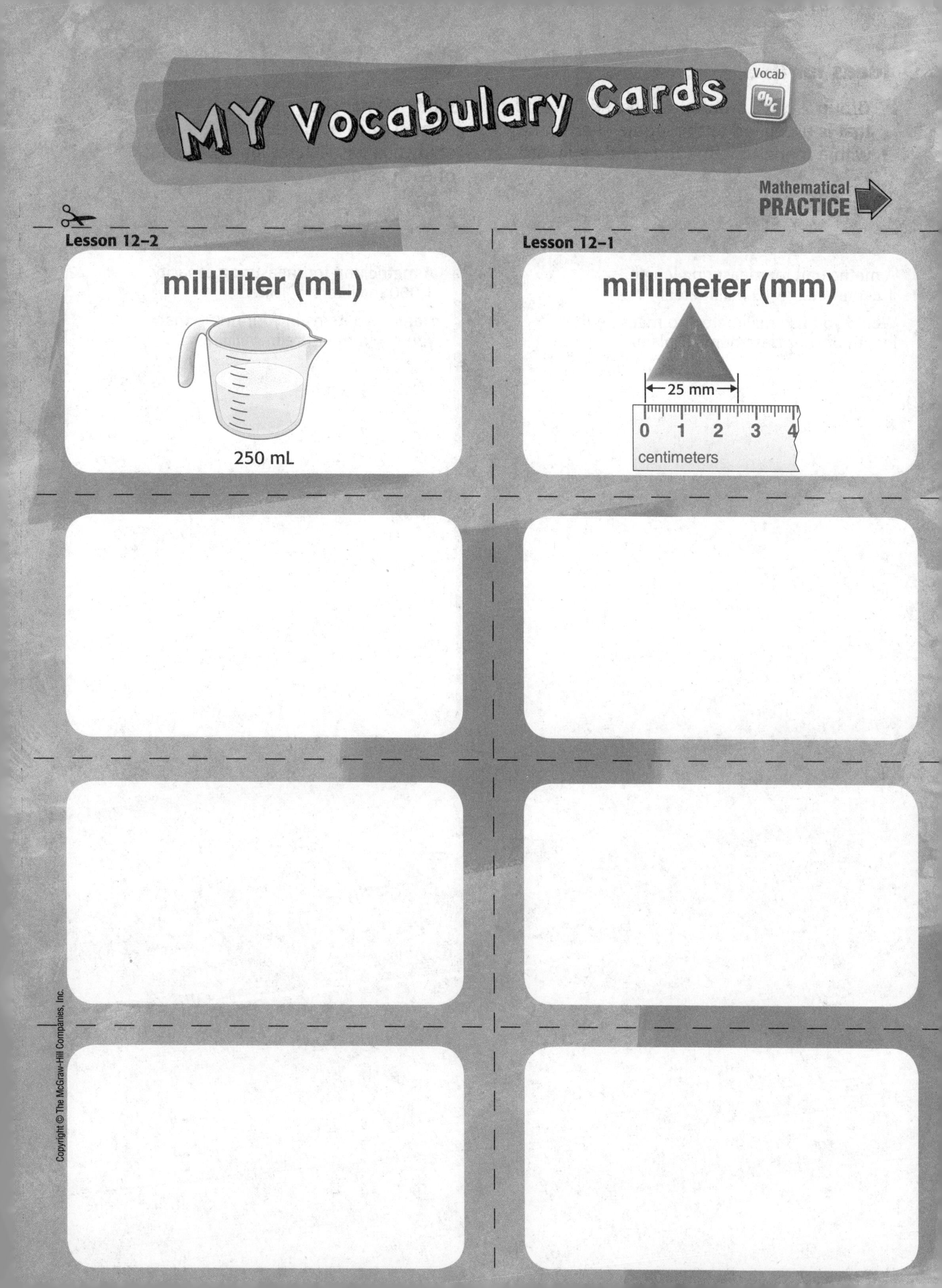
MY Vocabulary Cards
Vocab
abc
Mathematical
PRACTICE
Lesson 12-2
milliliter (mL)
250 mL
Lesson 12-1
millimeter (mm)
25 mm
0
1
2
3
4
centimeters

Ideas for Use

- Group 2 or 3 common words. Add a word that is unrelated to the group. Then work with a friend to name the unrelated word.
- Write the name of each lesson on the front of each blank card. Write a few study tips for each lesson on the back of each card.

A metric unit for measuring length.
1 centimeter = 10 millimeters

Would you use millimeters to measure the length of your classroom? Explain.

A metric unit for measuring capacity.
1,000 milliliters = 1 liter

Name two items in your home that you could measure in milliliters.

MY Foldable

Follow the steps on the back to make your Foldable.

Metric Conversions

Length
1 centimeter (cm) = 10 millimeters (mm)
1 meter (m) = 100 centimeters (cm)
1 kilometer (km) = 1,000 meters (m)

Capacity
1 liter (L) = 1,000 milliliters (mL)

Mass
1 kilogram (kg) = 1,000 grams (g)

Metric Measurements

Mass

gram (g)

meter (m)

kilometer (km)

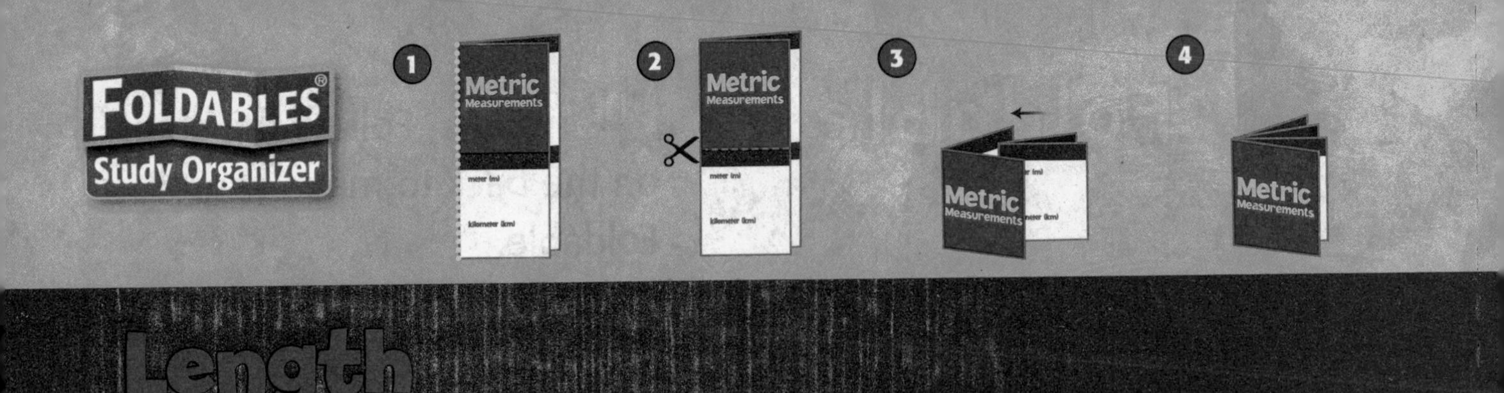

Length

millimeter (mm)

centimeter (cm)

kilogram (kg)

Capacity

milliliter (mL)

liter (L)

Name ______________________

Metric Units of Length

Lesson 1

ESSENTIAL QUESTION
How can conversion of measurements help me solve real-world problems?

Length is the measurement of a line between two points. Millimeter, centimeter, meter, and kilometer are units that are part of the **metric system** of measure for length.

A **millimeter (mm)** is about as thick as 6 sheets of notebook paper.

A **centimeter (cm)** is about the length of a ladybug.

A **meter (m)** is about the height of a chair.

A **kilometer (km)** is about six city blocks.

Example 1

Doug is growing carrots in his garden. He pulled out a carrot for lunch. Measure the carrot to the nearest centimeter.

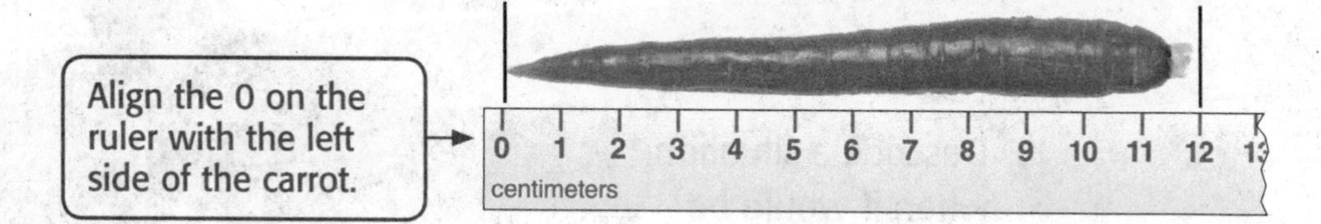

The carrot is closer to the 12–centimeter mark than the 11-centimeter mark.

So, the carrot is almost ________ centimeters long.

Before measuring the length of an object, always estimate the length to decide which unit of measurement is best to use.

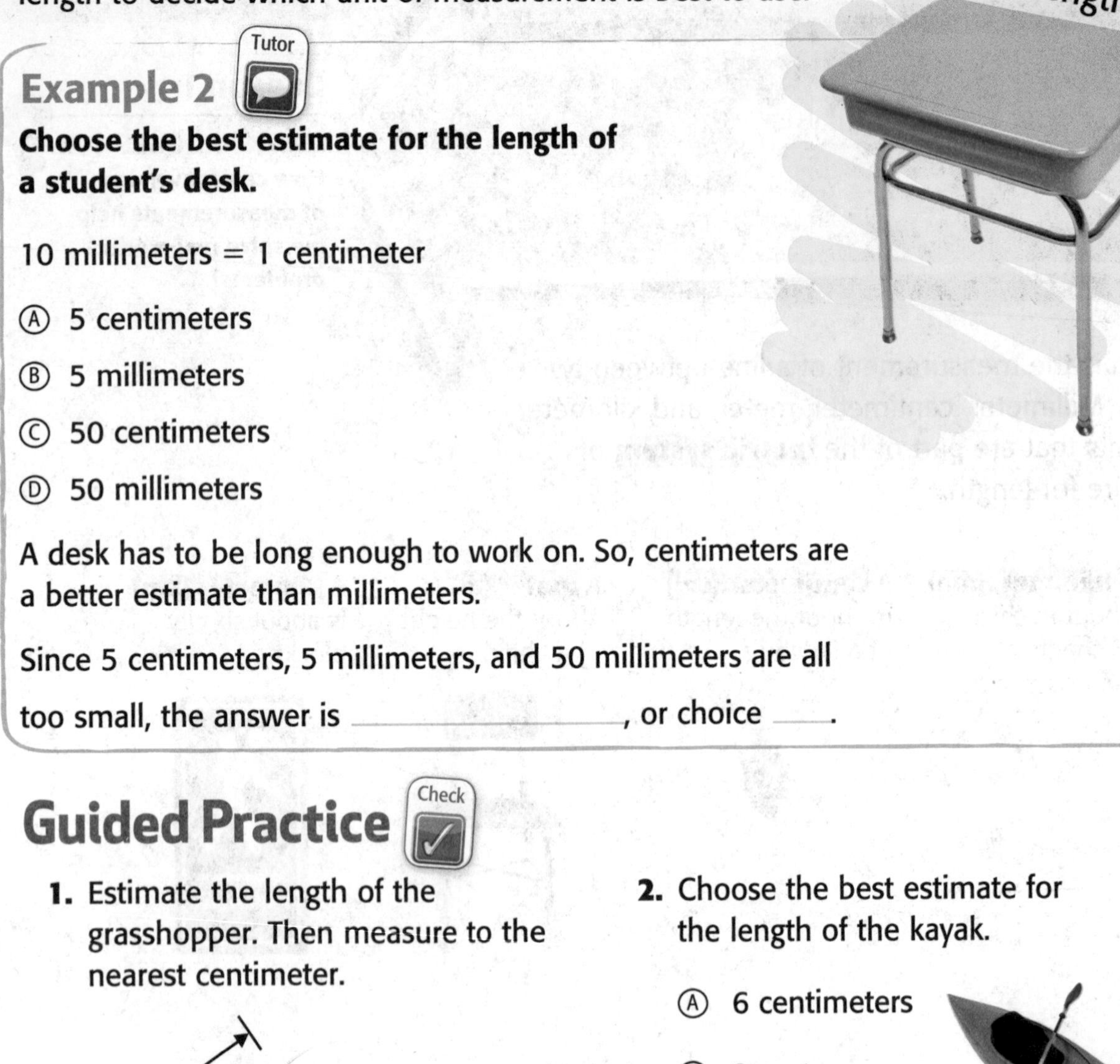

Example 2 Tutor

Choose the best estimate for the length of a student's desk.

10 millimeters = 1 centimeter

Ⓐ 5 centimeters

Ⓑ 5 millimeters

Ⓒ 50 centimeters

Ⓓ 50 millimeters

A desk has to be long enough to work on. So, centimeters are a better estimate than millimeters.

Since 5 centimeters, 5 millimeters, and 50 millimeters are all too small, the answer is ______________, or choice ____.

Guided Practice Check

1. Estimate the length of the grasshopper. Then measure to the nearest centimeter.

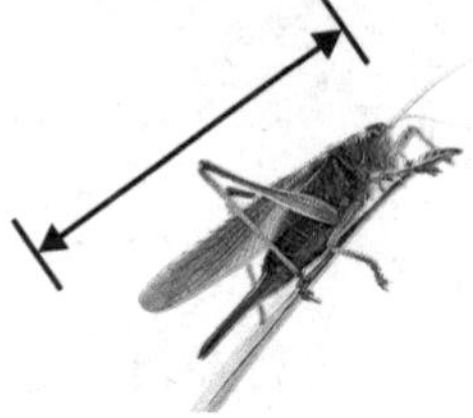

____

2. Choose the best estimate for the length of the kayak.

Ⓐ 6 centimeters

Ⓑ 2 meters

Ⓒ 6 meters

Ⓓ 2 kilometers

Talk MATH

Describe a situation when it would be appropriate to measure an object using millimeters.

Name ....

Independent Practice

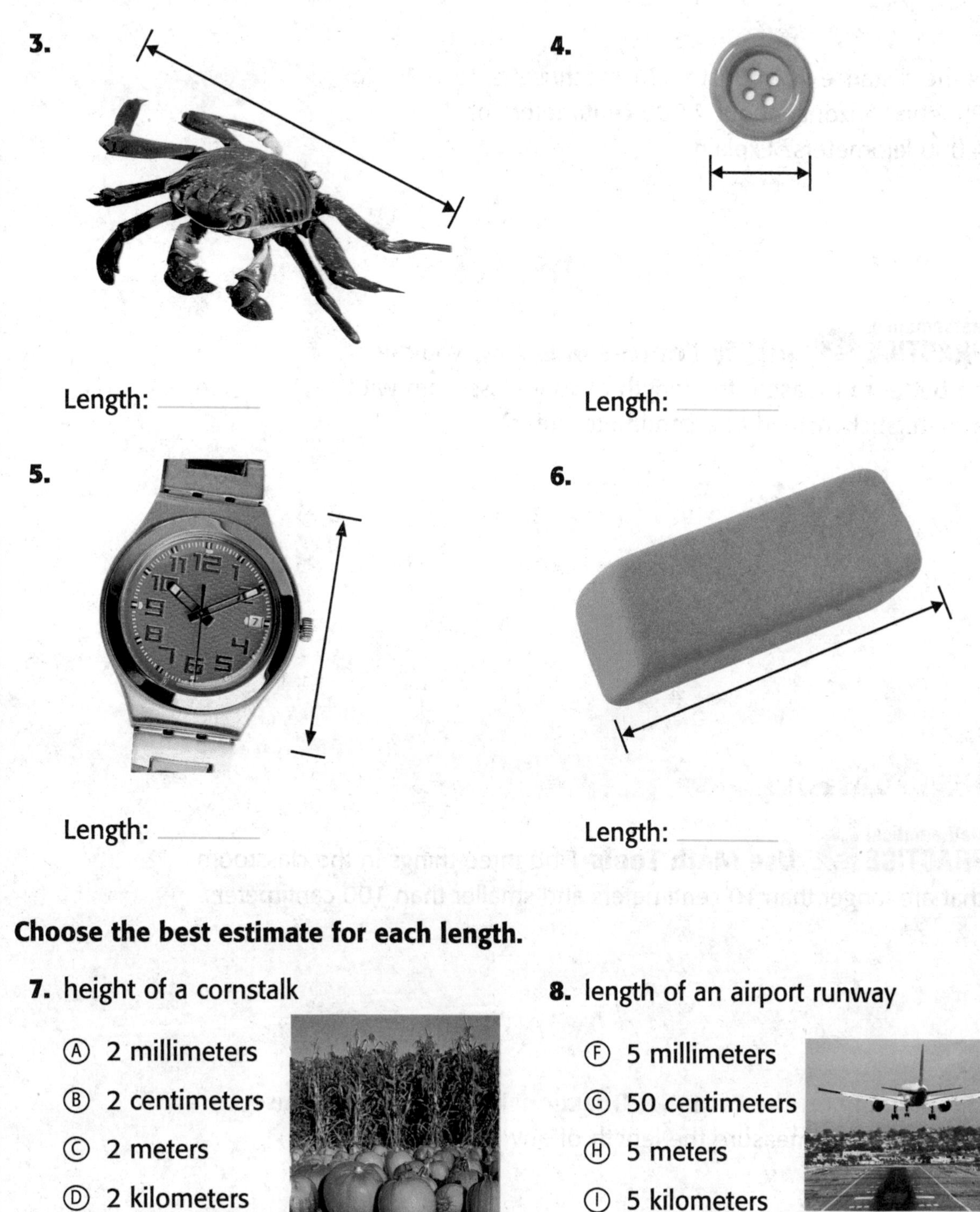

Estimate each length. Then measure each object to the nearest centimeter.

3. Length: ______

4. Length: ______

5. Length: ______

6. Length: ______

Choose the best estimate for each length.

7. height of a cornstalk

Ⓐ 2 millimeters
Ⓑ 2 centimeters
Ⓒ 2 meters
Ⓓ 2 kilometers

8. length of an airport runway

Ⓕ 5 millimeters
Ⓖ 50 centimeters
Ⓗ 5 meters
Ⓘ 5 kilometers

Problem Solving

9. A giraffe at the zoo is 5 meters tall. Name something else that is about 5 meters tall.

My Work!

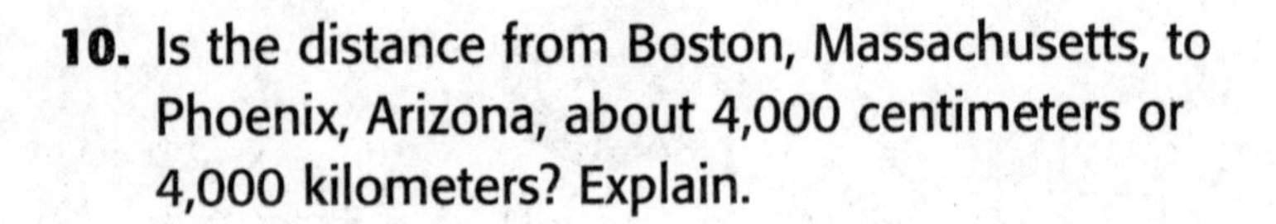

10. Is the distance from Boston, Massachusetts, to Phoenix, Arizona, about 4,000 centimeters or 4,000 kilometers? Explain.

11. Mathematical PRACTICE 3 **Justify Conclusions** Why would it be better to measure the length of your classroom with a meterstick instead of a centimeter ruler?

HOT Problems

12. Mathematical PRACTICE 5 **Use Math Tools** Find three things in the classroom that are longer than 10 centimeters and smaller than 100 centimeters.

13. **Building on the Essential Question** Is it reasonable to use centimeters to measure the length of any object? Explain.

MY Homework

Lesson 1
Metric Units of Length

Homework Helper

Need help? connectED.mcgraw-hill.com

Estimate the length. Then measure to the nearest centimeter.

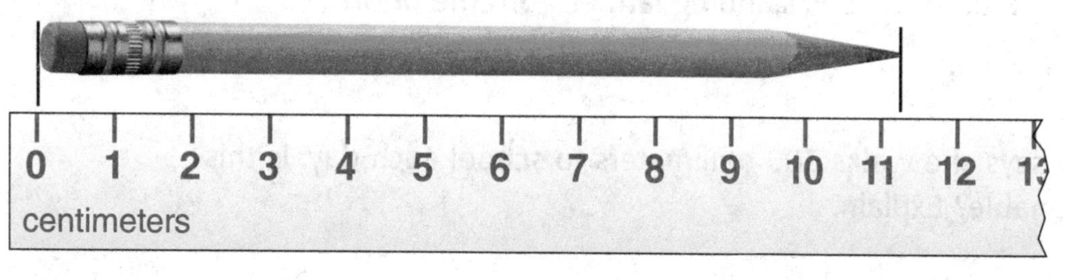

1 Estimate.
You know that the length of a ladybug is about 1 centimeter. You can estimate that the pencil is about 10 ladybugs, or 10 centimeters, long.

2 Measure.
Use a centimeter ruler. Line up the 0 on the ruler with the end of the pencil. The pencil ends just after the 11-centimeter mark.

So, the pencil is about 11 centimeters long.

Practice

Estimate each length. Then measure each object to the nearest centimeter.

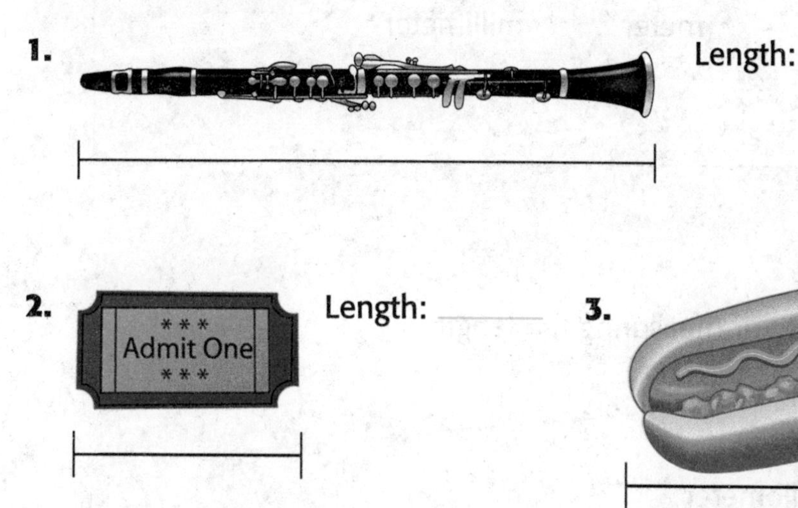

1. Length: ______

2. Length: ______

3. Length: ______

Choose the best estimate for each length.

4. length of a river

Ⓐ 27 km   Ⓒ 170 cm

Ⓑ 7 m   Ⓓ 270 mm

5. length of a sunflower seed

Ⓕ 90 cm   Ⓗ 90 mm

Ⓖ 9 cm   Ⓘ 9 mm

Real World

Problem Solving

6. Mathematical PRACTICE 3 **Draw a Conclusion** Sonia is standing 20 centimeters from the door. Brice is standing 20 meters from the door. Who is standing farther from the door?

____

7. Carly says she walks 300 millimeters to school each day. Is this reasonable? Explain.

____

____

8. At his aunt's farm, Benjamin sees a horse that is 2 meters long. Name two other things that are about 2 meters long.

____

Vocabulary Check

Vocab

9. List the metric system units for measuring length in order from *greatest* to *least*.

centimeter   kilometer   meter   millimeter

____

Test Practice

10. Which is the best unit to use for measuring the length of an eyelash?

Ⓐ millimeter   Ⓒ meter

Ⓑ centimeter   Ⓓ kilometer

Name ....

Metric Units of Capacity

Lesson 2

ESSENTIAL QUESTION
How can conversion of measurements help me solve real-world problems?

The amount of liquid a container can hold is its capacity. The liter (L) and milliliter (mL) are units of measurement for capacity in the metric system.

liter (L)

A bottle this size can hold a liter.

milliliter (mL)

An eyedropper holds about one milliliter.

Math in My World

Real World

Watch Tutor

Example 1

Decide whether 300 milliliters or 300 liters is the more reasonable estimate for the capacity of the mug.

Use logic to estimate the capacity.

300 mL

Helpful Hint
300 eyedroppers are reasonable.

300 L

Helpful Hint
300 bottles are too much.

So, ______________ is the more reasonable estimate.

Example 2 

Decide whether 600 milliliters or 600 liters is the more reasonable estimate for the capacity of the swimming pool.

The pool is a large object. So, 600 milliliters is too small.

So, ______________ is the more reasonable estimate.

Guided Practice

Check

Circle the more reasonable estimate for each capacity.

1. 

1 mL

1 L

2. 

38 mL

38 L

Talk MATH

Describe the unit of capacity you would use to measure the capacity of a bottle of medicine.

3. 

220 mL

220 L

Name ....

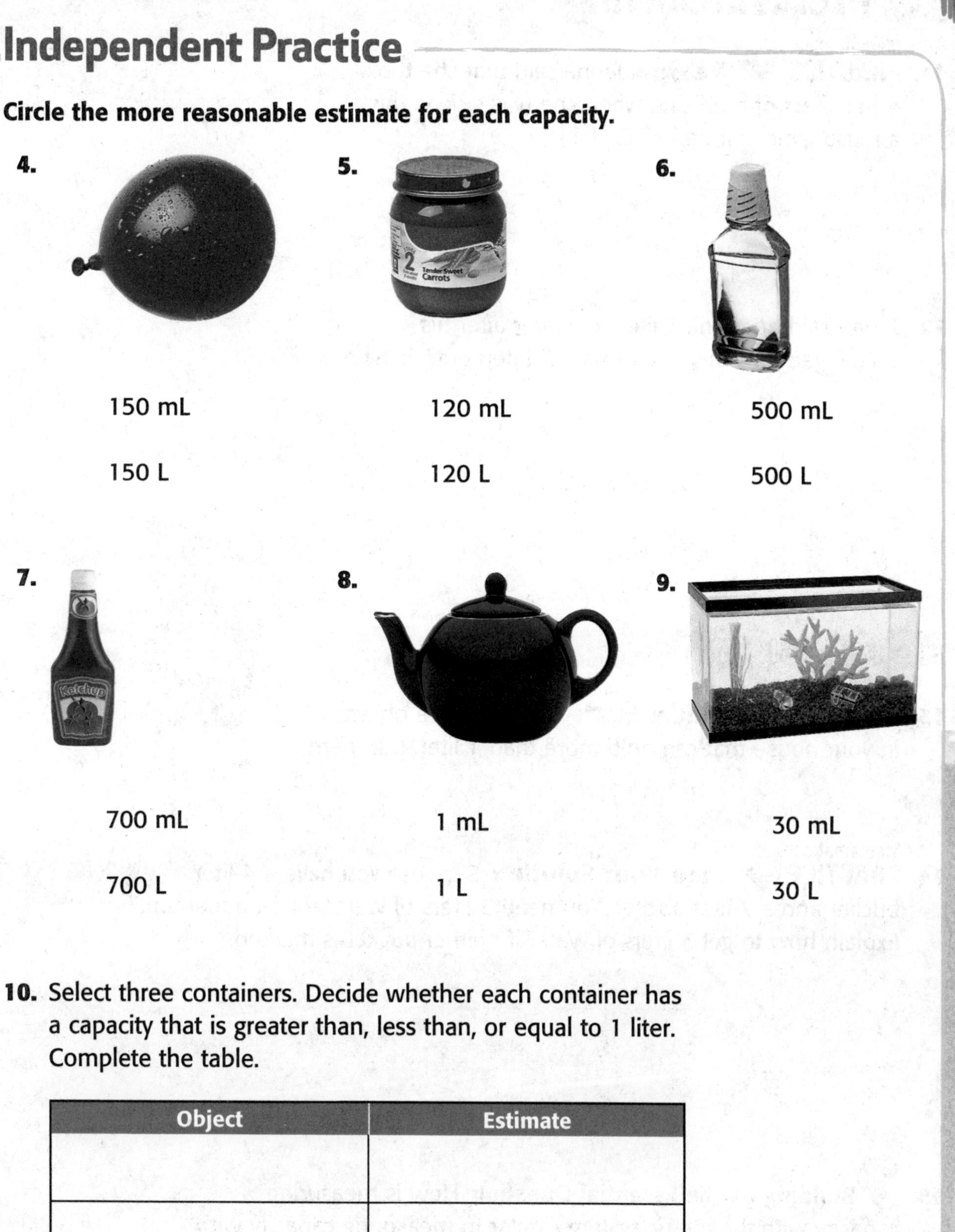

Independent Practice

Circle the more reasonable estimate for each capacity.

4.

150 mL

150 L

5.

120 mL

120 L

6.

500 mL

500 L

7.

700 mL

700 L

8.

1 mL

1 L

9.

30 mL

30 L

10. Select three containers. Decide whether each container has a capacity that is greater than, less than, or equal to 1 liter. Complete the table.

Object	Estimate

Problem Solving

11. Mathematical PRACTICE 2 **Reason** Jenna said that she took 4 milliliters of medicine when she was sick. Is this a reasonable statement? Explain.

My Work!

12. Jonah said he drank 3 liters of water after his soccer game. Is this a reasonable statement? Explain.

HOT Problems

13. Mathematical PRACTICE 4 **Model Math** Think of three objects in your house that can hold more than 1 liter. List them.

14. Mathematical PRACTICE 1 **Plan Your Solution** Suppose you have a 4-liter bucket and a 7-liter bucket. You need 3 liters of water for an aquarium. Explain how to get 3 liters of water if neither bucket is marked.

15. **Building on the Essential Question** How is measuring capacity with the metric system similar to measuring capacity with the customary system?

Name ....

MY Homework

Lesson 2
Metric Units of Capacity

Homework Helper

Need help? connectED.mcgraw-hill.com

Thom is making stew for his family. Is it more reasonable to say the capacity of the stew pot is 5 liters or 5 milliliters?

You know that a milliliter is a tiny amount—about the capacity of an eyedropper.

You know that a liter is a greater amount—about the capacity of a large water bottle.

It would not be reasonable to estimate the capacity of a stew pot in milliliters.

So, it would be more reasonable to say the capacity of the stew pot is 5 liters.

Practice

Choose the most reasonable estimate for each capacity.

1. Ⓐ 40 liters
   Ⓑ 4 liters
   Ⓒ 40 milliliters
   Ⓓ 4 milliliters

2. Ⓕ 10 mL
   Ⓖ 100 mL
   Ⓗ 10 L
   Ⓘ 100 L

3. Ⓐ 1 liter
   Ⓑ 3 liters
   Ⓒ 7 liters
   Ⓓ 10 liters

4. Ⓕ 17 mL
   Ⓖ 170 mL
   Ⓗ 170 L
   Ⓘ 17 L

Match each object to its reasonable capacity.

5. bottle of nail polish
6. bathtub
7. large pitcher

- 300 liters
- 2 liters
- 15 milliliters

Problem Solving

8. **Mathematical PRACTICE 1** **Check for Reasonableness** Emerson needs to use eye drops. Is it reasonable for her to put 1 milliliter of drops in each eye? Explain.

____

9. Ryan fills his cat's water bowl. Is it reasonable to say he uses 1 milliliter of water? Explain.

____

10. Identify 2 objects you could find in a grocery store that hold less than 100 milliliters.

____

Vocabulary Check

Vocab

Write a vocabulary term to complete each sentence.

liters   milliliters

11. The capacity of a baby's bottle would be measured in __________.

12. The capacity of a fish tank would be measured in __________.

Test Practice

13. Which is a reasonable estimate for the capacity of a bottle of mouthwash?

Ⓐ 1 milliliter   Ⓒ 1 liter

Ⓑ 20 milliliters   Ⓓ 20 liters

Name

Metric Units of Mass

Lesson 3

ESSENTIAL QUESTION
How can conversion of measurements help me solve real-world problems?

Mass is the amount of matter an object has. The mass of an object is not affected by gravity. However, an object's weight differs depending on gravity.

gram (g)

The mass of a penny is about 1 gram.

kilogram (kg)

The mass of six medium apples is about 1 kilogram.

1,000 grams (g) = 1 kilogram (kg)

Math in My World

Watch Tutor

Example 1

Which is the more reasonable estimate for the mass of the laptop, 2 grams or 2 kilograms?

Use logic to estimate the mass.

2 grams | 2 kilograms

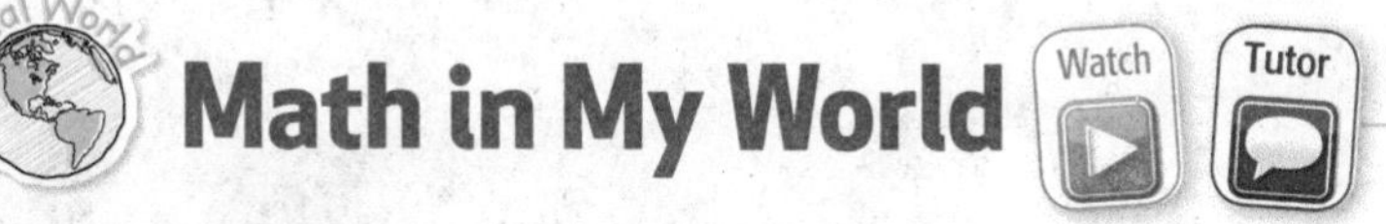

Helpful Hint
2 grams would have the same mass as about 2 pennies.

Helpful Hint
2 kilograms would have the same mass as about 12 medium apples.

So, ________ is the more reasonable estimate.

Example 2

Is it more reasonable to say that a rabbit's mass is 3 grams or 3 kilograms?

3 grams

Helpful Hint
3 grams is too small.

3 kilograms

Helpful Hint
3 kilograms is a reasonable estimate.

So, ________________ is the more reasonable estimate.

Guided Practice

Talk MATH
Explain the difference between weight and mass.

Circle the more reasonable estimate for each mass.

1.

25 grams

25 kilograms

2.

450 grams

450 kilograms

Name

Independent Practice

Circle the more reasonable estimate for each mass.

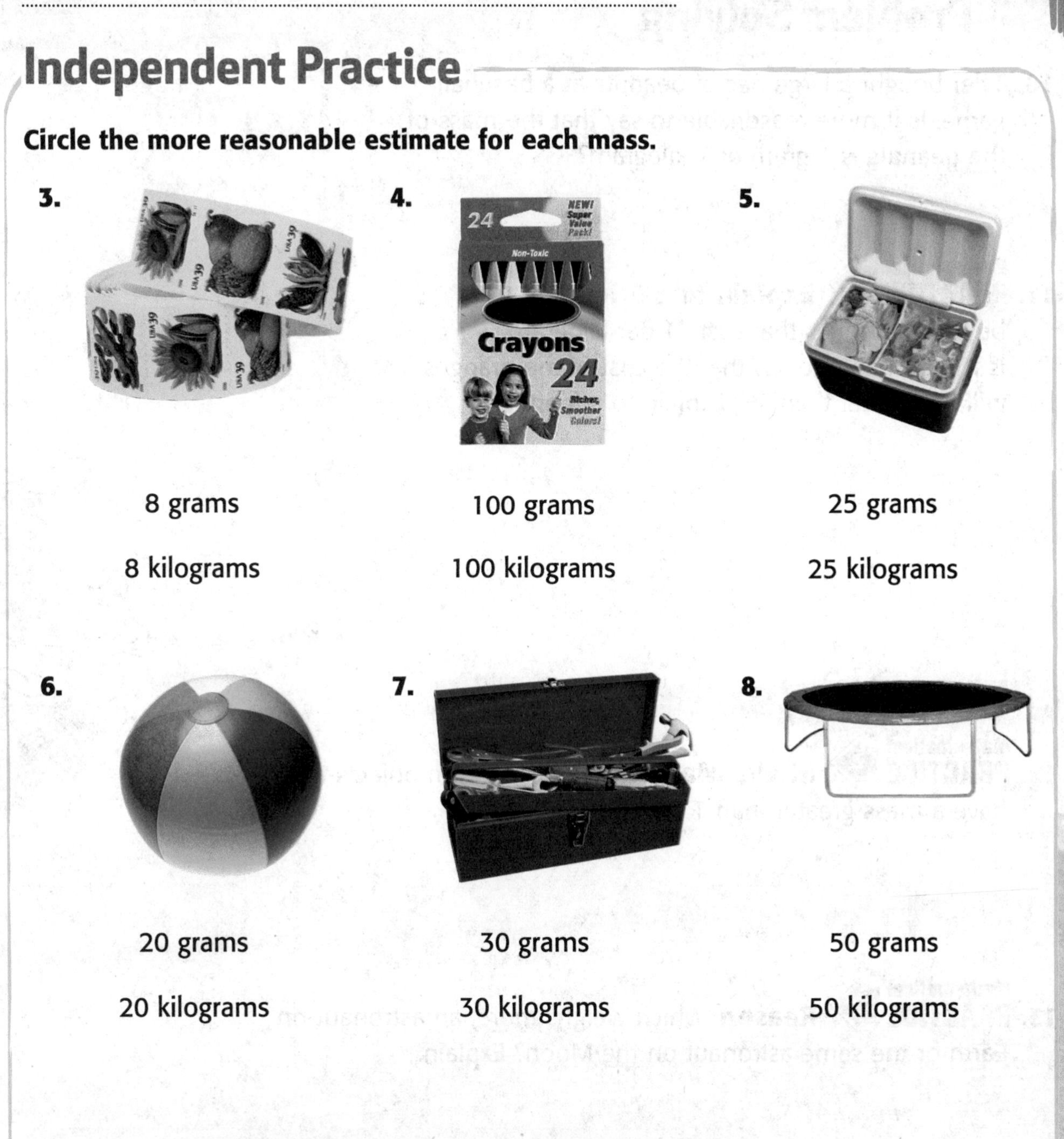

3. 8 grams / 8 kilograms

4. 100 grams / 100 kilograms

5. 25 grams / 25 kilograms

6. 20 grams / 20 kilograms

7. 30 grams / 30 kilograms

8. 50 grams / 50 kilograms

9. The table lists items that can be found in the classroom. Estimate the mass of each object. Record your estimates in the table.

Object	Estimate
glue bottle	
paper clip	
pencil	
stapler	

Problem Solving

My Work!

10. Tyler bought a large bag of peanuts at a baseball game. Is it more reasonable to say that the mass of the peanuts is 1 gram or 1 kilogram?

11. Mathematical PRACTICE 6 **Explain to a Friend** Alicia is buying 6 oranges that cost $1 per kilogram. Is it reasonable to say that the cost of the oranges will be greater than $6? Explain to a friend.

HOT Problems

12. Mathematical PRACTICE 4 **Model Math** List five classroom objects that have a mass greater than 1 kilogram.

13. Mathematical PRACTICE 2 **Reason** Which weighs more, an astronaut on Earth or the same astronaut on the Moon? Explain.

14. **Building on the Essential Question** Name a real-world example of something that has a mass that can be measured with a metric unit.

Name

MY Homework

Lesson 3
Metric Units of Mass

Homework Helper

Need help? connectED.mcgraw-hill.com

Geneva subscribes to a nature magazine. Is it reasonable to estimate that the mass of one issue of the magazine is 25 grams or 25 kilograms?

Twenty-five kilograms is too much.

You know that the mass of 1 penny is about 1 gram. Imagine holding 25 pennies in one hand and a magazine in the other. They would probably feel about the same.

So, it is reasonable to say the mass of one magazine is about 25 grams.

Practice

Circle the more reasonable estimate for each mass.

1.

1,500 grams 1,500 kilograms

2.

5 grams 5 kilograms

3.

3 grams 3 kilograms

4.

14 grams 14 kilograms

Complete the table by writing a reasonable unit of mass, grams or kilograms, for each object.

	Mass of Fruits and Vegetables	
	Object	**Mass (g or kg)**
5.	grape	1 ☐
6.	pumpkin	2 ☐
7.	apple	150 ☐
8.	cantaloupe	1 ☐
9.	potato	1 ☐

Problem Solving

10. The mass of a pen cap is 1 unit. What metric unit, gram or kilogram, was used to measure the mass of the pen cap?

____

11. Mathematical PRACTICE 6 **Explain to a Friend** Julio is buying a carton of blueberries that has a mass of 100 grams. Is it reasonable to say that there are 250 blueberries in the carton? Explain.

____

Vocabulary Check

Match each vocabulary term to its definition or example.

12. kilogram • the amount of matter an object has

13. mass • a metric unit of mass equal to about 1 penny

14. gram • 1,000 grams

Test Practice

15. Which is a reasonable estimate for the mass of a toothbrush?

Ⓐ 2 grams Ⓑ 20 grams Ⓒ 200 grams Ⓓ 2,000 grams

Check My Progress

Vocabulary Check

Vocab

1. Use the word bank to complete the charts about the **metric system** of measurement.

centimeter **gram** **kilogram** **kilometer** **mass**

milliliter **millimeter** **meter** **liter**

length

mass

capacity

Concept Check

2. Use the word bank to write each vocabulary word next to its abbreviation.

mm ______________ cm ______________

mL ______________ km ______________

g ______________ kg ______________

L ______________ m ______________

Problem Solving

3. Adrianna went on a hiking trip. Which measurement best describes how far she hiked, 10 kilometers or 10 meters?

____

4. Which is the more reasonable estimate for the mass of a dog, 20 grams or 20 kilograms?

____

5. Raul has a bottle of salad dressing. Is 700 milliliters or 700 liters a more reasonable estimate for the capacity of the bottle of salad dressing?

____

My Work!

Test Practice

6. Which of the following holds about 800 milliliters of water?

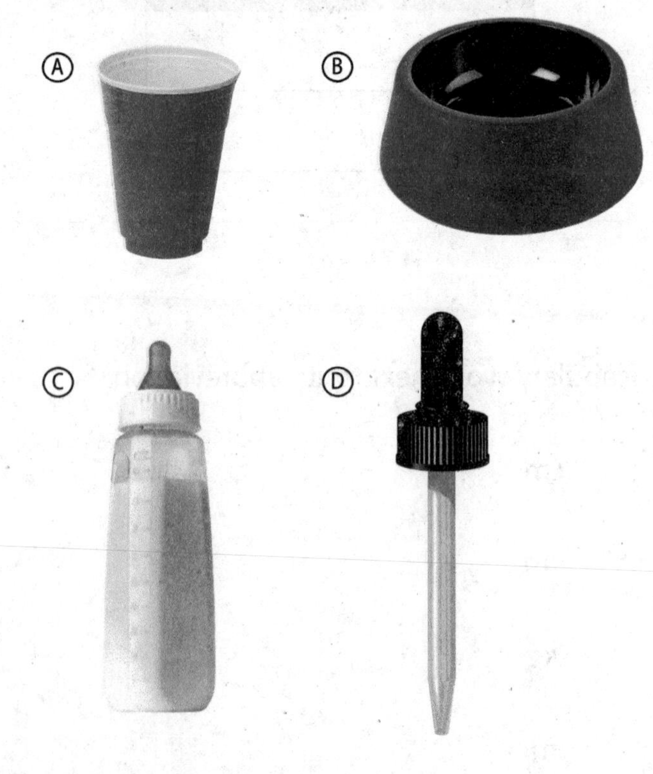

Name

Problem-Solving Investigation

STRATEGY: Make an Organized List

Lesson 4

ESSENTIAL QUESTION
How can conversion of measurements help me solve real-world problems?

Learn the Strategy

Watch Tutor

Sebastian has 0.24 of a dollar. How many different combinations of coins could he have?

1 Understand

What facts do you know?

Sebastian has ______ of a dollar.

What do you need to find?

the number of possible coin combinations

2 Plan

I will make an organized list to solve the problem.

3 Solve

0.24 of a dollar = 24¢

- 2 dimes, 4 pennies
- 1 dime, 2 nickels, 4 pennies
- 1 dime, 1 nickel, 9 pennies
- 4 nickels, 4 pennies
- 3 nickels, 9 pennies
- 2 nickels, 14 pennies
- 24 pennies
- 1 dime, 14 pennies
- 1 nickel, 19 pennies

There are ______ different combinations.

4 Check

Does your answer make sense? Explain.

______

Practice the Strategy

Brody has three cats. One has a mass of 4,523 grams. One has a mass of 5,012 grams. One has a mass of 4,702 grams. If Brody picks up two of the cats at once, what are the possible total masses that Brody is carrying?

1 Understand

What facts do you know?

____

____

What do you need to find?

____

2 Plan

____

3 Solve

4 Check

Does your answer make sense? Explain.

____

____

Name

Apply the Strategy

Solve each problem by making an organized list.

1. Mathematical PRACTICE 1 **Make a Plan** Brianna has 0.16 of a dollar. How many different combinations of coins could she have?

2. There were three races at the track meet. The distances were 100 meters long, 800 meters long, and 3,200 meters long. Suppose Lucy ran two of the races. What are the possible total distances that she ran?

3. Aaron has 3,700 milliliters of lemonade in a pitcher. He has three cups. Their capacities are 320 milliliters, 495 milliliters, and 583 milliliters. Suppose Aaron fills two of the cups. What are the possible capacities of lemonade that he could have left in the pitcher?

4. Michael has 0.18 of a dollar. How many possible combinations of coins could he have?

5. Dean has four pieces of clay to make a clay pot. The masses of the pieces are 10 grams, 15 grams, 20 grams, and 14 grams. If he uses three of the pieces, what are the possible total masses of the clay pot?

My Work!

Review the Strategies

Use any strategy to solve each problem.

- Make an organized list.
- Guess, check, and revise.
- Find extra or missing information.
- Use logical reasoning.

6. There are three trees in a backyard. The second tree is half as tall as the first. The third tree is taller than the second tree and shorter than the first tree. The total height of the trees is 24 feet. What is the height of each tree?

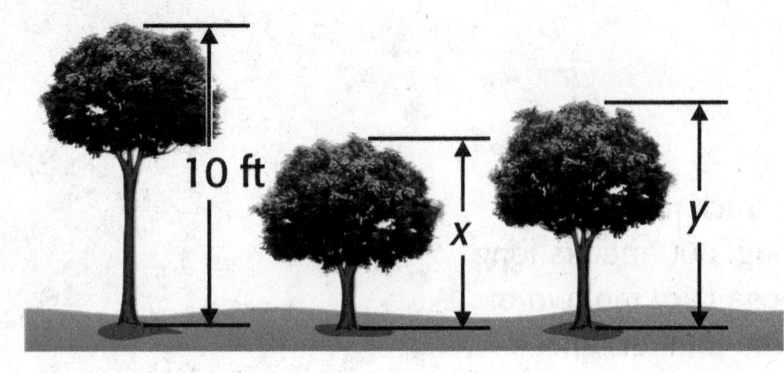

____

7. There are three lines. The first line is 3 times as long as the second. The second line is 4 meters longer than the third. The third line is 2 meters long. How long is the first line?

____

8. Darin has 5 coins that total 62¢. What are the coins?

____

____

9. Mathematical PRACTICE 4 **Model Math** Alfonso, Erik, Owen, and Alek are going hiking in pairs. How many different pairs of hiking partners are possible? List them.

____

____

____

My Work!

Name ....

MY Homework

Lesson 4
Problem Solving: Make an Organized List

Homework Helper

eHelp Need help? connectED.mcgraw-hill.com

Quinn's backpack can carry 5 kilograms of mass. Look at the items in the chart. What possible combinations of items can Quinn carry in her backpack without exceeding 5 kilograms?

Item	Mass
math book	3 kg
art supply kit	2 kg
lunch box	2 kg
water bottle	1 kg

1 Understand

What facts do you know?

Quinn's backpack can carry 5 kilograms of mass.

What do you need to find?

I need to find the possible combinations of items Quinn can carry in her backpack.

2 Plan

I will make an organized list of the possible combinations

3 Solve

- math book, art supply kit–5 kg
- math book, lunch box–5 kg
- math book, water bottle–4 kg
- art supply kit, lunch box, water bottle–5 kg
- art supply kit, lunch box–4 kg
- art supply kit, water bottle–3 kg
- lunch box, water bottle–3 kg

So, there are seven possible combinations.

4 Check

Does the answer make sense?

Yes. I listed each combination and its total mass. None of the masses are greater than 5 kilograms.

So, the answer is reasonable.

Problem Solving

Solve each problem by making an organized list.

My Work!

1. Paul's bathtub is clogged. He has to empty 30 liters of water by hand. Paul has a 3-liter, a 4-liter, and a 5-liter bucket. If Paul carries two buckets each trip, what combinations of sizes allow him to empty the bathtub in exactly four trips?

2. Tyra is training for a bicycle race. Each week she rides a total distance greater than 10 kilometers and less than or equal to 30 kilometers. If the distance is always an even number and a multiple of 3, what are the possible distances Tyra rides in one week?

3. Mathematical PRACTICE 1 **Keep Trying** Lexi's bulletin board is 40 centimeters wide. Each of her ribbons is 4 centimeters wide, and her photos are 12 centimeters wide. What combinations of ribbons and photos will fit side by side with no overlap on Lexi's bulletin board?

4. Carmen buys a pack of crackers for 75 cents from the vending machine. She puts a $1-bill in the machine. What combination of coins, excluding pennies, could Carmen get in change?

Name

Convert Metric Units

Lesson 5

ESSENTIAL QUESTION
How can conversion of measurements help me solve real-world problems?

You can multiply to convert, or change between, units.

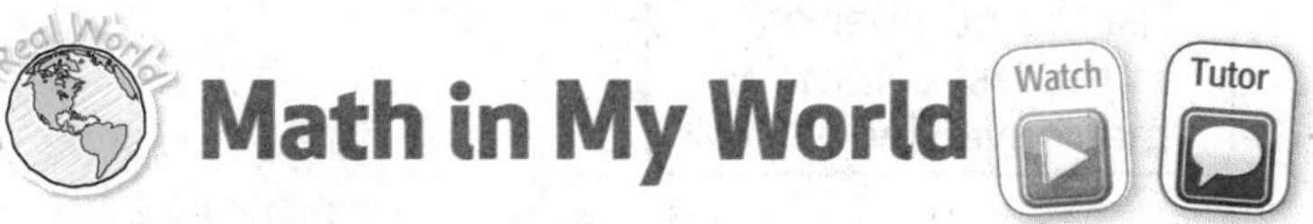

Example 1

The tree in Camryn's front yard is 4 meters tall. How many centimeters tall is the tree?

Since meters are larger than centimeters, multiply.

$4 \times 100 = 400$

Multiply by 100 because there are 100 centimeters in each meter.

4 meters = ______ centimeters

So, the tree is ______ centimeters tall.

Metric Units of Length
1 centimeter (cm) = 10 millimeters (mm)
1 meter (m) = 100 centimeters (cm)
1 kilometer (km) = 1,000 meters (m)

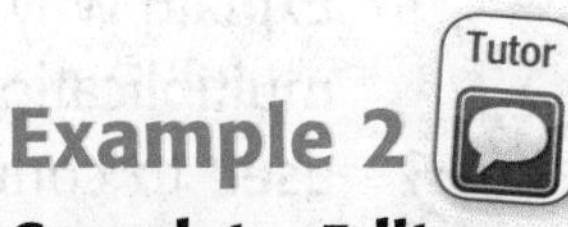

Example 2

Complete. 5 liters = ■ milliliters

Since liters are larger than milliliters, multiply.

$5 \times 1{,}000 = 5{,}000$

Multiply by 1,000 because there are 1,000 milliliters in each liter.

So, 5 liters = ______ milliliters.

Metric Units of Capacity
1 liter (L) = 1,000 milliliters (mL)

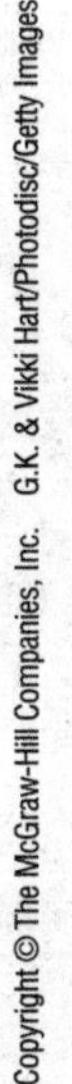

Example 3

Convert 7 kilograms to grams.

Metric Units of Mass
1 kilogram (kg) = 1,000 grams (g)

7 kilograms = ? grams

Kilograms are larger than grams. So, use multiplication.

Multiply by 1,000 because 1 kilogram = 1,000 grams.

1,000 × 7 = ______

So, 7 kilograms = ______ grams.

Guided Practice

Complete each conversion table.

List the number pairs in the last column of the conversion table.

1.

kilometers (km)	meters (m)	(km, m)
1	1,000	(1, 1,000)
2		
3		
4		

2.

centimeters (cm)	millimeters (mm)	(cm, mm)
1		
2		
3		
4		

3.

meters (m)	centimeters (cm)	(m, cm)
5		
6		
7		
8		

4.

liters (L)	milliliters (mL)	(L, mL)
1	1,000	(1, 1,000)
2		
3		
4		

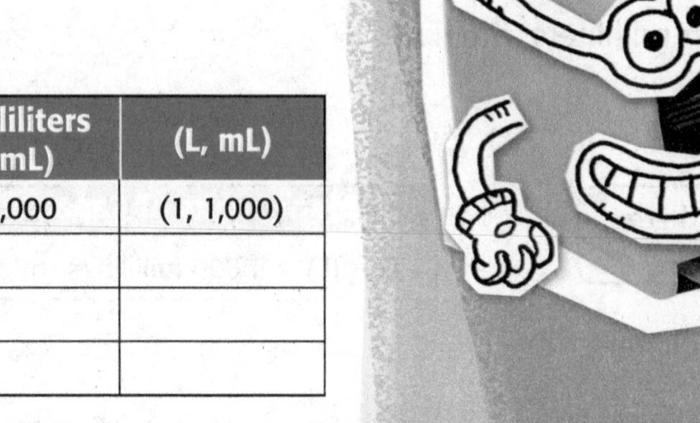

Name ....

Independent Practice

Complete each conversion table.

5.

meters (m)	centimeters (cm)	(m, cm)
4		
5		
8		
9		

6.

kilograms (kg)	grams (g)	(kg, g)
7		
9		
11		
13		

Algebra Find each unknown number.

7. 6 L = ■ mL

■ = ______

8. 5 m = ■ cm

■ = ______

9. 2 kg = ■ g

■ = ______

10. 5 cm = ■ mm

■ = ______

11. 12 kg = ■ g

■ = ______

12. 4 m = ■ mm

■ = ______

13. 5 L = ■ mL

■ = ______

14. 7 km = ■ m

■ = ______

15. 19 m = ■ cm

■ = ______

16. 9 kg = ■ g

■ = ______

17. 18 L = ■ mL

■ = ______

18. 22 cm = ■ mm

■ = ______

19. How many times larger is one kilogram than one gram? ______

20. Mathematical PRACTICE 2 **Use Number Sense** How many times longer is one kilometer than one meter? ______

21. How many times longer is one meter than one centimeter? ______

Problem Solving

22. The mass of Kendall's bicycle is 12 kilograms. What is the mass of the bicycle in grams?

23. Mrs. Liu's house is 7 meters tall. How tall is the house in centimeters?

24. Mathematical PRACTICE 2 **Use Number Sense** Javier needs 2 liters of iced tea for a picnic. How many milliliters of iced tea does he need?

25. Avery's dad is running a race that is 6 kilometers long. How many meters is that race?

My Work!

HOT Problems

26. Mathematical PRACTICE 3 **Which One Doesn't Belong?** Circle the measurement that does not belong. Explain.

300 grams	10 kilograms	10 pounds	600 grams

27. **Building on the Essential Question** When converting from a larger unit to a smaller unit, why does the value of the measurement increase?

Name

MY Homework

Lesson 5
Convert Metric Units

Homework Helper

eHelp Need help? connectED.mcgraw-hill.com

Liam is delivering trophies for the soccer team. The mass of one trophy is 2 kilograms. What is the mass in grams of the trophy?

Kilograms are larger than grams, so you will multiply.

2 kilograms × 1,000 = 2,000 grams

So, the mass of the trophy in grams is 2,000 grams.

Helpful Hint

1 centimeter (cm) = 10 millimeters (mm)
1 meter (m) = 100 centimeters (cm)
1 kilometer (km) = 1,000 meters (m)

1 liter (L) = 1,000 milliliters (mL)

1 kilogram (kg) = 1,000 grams (g)

Kymie said her driveway is 14 meters long. How long is her driveway in centimeters?

Meters are larger than centimeters, so you will multiply.

14 meters × 100 = 1,400 centimeters

So, Kymie's driveway is 1,400 centimeters long.

Practice

Algebra Find each unknown number.

1. 7 kg = ■ g

■ = ______

2. ■ mm = 9 cm

■ = ______

3. 5 L = ■ mL

■ = ______

4. 23 m = ■ cm

■ = ______

5. 17 kg = ■ g

■ = ______

6. 450 cm = ■ mm

■ = ______

Problem Solving

7. Molly measured the distance her paper airplane flew. The paper airplane traveled 5 meters. How many centimeters did the paper airplane travel?

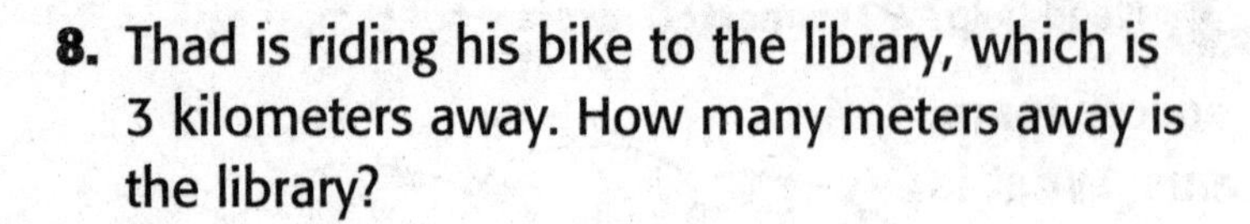

8. Thad is riding his bike to the library, which is 3 kilometers away. How many meters away is the library?

9. Patrick has 5 liters of water. How many milliliters of water does he have?

10. Maria's suitcase has a mass of 14 kilograms. How many grams is the mass of her suitcase?

11. Mathematical PRACTICE 2 **Use Number Sense** Minh is packing books into boxes. The mass of one of the boxes is 20 kilograms. What is the mass of the box in grams?

My Work!

Test Practice

12. Which is equivalent to 300 meters?

Ⓐ 30 kilometers
Ⓑ 3 kilometers
Ⓒ 30,000 centimeters
Ⓓ 3,000 centimeters

Name

Solve Measurement Problems

Lesson 6

ESSENTIAL QUESTION
How can conversion of measurements help me solve real-world problems?

Math in My World

Tools Tutor

Example 1

Lauren lives 0.2 kilometer from Alex. Alex lives three times as far from Colin's house than Lauren's house. How far away does Alex live from Colin?

Find 3×0.2.

You can use a number line to solve the problem.

Start at zero. Count by 0.2 three times.

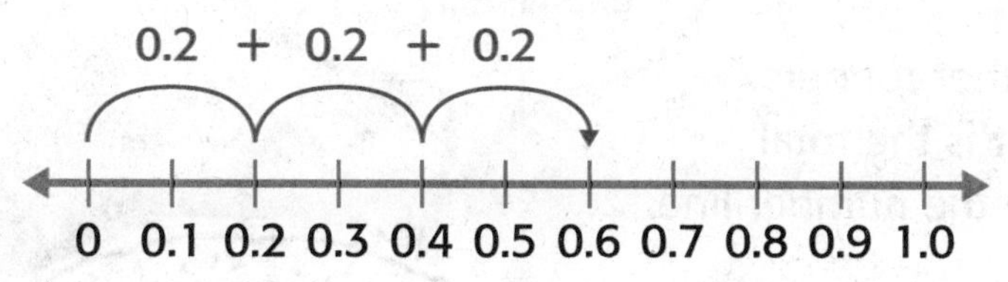

$3 \times 0.2 = 0.6$

So, Alex lives ______ kilometer from Colin.

Check

Convert 0.2 to a fraction. Then multiply the fraction by 3.

$0.2 = \textit{two tenths} = \frac{2}{10}$

$3 \times \frac{2}{10} = 3 \times \left(2 \times \frac{1}{10}\right)$ $\quad \frac{2}{10} = 2 \times \frac{1}{10}$

$= (3 \times 2) \times \frac{1}{10}$ $\quad$ Associative Property

$= 6 \times \frac{1}{10}$ $\quad$ Multiply. $3 \times 2 = 6$

$= \frac{6}{10}$ $\quad$ 6 groups of $\frac{1}{10}$ is $\frac{6}{10}$

Since $\frac{6}{10} = \textit{six tenths} = 0.6$, the answer is correct.

Helpful Hint
Think of $\frac{2}{10}$ as a multiple of $\frac{1}{10}$.

Example 2

Tutor

Javier poured 500 milliliters of lemon juice and 2 liters of water in a pitcher to make lemonade. How many milliliters of lemon juice and water did he pour into the pitcher in all?

1 Convert.

Convert 2 liters to milliliters.

Since 1 liter = 1,000 milliliters, multiply the number of liters by 1,000.

$2 \times 1{,}000 = 2{,}000$

So, 2 liters = 2,000 milliliters.

2 Add.

2,000 milliliters + 500 milliliters = 2,500 milliliters

So, Javier poured ________ milliliters of lemon juice and water into the pitcher.

Talk MATH

Explain how you can check your answer for Exercise 1.

Guided Practice

Check

1. Evelyn is in a relay race with three other runners. Each runner runs 0.1 kilometer. What is the total distance run by all four runners? Use the number line.

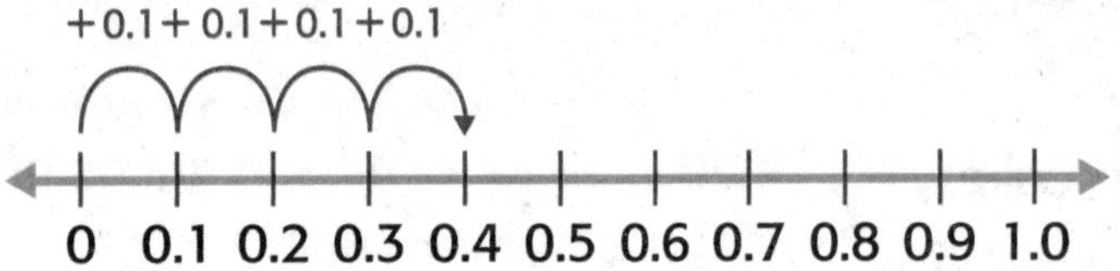

+0.1 + 0.1 + 0.1 + 0.1

0 0.1 0.2 0.3 0.4 0.5 0.6 0.7 0.8 0.9 1.0

____

2. A bag of potatoes has a mass of 4 kilograms. Some potatoes are taken out. The mass is now 2,305 grams. What is the mass of the potatoes in grams that were taken out of the bag?

____

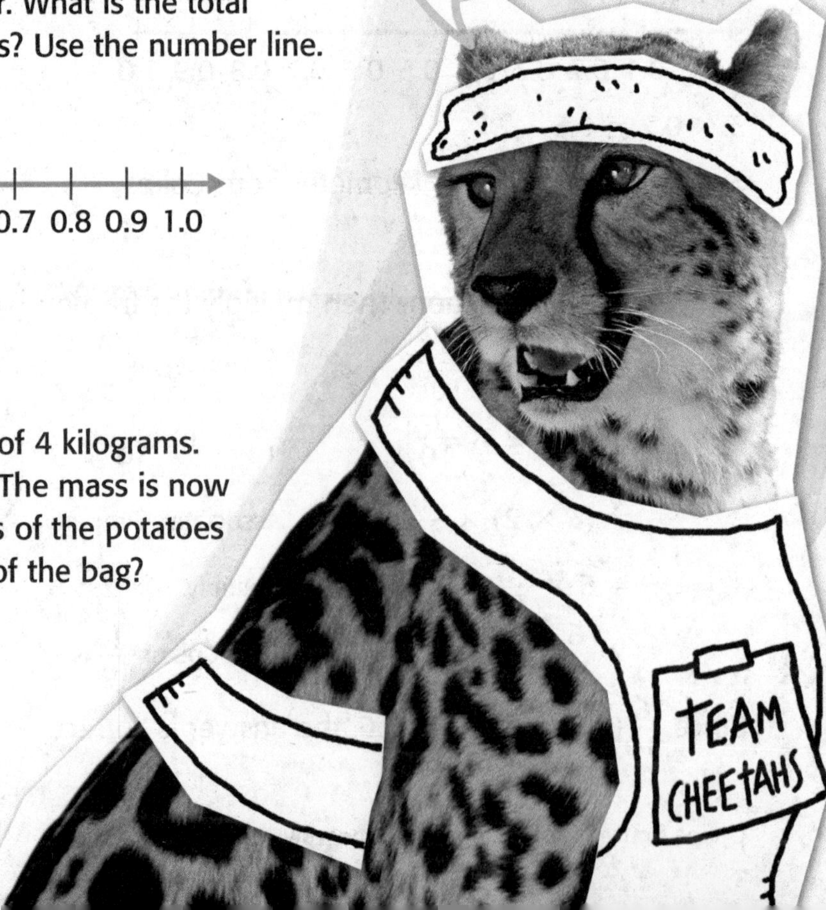

TEAM CHEETAHS

Name ....

Independent Practice

3. A ribbon is 1 meter long. Keira cut a piece of the ribbon. The piece she cut is 0.4 meter long. How long is the other piece? Use the number line.

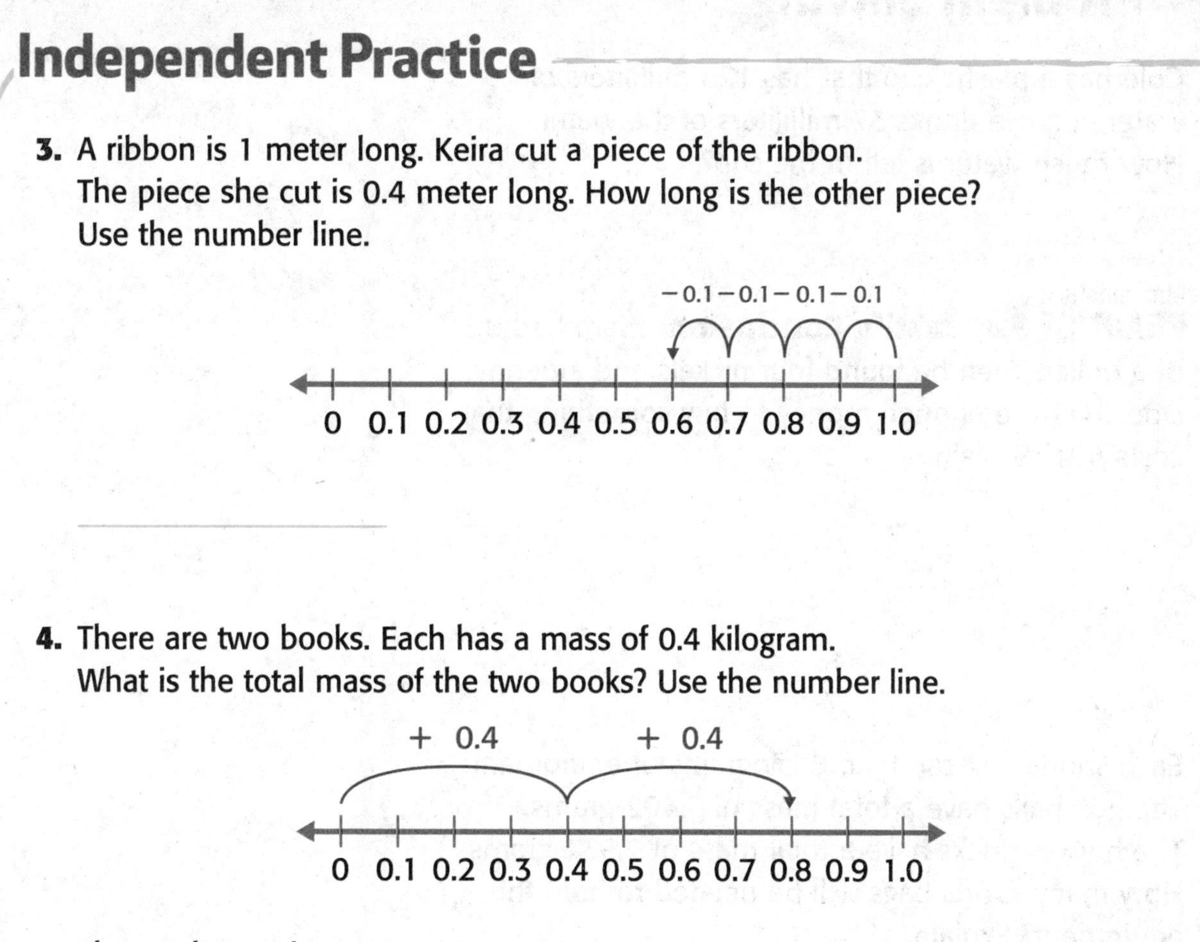

____

4. There are two books. Each has a mass of 0.4 kilogram. What is the total mass of the two books? Use the number line.

The total mass is ______________.

Convert to solve each problem. Draw a number line if needed.

5. One insect is 47 millimeters long. Another insect is 3 centimeters long. What is the total length in millimeters of the insects?

____

6. A table has a mass of 7 kilograms. A chair has a mass of 4,048 grams. What is the total mass in grams of the table and the chair?

____

Problem Solving

My Work!

7. Cole has a plastic cup that has 125 milliliters of water in it. He drinks 37 milliliters of the water. How much water is left in the cup?

8. Mathematical PRACTICE 3 **Justify Conclusions** Sam had 0.3 of a dollar. Then he found four nickels and a penny. Does he have enough money to buy something that costs 50¢? Explain.

9. Each sports bag can hold 6 kilograms of equipment. The golf balls have a total mass of 3,402 grams. The hockey pucks have a total mass of 2,932 grams. How many sports bags will be needed to hold the equipment? Explain.

HOT Problems

10. Mathematical PRACTICE 2 **Use Symbols** Compare. Write <, >, or =.

3 L + 2,492 mL ◯ 2 L + 1,301 mL + 2,191 mL

11. **Building on the Essential Question** How do I know when it is necessary to convert units before solving a problem?

MY Homework

Lesson 6
Solve Measurement Problems

Homework Helper

eHelp Need help? connectED.mcgraw-hill.com

Alfredo has a board that is 2 meters long. He needs 3 pieces that are each 50 centimeters long. How many centimeters of the original board will be left after he cuts the 3 pieces he needs?

1. Convert the meters to centimeters. Meters are larger than centimeters, so you will multiply.
   2 meters × 100 = 200 centimeters

2. Multiply to find the total length of the 3 pieces Alfredo needs.
   3 × 50 centimeters = 150 centimeters

3. Subtract to find the new length of the original board.
   200 − 150 = 50 centimeters

So, Alfredo will have 50 centimeters left of the original board.

Practice

1. Chloe buys 3 jars of peanut butter for the food drive. Each jar has a mass of 0.2 kilogram. What is the mass of all 3 jars of peanut butter? Use the number line.

   ____

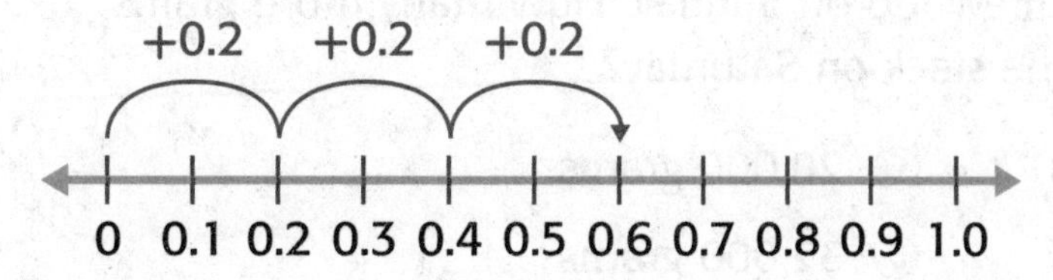

Problem Solving

2. Aidan mixes 630 milliliters of blue paint with 2 liters of red paint to make purple paint. How many milliliters of purple paint does Aidan have?

____

3. Celine is knitting a scarf. The finished length will be 1.2 meters. So far she has knitted 0.8 meters. How many more meters does Celine need to knit? Draw a number line to solve.

____

4. Joanie's bag of trail mix has a mass of 0.75 kilogram. She and some friends eat 0.5 kilogram of the mix. How many kilograms of trail mix are left? Draw a number line to solve.

____

5. Mathematical PRACTICE 5 **Use Math Tools** Ross found 5 ladybugs. Each is 0.8 centimeter long. If he lays the ladybugs in a row, what is the total length in millimeters? Draw a number line to solve.

____

6. Lucas is riding his bike to the park which is 2 kilometers from his house. When he is one-fourth of the way to the park, it starts to rain. Lucas turns around and rides back home. How many meters did Lucas ride?

____

Test Practice

7. Grayson stacked 17 kilograms of firewood on Saturday and 15 kilograms of firewood on Sunday. How many more grams of firewood did he stack on Saturday?

Ⓐ 2,000 grams
Ⓑ 3,200 grams
Ⓒ 20,000 grams
Ⓓ 32,000 grams

Review

Chapter 12
Metric Measurement

Vocabulary Check

Vocab

Draw a line to the sentence that each vocabulary word completes.

1. **capacity**
2. **convert**
3. **mass**
4. **metric system**
5. **length**

- Centimeter, gram, and liter are all examples of units of measure from the ____________.
- ____________ is the amount of liquid a container holds.
- ____________ is the measurement of a line between two points.
- When you change the unit of measure, you ____________ measurements.
- The amount of matter that an object has is known as its ____________.

Color the metric units of length red. Color the metric units of capacity blue. Color the metric units of mass green. Then write each abbreviation on the roof.

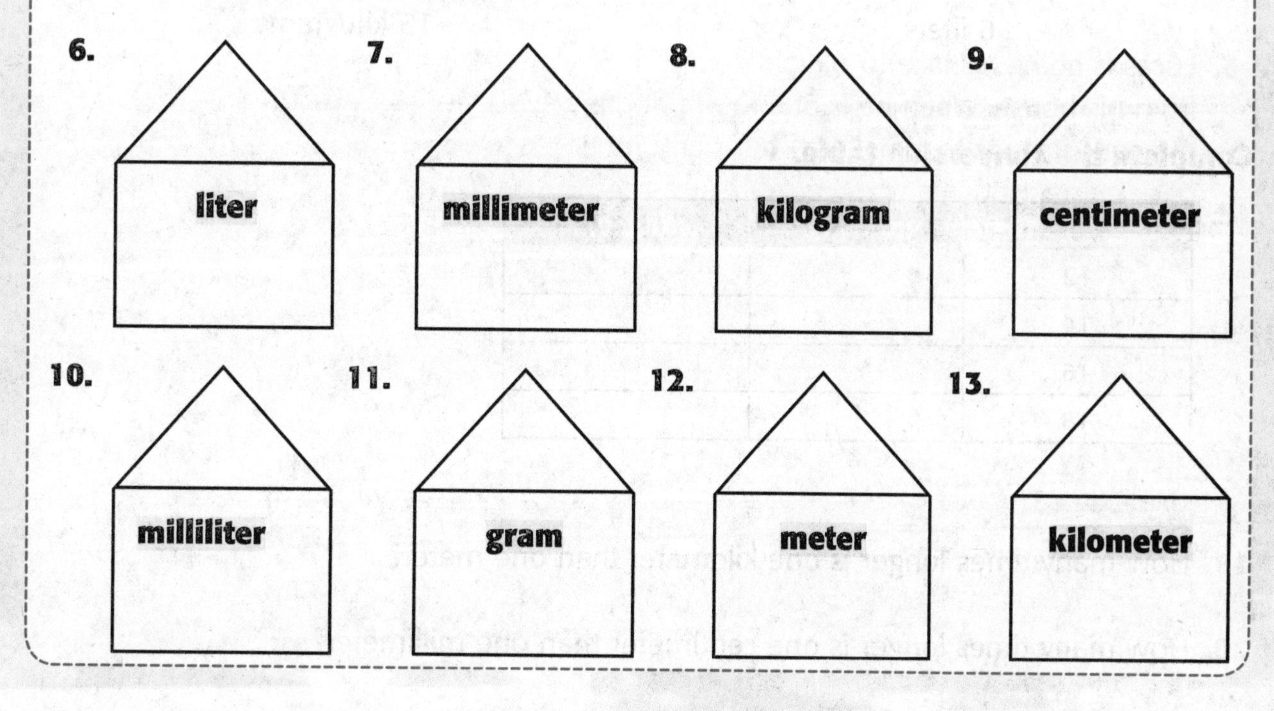

Concept Check

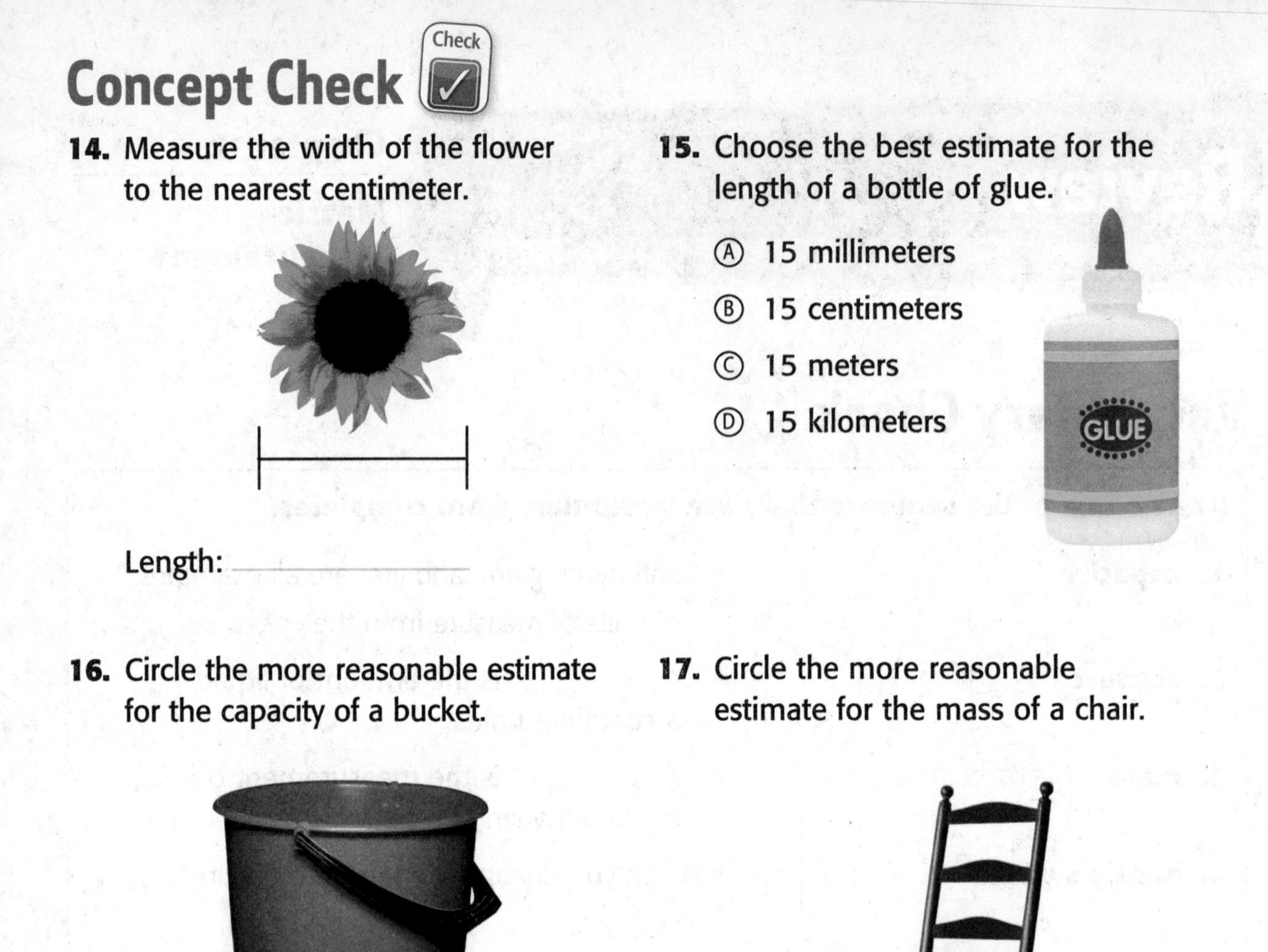

14. Measure the width of the flower to the nearest centimeter.

Length: ________

15. Choose the best estimate for the length of a bottle of glue.

Ⓐ 15 millimeters

Ⓑ 15 centimeters

Ⓒ 15 meters

Ⓓ 15 kilometers

16. Circle the more reasonable estimate for the capacity of a bucket.

6 milliliters

6 liters

17. Circle the more reasonable estimate for the mass of a chair.

15 grams

15 kilograms

Complete the conversion table.

18.

kilograms (kg)	grams (g)	(kg, g)
12		
14		
16		
18		

19. How many times longer is one kilometer than one meter? ________

20. How many times longer is one centimeter than one millimeter? ________

Name

Problem Solving

21. Carson has 0.21 of a dollar. How many different combinations of coins could he have?

22. There are three picture frames. Each has a mass of 0.2 kilogram. What is the total mass of the three picture frames? Use the number line.

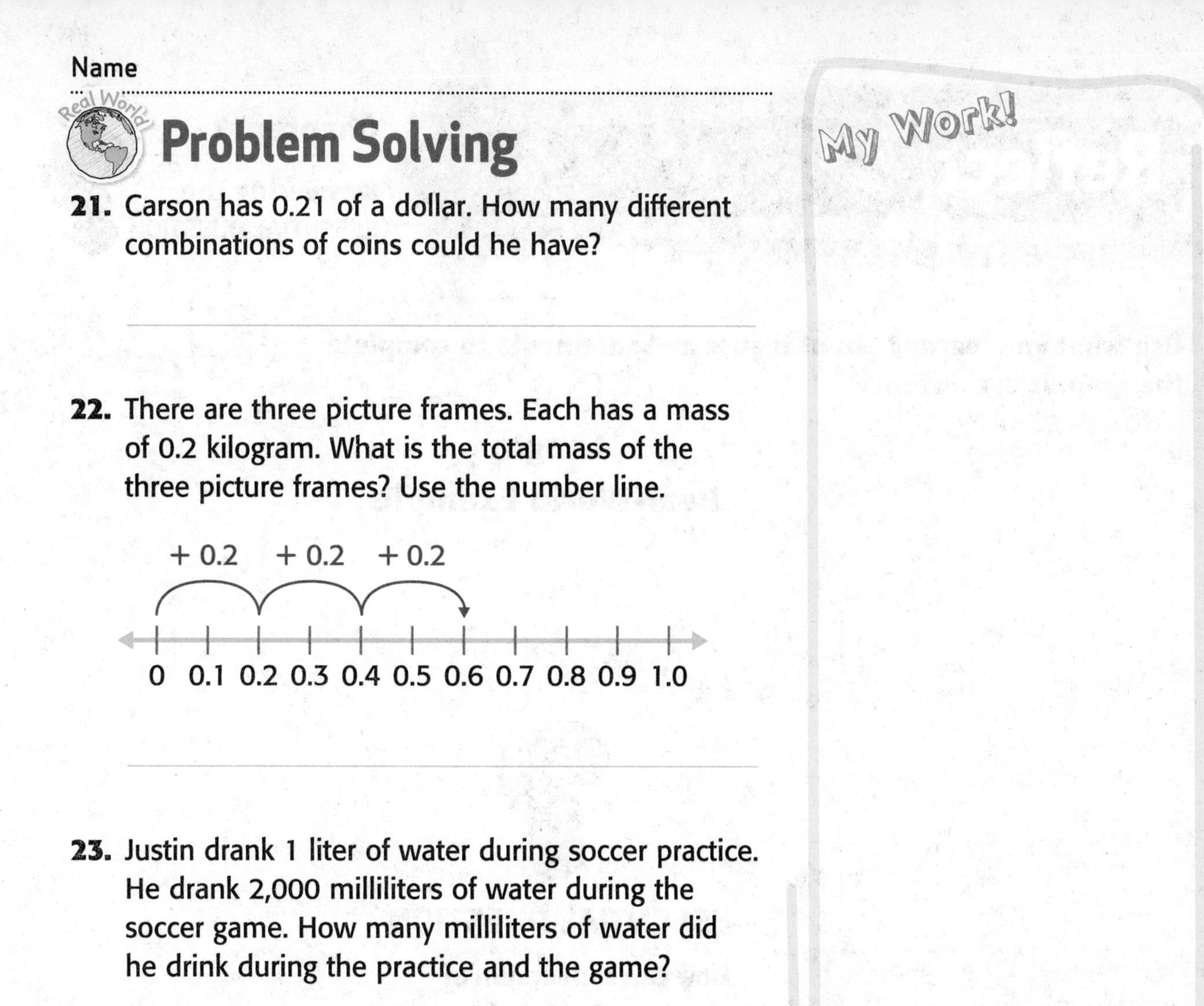

23. Justin drank 1 liter of water during soccer practice. He drank 2,000 milliliters of water during the soccer game. How many milliliters of water did he drink during the practice and the game?

24. Julian walked a distance of 2 meters. Keira walked a distance of 300 centimeters. Which distance is longer?

Test Practice

25. Henry's water bottle has a capacity of 1 liter. What is the capacity of Henry's water bottle in milliliters?

Ⓐ 1 milliliter
Ⓑ 10 milliliters
Ⓒ 100 milliliters
Ⓓ 1,000 milliliters

Chapter 12

Answering the ESSENTIAL QUESTION

Use what you learned about metric measurements to complete the graphic organizer.

Length
Real-World Example

ESSENTIAL QUESTION

How can conversion of measurements help me solve real-world problems?

Capacity
Real-World Example

Mass
Real-World Example

Reflect on the ESSENTIAL QUESTION  **Write your answer below.**

Chapter 13 Perimeter and Area

ESSENTIAL QUESTION

Why is it important to measure perimeter and area?

Let's Cheer OUR TEAM!

Watch a video!

Watch

MY Common Core State Standards

Measurement and Data

4.MD.3 Apply the area and perimeter formulas for rectangles in real world and mathematical problems.

I'll be able to get this—no problem!

Standards for Mathematical PRACTICE

1. Make sense of problems and persevere in solving them.
2. Reason abstractly and quantitatively.
3. Construct viable arguments and critique the reasoning of others.
4. Model with mathematics.
5. Use appropriate tools strategically.
6. Attend to precision.
7. Look for and make use of structure.
8. Look for and express regularity in repeated reasoning.

= focused on in this chapter

Name

Check

← Go online to take the Readiness Quiz

Draw an array to model each multiplication problem. Find the product.

1. $4 \times 8 =$ ______

2. $7 \times 7 =$ ______

3. $3 \times 12 =$ ______

4. $10 \times 4 =$ ______

5. $9 \times 11 =$ ______

6. $4 \times 15 =$ ______

7. Cora is making a picture. It will be a square. How many sides are the same length?

______

How Did I Do?

Shade the boxes to show the problems you answered correctly.

1	2	3	4	5	6	7

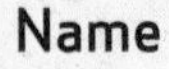

Name

Review Vocabulary

length   product

Making Connections

Use the review vocabulary to complete the chart.

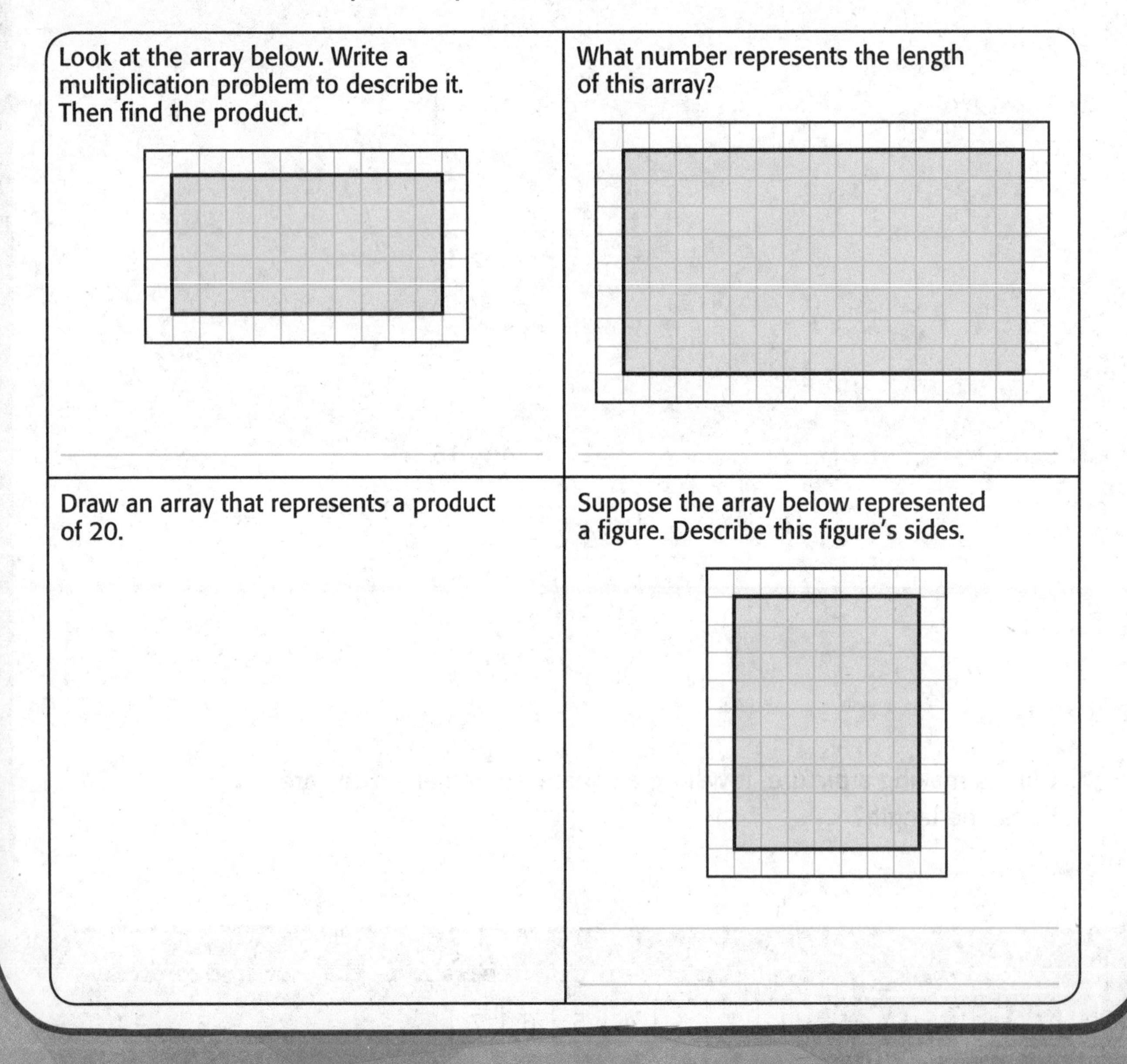

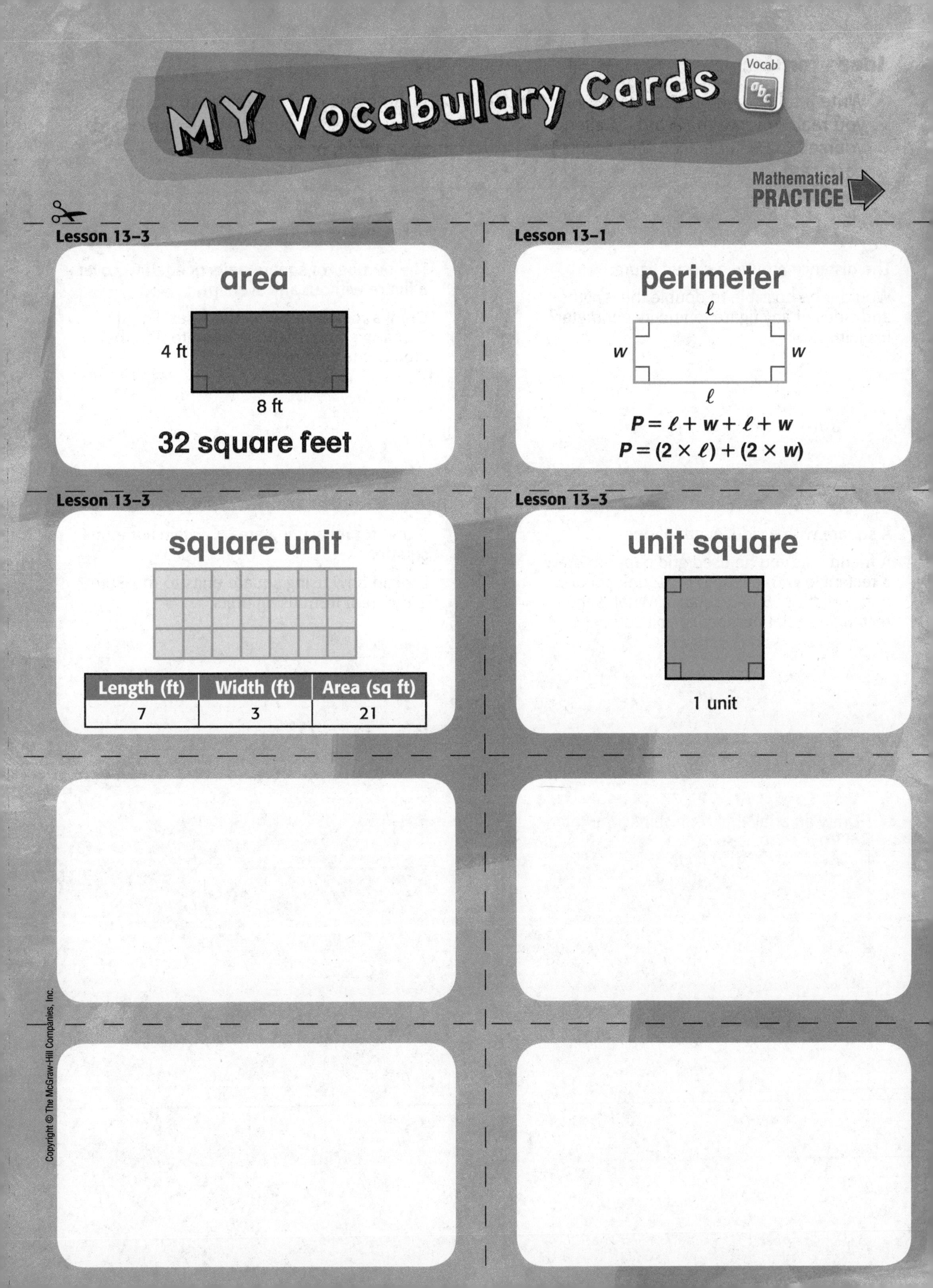
MY Vocabulary Cards
Vocab
abc
Mathematical PRACTICE
Lesson 13-3
area
4 ft
8 ft
32 square feet
Lesson 13-1
perimeter
ℓ
w
w
ℓ
P = ℓ + w + ℓ + w
P = (2 × ℓ) + (2 × w)
Lesson 13-3
square unit
Length (ft) | Width (ft) | Area (sq ft)
7 | 3 | 21
Lesson 13-3
unit square
1 unit

Ideas for Use

- Write a tally mark on each card every time you read or write the word. Challenge yourself to use at least 3 tally marks for each card.
- Use the blank cards to review problem-solving strategies, such as work backward, make a table, or draw a picture.

The distance around a closed figure.

Would it be possible to double the lengths and sides of any figure to find its perimeter? Explain.

The number of square units needed to cover a figure without any overlap.

Draw a rectangle. Label the sides' lengths. Exchange papers with a friend to find the area of the rectangle.

A square with a side length of one unit.

A friend tells you he used grid paper to draw a rectangle with a length of 12 unit squares and a width of 5 unit squares. What is the rectangle's total number of unit squares?

A unit for measuring area. Contains one unit square.

Explain how using square units to measure is different from using units.

MY Foldable

FOLDABLES® Follow the steps on the back to make your Foldable.

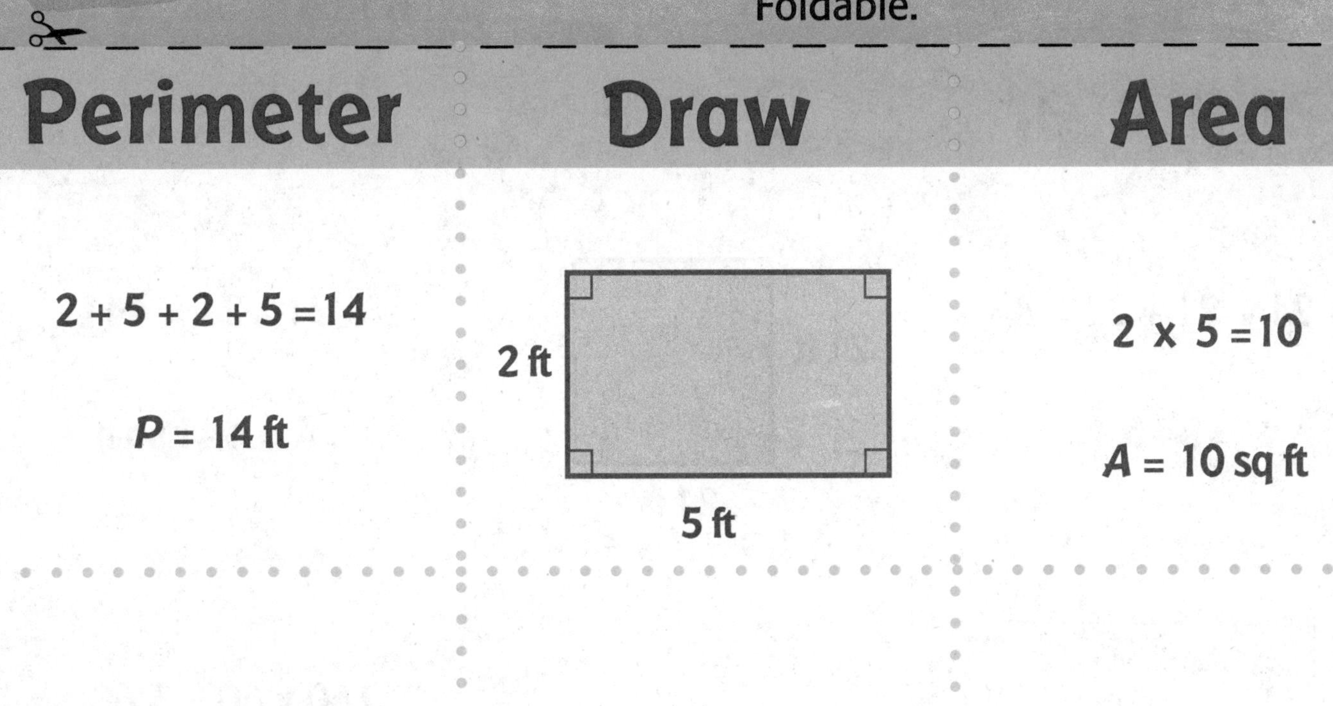

Perimeter	Draw	Area
2 + 5 + 2 + 5 = 14 *P* = 14 ft	2 ft 5 ft	2 x 5 = 10 *A* = 10 sq ft
20 + 23 + 20 + 23 = 86 *P* = 86 in.		
	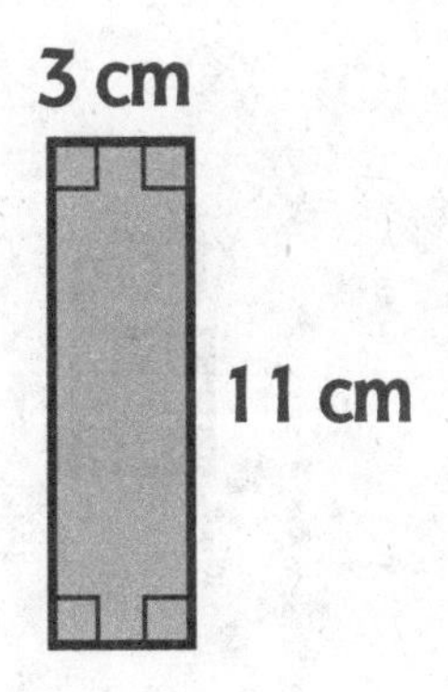	
		26 x 4 = 104 *A* = 104 sq in.

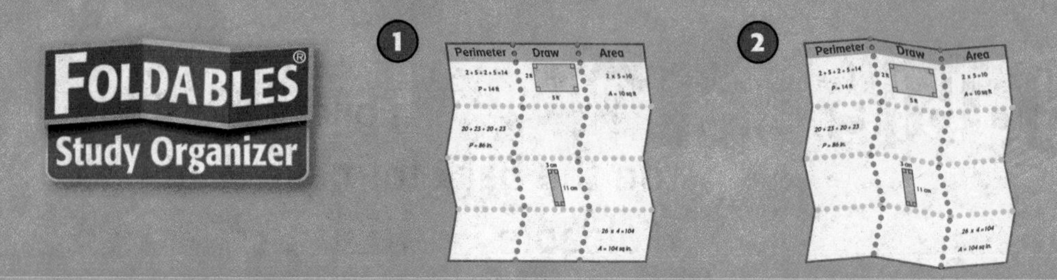

Perimeter Draw Area

$21 + 21 + 21 + 21 = 84$

$P = 84$ ft

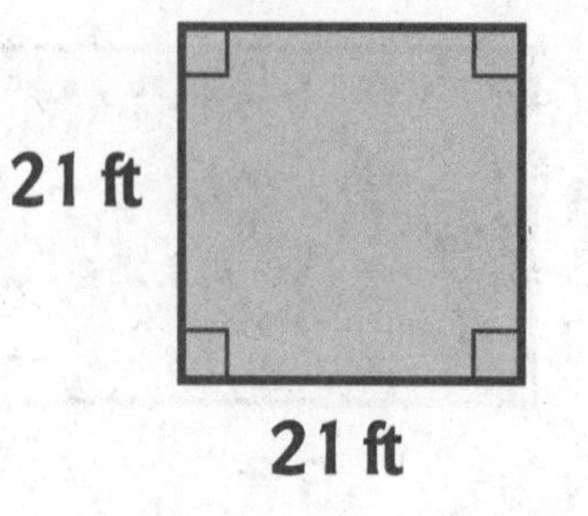

$21 \times 21 = 441$

$A = 441$ sq ft

$60 \times 60 = 3{,}600$

$A = 3{,}600$ sq yd

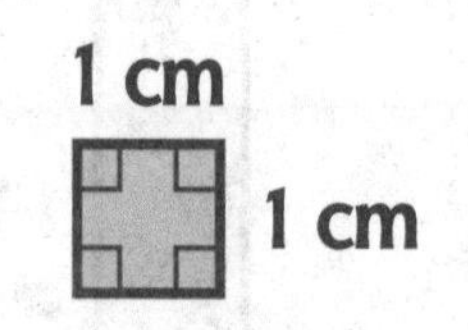

$34 + 34 + 34 + 34 = 136$

$P = 136$ in.

Measure Perimeter

Lesson 1

ESSENTIAL QUESTION
Why is it important to measure perimeter and area?

The distance around a closed figure is called the **perimeter**.

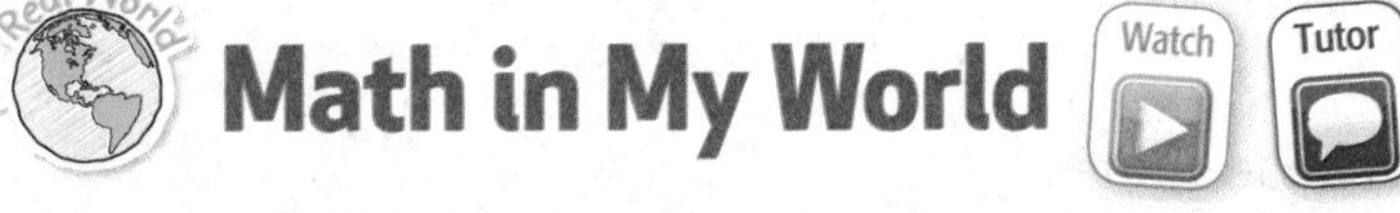

Math in My World

Watch Tutor

Example 1

Berto walked around a park on the rectangular path shown. How far did Berto walk?

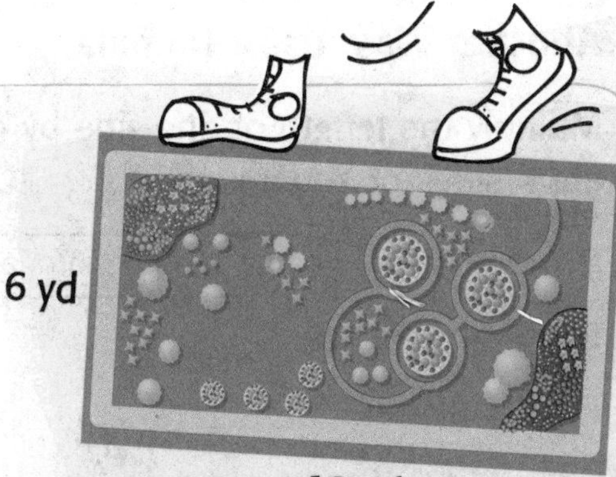

Opposite sides of a rectangle are equal. So, the side lengths are 12 yards, 12 yards, 6 yards, and 6 yards.

Add the measures of all of the sides of the figure.

Perimeter = 12 yards + 12 yards + 6 yards + 6 yards

Perimeter = ______ yards

So, Berto walked ______ yards.

Key Concept Perimeter of a Rectangle

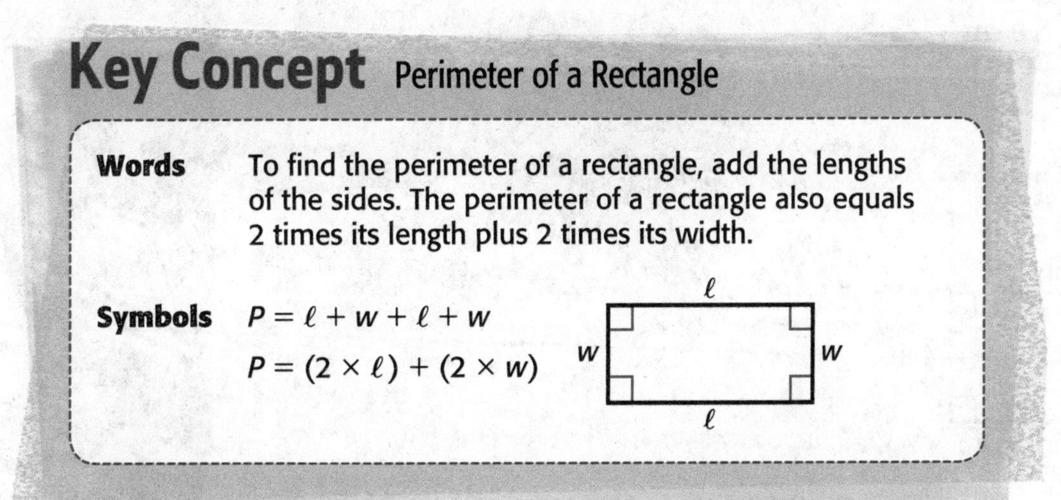

Words To find the perimeter of a rectangle, add the lengths of the sides. The perimeter of a rectangle also equals 2 times its length plus 2 times its width.

Symbols $P = \ell + w + \ell + w$

$P = (2 \times \ell) + (2 \times w)$

A square has four sides of equal length. To find the perimeter of a square, multiply the length of one side by four.

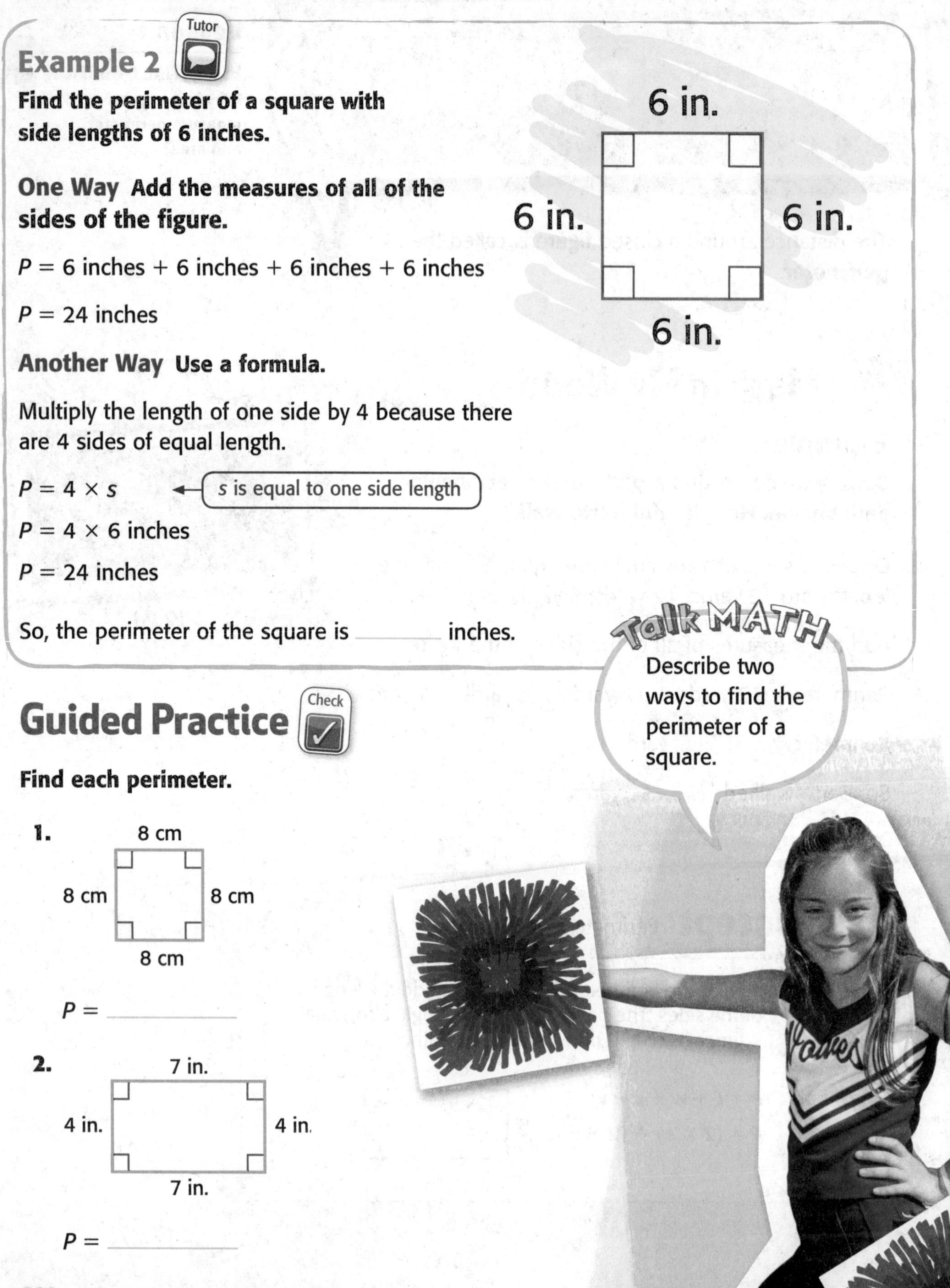

Example 2

Find the perimeter of a square with side lengths of 6 inches.

One Way **Add the measures of all of the sides of the figure.**

$P = 6$ inches $+ 6$ inches $+ 6$ inches $+ 6$ inches

$P = 24$ inches

Another Way **Use a formula.**

Multiply the length of one side by 4 because there are 4 sides of equal length.

$P = 4 \times s$ ← *s* is equal to one side length

$P = 4 \times 6$ inches

$P = 24$ inches

So, the perimeter of the square is ______ inches.

Talk MATH

Describe two ways to find the perimeter of a square.

Guided Practice

Find each perimeter.

1. 8 cm, 8 cm, 8 cm, 8 cm

$P =$ ______

2. 7 in., 4 in., 4 in., 7 in.

$P =$ ______

Name

Independent Practice

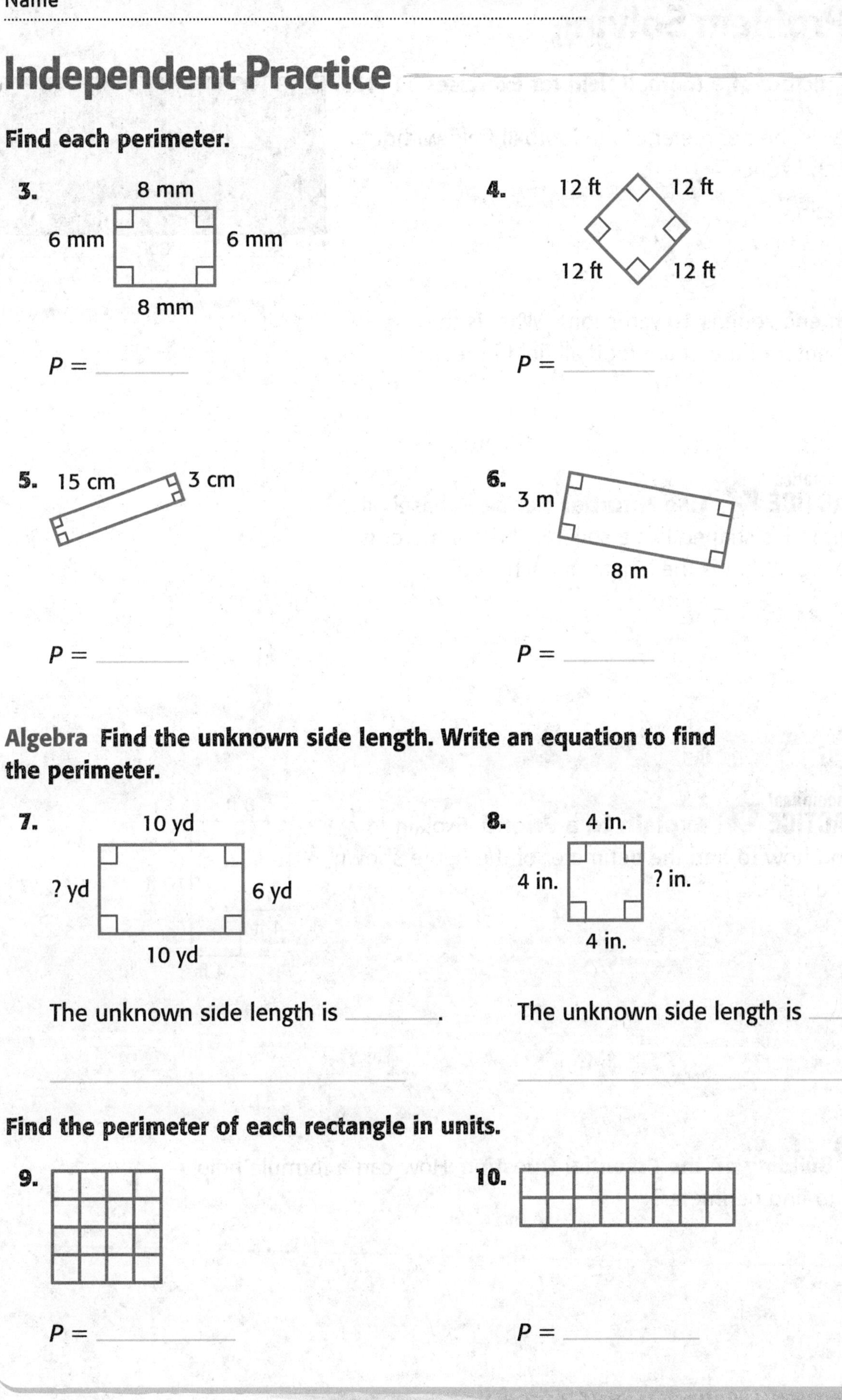

Find each perimeter.

3. 8 mm, 6 mm, 6 mm, 8 mm

$P =$ ______

4. 12 ft, 12 ft, 12 ft, 12 ft

$P =$ ______

5. 15 cm, 3 cm

$P =$ ______

6. 3 m, 8 m

$P =$ ______

Algebra Find the unknown side length. Write an equation to find the perimeter.

7. 10 yd, ? yd, 6 yd, 10 yd

The unknown side length is ______.

______

8. 4 in., 4 in., ? in., 4 in.

The unknown side length is ______.

______

Find the perimeter of each rectangle in units.

9.

$P =$ ______

10.

$P =$ ______

Use the picture of a football field for Exercises 11–12.

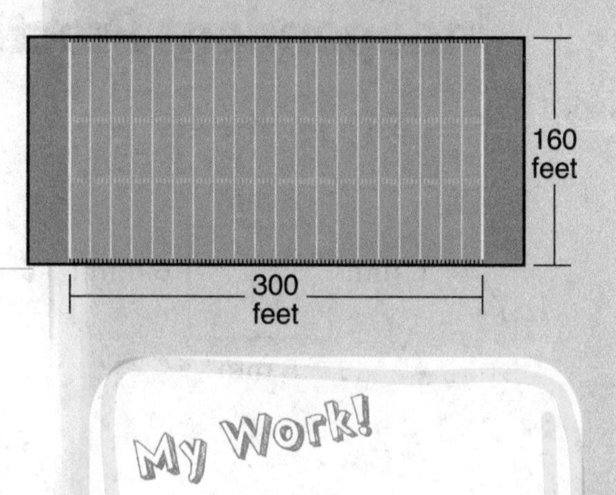

11. What is the perimeter of the football field without the end zones?

12. Each end zone is 10 yards long. What is the perimeter of the entire football field in feet?

My Work!

13. Mathematical PRACTICE 2 **Use Number Sense** A baseball diamond is shaped like a square. The perimeter is 360 feet. What is the length of each side?

HOT Problems

14. Mathematical PRACTICE 6 **Explain to a Friend** Explain to a friend how to find the perimeter of the figure shown to the right.

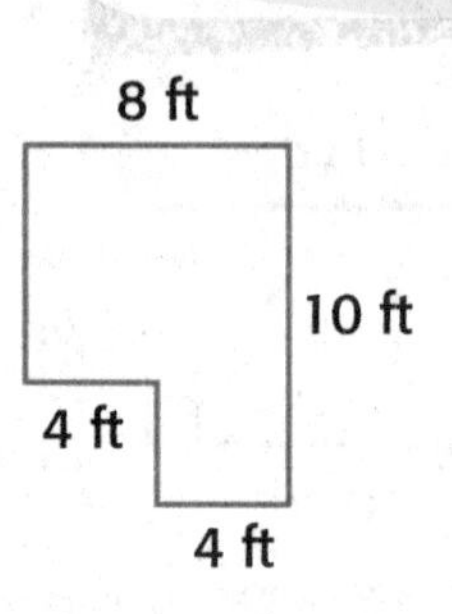

15. **Building on the Essential Question** How can a formula help me to find perimeter?

Name ....

MY Homework

Lesson 1
Measure Perimeter

Homework Helper

eHelp Need help? connectED.mcgraw-hill.com

Claire plans to glue ribbon around the edges of the picture frame. How much ribbon will she need?

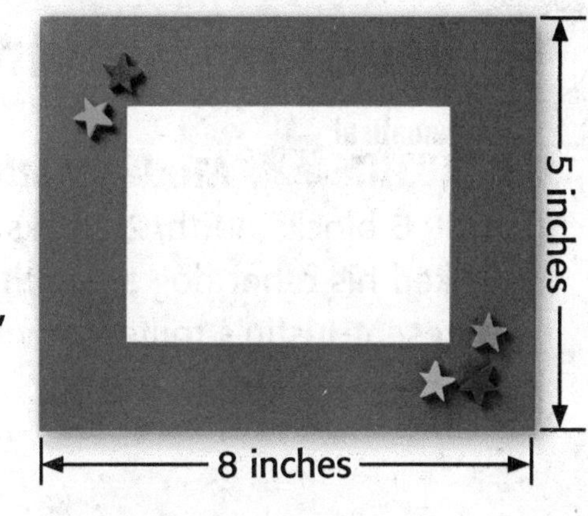

One Way **Add the measures of all sides of the figure.**

You know that opposite sides of a rectangle are equal, so the measures are 5 inches, 5 inches, 8 inches, and 8 inches.

P = 5 inches + 5 inches + 8 inches + 8 inches
P = 26 inches

Another Way **Use a formula.**

P = (2 × 8 inches) + (2 × 5 inches)
P = 16 inches + 10 inches
P = 26 inches

So, Claire will need 26 inches of ribbon.

Helpful Hint
The perimeter of a rectangle equals 2 times the length plus 2 times the width.
$P = (2 \times \ell) + (2 \times w)$

Practice

Find each perimeter.

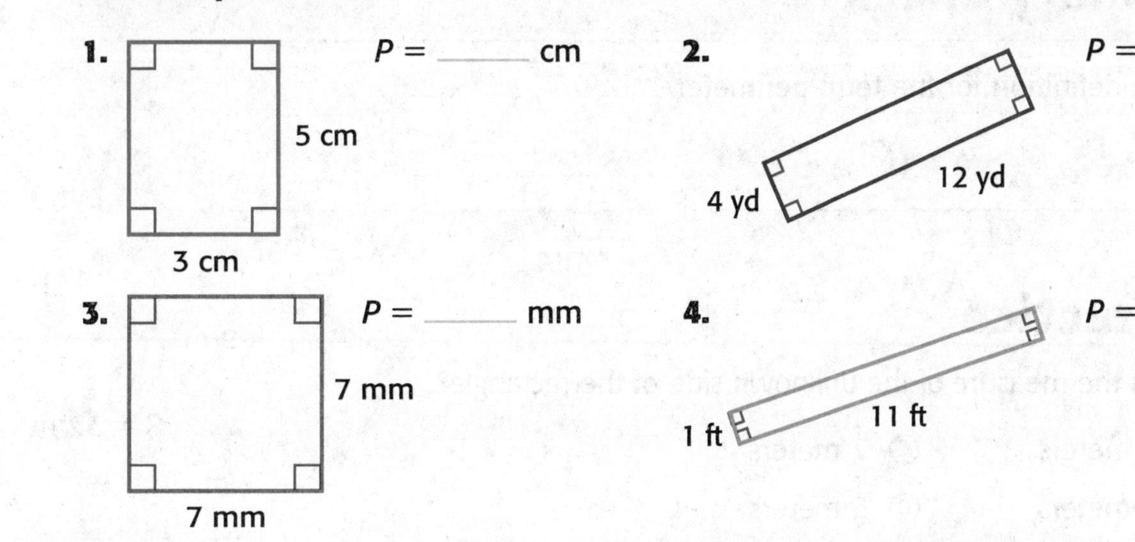

Find the perimeter of each rectangle in units.

5.

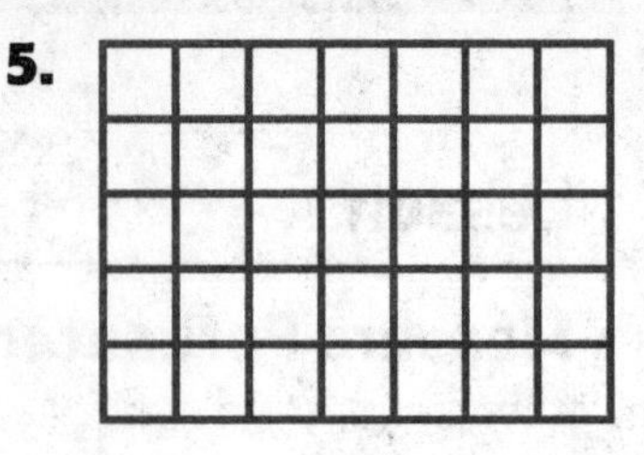

$P =$ ______ units

6.

$P =$ ______ units

7. Mathematical PRACTICE 4 **Model Math** Justin walked his dog 2 blocks west, 6 blocks north, 2 blocks east, and 6 blocks south. Then he walked his other dog along the same route. Draw a picture to represent Justin's route. How many blocks did Justin walk in all?

______

8. A rectangle has a perimeter of 30 inches. One side measures 5 inches. What are the measures of the other three sides?

______

9. Carolla is putting a border around the edge of a rectangular bulletin board. One side of the bulletin board measures 2 feet and another side measures 4 feet. Is 10 feet of border enough? Explain.

______

Vocabulary Check

10. Write a definition for the term perimeter. ______

______

Test Practice

11. What is the measure of the unknown side of the rectangle?

9 m

? m

$P = 32$ m

Ⓐ 23 meters
Ⓑ 14 meters
Ⓒ 7 meters
Ⓓ 5 meters

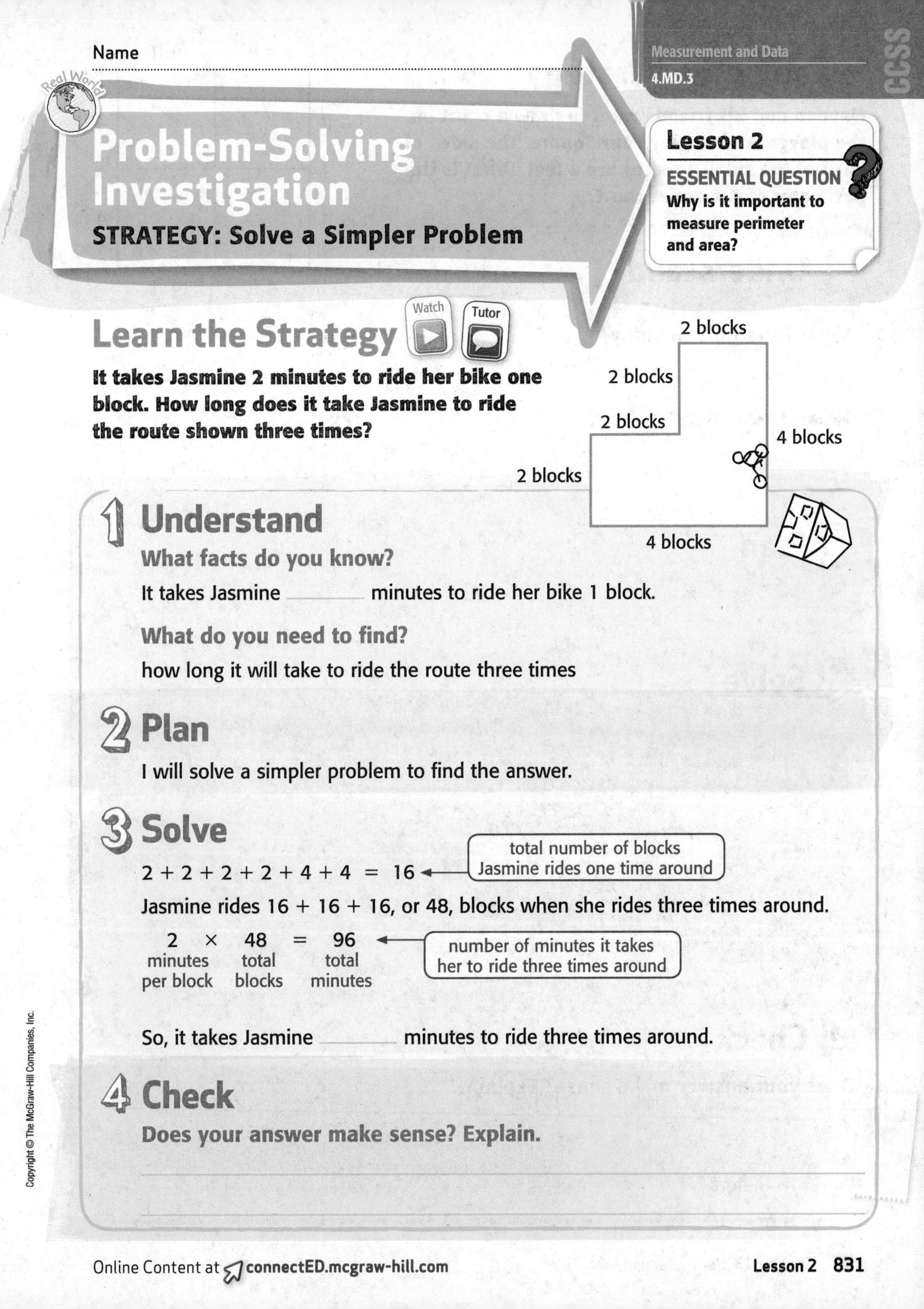

Problem-Solving Investigation

STRATEGY: Solve a Simpler Problem

Lesson 2

ESSENTIAL QUESTION

Why is it important to measure perimeter and area?

Learn the Strategy

It takes Jasmine 2 minutes to ride her bike one block. How long does it take Jasmine to ride the route shown three times?

1 Understand

What facts do you know?

It takes Jasmine ______ minutes to ride her bike 1 block.

What do you need to find?

how long it will take to ride the route three times

2 Plan

I will solve a simpler problem to find the answer.

3 Solve

$2 + 2 + 2 + 2 + 4 + 4 = 16$ ← total number of blocks Jasmine rides one time around

Jasmine rides 16 + 16 + 16, or 48, blocks when she rides three times around.

$2 \times 48 = 96$ ← number of minutes it takes her to ride three times around

(2 = minutes per block; 48 = total blocks; 96 = total minutes)

So, it takes Jasmine ______ minutes to ride three times around.

4 Check

Does your answer make sense? Explain.

______

______

Practice the Strategy

Hayden and his friends want to draw a court on the playground to play Four Square. The sides of each of the small squares are 4 feet. What is the perimeter of the entire court?

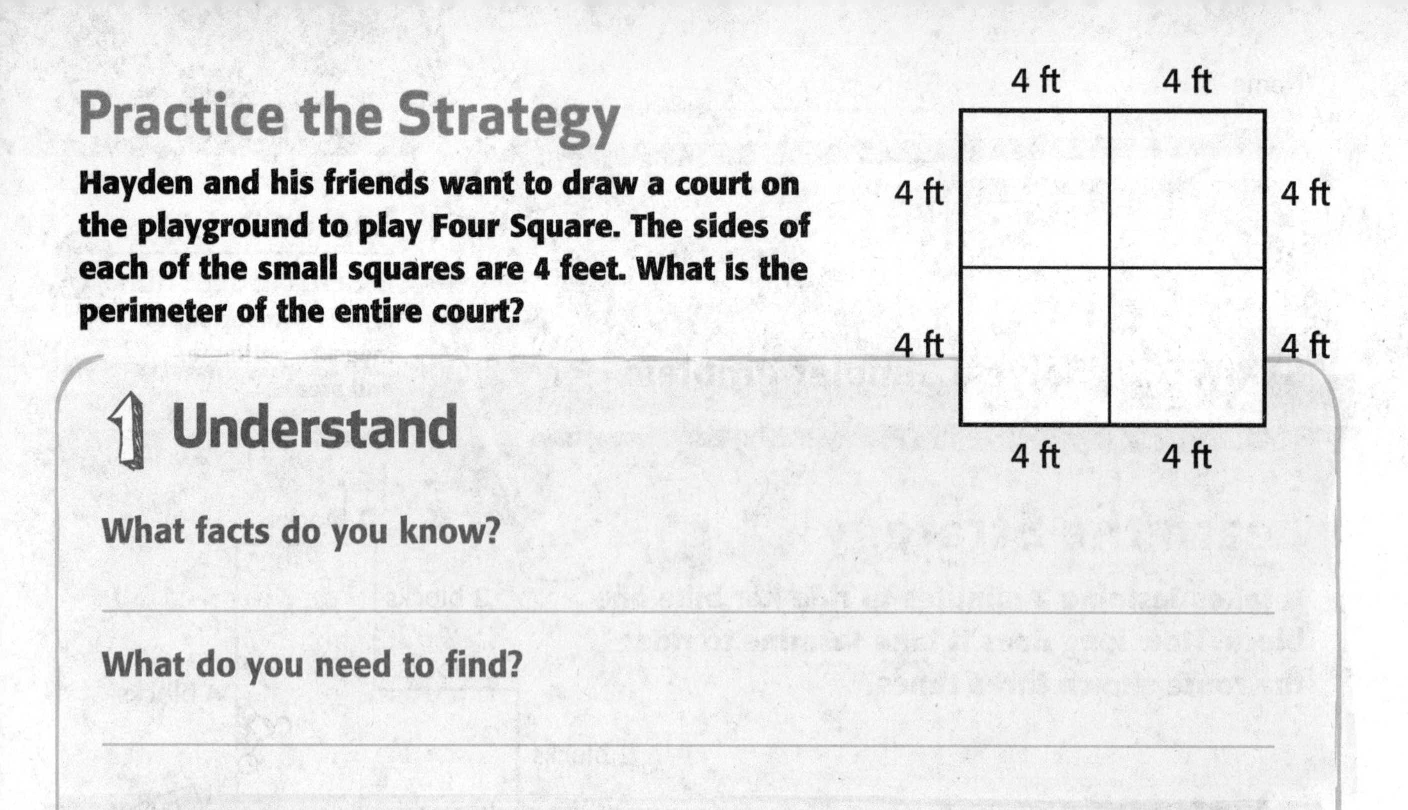

1 Understand

What facts do you know?

____

What do you need to find?

____

2 Plan

____

3 Solve

4 Check

Does your answer make sense? Explain.

____

____

Name

Apply the Strategy

Solve each problem by solving a simpler problem.

GO TEAM!

My Work!

1. Clarissa has four pictures that are each the size of the one shown. What will be the perimeter of the rectangle formed if the four pictures are laid end to end as shown?

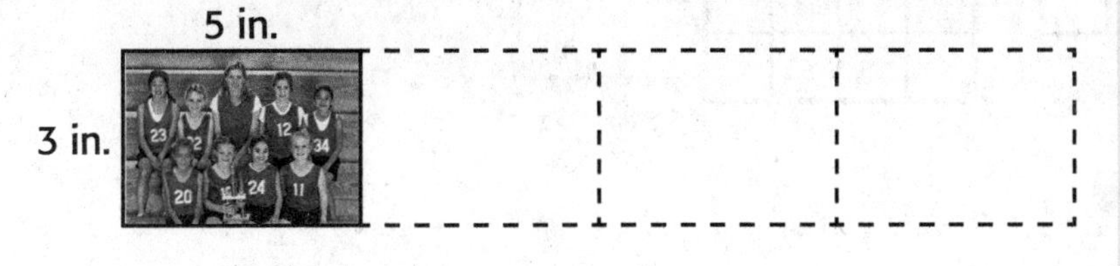

2. Mr. and Mrs. Lopez are putting square tiles on the floor in their bathroom. They can fit 6 rows of 4 tiles in the bathroom. How many tiles do they need to buy? If each tile costs $5, what is the total cost?

3. Ling is putting up a wallpaper border on three walls that are each 14 feet long and 12 feet tall. How many feet of wallpaper border will she use if she puts the border only at the top of the wall?

4. **Mathematical PRACTICE 1 Plan Your Solution** Takeisha is placing 72 photographs in an album. She will put the same number of photos on each of 6 pages. She can put 4 pictures in each row. How many rows will be on each page?

Review the Strategies

Use any strategy to solve each problem.

- Solve a simpler problem.
- Draw a picture.
- Make a table.
- Guess, check, and revise.

My Work!

5. **Mathematical PRACTICE 8 Look for a Pattern** What is the perimeter of the eighth figure if this pattern continues?

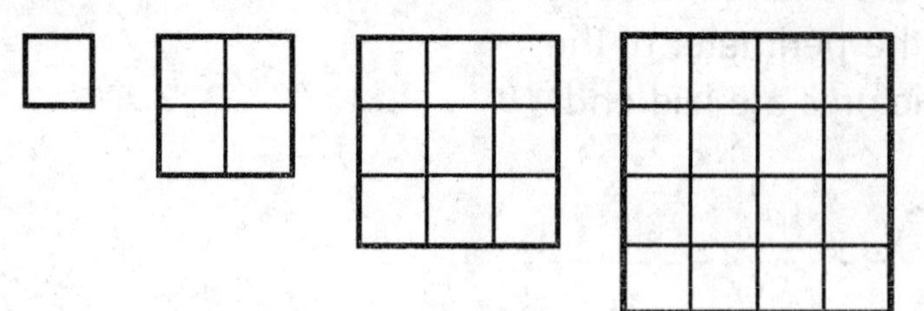

6. Marcos is making three tile pictures. He uses 310 green tiles to make each picture. He uses 50 fewer red tiles than green tiles for each picture. How many red and green tiles does he use in all?

7. A type of bacteria doubles in number every 12 hours. After 2 days, there are 48 bacteria. How many bacteria were there at the beginning of the first day?

8. Martell had boxes he was stacking. Each was 2 feet high. If he stacks 3 boxes on top of a table that is 3 feet high, what will be the total height of the boxes and the table?

9. Mr. Waters is building a rectangular sandbox. He needs to buy enough wood to go around the perimeter of the sandbox. If the length is 6 feet and the width is 5 feet, how many feet of wood does Mr. Waters need to buy?

Lesson 2
Problem Solving: Solve a Simpler Problem

Homework Helper

eHelp Need help? connectED.mcgraw-hill.com

Aaliyah and her 5 friends are picking up litter at the park. Each will clean a rectangular area with one side that measures 2 yards and another side that measures 6 yards. What is the total perimeter of all 6 areas they will clean?

1 Understand

What facts do you know?

Six people will each clean an area that measures 2 yards × 6 yards.

What do you need to find?

the total perimeter of all 6 areas

2 Plan

Solve a simpler problem.

3 Solve

Make a drawing and find the perimeter for one area. One side measures 2 yards and another side measures 6 yards.

2 yd
6 yd

Perimeter = 2 yd + 2 yd + 6 yd + 6 yd = 16 yd

Multiply 16 by the number of areas.

6 areas × 16 yards = 96 yards

So, the total perimeter of all six areas is 96 yards.

4 Check

Use addition to check the answer.

16 yd + 16 yd + 16 yd + 16 yd + 16 yd + 16 yd = 96 yards

So, the answer is correct.

Real World!

Problem Solving

Solve each problem by solving a simpler problem.

1. There are two identical figures with all sides of equal length. The combined perimeter of the figures is 80 centimeters. What shape are the figures? What is the length of one side?

2. Miranda packed 19 glass ornaments into each box. She filled 5 boxes. What is the total number of ornaments in the boxes?

3. Carrie is making a bedskirt for her twin beds. One side of a mattress measures 39 inches, and another side measures 75 inches. How many feet of fabric will Carrie need to make skirts for both beds?

4. Dalton's dad rode his bike 1 kilometer north, 1 kilometer west, 1 kilometer south, and 2 kilometers east. How many total kilometers will Dalton's dad ride if he follows this path 9 times?

5. Mathematical PRACTICE 1 **Make Sense of Problems** For each mile that Jason runs, Megan runs $\frac{1}{4}$ mile farther. If Jason ran 6 miles, how far did Megan run?

My Work!

Check My Progress

Vocabulary Check

The distance around a closed figure is called the **perimeter.**

1. Which of the following is a formula for finding the perimeter of a rectangle? Circle the correct response.

$P = \ell + w$

$P = 4 \times \ell \times w$

$P = \ell \times w$

$P = (2 \times \ell) + (2 \times w)$

Concept Check

Find each perimeter.

2.

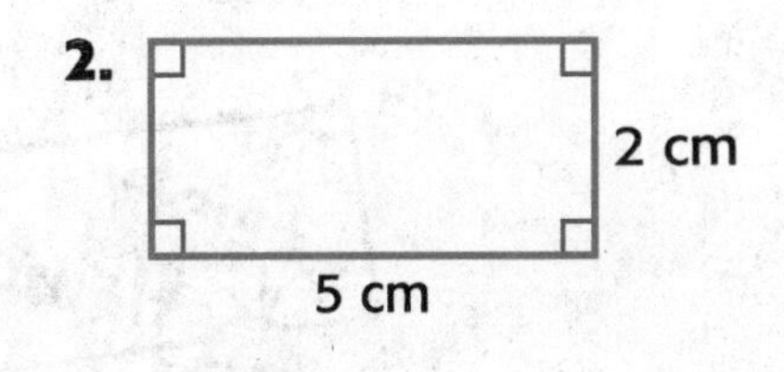

$P =$ ______

3.

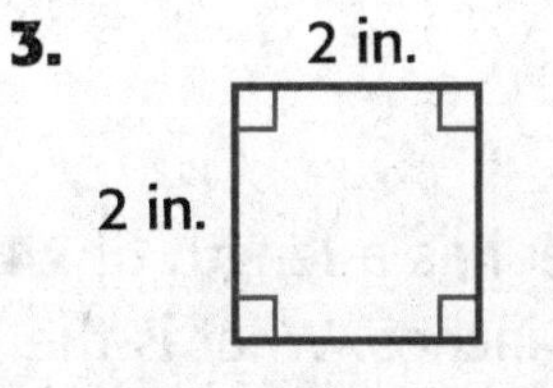

$P =$ ______

4.

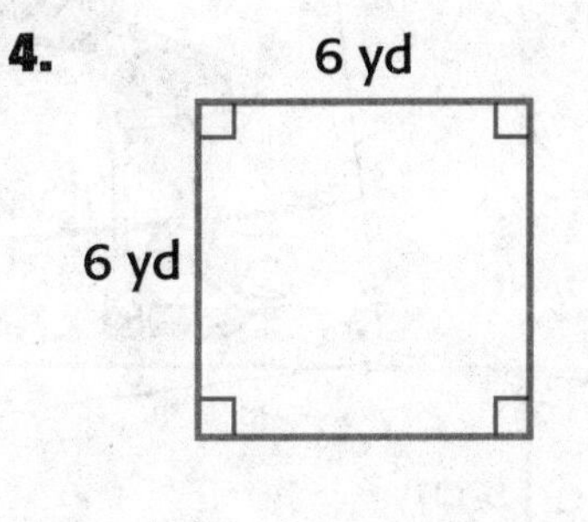

$P =$ ______

5.

$P =$ ______

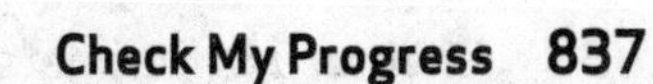

Problem Solving

My Work!

6. Byron made a drawing of his room. His drawing is shown. What is the perimeter of Byron's room?

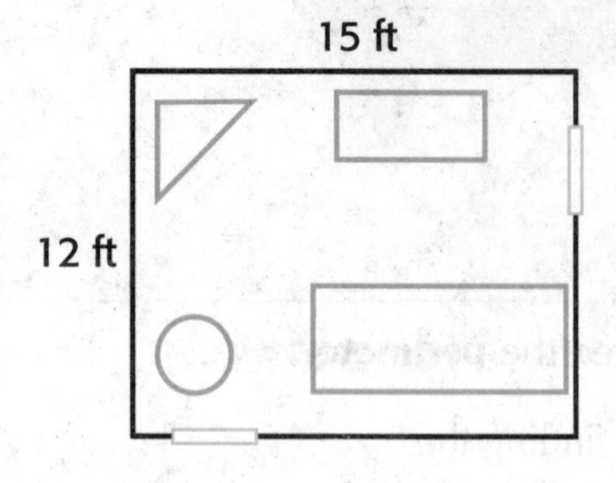

7. What is the perimeter of a square with side lengths of 4 inches?

8. Felicia is building a rectangular garden. The garden will have a perimeter of 20 meters. Give three pairs of possible side lengths.

9. A rectangular poster has a length of 24 inches, and its width is 12 inches. What is the perimeter of the poster?

Test Practice

10. Which of the following is the perimeter of the square?

Ⓐ 10 centimeters
Ⓑ 15 centimeters
Ⓒ 20 centimeters
Ⓓ 25 centimeters

5 cm
5 cm
5 cm
5 cm

Name ....

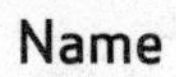

Hands On
Model Area

Lesson 3

ESSENTIAL QUESTION
Why is it important to measure perimeter and area?

A square with a side length of one unit is called a **unit square**.

A unit square has one **square unit** of area and can be used to measure area. **Area** is the number of square units needed to cover a figure without overlapping.

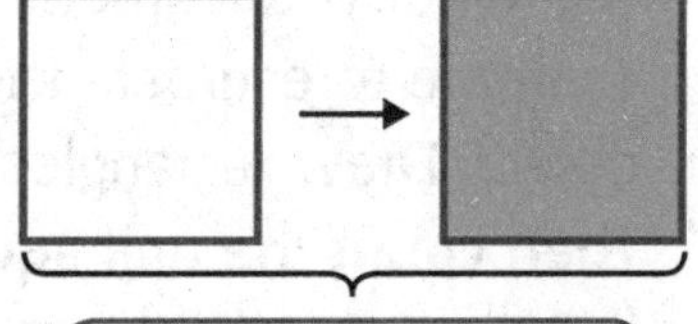

Shading, or covering, a unit square results in one square unit.

Draw It

Find the area of the rectangles shown in the table.

Rectangle	Length (units)	Width (units)	Area (sq units)

1. **Draw each rectangle.**
   Use grid paper to draw each rectangle.

2. **Find the length and width of each rectangle.**
   Count the number of unit squares that cover the length and width of the rectangle. Record each in the table.

3. **Determine the area of each rectangle.**
   Count the number of whole squares that cover the rectangle. Each whole square is 1 square unit.

Try It

Find a formula that can be used to find the area of a rectangle.

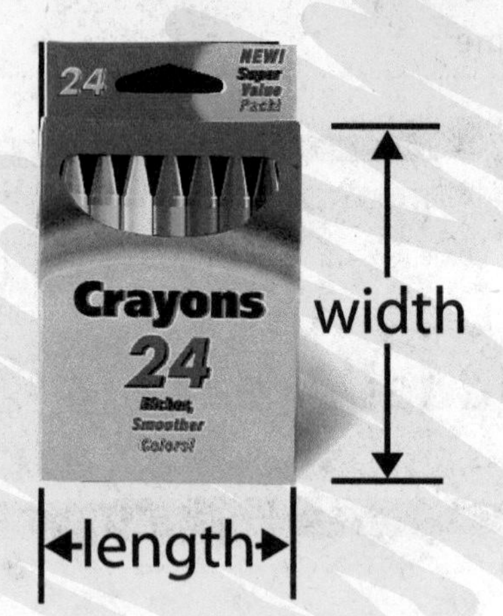

1 Measure the length and width of each object listed in the table.
Use a centimeter ruler to measure the length and the width of each object to the nearest centimeter. Record the results in the table.

2 Find the area of each object.
Use what you learned in the first example to estimate the area of each object. Draw rectangles on grid paper. Then count the unit squares to find the area. Record the results.

Object	Length (cm)	Width (cm)	Area (sq cm)
sticky note			
crayon box			
book			

3 Justify the formula for area.
Look for a pattern to find how length and width relate to area.

The area of each object is the product of the

______________ and ______________.

Talk About It

1. How did you make your estimates for the areas of the objects in Activity 2? How close were the estimates to the actual areas?

____

____

2. What operation can you use with the length and the width to equal the area of a rectangle? Explain.

____

____

3. Mathematical PRACTICE 2 **Use Symbols** What is the formula for the area of a rectangle? Use A for area, ℓ for length, and w for width.

____

Name ....

CCSS

Practice It

Complete the table below.

Rectangle	Length (units)	Width (units)	Area (sq units)
4.		1	
5.			
6.	6	3	
7.			
8.	7	6	

Apply It

Use the area formula you wrote in Exercise 3 to solve each problem.

9. Mr. Hart is hanging a picture on a wall. The picture frame has a length of 12 inches and a width of 9 inches. How much wall space will the picture need?

My Work!

10. What is the area of a classroom with a length of 30 feet and a width of 15 feet?

11. Miss Foster wants to buy carpet for her living room. The living room has a length of 15 feet and a width of 10 feet. How much carpet will she need?

12. Mathematical PRACTICE 5 **Use Math Tools** Ricky's computer monitor is a rectangle. The length is 15 inches and the width is 12 inches. Estimate the area of the monitor.

Write About It

13. Suppose two rectangles have the same area. Must they have the same length and width? Explain.

Name

MY Homework

Lesson 3
Hands On: Model Area

Homework Helper

eHelp Need help? connectED.mcgraw-hill.com

Find the area of a parking lot that is 9 yards by 12 yards.

One Way **Use a model.**

Use grid paper to model the parking lot. Each square equals 1 square yard. Count the number of squares it takes to fill a rectangle that measures 12 units by 9 units.

There are 108 squares, or square units, in all.

Another Way **Multiply.**

To find the area of a rectangle, multiply the length by the width.

$A = \ell \times w$

$A = 9$ yards $\times$ 12 yards

$A = 108$ square yards

So, the area of the parking lot is 108 square yards.

Practice

Find the area of each rectangle.

1.

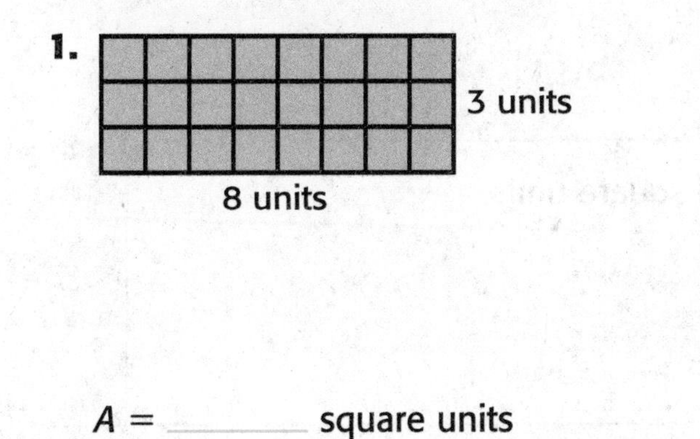

$A =$ ______ square units

2.

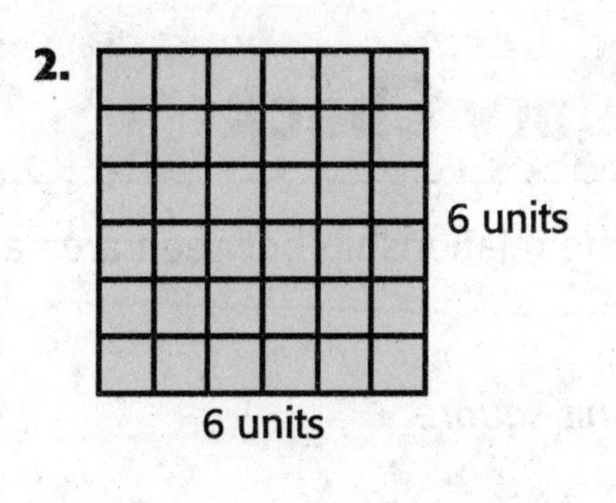

$A =$ ______ square units

Find the area of each rectangle.

3. **4.**

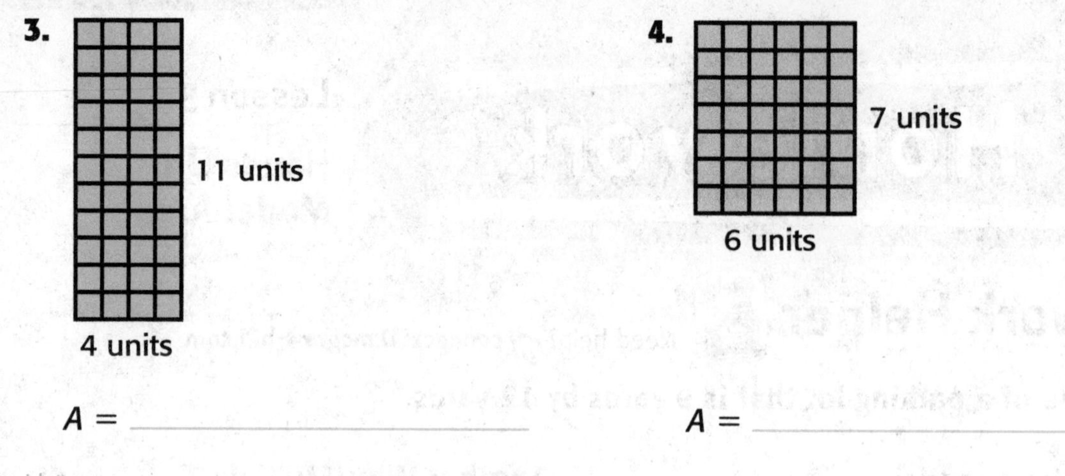

$A =$ __________ $A =$ __________

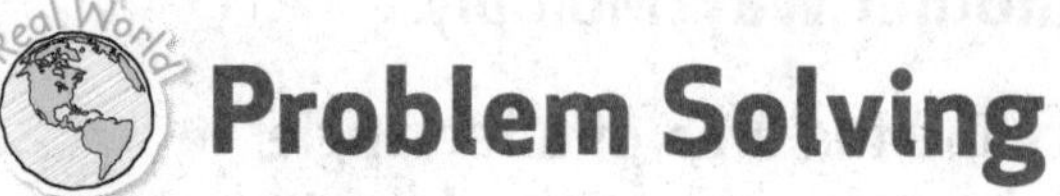

Problem Solving

Use the area formula $A = \ell \times w$ you discovered in the lesson to solve each problem.

5. Mathematical PRACTICE 3 **Justify Conclusions** Three guinea pigs living in the same cage need at least 12 square feet of living space. Is a cage that measures 3 feet by 5 feet big enough for three guinea pigs? Explain.

____

____

6. Jeanette drew a rectangle with an area of 6 square centimeters. Identify a possible length and width for her rectangle.

____

7. Marcus wants to carpet his clubhouse. One side of the rectangular floor measures 11 feet. Another side measures 8 feet. How many square feet of carpet does Marcus need to completely cover the floor?

____

Vocabulary Check

Vocab

8. Explain the relationship between area and square units.

____

9. Define *unit square*.

____

Measure Area

Lesson 4

ESSENTIAL QUESTION
Why is it important to measure perimeter and area?

You know that area is the number of square units needed to cover a region or figure without any overlap.

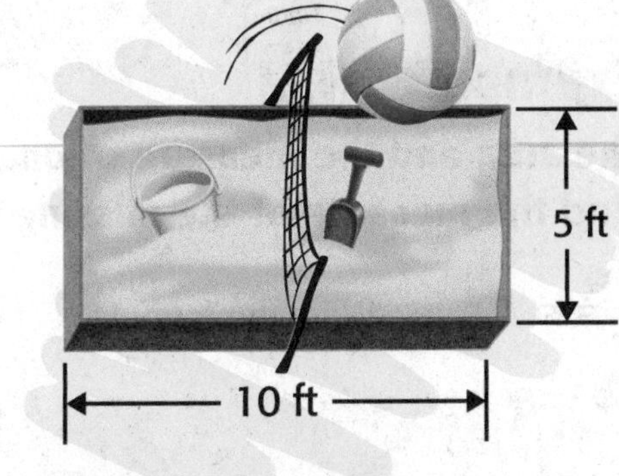

Math in My World

Watch Tutor

Example 1

The Perez family wants to put the sandbox shown in their backyard. What is the area of the sandbox?

One Way **Count unit squares.**

Tile the rectangle with unit squares. Each unit square has an area of one square foot.

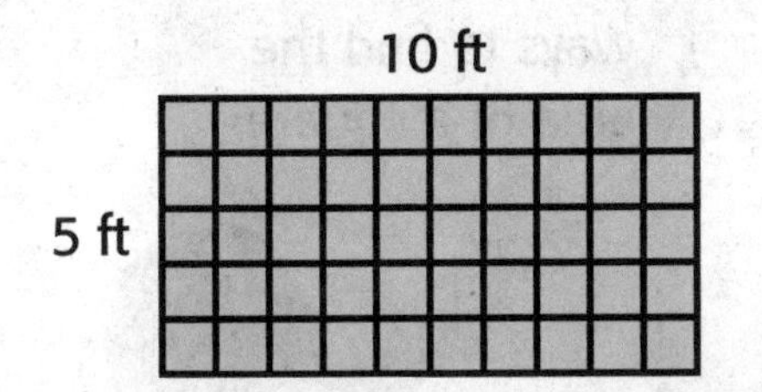

There are ______ unit squares.

There are ______ square feet.

Another Way **Multiply.**

Multiply the length times the width to find the area.

$A = \text{length} \times \text{width}$

$A = \ell \times w$

$A = 10 \text{ feet} \times 5 \text{ feet}$

$A =$ ______ square feet

So, the area of the sandbox is ______ square feet.

Key Concept Area of a Rectangle

Words To find the area A of a rectangle, multiply the length ℓ by the width w.

Symbols $A = \ell \times w$

ℓ

w w

ℓ

You can also find the area of a square.

Key Concept Area of a Square

Words To find the area A of a square, multiply the length of one side s by itself.

Symbols $A = s \times s$

Example 2

The area and the measure of one side of the square is given. Find the measure of the missing side.

8 m

Area = 64 sq m

$A = s \times s$ Write the formula.

$64 = 8 \times s$

Think: 8 times what number equals 64?

$s =$ ______ meters

The measure of the missing side is ______ meters.

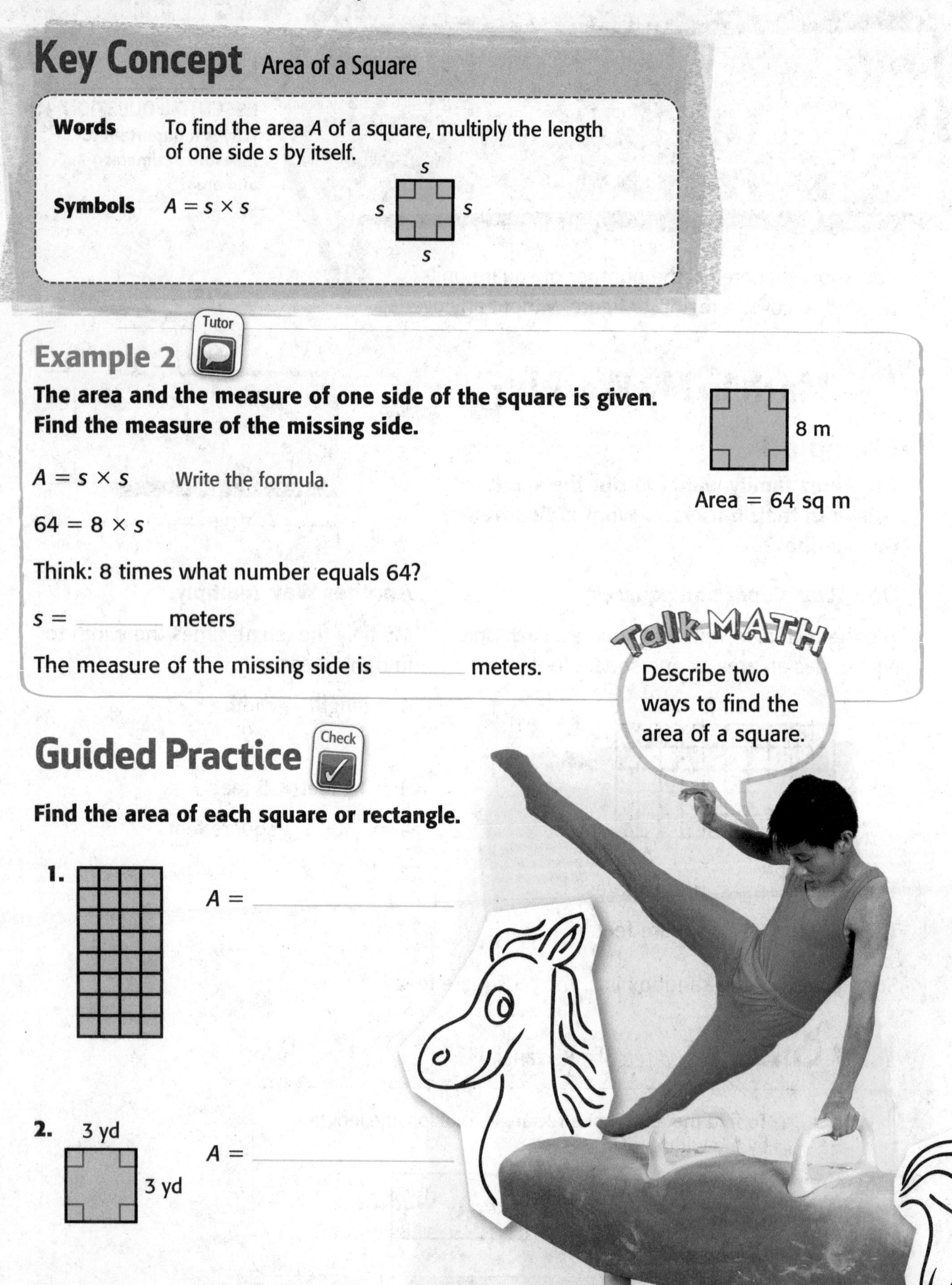

Guided Practice

Find the area of each square or rectangle.

1. $A =$ ______

2. 3 yd, 3 yd $A =$ ______

Name

Independent Practice

Find the area of each square or rectangle.

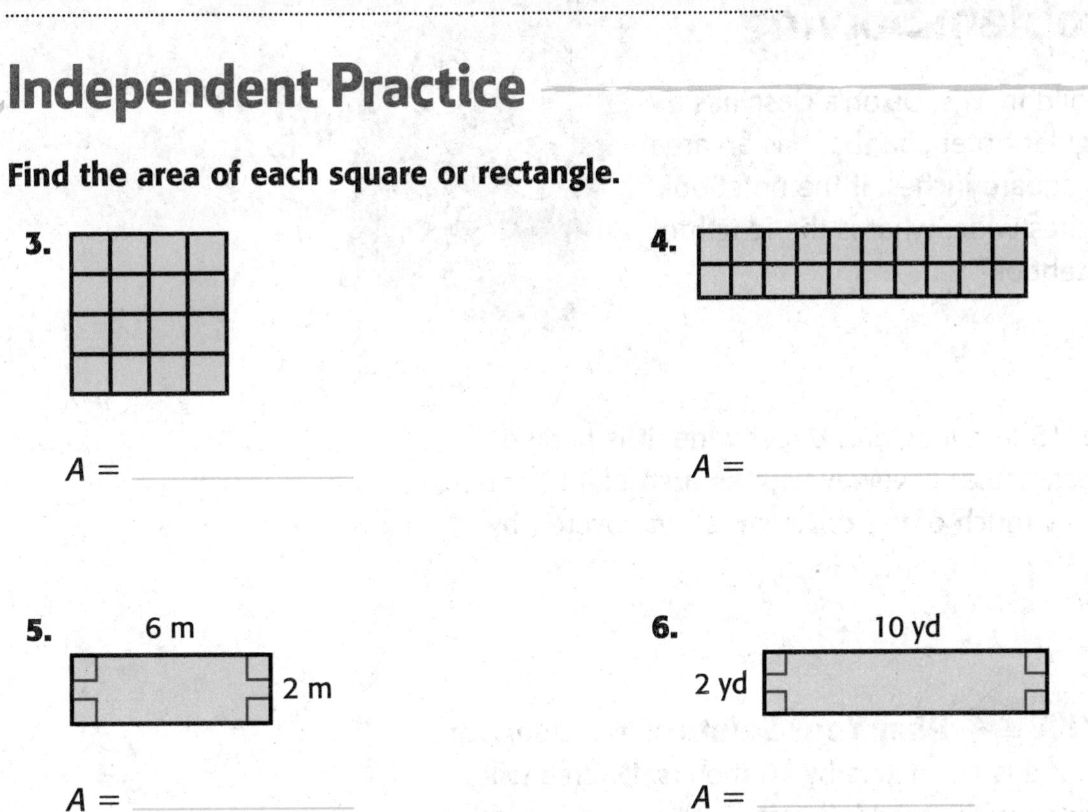

3. $A =$ ______

4. $A =$ ______

5. $A =$ ______

6. $A =$ ______

Algebra The area and the measure of one side of each square or rectangle are given. Label the missing sides.

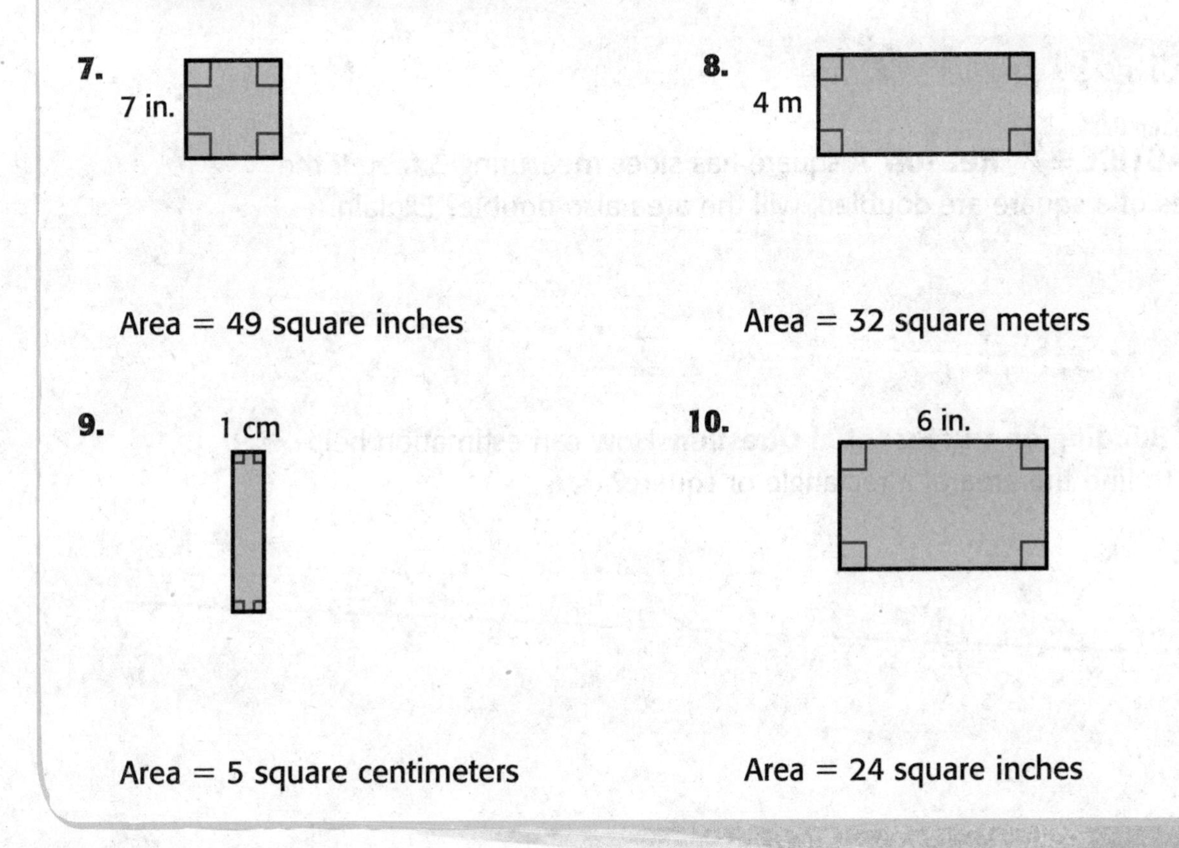

7. Area = 49 square inches

8. Area = 32 square meters

9. Area = 5 square centimeters

10. Area = 24 square inches

Problem Solving

11. Each child in Mrs. Dixon's class has a rectangular notebook that has an area of 108 square inches. If the notebook is 9 inches wide, what is the length of the notebook?

My Work!

12. A car is 15 feet long and 6 feet wide. It is parked on a rectangular driveway with an area of 112 square feet. How much of the driveway is not covered by the car?

13. Mathematical PRACTICE 1 **Plan Your Solution** A rectangular playground is 40 meters by 10 meters. Its area will be covered with shredded tires. Each bag of shredded tires covers 200 square meters and costs $30. Find the total cost for this project.

HOT Problems

14. Mathematical PRACTICE 2 **Reason** A square has sides measuring 3 feet. If the sides of a square are doubled, will the area also double? Explain.

15. **Building on the Essential Question** How can estimation help me to find the area of a rectangle or square?

Name ....

MY Homework

Lesson 4
Measure Area

Homework Helper

eHelp Need help? connectED.mcgraw-hill.com

Kristin is making a scrapbook about her pets. She covers one page of the scrapbook with decorative paw print paper. What is the area of this page?

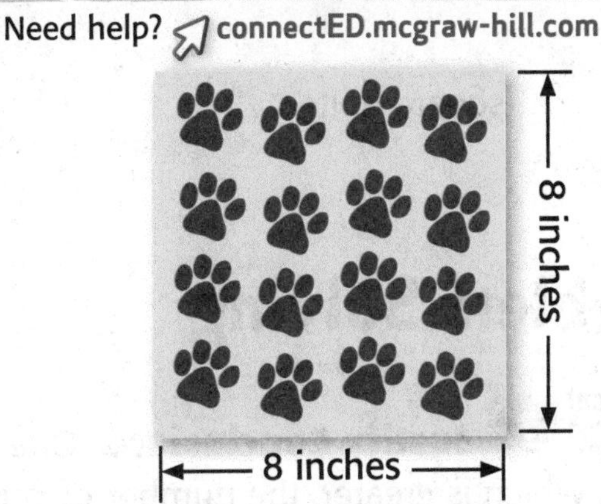

One Way **Count the units.**

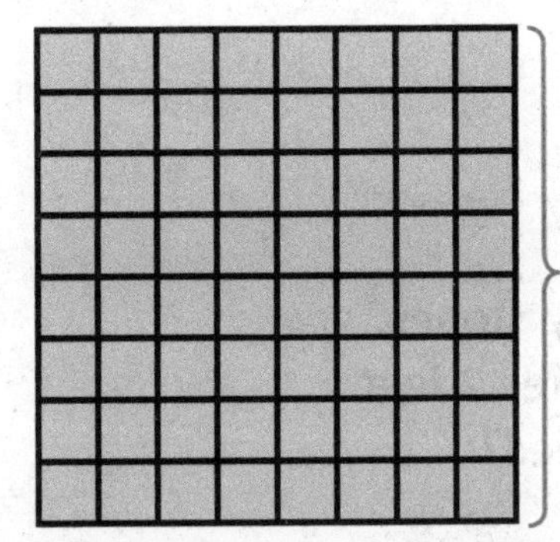

There are 64 square units in all.

So, the area of the scrapbook page is 64 square inches.

Another Way **Multiply.**

To find the area of a square, multiply the length of one side by itself.

$A = s \times s$

$A = 8 \text{ inches} \times 8 \text{ inches}$

$A = 64 \text{ square inches}$

Helpful Hint

To find the area of a rectangle, multiply the length by the width.

Practice

Find the area of each figure.

1.

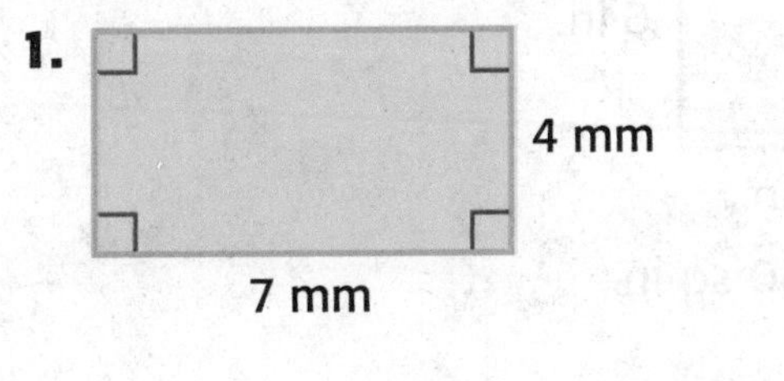

$A =$ _______ square millimeters

2.

$A =$ _______ square units

Find the area of each figure.

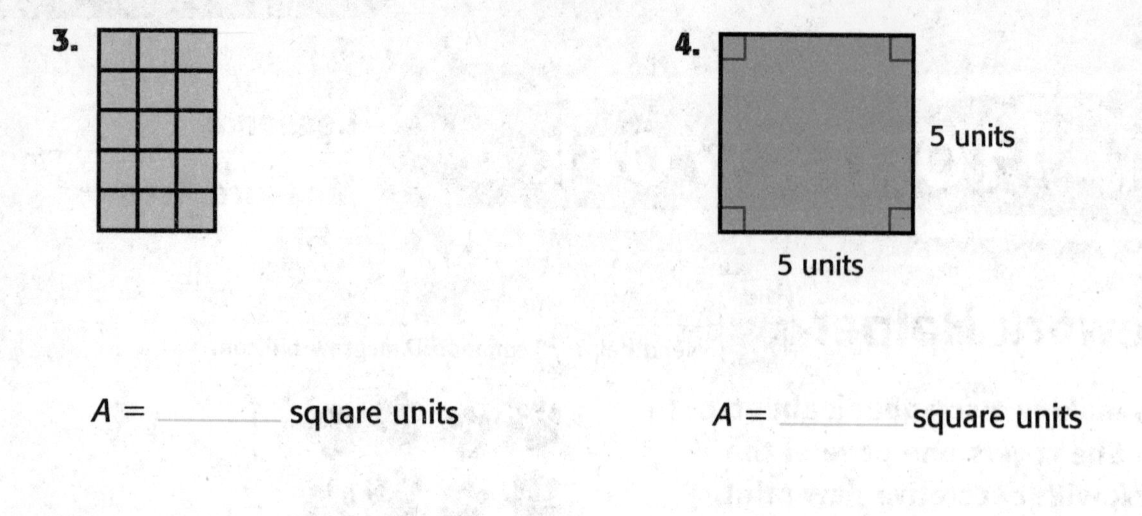

3. $A =$ ______ square units

4. $A =$ ______ square units

Problem Solving

5. **Mathematical PRACTICE 3 Justify Conclusions** One side of a square is 10 units. Which is greater, the number of square units for the area of the square or the number of units for the perimeter? Explain.

______

______

6. Eric created a rectangular patio using 1-foot square paving stones, which are sold in batches by the dozen. The patio measures 7 feet by 8 feet. How many batches of paving stones did Eric need? (*Hint:* 1 dozen = 12)

______

Test Practice

7. What is the perimeter of the rectangle?

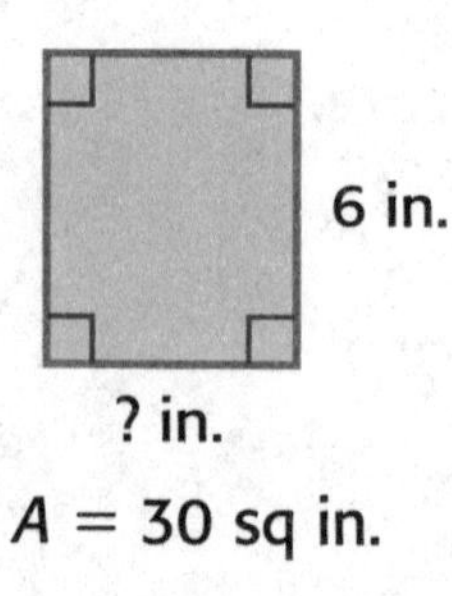

$A = 30$ sq in.

Ⓐ 22 inches

Ⓑ 24 inches

Ⓒ 26 inches

Ⓓ 28 inches

Name

Relate Area and Perimeter

Lesson 5

ESSENTIAL QUESTION
Why is it important to measure perimeter and area?

Math in My World

Tutor

Example 1

The swim team put its trophy on a table that has an area of 12 square feet. List all of the possible lengths and widths of rectangles with an area of 12 square feet.

The models show all of the possible rectangles. Label each model.

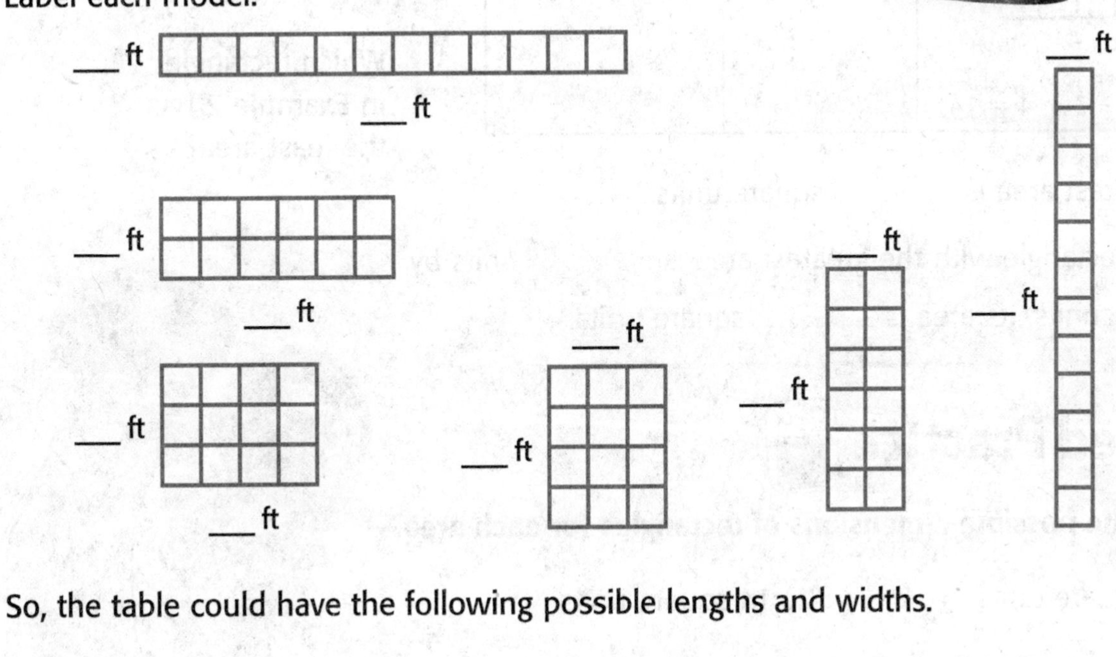

So, the table could have the following possible lengths and widths.

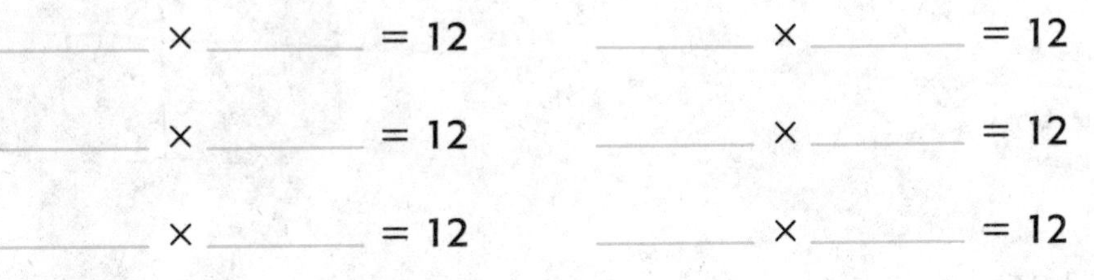

_____ × _____ = 12   _____ × _____ = 12

_____ × _____ = 12   _____ × _____ = 12

_____ × _____ = 12   _____ × _____ = 12

Example 2

Find the rectangle with the greatest area whose perimeter is 14 units.

The table shows each rectangle that has a perimeter of 14 units. Complete the table.

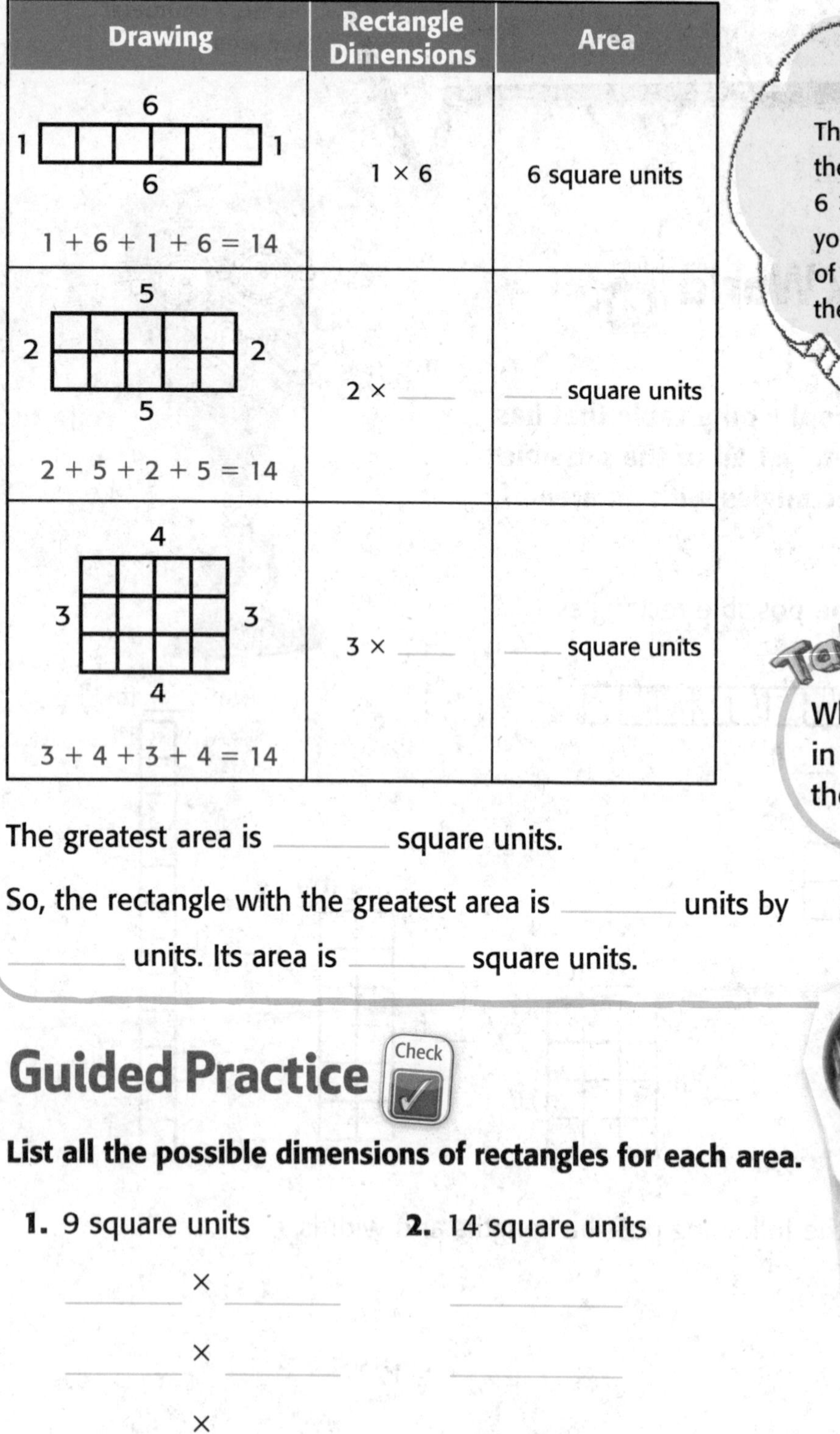

Drawing	Rectangle Dimensions	Area
6 / 1 [grid] 1 / 6 $1 + 6 + 1 + 6 = 14$	1×6	6 square units
5 / 2 [grid] 2 / 5 $2 + 5 + 2 + 5 = 14$	$2 \times$ ____	____ square units
4 / 3 [grid] 3 / 4 $3 + 4 + 3 + 4 = 14$	$3 \times$ ____	____ square units

Helpful Hint

The rectangles not listed in the table have dimensions of 6×1, 5×2, and 4×3. If you reverse the dimensions of a rectangle, it will still have the same area.

The greatest area is ________ square units.

So, the rectangle with the greatest area is ________ units by ________ units. Its area is ________ square units.

Talk MATH

Which rectangle in Example 2 has the least area?

Guided Practice

Check

List all the possible dimensions of rectangles for each area.

1. 9 square units

____ × ____

____ × ____

____ × ____

2. 14 square units

____

____

____

____

Name

Independent Practice

List all the possible dimensions of rectangles for each area.

3. 16 square units

4. 20 square units

Find the perimeter and area for each square or rectangle.

5.

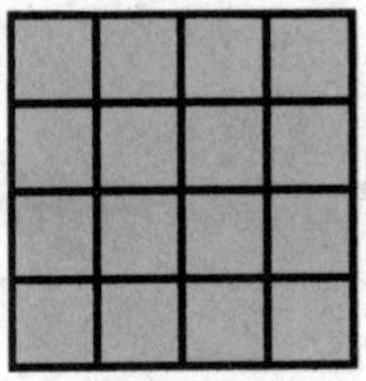

Perimeter: ______

Area: ______

6.

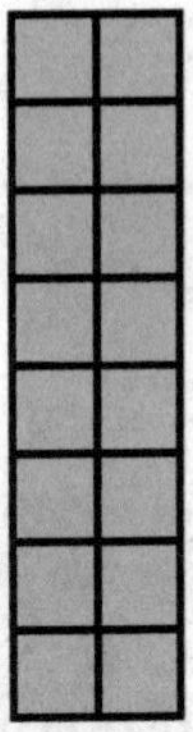

Perimeter: ______

Area: ______

7. What do the figures in Exercises 5 and 6 have in common? How do these figures differ?

______

______

8. Mathematical PRACTICE 1 **Plan Your Solution** Violet is making a rectangular banner for the basketball team to run through before the start of the game. She has 24 square feet of paper. List all of the possible dimensions of rectangles with an area of 24 square feet.

My Work!

9. Which of the dimensions found in Exercise 8 has the greatest perimeter?

10. If a rectangle has a greater perimeter than another rectangle, does it also have a greater area? Explain.

11. Mathematical PRACTICE 2 **Reason** Is it possible to draw a rectangle that has an area of 24 square units and a perimeter of 24 units? Explain.

12. **Building on the Essential Question** What is the difference between area and perimeter?

MY Homework

Lesson 5
Relate Area and Perimeter

Homework Helper

eHelp Need help? connectED.mcgraw-hill.com

A rectangle has a perimeter of 16 units. What is its greatest possible area?

1 Draw all of the possible rectangles with a perimeter of 16 units.

Drawing	Rectangle Dimensions	Area
1 by 7 rectangle; $1 + 7 + 1 + 7 = 16$	1×7 (or 7×1)	$A = 7$ square units
2 by 6 rectangle; $2 + 6 + 2 + 6 = 16$	2×6 (or 6×2)	$A = 12$ square units
3 by 5 rectangle; $3 + 5 + 3 + 5 = 16$	3×5 (or 5×3)	$A = 15$ square units
4 by 4 rectangle; $4 + 4 + 4 + 4 = 16$	4×4	$A = 16$ square units

If you reverse the dimensions of a rectangle, it will still have the same area.

2 Compare the areas of the rectangles.
The greatest area is 16 square units.

So, 16 square units is the greatest possible area for a rectangle whose perimeter is 16 units.

Practice

Draw two possible rectangles for each perimeter. Find the area of each.

1. 20 units

2. 8 units

Problem Solving

3. Mathematical PRACTICE 2 **Use Number Sense** Tomás drew a rectangle with an area of 6 square centimeters. What is the greatest possible perimeter for this rectangle?

My Work!

4. Danica has laid out floor tiles so they form a rectangle with a perimeter of 18 inches. What is the difference between the greatest and least possible areas of the rectangle?

5. A rectangle has an area of 30 square meters and a perimeter of 34 meters. What are the dimensions of the rectangle?

Test Practice

6. A square has a perimeter of 28 feet. What is its area?

Ⓐ 45 square feet

Ⓑ 48 square feet

Ⓒ 49 square feet

Ⓓ 50 square feet

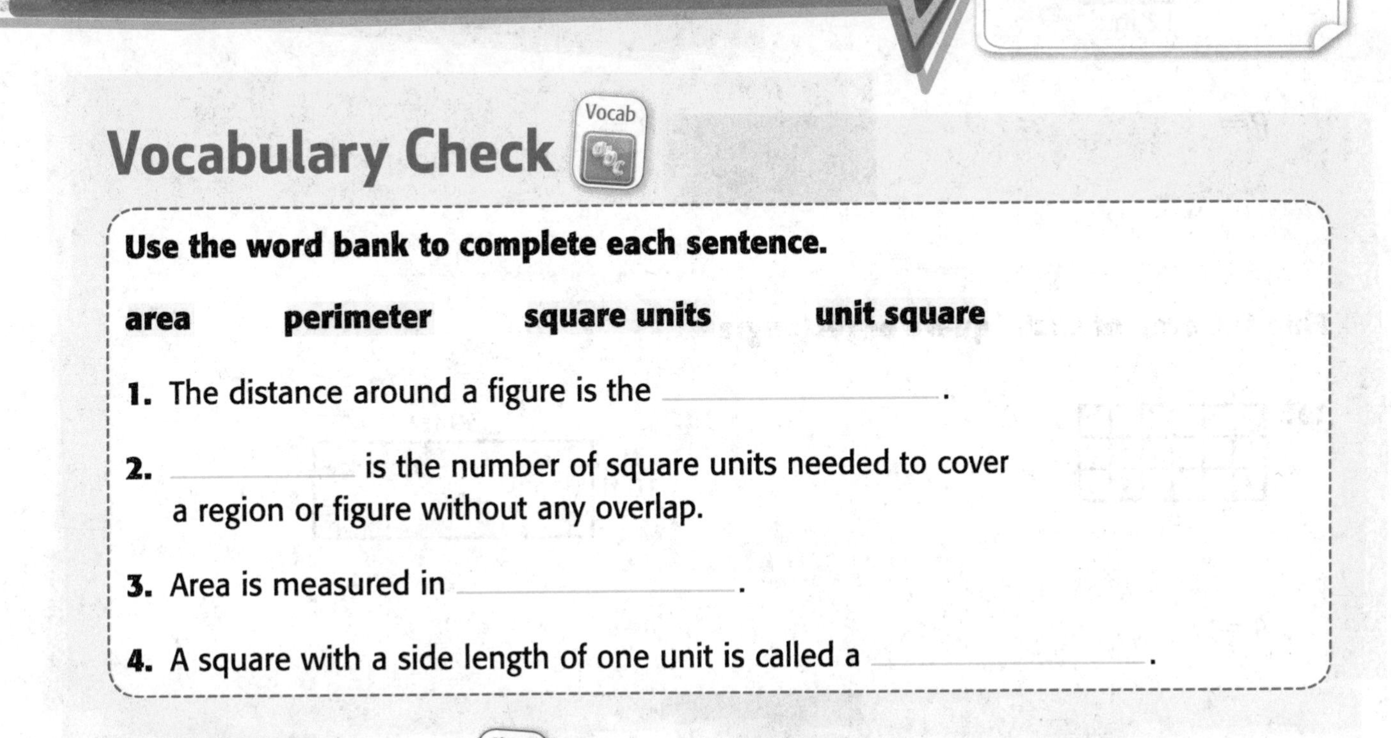

Vocabulary Check

Use the word bank to complete each sentence.

area **perimeter** **square units** **unit square**

1. The distance around a figure is the ______________.

2. ______________ is the number of square units needed to cover a region or figure without any overlap.

3. Area is measured in ______________.

4. A square with a side length of one unit is called a ______________.

Concept Check

Look at the tennis court below. Find the perimeter and area.

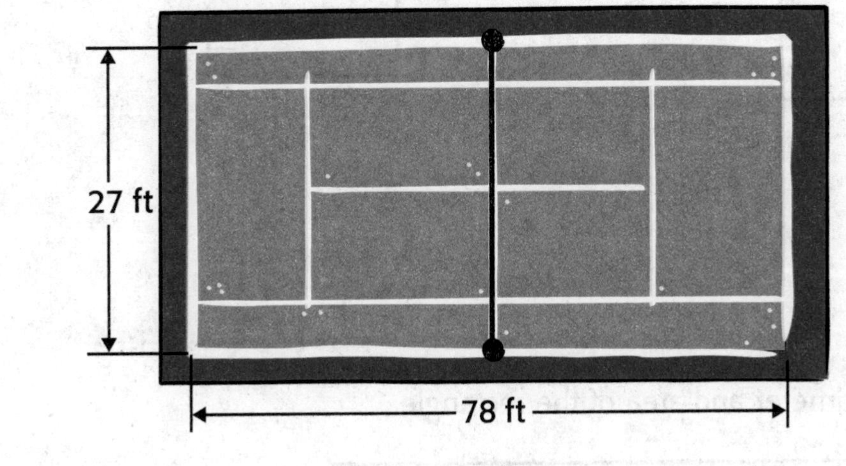

5. Perimeter = ________

6. Area = ________

Find each perimeter.

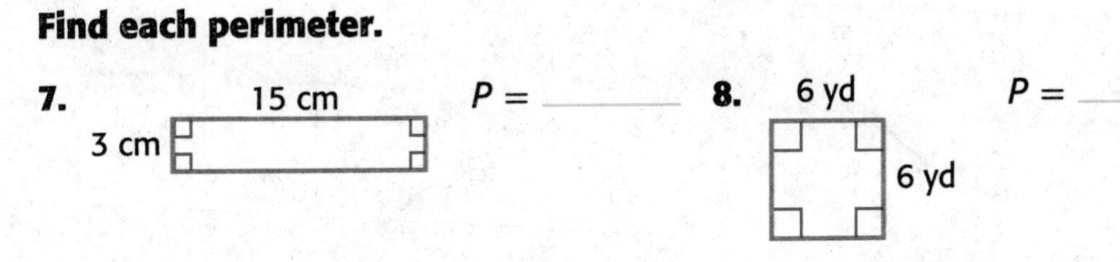

7. $P =$ ________

8. $P =$ ________

Find each perimeter.

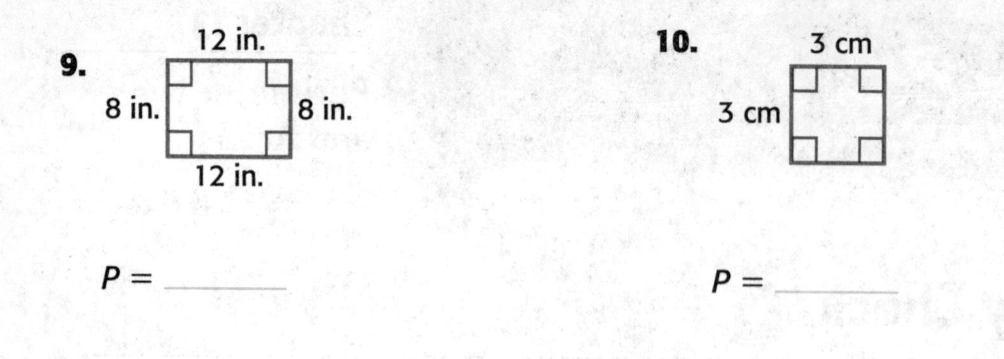

9. $P =$ ______

10. $P =$ ______

Find the area of each square or rectangle.

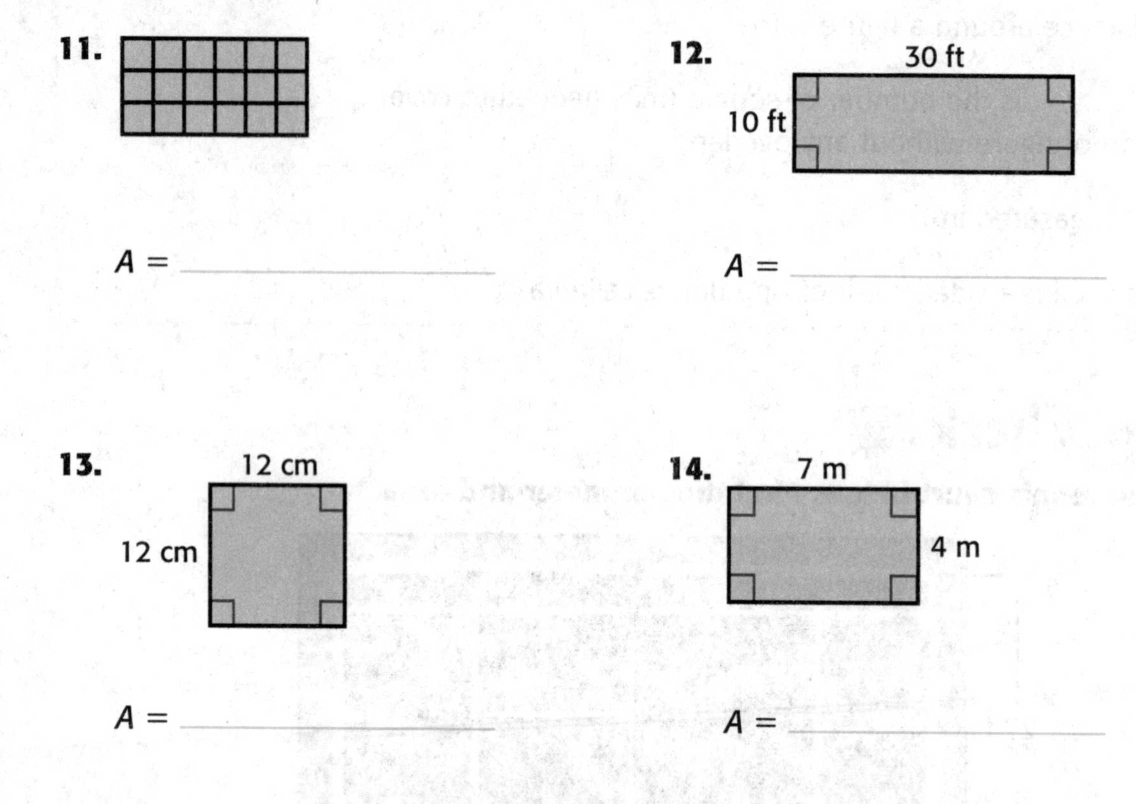

11. $A =$ ______

12. $A =$ ______

13. $A =$ ______

14. $A =$ ______

15. Find the perimeter and area of the rectangle.

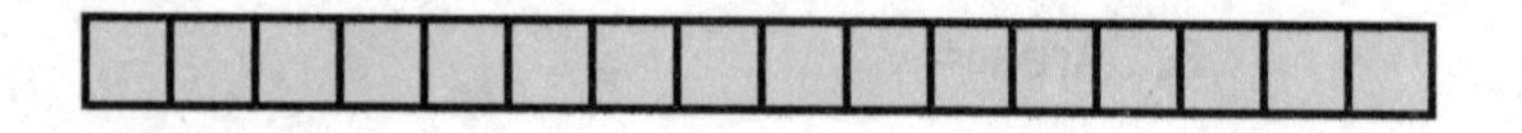

Perimeter: ______

Area: ______

Name

Problem Solving

16. Rodolfo's ping pong table has an area of 45 square feet. The length is 9 feet. What is the perimeter of the ping pong table?

17. Mr. Lobo is building a fence around his rectangular yard. It is 16 feet long and 14 feet wide. How many feet of fencing will he need?

18. Brett painted 3 walls. Each wall was 9 feet tall and 12 feet long. How much wall area did he paint?

19. Is there a relationship between the area and the perimeter of a rectangle? Explain.

Test Practice

20. Heidi ran two laps around the city block shown. How many feet did she run?

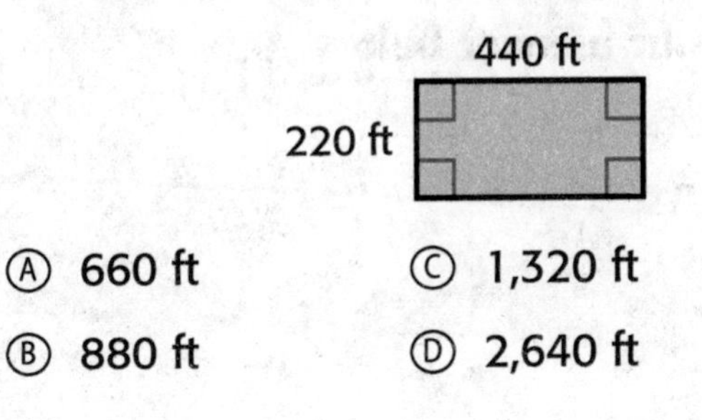

Ⓐ 660 ft   Ⓒ 1,320 ft

Ⓑ 880 ft   Ⓓ 2,640 ft

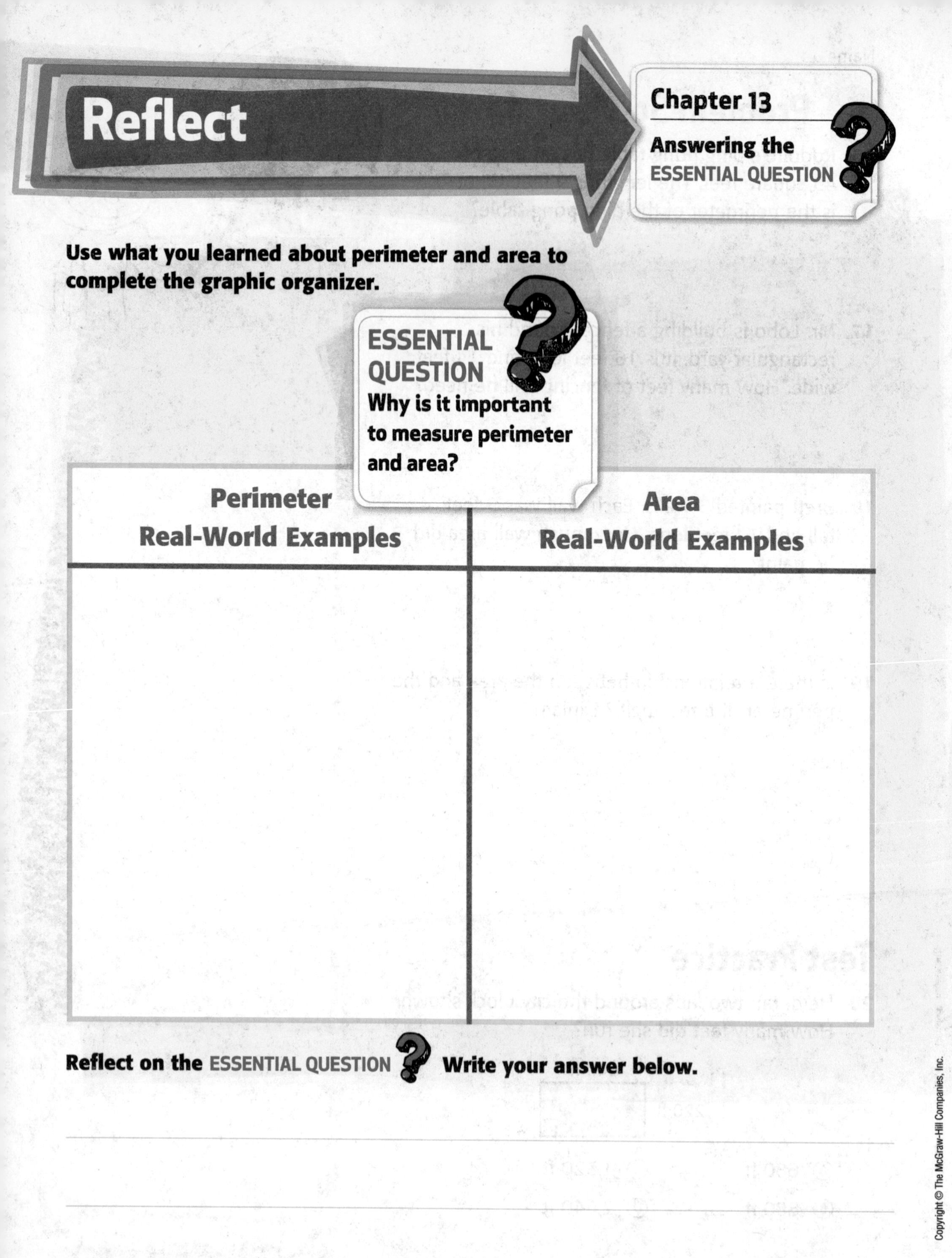

Use what you learned about perimeter and area to complete the graphic organizer.

Perimeter Real-World Examples	Area Real-World Examples

Reflect on the ESSENTIAL QUESTION Write your answer below.

Chapter 14 Geometry

ESSENTIAL QUESTION

How are different ideas about geometry connected?

Sign Me Up!

SLOW

STOP

Watch

Watch a video!

MY Common Core State Standards

Geometry

4.G.1 Draw points, lines, line segments, rays, angles (right, acute, obtuse), and perpendicular and parallel lines. Identify these in two-dimensional figures.

4.G.2 Classify two-dimensional figures based on the presence or absence of parallel or perpendicular lines, or the presence or absence of angles of a specified size. Recognize right triangles as a category, and identify right triangles.

4.G.3 Recognize a line of symmetry for a two-dimensional figure as a line across the figure such that the figure can be folded along the line into matching parts. Identify line-symmetric figures and draw lines of symmetry.

Measurement and Data *This chapter also addresses these standards:*

4.MD.5 Recognize angles as geometric shapes that are formed wherever two rays share a common endpoint, and understand concepts of angle measurement:

4.MD.5a An angle is measured with reference to a circle with its center at the common endpoint of the rays, by considering the fraction of the circular arc between the points where the two rays intersect the circle. An angle that turns through 1/360 of a circle is called a "one-degree angle," and can be used to measure angles.

4.MD.5b An angle that turns through *n* one-degree angles is said to have an angle measure of *n* degrees.

4.MD.6 Measure angles in whole-number degrees using a protractor. Sketch angles of specified measure.

4.MD.7 Recognize angle measures as additive. When an angle is decomposed into non-overlapping parts, the angle measure of the whole is the sum of the angle measures of the parts. Solve addition and subtraction problems to find unknown angles on a diagram in real world and mathematical problems, e.g., by using an equation with a symbol for the unknown angle measure.

Standards for Mathematical PRACTICE

1. Make sense of problems and persevere in solving them.
2. Reason abstractly and quantitatively.
3. Construct viable arguments and critique the reasoning of others.
4. Model with mathematics.
5. Use appropriate tools strategically.
6. Attend to precision.
7. Look for and make use of structure.
8. Look for and express regularity in repeated reasoning.

= focused on in this chapter

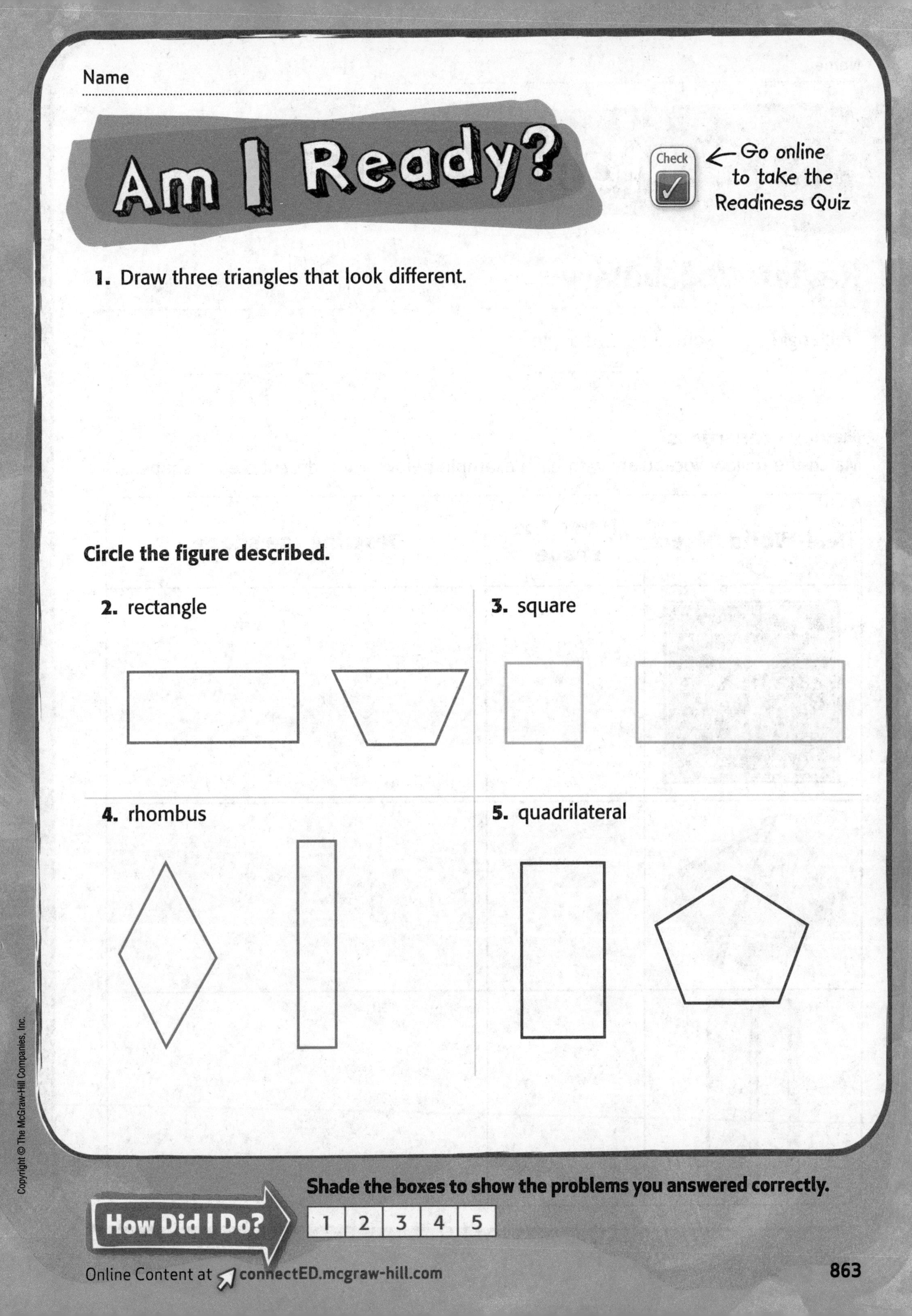

Name

Am I Ready?

Check

Go online to take the Readiness Quiz

1. Draw three triangles that look different.

Circle the figure described.

2. rectangle

3. square

4. rhombus

5. quadrilateral

How Did I Do?

Shade the boxes to show the problems you answered correctly.

1	2	3	4	5

Name

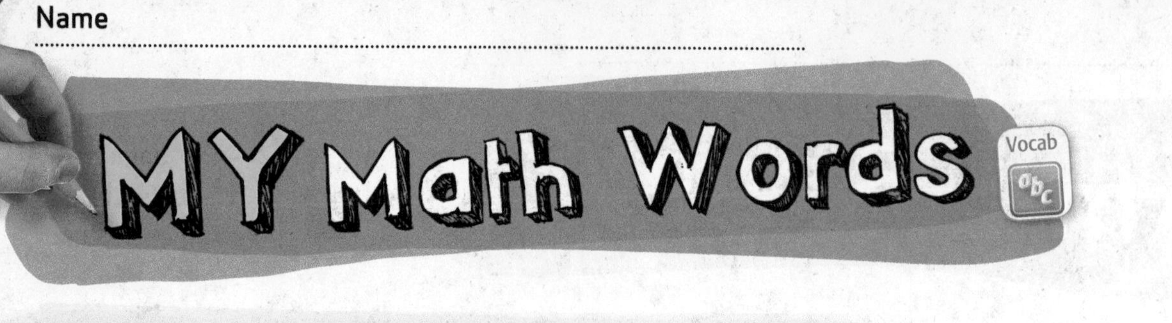

Review Vocabulary

rectangle square triangle

Making Connections

Match the review vocabulary with each example below. Then describe each shape.

Real-World Object	Name the shape.	Describe the shape.

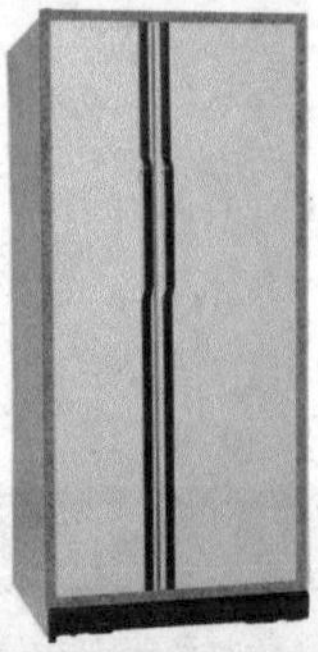		

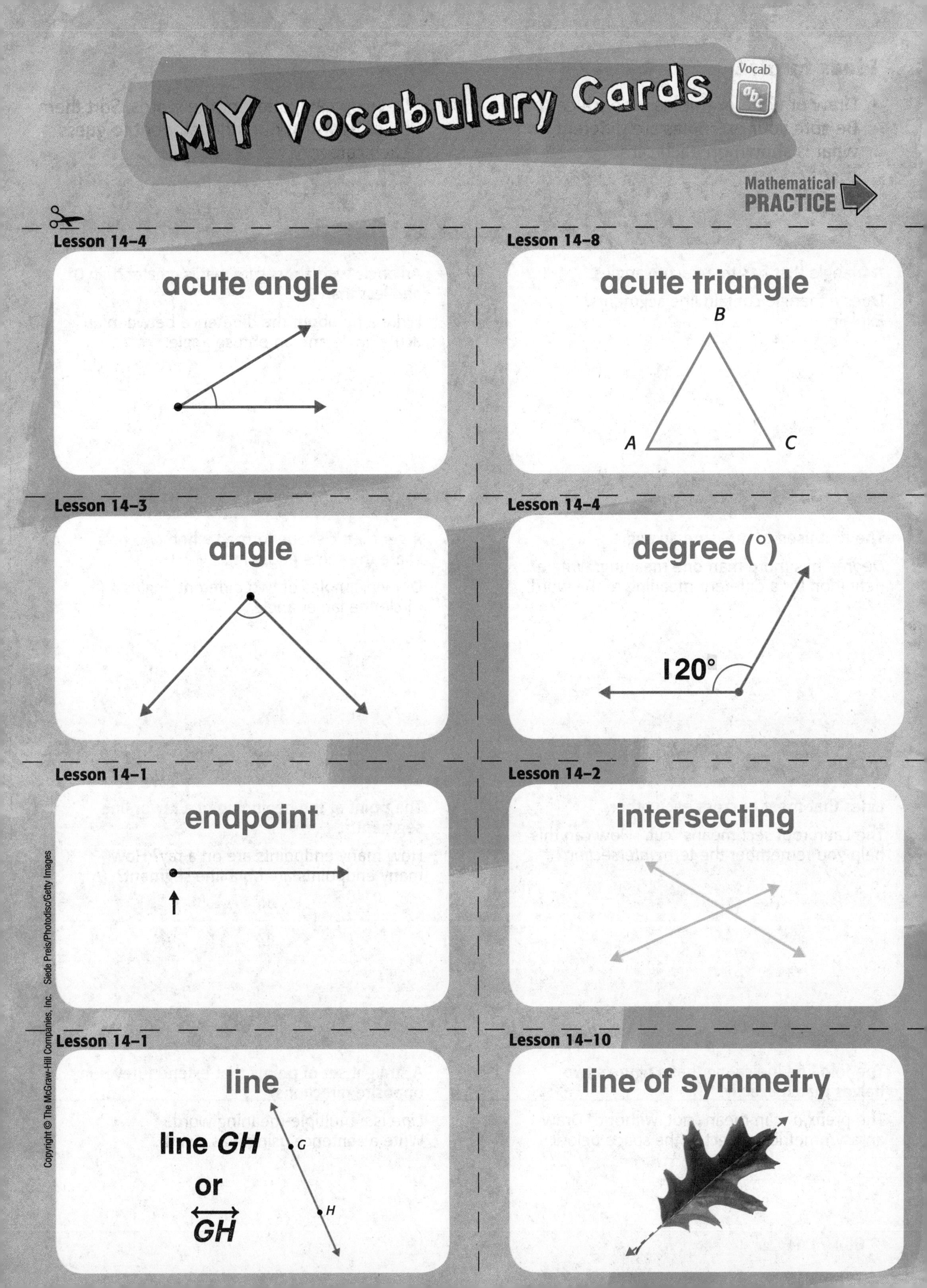
MY Vocabulary Cards
Vocab
Mathematical PRACTICE
Lesson 14–4
acute angle
Lesson 14–8
acute triangle
B
A
C
Lesson 14–3
angle
Lesson 14–4
degree (°)
120°
Lesson 14–1
endpoint
Lesson 14–2
intersecting
Lesson 14–1
line
line GH
or
GH
G
H
Lesson 14–10
line of symmetry

Ideas for Use

- Draw or write examples for each card. Be sure your examples are different from what is shown on each card.
- Develop categories for the words. Sort them by category. Ask another student to guess each category.

A triangle that has three acute angles.

Does a triangle contain line segments? Explain.

An angle with a measure that is greater than 0° and less than 90°.

Write a tip about the difference between an acute angle and an obtuse angle.

The unit used to measure an angle.

Degree has more than one meaning. Write a definition for a different meaning of the word.

A geometric shape formed when two rays share the same endpoint.

Draw examples of two different angles. Circle the larger angle.

Lines that meet or cross each other.

The Latin root *sect* means "cut." How can this help you remember the term *intersecting*?

The point at the beginning of a ray or line segment.

How many endpoints are on a ray? How many endpoints are on a line segment?

The fold line indicating that a figure's two halves match exactly.

The prefix *a-* can mean "not, without." Draw an asymmetrical object in the space below.

A straight set of points that extend forever in opposite directions.

Line is a multiple-meaning word. Write a sentence using it as a verb.

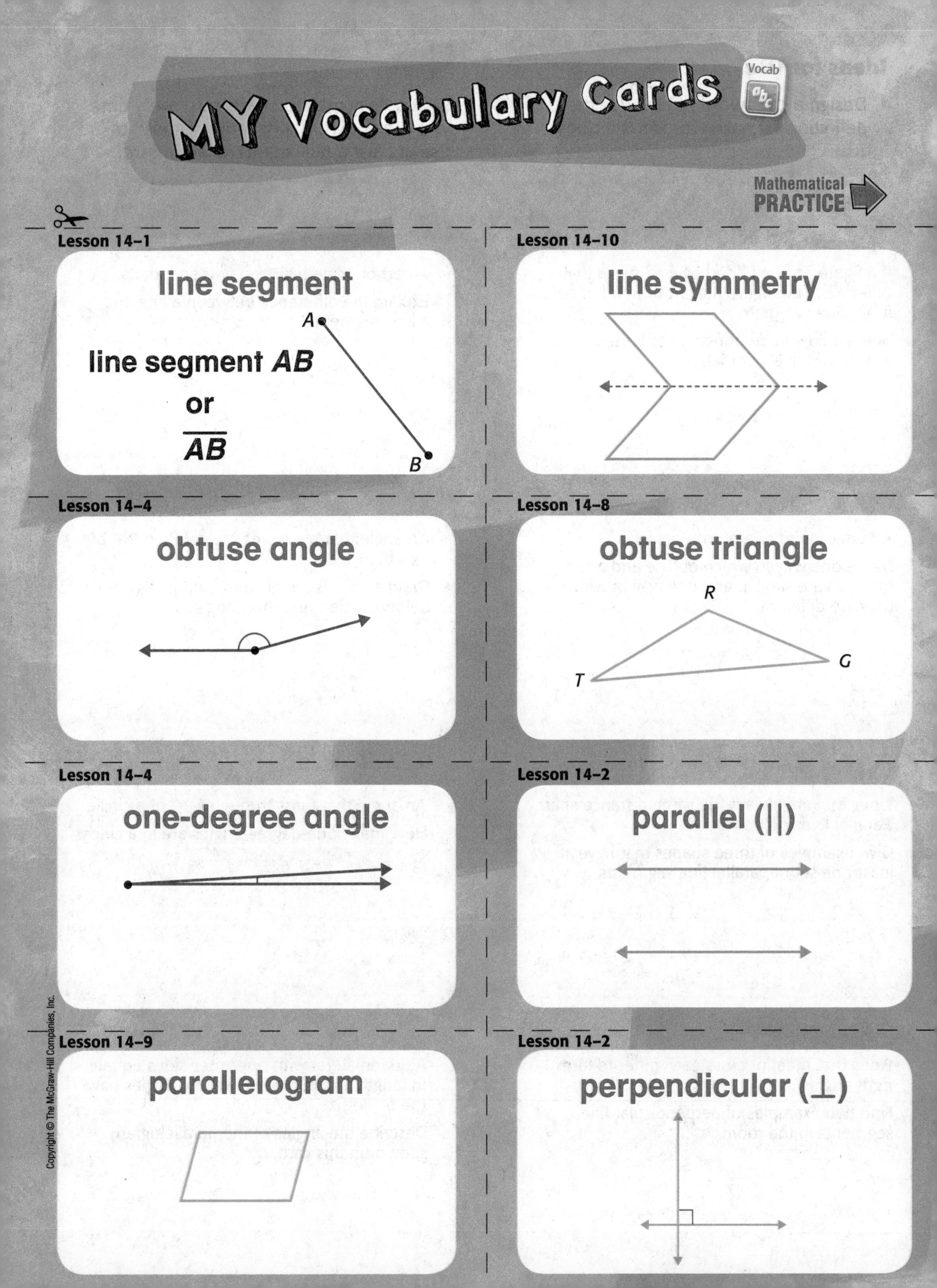
MY Vocabulary Cards
Vocab
abc
Mathematical PRACTICE
Lesson 14-1
line segment
A
line segment AB
or
AB
B
Lesson 14-10
line symmetry
Lesson 14-4
obtuse angle
Lesson 14-8
obtuse triangle
R
G
T
Lesson 14-4
one-degree angle
Lesson 14-2
parallel (||)
Lesson 14-9
parallelogram
Lesson 14-2
perpendicular (⊥)
Copyright © The McGraw-Hill Companies, Inc.

Ideas for Use

- Design a crossword puzzle. Use the definition for each word as the clues.
- Write a tally mark on each card every time you use the word. Challenge yourself to use at least 3 tally marks for each card.

If a figure can be folded over a line so that one half of the figure matches the other half, it has line symmetry.

How could you demonstrate to a friend that a circle has line symmetry?

A part of a line between two endpoints.

Explain the difference between a line and a line segment.

A triangle with one obtuse angle.

Name one way in which obtuse and acute triangles are similar, and one way in which they are different.

An angle that measures greater than 90° but less than 180°.

Draw an obtuse angle and a right angle below. Circle the obtuse angle.

Lines that are always the same distance apart. Parallel lines do not meet.

Give examples of three shapes that have at least one set of parallel line segments.

An angle that turns through $\frac{1}{360}$ of a circle.

How many one-degree angles are in a cirlce?

Lines that meet or cross each other to form right angles.

Find two examples of perpendicular line segments in the room.

A quadrilateral with opposites sides equal in length and parallel. Opposite angles have the same size.

Describe the angles of the parallelogram shown on this card.

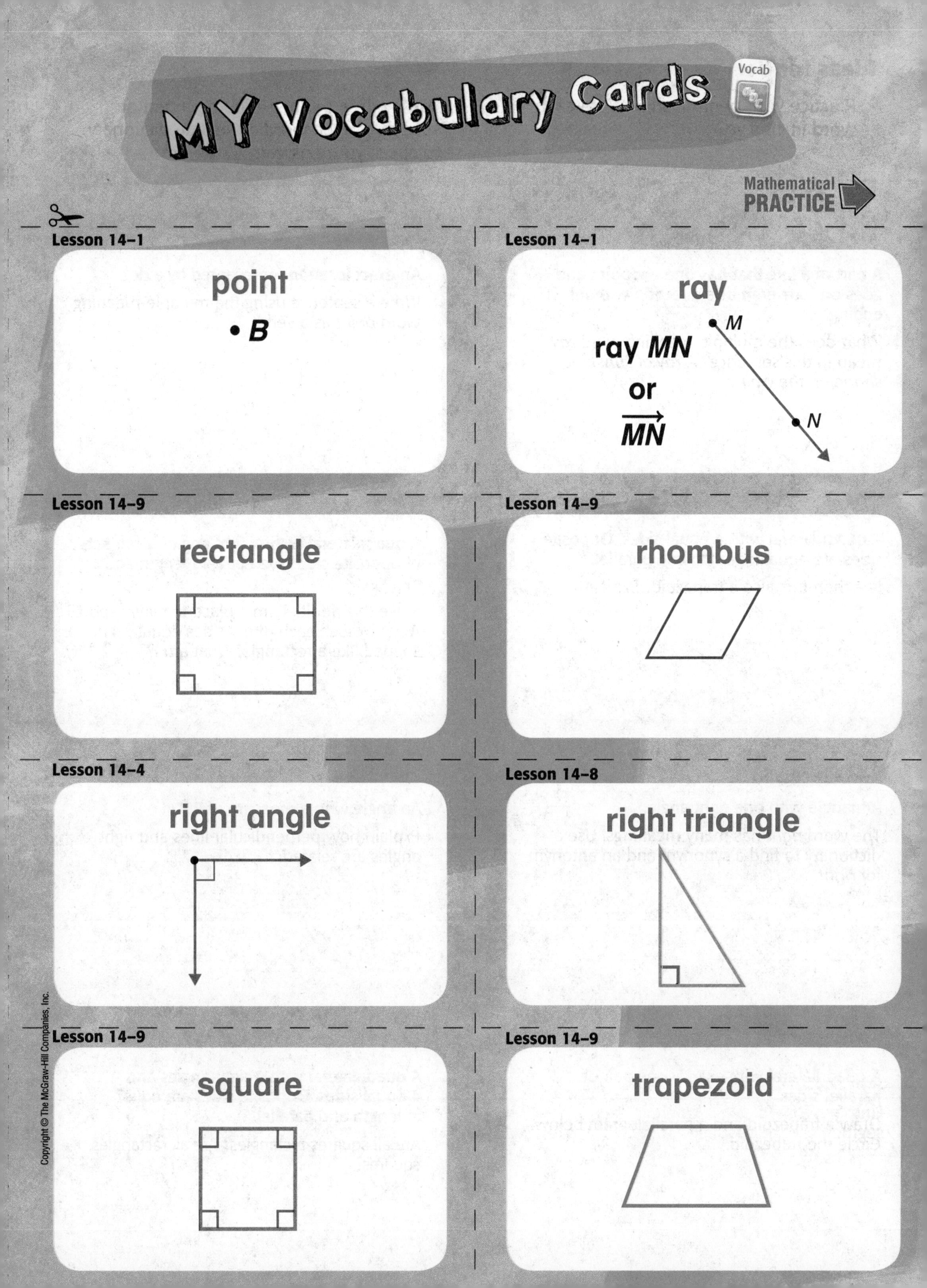
MY Vocabulary Cards
Vocab
Mathematical PRACTICE
Lesson 14-1
point
B
Lesson 14-1
ray
M
ray MN
or
MN
N
Lesson 14-9
rectangle
Lesson 14-9
rhombus
Lesson 14-4
right angle
Lesson 14-8
right triangle
Lesson 14-9
square
Lesson 14-9
trapezoid

Ideas for Use

- Practice your penmanship! Write each word in cursive.
- Work with a partner to name a part of speech of each word. Consult a dictionary to check your answers.

A part of a line that has one endpoint and goes on forever in one direction without ending.

What does the multiple-meaning word *ray* mean in this sentence? *A ray of sunshine shone on the wall.*

An exact location represented by a dot.

Write a sentence using the multiple-meaning word *point* as a verb.

A quadrilateral with 4 equal sides. Opposite sides are equal in length and parallel.

Is a rhombus also a trapezoid? Explain.

A quadrilateral with 4 right angles. Both sets of opposite sides are equal in length and parallel.

Solve this riddle: I am a place to play a sport. A net divides each of my sides equally. I'm shaped like a rectangle. What am I?

A triangle with one right angle.

The word *right* has many meanings. Use a dictionary to find a synonym and an antonym for *right*.

An angle with a measure of 90°.

Explain how perpendicular lines and right angles are related.

A quadrilateral with exactly one pair of parallel sides.

Draw a trapezoid and a parallelogram below. Circle the trapezoid.

A quadrilateral with 4 right angles and 4 equal sides. Opposite sides are equal in length and parallel.

Are all squares rectangles? Are all rectangles squares?

MY Foldable

FOLDABLES® Follow the steps on the back to make your Foldable.

FOLDABLES®
Study Organizer
1
2

Name ....

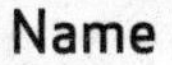

Draw Points, Lines, and Rays

Lesson 1

ESSENTIAL QUESTION
How are different ideas about geometry connected?

A **point** is an exact location that is represented by a dot. A **line** is a straight set of points that extends in opposite directions without ending.

Math in My World

Tools Tutor

Example 1

Molly drew the figure shown. Identify the figure she drew.

The figure extends in opposite directions. The arrows indicate that it extends without ending. It is a line.

This line is labeled with point X and point Y. There are different ways to represent this line, such as line XY or $\overleftrightarrow{XY}$.

So, Molly drew ______________________.

Key Concept Lines, Rays, Line Segments

Words	Models
A line is a straight set of points that extends in opposite directions without ending.	line AB or $\overleftrightarrow{AB}$
A **ray** is a part of a line that has one **endpoint** and extends in one direction without ending.	endpoint; ray AB or $\overrightarrow{AB}$
A **line segment** is a part of a line between two endpoints.	endpoint; line segment AB or $\overline{AB}$

Example 2

Tutor

Draw a figure that could be represented by $\overline{CD}$.

$\overline{CD}$ represents a line segment with endpoints C and D.

My Drawing!

Example 3

Identify the figure at the right. A B

The figure has one endpoint and extends in one direction without ending. It is a ray.

The endpoint is A. The ray extends in the direction of point B.

So, the figure is ______________________.

Guided Practice

Identify each figure.

1. 

____

2. 

____

3. 

____

4. 

____

Talk MATH

How are lines and line segments alike? How are they different?

Name ....

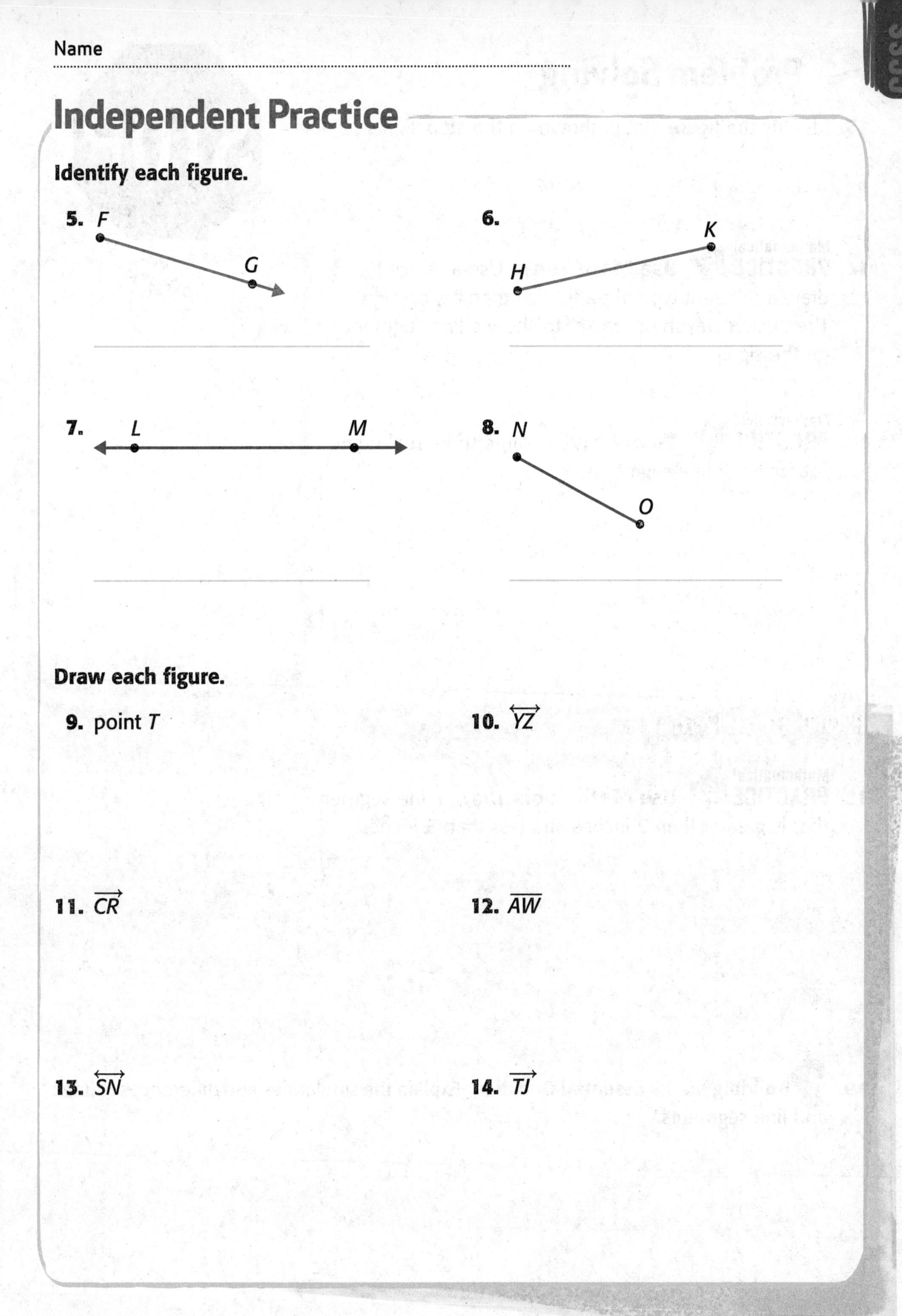

Independent Practice

Identify each figure.

5.

6.

7.

8.

Draw each figure.

9. point T

10. $\overleftrightarrow{YZ}$

11. $\overrightarrow{CR}$

12. $\overline{AW}$

13. $\overleftrightarrow{SN}$

14. $\overrightarrow{TJ}$

Problem Solving

15. Identify the figure that is shown on the stop sign.

16. Mathematical **PRACTICE 5** **Use Math Tools** Use a pencil to draw a different type of traffic sign than a stop sign. Then use a crayon or marker to show a line segment on the sign.

My Drawing!

17. Mathematical **PRACTICE 4** **Model Math** Name three real-world examples of line segments.

HOT Problems

18. Mathematical **PRACTICE 5** **Use Math Tools** Draw a line segment that is greater than 2 inches and less than 5 inches.

19. **Building on the Essential Question** Explain the similarities and differences of lines and line segments.

Name

MY Homework

Lesson 1
Draw Points, Lines, and Rays

Homework Helper

eHelp

Need help? connectED.mcgraw-hill.com

Abbey is reading the instructions in her drawing book. The instructions are to draw $\overleftrightarrow{AB}$, $\overrightarrow{EK}$, and $\overline{JT}$. What figures should she draw?

A line is a straight set of points that extend in opposite directions without ending. You can represent it as line *AB* or $\overleftrightarrow{AB}$.

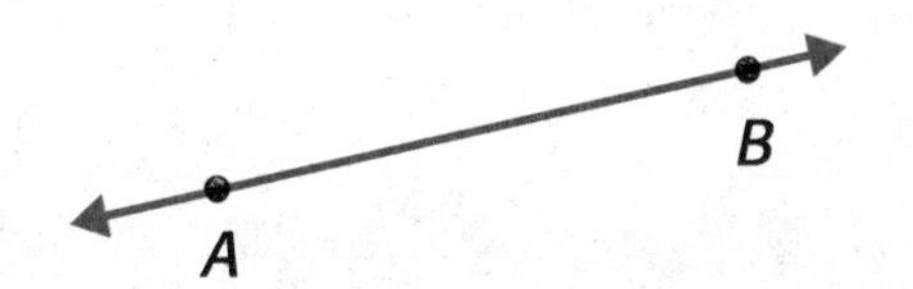

A ray is a figure that has one endpoint and extends in one direction without ending. You can represent it as ray *EK* or $\overrightarrow{EK}$.

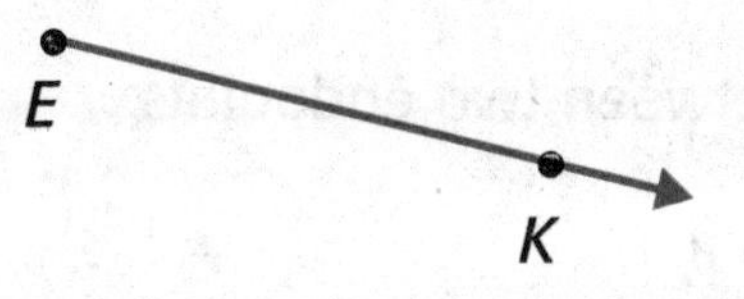

A line segment is a part of line between two endpoints. You can represent it as line segment *JT* or $\overline{JT}$.

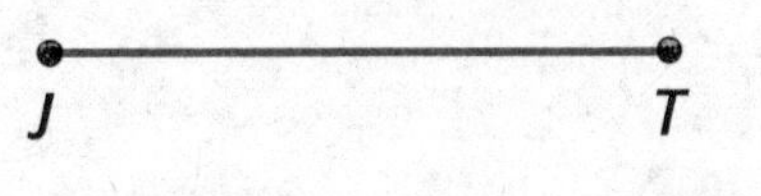

Practice

Identify each figure.

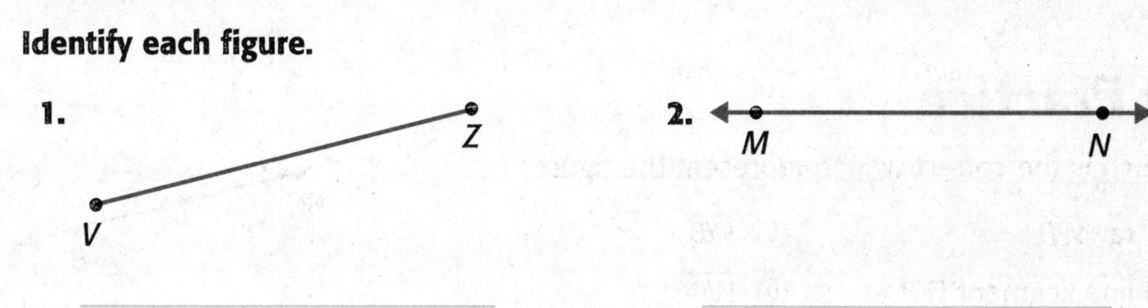

Draw each figure.

3. $\overline{YB}$

4. $\overrightarrow{LR}$

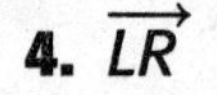

Problem Solving

5. How many line segments are represented in this figure?

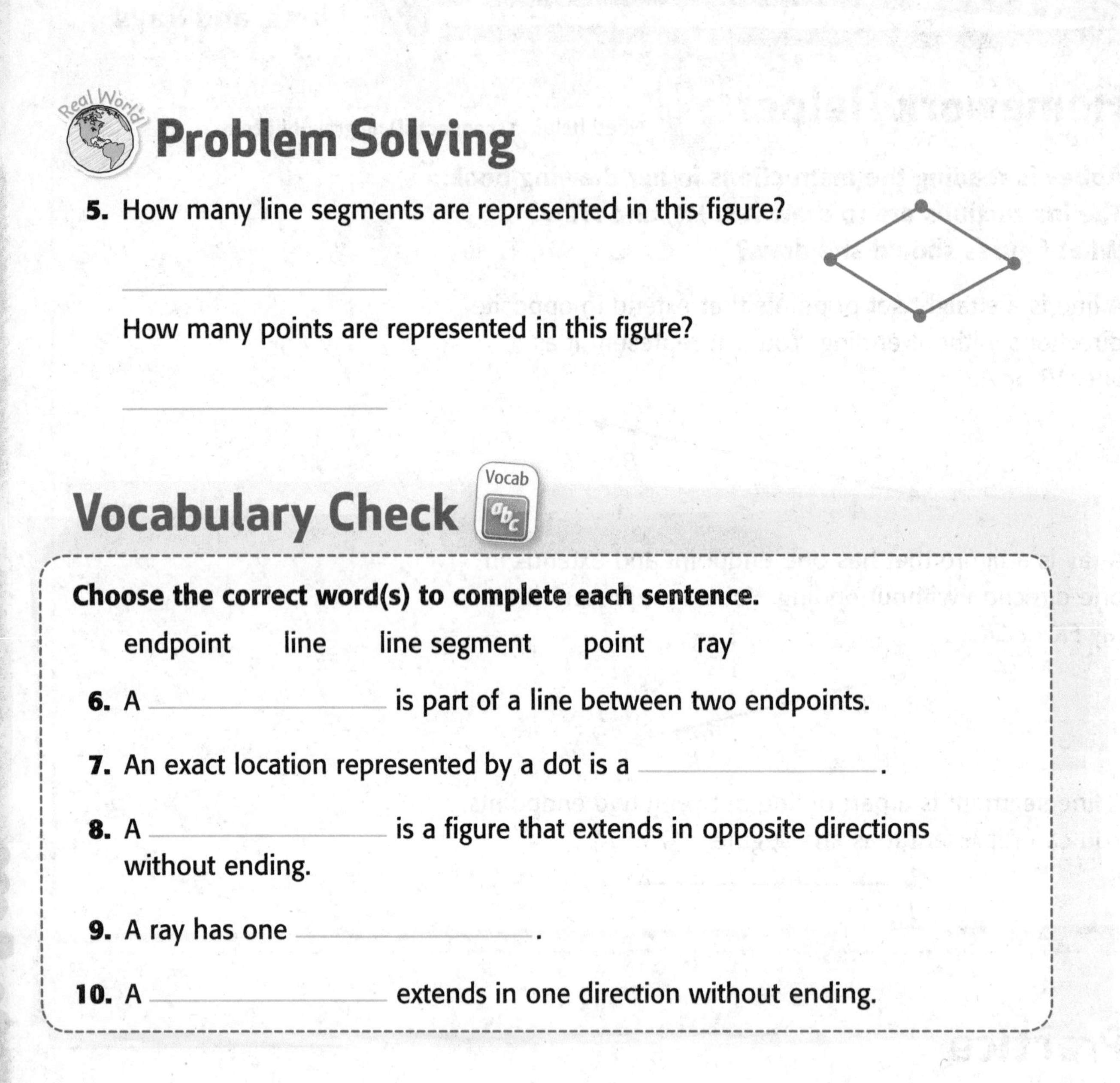

____

How many points are represented in this figure?

____

Vocabulary Check

Choose the correct word(s) to complete each sentence.

endpoint line line segment point ray

6. A ______________ is part of a line between two endpoints.

7. An exact location represented by a dot is a ______________.

8. A ______________ is a figure that extends in opposite directions without ending.

9. A ray has one ______________.

10. A ______________ extends in one direction without ending.

Test Practice

11. Which is the correct way to represent the figure?

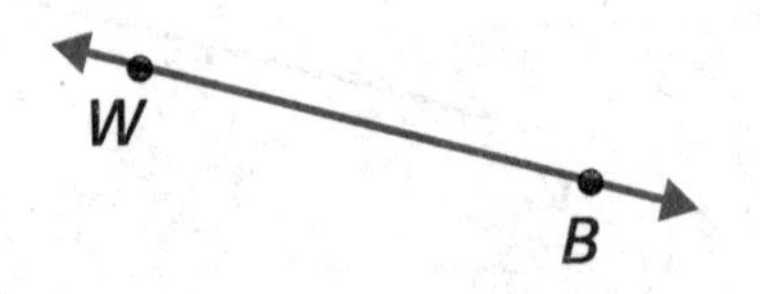

Ⓐ ray *WB*

Ⓑ line segment *WB*

Ⓒ $\overline{WB}$

Ⓓ $\overleftrightarrow{WB}$

Name ....

Draw Parallel and Perpendicular Lines

Lesson 2

ESSENTIAL QUESTION
How are different ideas about geometry connected?

You can describe lines, rays, and line segments by the way they cross each other or do not cross each other.

Example 1

Oliver was riding in the car and saw this sign. Describe how the outlined line segments cross each other or do not cross each other.

Parallel lines are always the same distance apart. They do not meet or cross each other.

So, Oliver saw a figure with ______________ line segments.

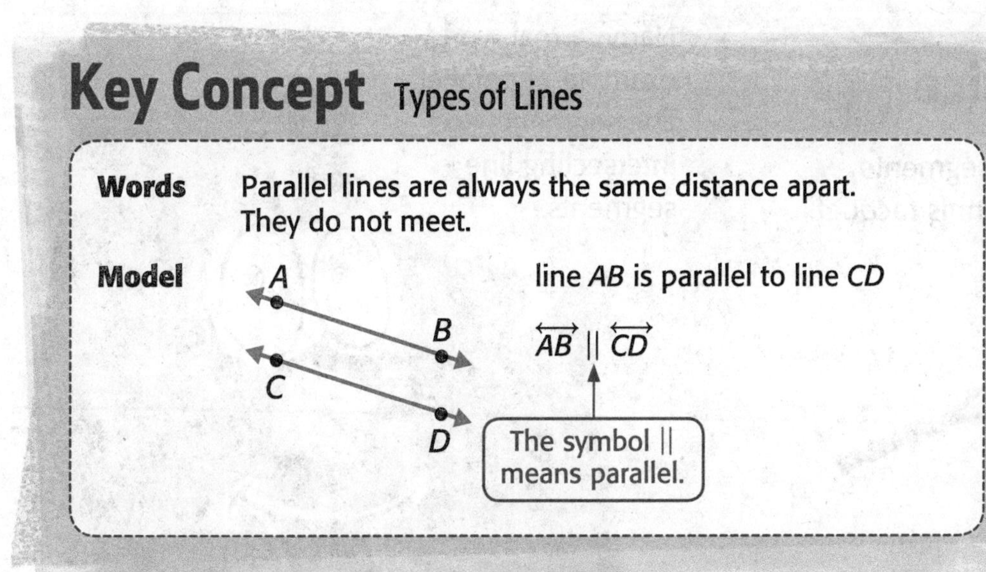

Key Concept Types of Lines

Words Parallel lines are always the same distance apart. They do not meet.

Model

line *AB* is parallel to line *CD*

$\overleftrightarrow{AB} \parallel \overleftrightarrow{CD}$

The symbol || means parallel.

Key Concept Types of Lines

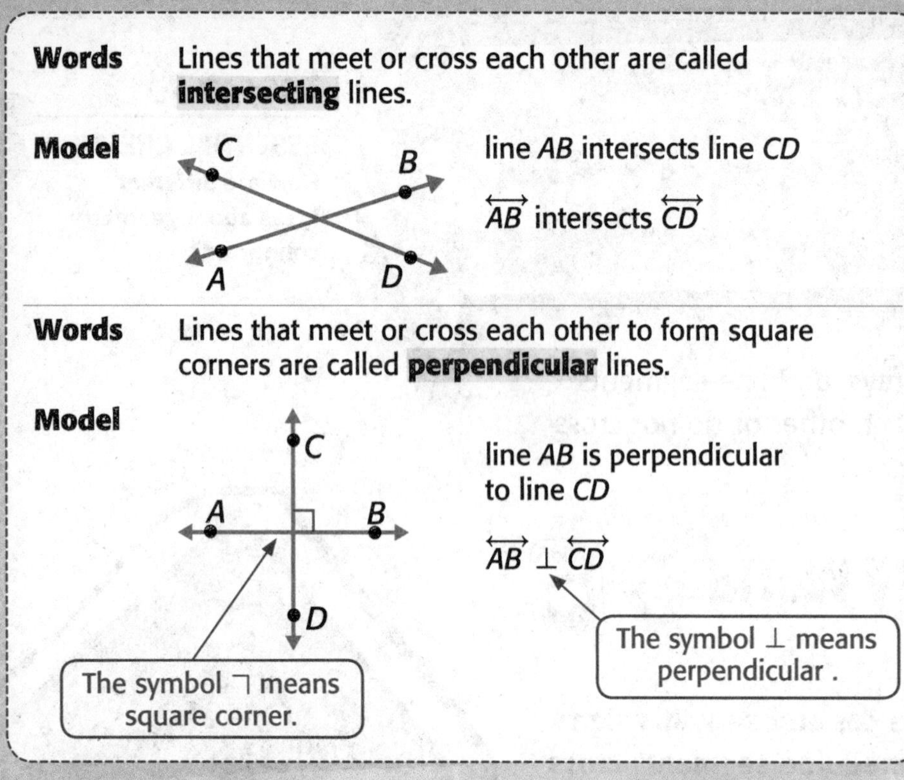

Words Lines that meet or cross each other are called **intersecting** lines.

Model line *AB* intersects line *CD*

$\overleftrightarrow{AB}$ intersects $\overleftrightarrow{CD}$

Words Lines that meet or cross each other to form square corners are called **perpendicular** lines.

Model line *AB* is perpendicular to line *CD*

$\overleftrightarrow{AB} \perp \overleftrightarrow{CD}$

The symbol ⅂ means square corner.

The symbol ⊥ means perpendicular .

Example 2

Describe the figure. Use *parallel, perpendicular,* or *intersecting*. Use the most specific term.

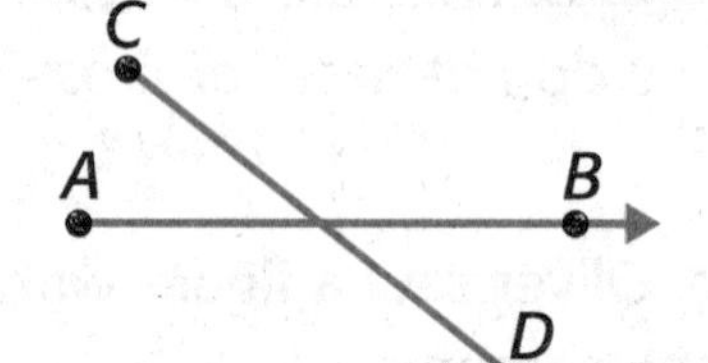

The figure shows ray *AB* and line segment *CD*.

The figures cross each other but do not form square corners.

$\overrightarrow{AB}$ and $\overline{CD}$ are ______________.

Talk MATH

Name a real-world example of parallel line segments and intersecting line segments.

Guided Practice

1. Describe the line segments outlined on the tennis racquet.

____

Name

Independent Practice

Describe each figure. Use *parallel, perpendicular,* or *intersecting.* Use the most specific term.

2.

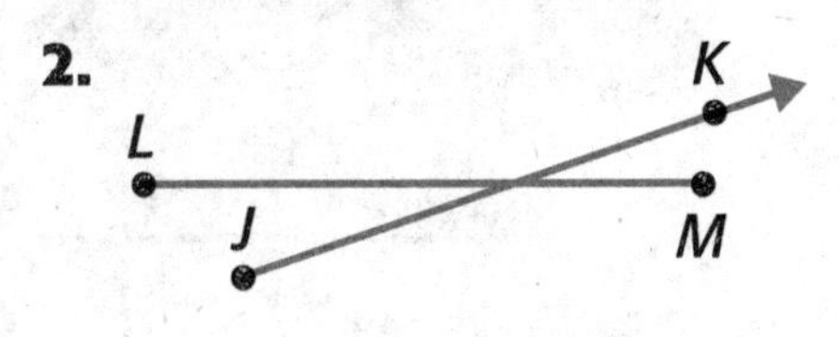

$\overline{LM}$ and $\overrightarrow{JK}$ are ______________.

3.

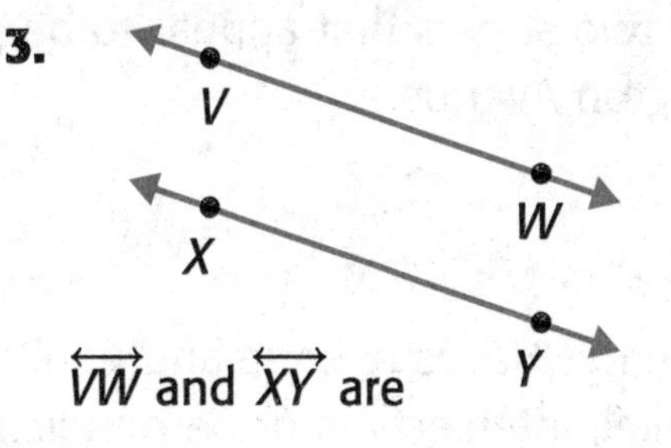

$\overleftrightarrow{VW}$ and $\overleftrightarrow{XY}$ are ______________.

Draw an example of each figure.

4. $\overleftrightarrow{DE} \parallel \overleftrightarrow{FG}$

5. $\overleftrightarrow{RS}$ intersects $\overleftrightarrow{TU}$

6. $\overleftrightarrow{NO} \perp \overleftrightarrow{PQ}$

7. $\overline{JK} \parallel \overrightarrow{LM}$

8. Circle the statement that is true about the figure below.

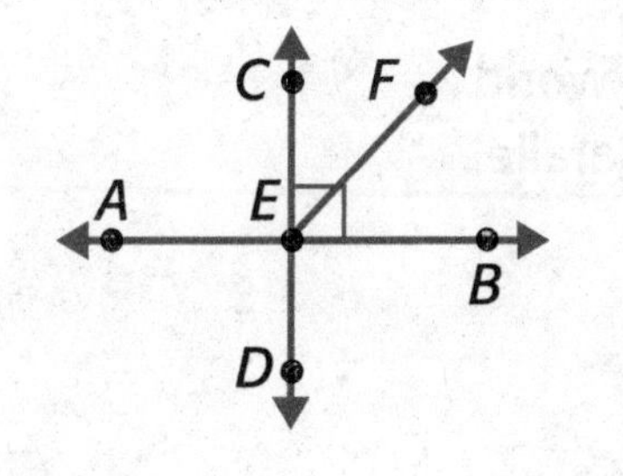

Line *AB* is parallel to ray *EF*.

Line *AB* is perpendicular to line *CD*.

Line *CD* is parallel to ray *EF*.

Line *CD* is parallel to line *AB*.

Problem Solving

Mathematical PRACTICE 4 Model Math On a map, streets can be represented by line segments. Use the map to answer Exercises 9–11.

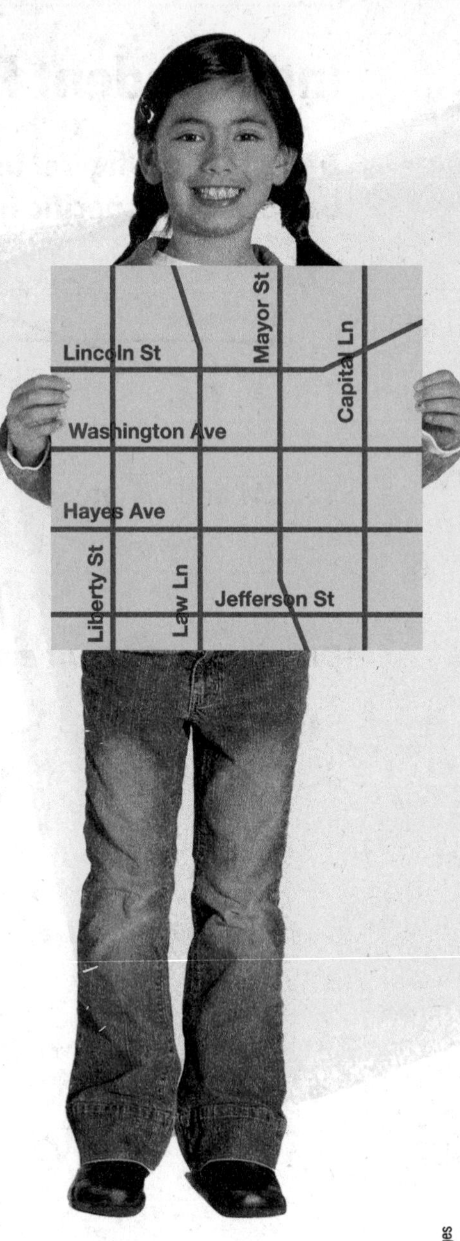

9. Identify two streets that appear to be parallel to Washington Avenue.

10. Tell whether Hayes Avenue and Capital Lane appear to be parallel, intersecting, or perpendicular lines. Explain.

11. Are there any streets that are intersecting but not perpendicular? Explain.

HOT Problems

12. Mathematical **PRACTICE 2** **Stop and Reflect** Tell whether each statement is *true* or *false*.

- If two lines are parallel, they are always the same distance apart. ______
- If two lines are parallel, they are also perpendicular. ______

13. **Building on the Essential Question** Describe a real-world example of when it is necessary that line segments are parallel.

Name ....

MY Homework

Lesson 2
Draw Parallel and Perpendicular Lines

Homework Helper

eHelp

Need help? connectED.mcgraw-hill.com

Describe each figure. Use *parallel, perpendicular,* or *intersecting*. Use the most specific term.

Lines or line segments that are always the same distance apart and do not meet are parallel. So, $\overline{RK} \parallel \overline{DM}$.

Lines or line segments that meet to form square corners are perpendicular.

So, $\overleftrightarrow{PB} \perp \overline{EH}$.

Lines or line segments that meet or cross are intersecting. So, $\overleftrightarrow{XY}$ intersects $\overleftrightarrow{CF}$.

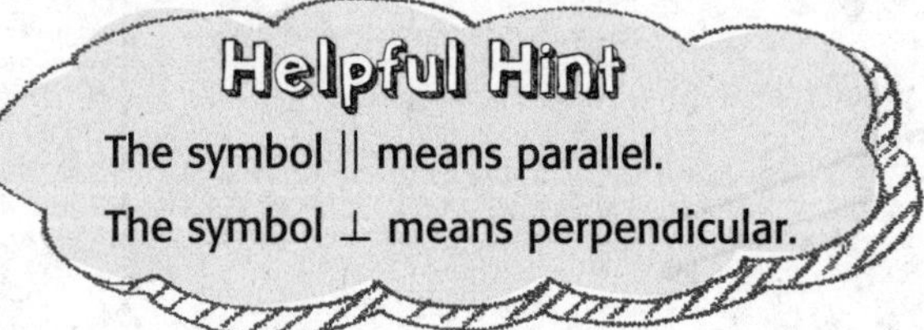

Practice

1. Describe the figure. Use *parallel, perpendicular,* or *intersecting*. Use the most specific term.

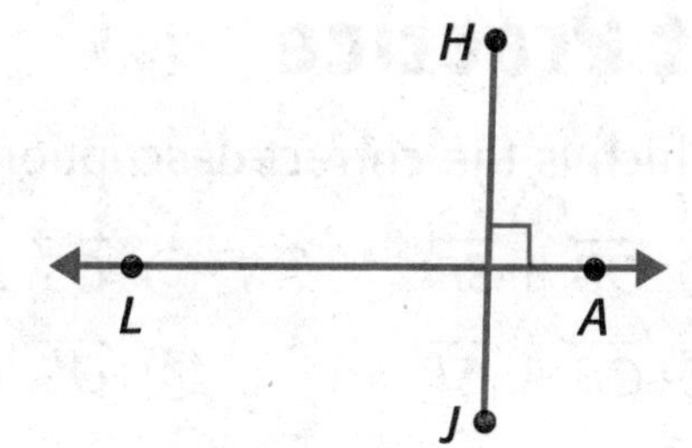

____

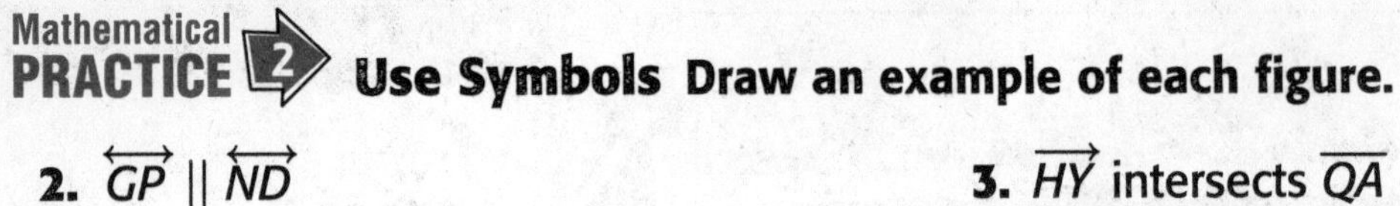

Mathematical PRACTICE 2 **Use Symbols** **Draw an example of each figure.**

2. $\overleftrightarrow{GP} \parallel \overleftrightarrow{ND}$

3. $\overrightarrow{HY}$ intersects $\overline{QA}$

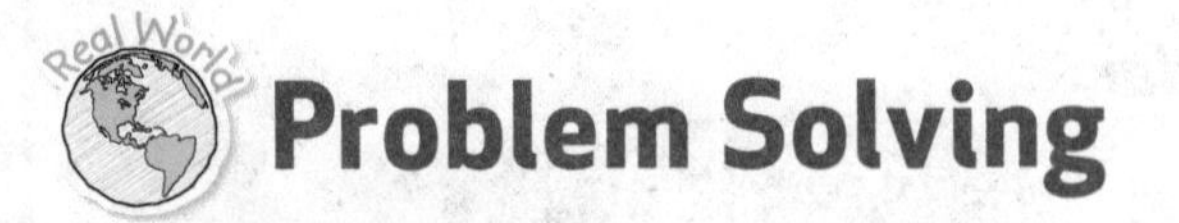

Problem Solving

4. **Mathematical PRACTICE 4** **Model Math** Martin is washing windows. First he must raise the blinds. Describe the kind of line segments formed by the horizontal blinds.

____

Vocabulary Check

Draw a line to match each vocabulary term to its example.

5. intersecting, but not perpendicular

6. parallel

7. perpendicular

Test Practice

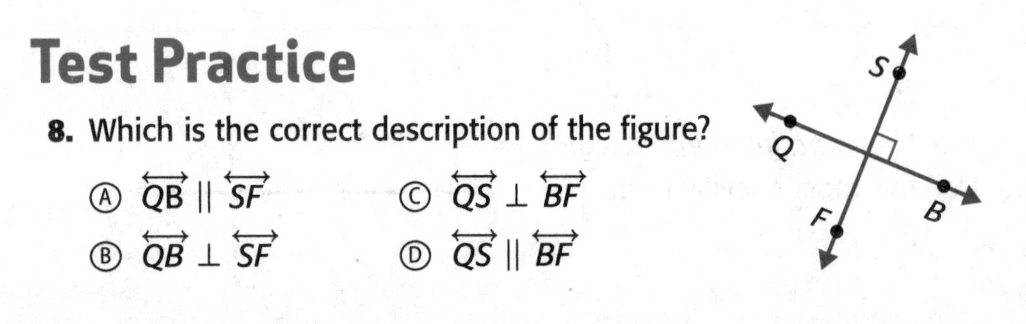

8. Which is the correct description of the figure?

Ⓐ $\overleftrightarrow{QB} \parallel \overleftrightarrow{SF}$

Ⓑ $\overleftrightarrow{QB} \perp \overleftrightarrow{SF}$

Ⓒ $\overleftrightarrow{QS} \perp \overleftrightarrow{BF}$

Ⓓ $\overleftrightarrow{QS} \parallel \overleftrightarrow{BF}$

Check My Progress

Vocabulary Check

Vocab

Use the word bank to complete each sentence.

endpoint **line** **line segment** **point** **ray**

1. A ______________ is part of a line between two endpoints.
2. A ______________ is a part of a line that has one ______________ and extends in one direction without ending.
3. A ______________ is a straight set of points that extends in opposite directions without ending.
4. A ______________ is an exact location that is represented by a dot.

Match each vocabulary word to its definition.

5. **intersecting** • lines that meet or cross each other to form square corners
6. **parallel** • lines that meet or cross each other, but do not necessarily form square corners
7. **perpendicular** • lines that are always the same distance apart and do not meet

Concept Check

Check

Circle the correct description of each figure.

8. M N

line
line segment
ray

9. P O

line
line segment
ray

10. Q R

line
line segment
ray

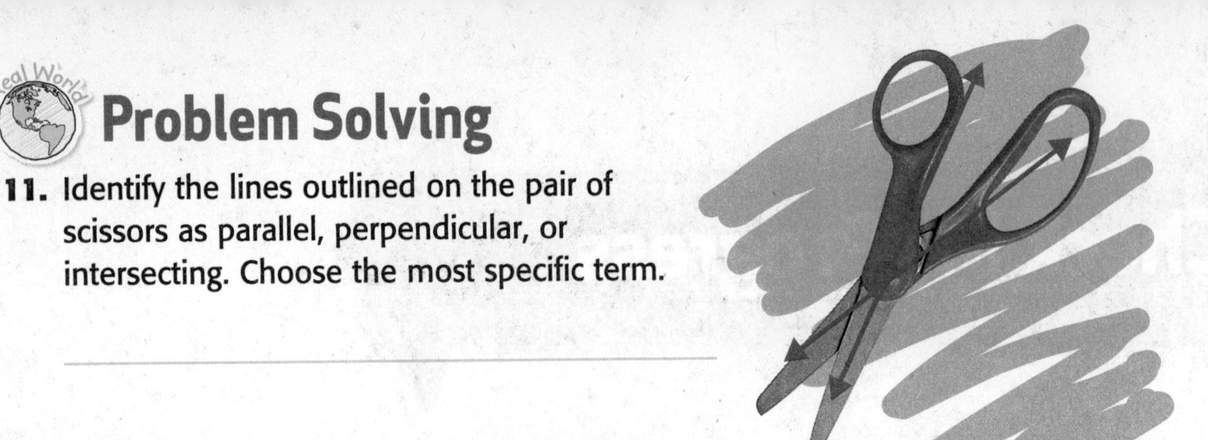

Problem Solving

11. Identify the lines outlined on the pair of scissors as parallel, perpendicular, or intersecting. Choose the most specific term.

____

12. Sandy is driving on Broadway Avenue. Which street appears to be perpendicular to Broadway Avenue?

____

13. Nathan practiced his handwriting by writing the alphabet in capital letters. He stopped at the first letter that contains parallel line segments. At which letter did Nathan stop writing?

____

Test Practice

14. Which figure shows parallel lines?

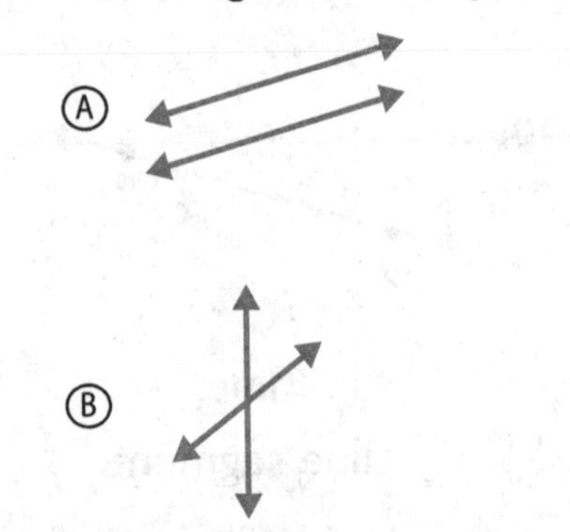

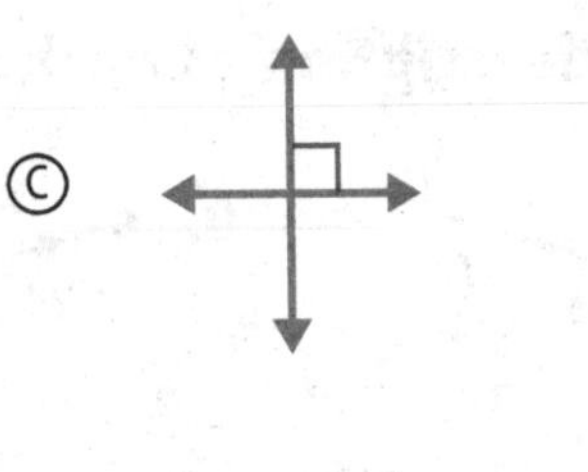

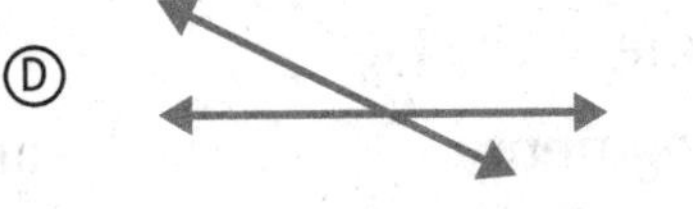

Name ....

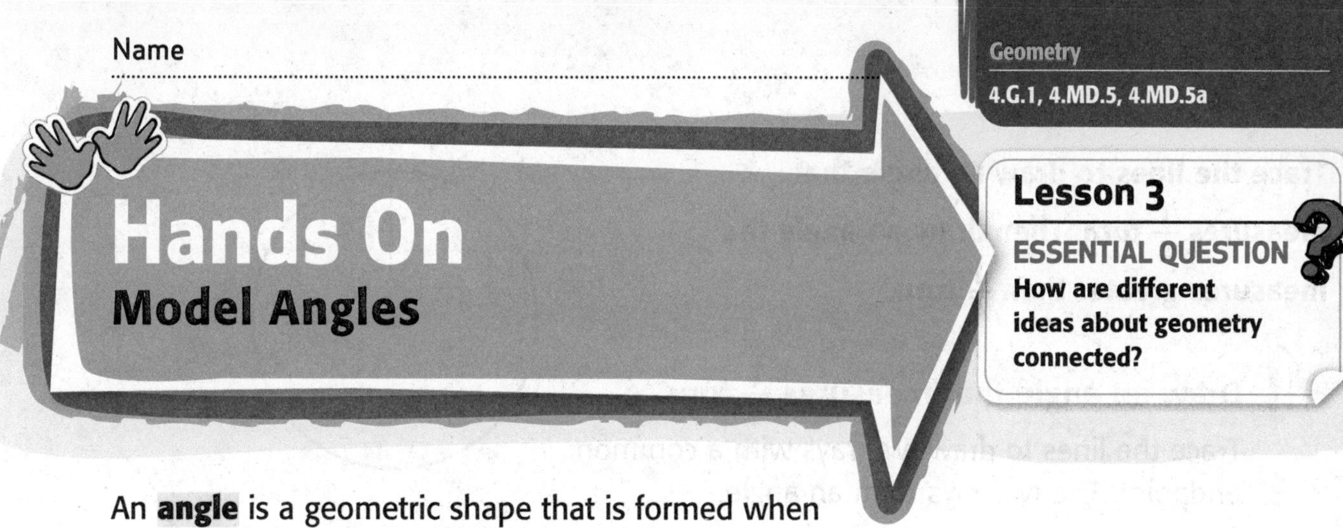

Hands On
Model Angles

Lesson 3

ESSENTIAL QUESTION
How are different ideas about geometry connected?

An **angle** is a geometric shape that is formed when two rays have the same endpoint.

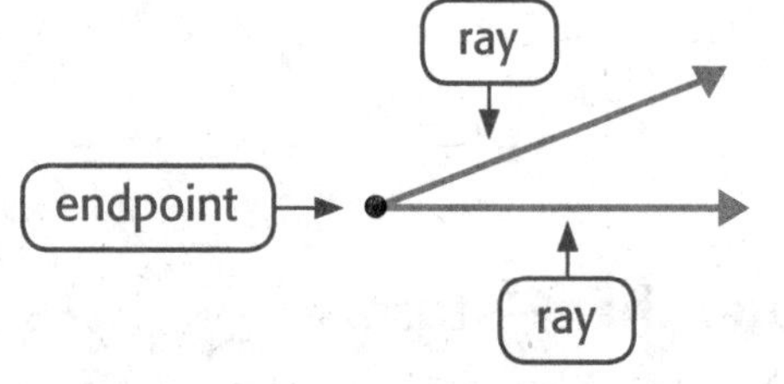

Angles are measured by the amount of rotation, or turning, from one ray to another.

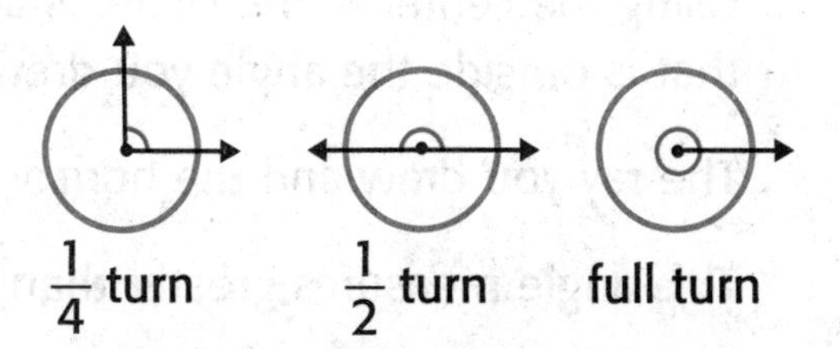

Draw It

Trace the lines to draw an angle that measures $\frac{1}{4}$ turn. Then draw an angle that measures less than $\frac{1}{4}$ turn.

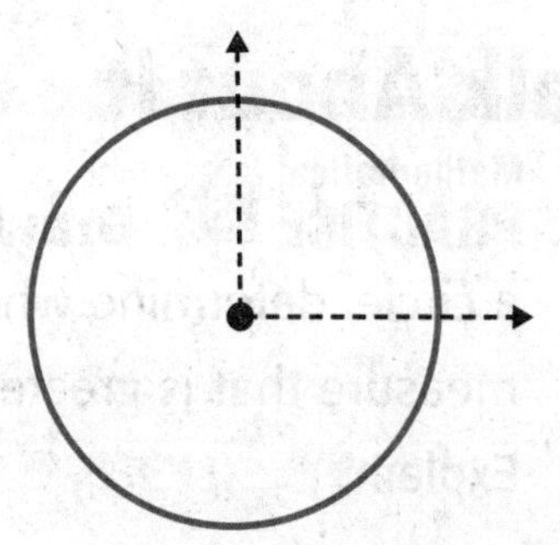

1 Draw an angle that measures $\frac{1}{4}$ turn.

Trace the lines to draw two rays with a common endpoint. The two rays form an angle.

The center of the circle is at the same point as the endpoint of the two rays. The angle you drew measures $\frac{1}{4}$ turn.

2 Draw an angle that measures less than $\frac{1}{4}$ turn.

Using the center of the circle as an endpoint, draw a ray that is inside the angle you drew in Step 1.

The ray you drew and the horizontal ray form an angle. This angle measures less than $\frac{1}{4}$ turn.

Try It

Trace the lines to draw an angle that measures $\frac{1}{4}$ turn. Then draw an angle that measures greater than $\frac{1}{4}$ turn.

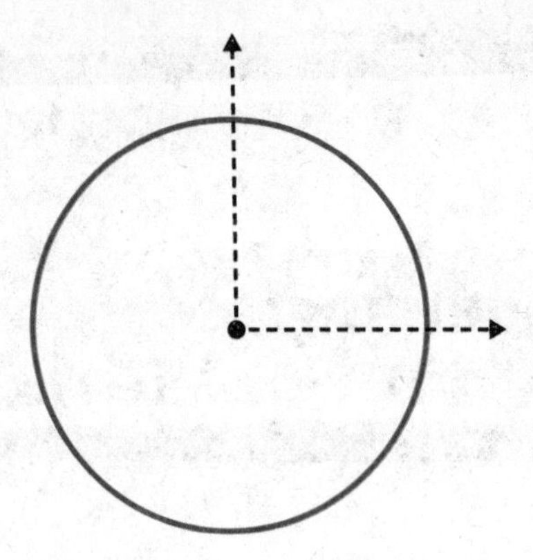

1 Draw an angle that measures $\frac{1}{4}$ turn.

Trace the lines to draw two rays with a common endpoint. The two rays form an angle.

The center of the circle is at the same point as the endpoint of the two rays.

The angle you drew measures $\frac{1}{4}$ turn.

2 Draw an angle that measures greater than $\frac{1}{4}$ turn.

Using the center of the circle as an endpoint, draw a ray that is outside the angle you drew in Step 1.

The ray you drew and the horizontal ray form an angle.

This angle measures greater than $\frac{1}{4}$ turn.

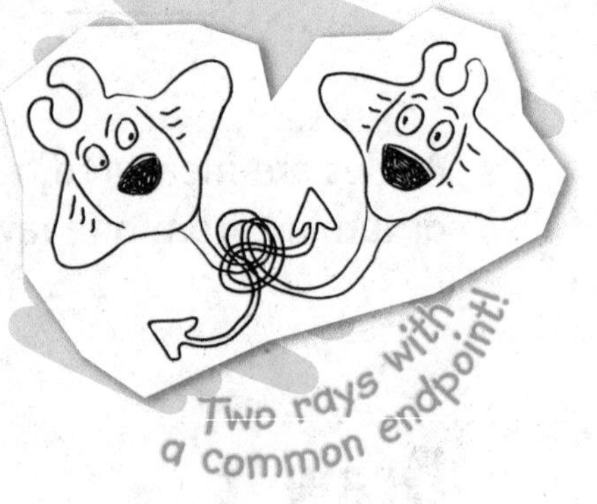

Talk About It

1. **Mathematical PRACTICE 3 Draw a Conclusion** Without drawing a circle, determine whether the angle at the right has a measure that is greater than, less than, or equal to $\frac{1}{2}$ turn. Explain.

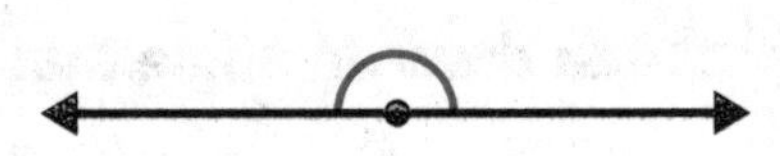

____

____

2. **Mathematical PRACTICE 6 Explain to a Friend** Refer to the angle you drew in the activity above. State whether the angle's measure is greater than, less than, or equal to $\frac{1}{2}$ turn. Explain.

____

____

____

Name

CCSS

Practice It

3. Draw an angle with a measure less than $\frac{1}{4}$ turn.

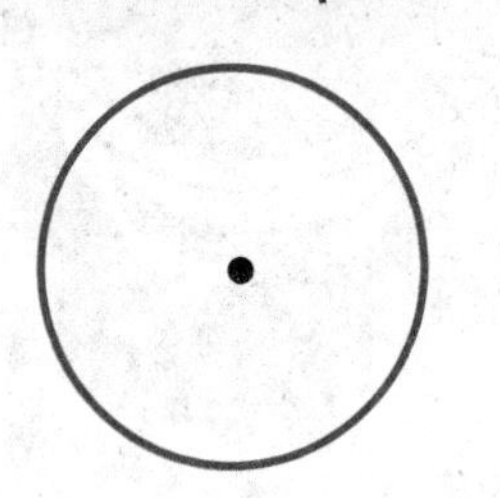

4. Draw an angle with a measure greater than $\frac{1}{4}$ turn.

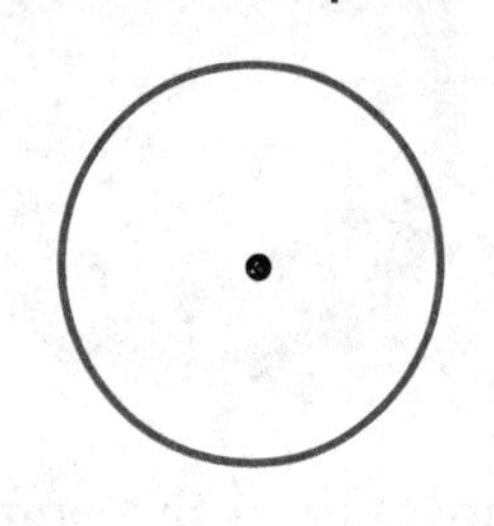

Draw lines to match each figure to its description.

5.

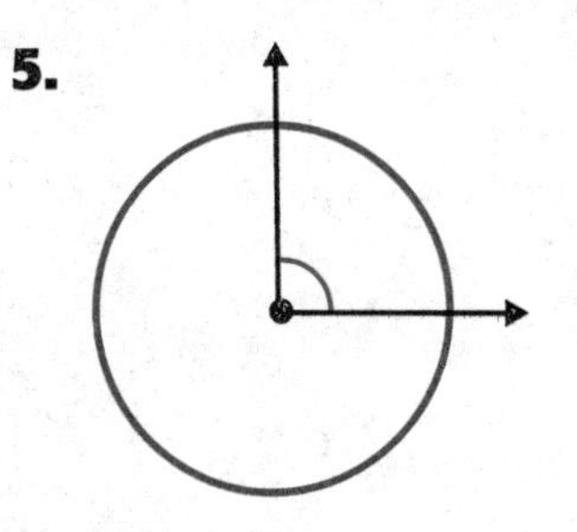

- An angle with a measure greater than $\frac{1}{4}$ turn, but less than $\frac{1}{2}$ turn.

6.

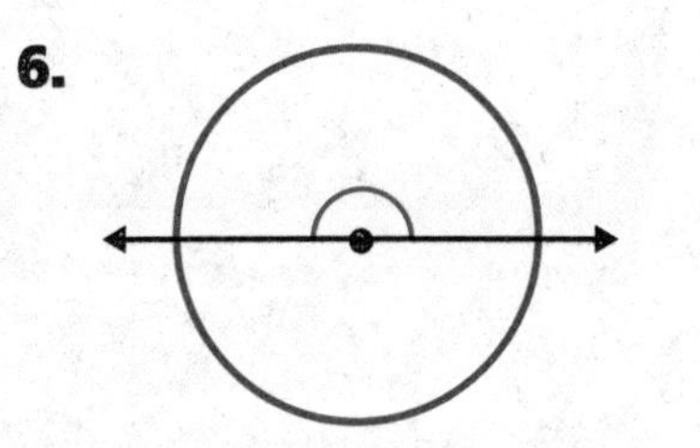

- An angle with a measure less than $\frac{1}{4}$ turn.

7.

- An angle with a measure of $\frac{1}{4}$ turn.

8.

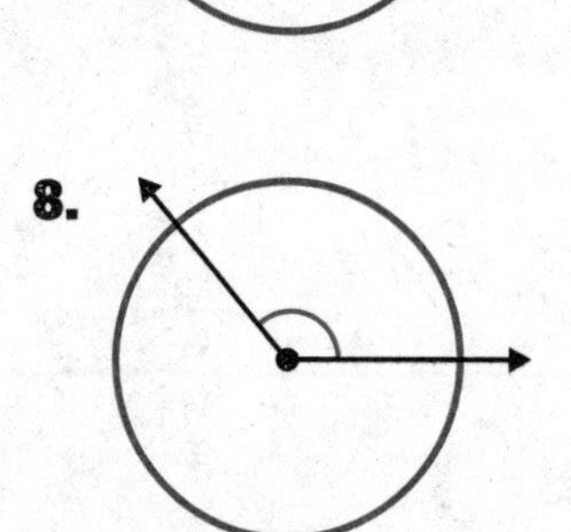

- An angle with a measure of $\frac{1}{2}$ turn.

Apply It

9. Draw the hands on the clock below to show 5:00.

Describe the measure of the angle formed by the hands of the clock.

____

____

____

10. Draw a real-world object that shows an angle with a measure of $\frac{1}{4}$ turn.

11. Mathematical PRACTICE 4 **Model Math** Draw two angles that share a common endpoint and a common ray. Together, they should form an angle with a measure of $\frac{1}{4}$ turn.

Write About It

12. How can I describe an angle's measure?

____

____

____

Name

MY Homework

Lesson 3

Hands On: Model Angles

Homework Helper

eHelp

Need help? connectED.mcgraw-hill.com

Draw an angle with a measure greater than $\frac{1}{4}$ turn.

Draw two rays with a common endpoint at the center of the circle.
Be sure that the angle created by the rays has a measure greater than $\frac{1}{4}$ turn.

How do you know the measure of the angle is greater than $\frac{1}{4}$ turn?

An angle with a measure of $\frac{1}{4}$ turn forms a square corner.
The angle drawn has a greater measure than $\frac{1}{4}$ turn.

Practice

Circle the correct description for the measure of each angle.

1.

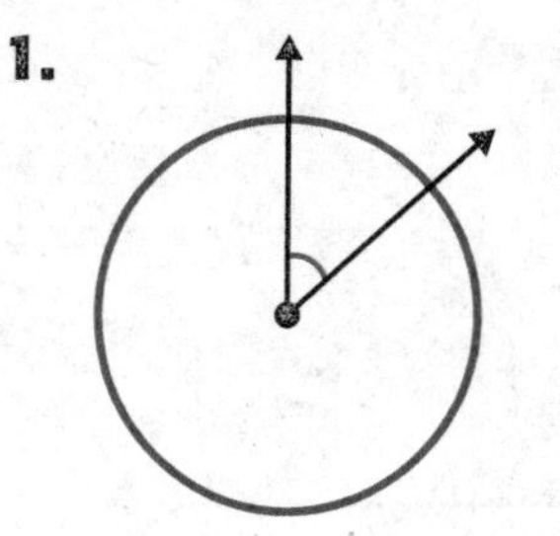

less than $\frac{1}{4}$ turn

greater than $\frac{1}{4}$ turn

2.

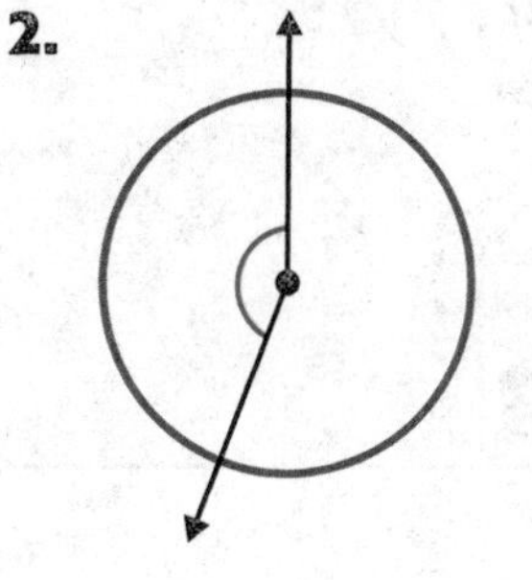

less than $\frac{1}{4}$ turn

greater than $\frac{1}{4}$ turn

3.

less than $\frac{1}{2}$ turn

greater than $\frac{1}{2}$ turn

Draw an angle that has each measure.

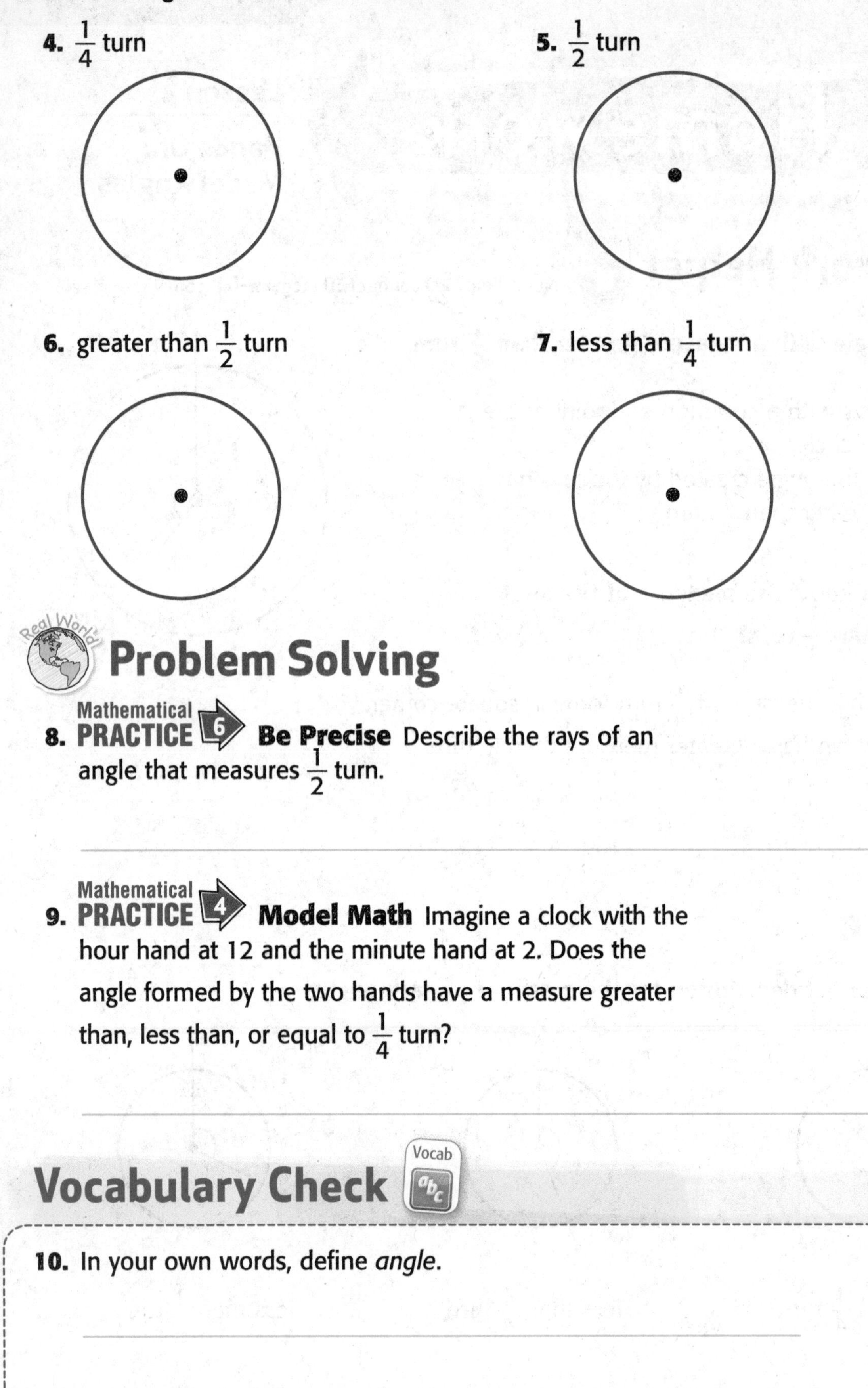

4. $\frac{1}{4}$ turn

5. $\frac{1}{2}$ turn

6. greater than $\frac{1}{2}$ turn

7. less than $\frac{1}{4}$ turn

Problem Solving

8. **Mathematical PRACTICE 6** **Be Precise** Describe the rays of an angle that measures $\frac{1}{2}$ turn.

---

9. **Mathematical PRACTICE 4** **Model Math** Imagine a clock with the hour hand at 12 and the minute hand at 2. Does the angle formed by the two hands have a measure greater than, less than, or equal to $\frac{1}{4}$ turn?

Vocabulary Check

Vocab

10. In your own words, define *angle*.

Name

Classify Angles

Lesson 4

ESSENTIAL QUESTION
How are different ideas about geometry connected?

Angles can be measured in a more precise way than turns. The unit used to measure an angle is called a **degree (°)**. A circle is made up of 360°.

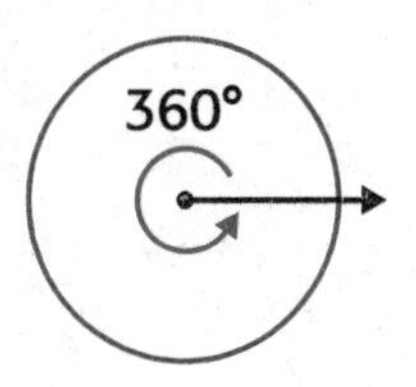

An angle that turns through $\frac{1}{360}$ of a circle is called a **one-degree angle**. That means that 360 one-degree angles sharing the same endpoint make a circle. The angle below turns through 3 one-degree angles. So, it measures 3°.

Math in My World

Watch Tutor

Example 1

David waits by the crosswalk sign on his way to school. The angle outlined on the sign turns through 50 one-degree angles. Find the measure of the angle.

The angle turns through 50 one-degree angles.

That means that 50 one-degree angles sharing the same endpoint make the angle.

So, the angle has a measure of ______°.

Angles can be classified as *right*, *acute*, or *obtuse*.

Key Concept Types of Angles

A **right angle** measures 90°.

This symbol means right angle.

An **acute angle** measures greater than 0° and less than 90°.

An **obtuse angle** measures greater than 90° but less than 180°.

Example 2 Tutor

Classify the angle as *right*, *acute*, or *obtuse*.

The angle is 90°.

So, it is a ________ angle.

Example 3 Tutor

Classify the angle as *right*, *acute*, or *obtuse*.

The angle is greater than 90° and less than 180°.

So, it is a(n) ________ angle.

Talk MATH

How many one-degree angles does a right angle turn through?

Guided Practice Check

1. The angle shown turns through 94 one-degree angles. Find the measure of the angle.

____

2. Classify the angle shown as *right*, *acute*, or *obtuse*.

____

Name

Independent Practice

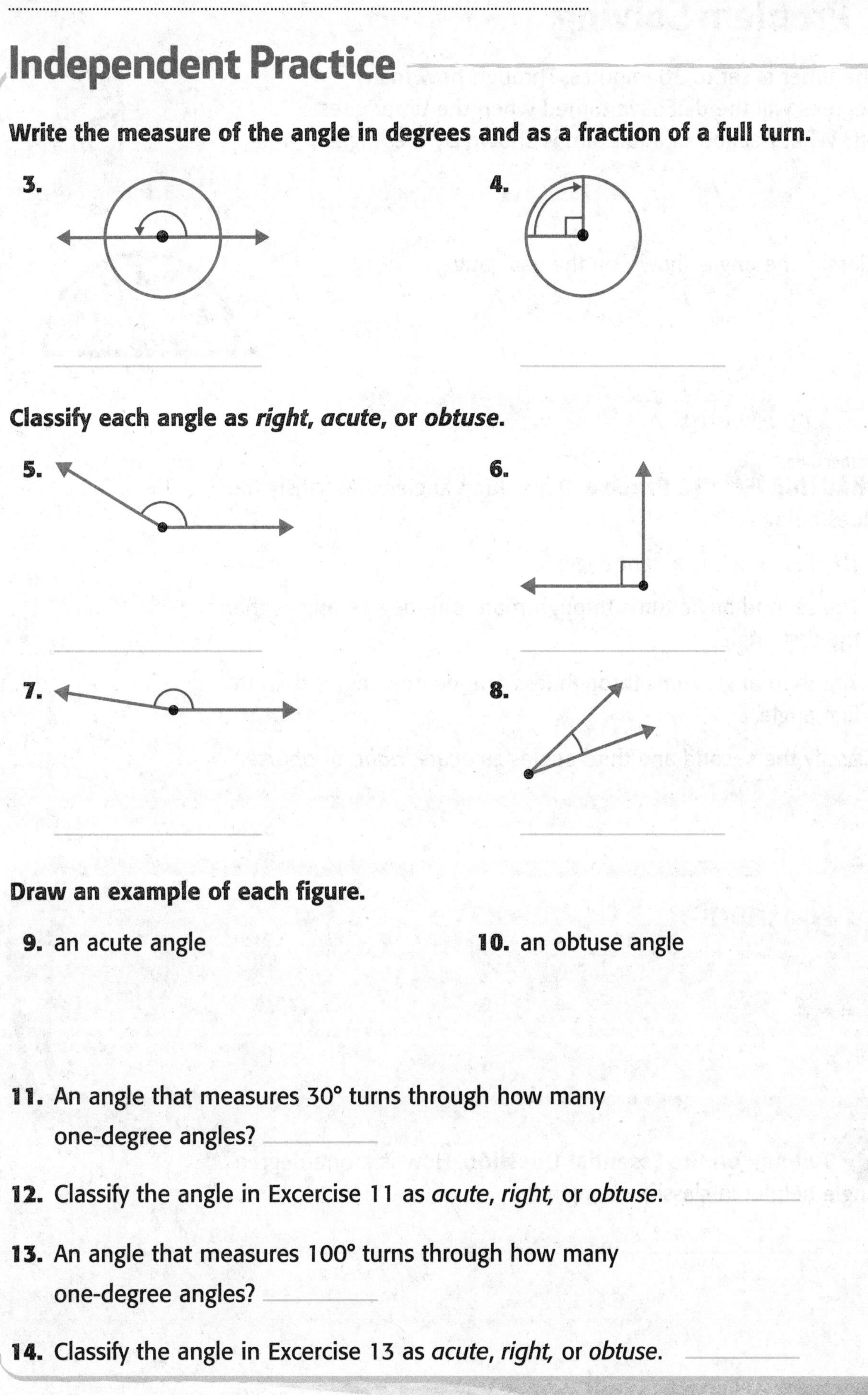

Write the measure of the angle in degrees and as a fraction of a full turn.

3.

4.

Classify each angle as *right, acute,* or *obtuse.*

5.

6.

7.

8.

Draw an example of each figure.

9. an acute angle

10. an obtuse angle

11. An angle that measures 30° turns through how many one-degree angles? ______

12. Classify the angle in Excercise 11 as *acute, right,* or *obtuse.* ______

13. An angle that measures 100° turns through how many one-degree angles? ______

14. Classify the angle in Excercise 13 as *acute, right,* or *obtuse.* ______

Problem Solving

15. The timer is set to 30 minutes. Through how many degrees will the dial have turned when the timer goes off? What fraction of a full turn is shown by the angle?

16. Classify the angle shown on the gas gauge.

HOT Problems

17. Mathematical PRACTICE 6 **Be Precise** Draw three angles that satisfy the clues below.

- The first angle is a right angle.
- The second angle turns through more one-degree angles than the first angle.
- The third angle turns through less one-degree angles than the first angle.

Classify the second and third angles as *acute*, *right*, or *obtuse*.

My Drawing!

18. **Building on the Essential Question** How is a one-degree angle helpful in classifying angles?

Name ....

Lesson 4
Classify Angles

Homework Helper

eHelp

Need help? connectED.mcgraw-hill.com

Classify the angle on the inside of the goalpost as *right*, *acute*, or *obtuse*.

The angle on the inside of the goalpost forms a square and measures 90°.

So, it is a right angle.

Practice

Write the measure of each angle in degrees and as a fraction of a full turn.

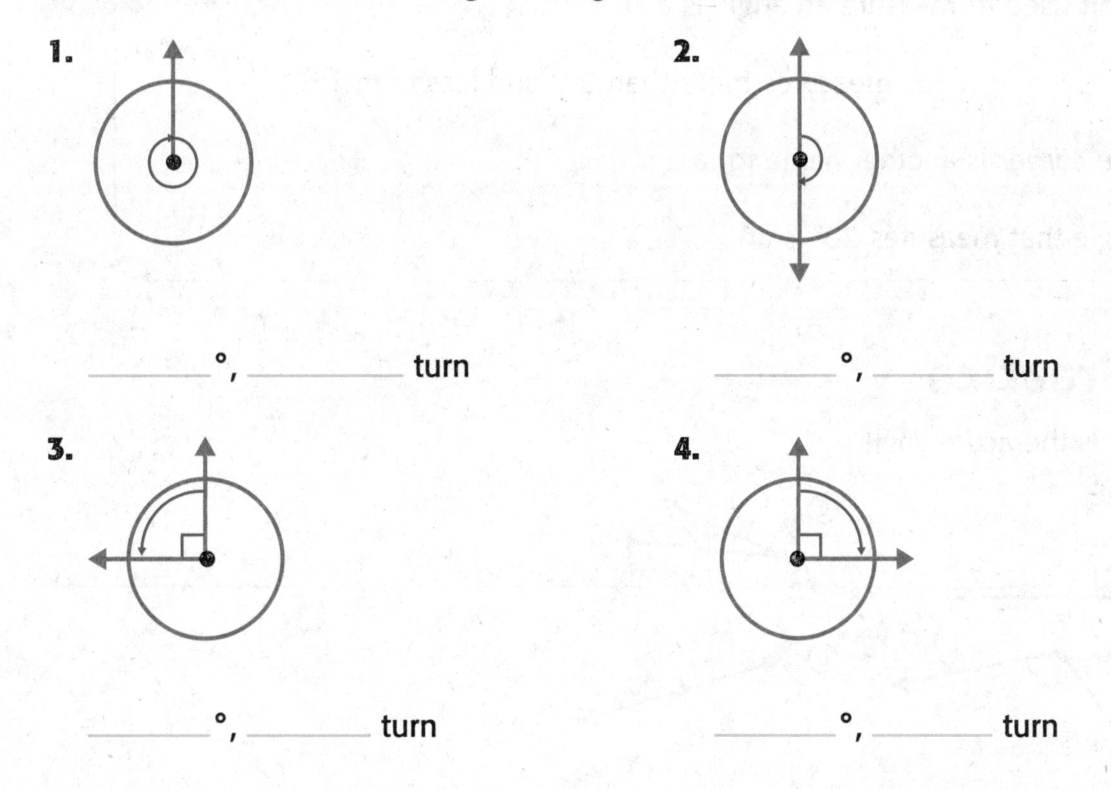

1. ________°, ________ turn

2. ________°, ________ turn

3. ________°, ________ turn

4. ________°, ________ turn

Classify each angle as *right, acute,* or *obtuse*.

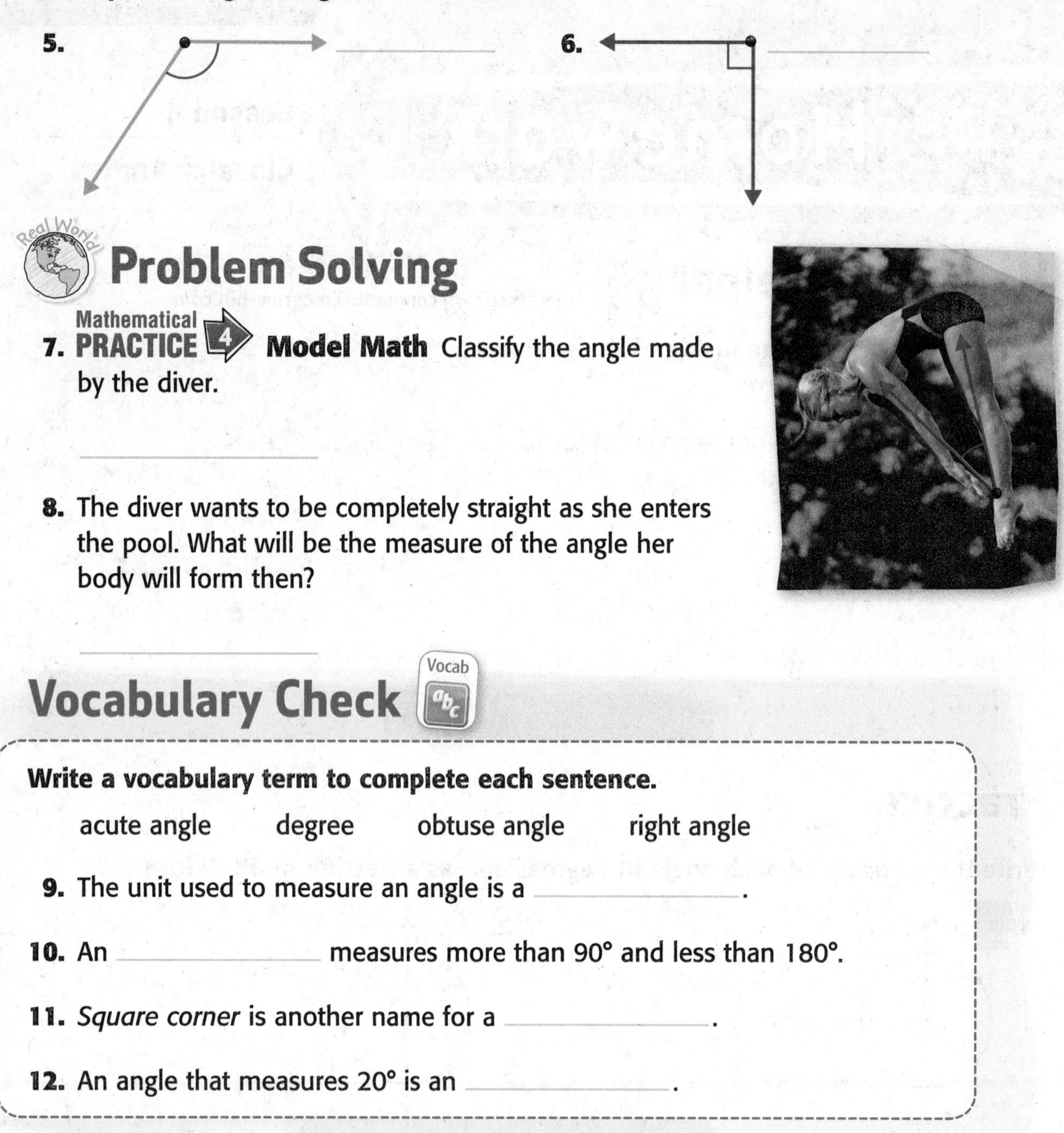

5. ______________

6. ______________

Problem Solving

7. Mathematical PRACTICE 4 **Model Math** Classify the angle made by the diver.

____

8. The diver wants to be completely straight as she enters the pool. What will be the measure of the angle her body will form then?

____

Vocabulary Check

Write a vocabulary term to complete each sentence.

acute angle degree obtuse angle right angle

9. The unit used to measure an angle is a ______________.

10. An ______________ measures more than 90° and less than 180°.

11. *Square corner* is another name for a ______________.

12. An angle that measures 20° is an ______________.

Test Practice

13. Which is the acute angle?

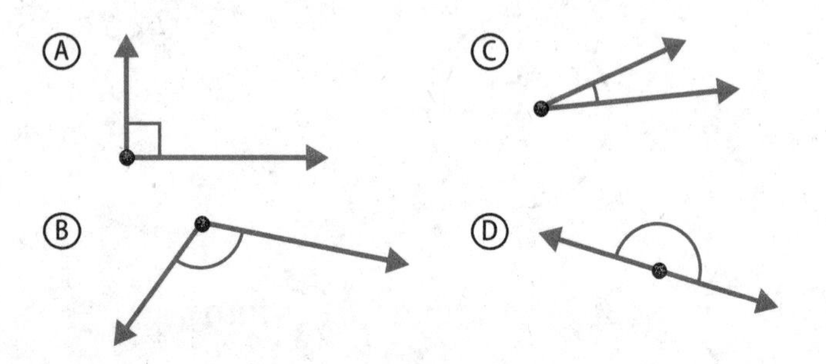

Name ....

Measure Angles

Lesson 5

ESSENTIAL QUESTION
How are different ideas about geometry connected?

The protractor is a tool used to measure angles. The length of the rays does not affect the measure of the angle.

Example 1

Henry drew the angle shown. Measure the angle.

Line up the protractor.
Place the center of the protractor on the endpoint of the angle with the straightedge along one ray.

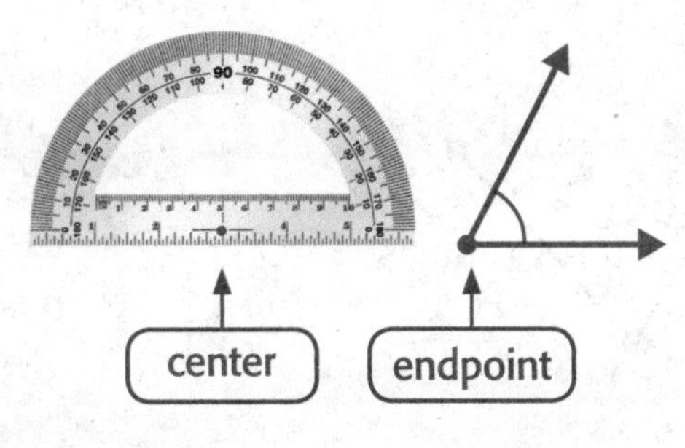

2 Line up the angle.
Line up one ray of the angle with the zero on the protractor.

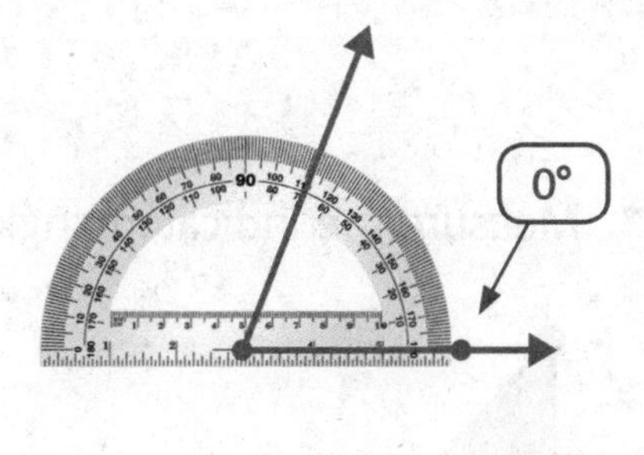

3 Measure the angle.
Find the tic mark on the protractor that aligns with the second ray of the angle.

70°

So, the measure of the angle is ______.

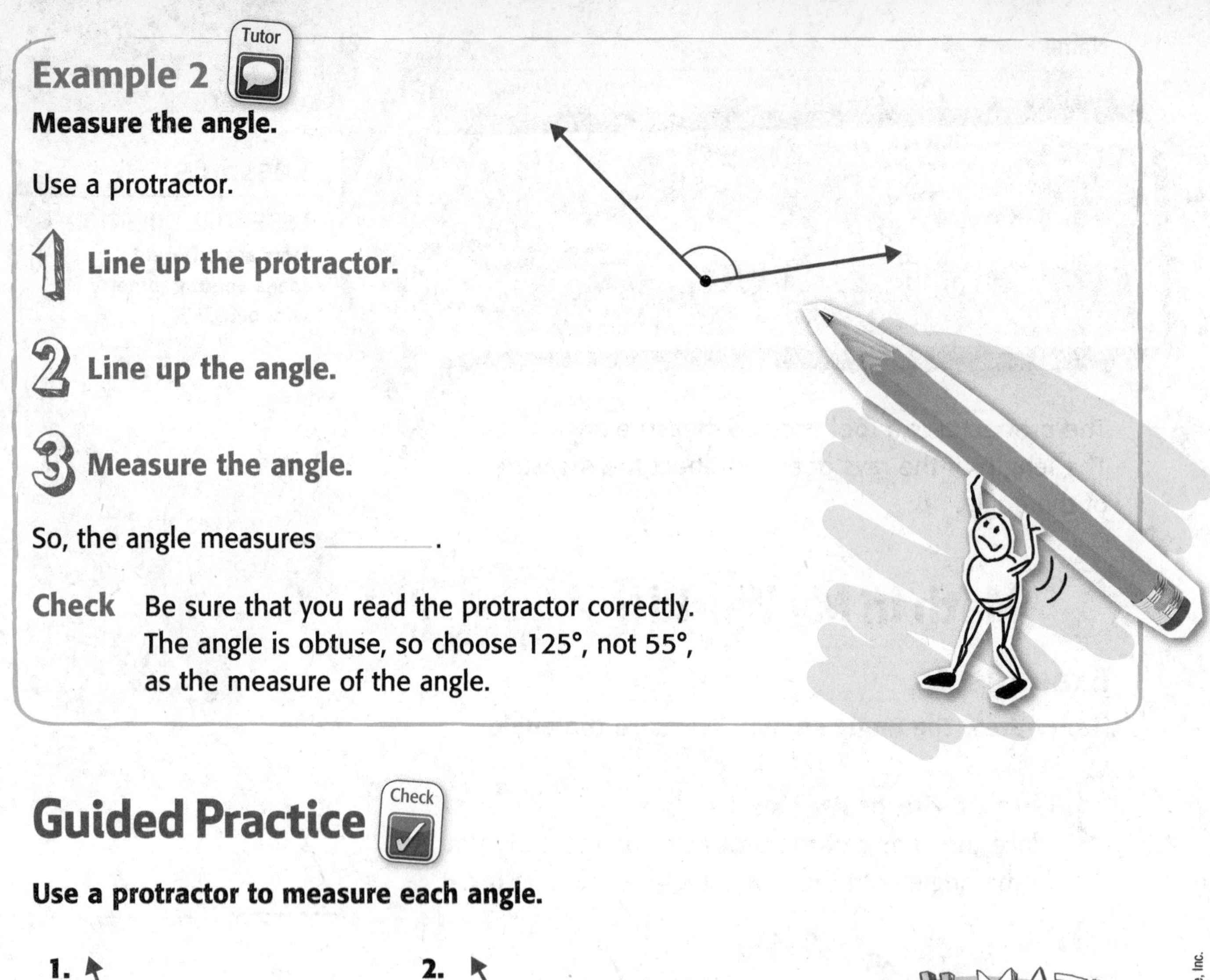

Example 2

Measure the angle.

Use a protractor.

1 **Line up the protractor.**

2 **Line up the angle.**

3 **Measure the angle.**

So, the angle measures ______.

Check Be sure that you read the protractor correctly. The angle is obtuse, so choose 125°, not 55°, as the measure of the angle.

Guided Practice

Use a protractor to measure each angle.

1.

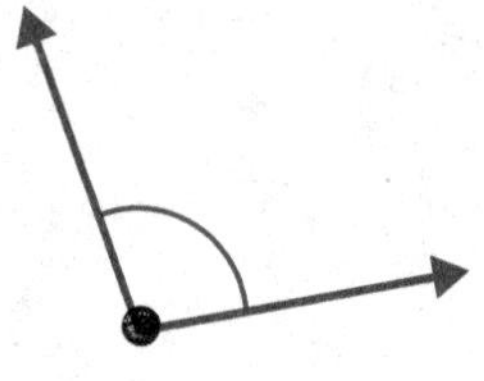

______

2.

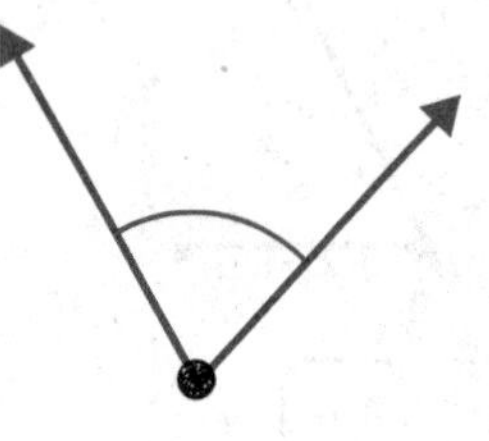

______

Talk MATH

Explain how to use a protractor.

3. Measure the indicated angle on the triangle.

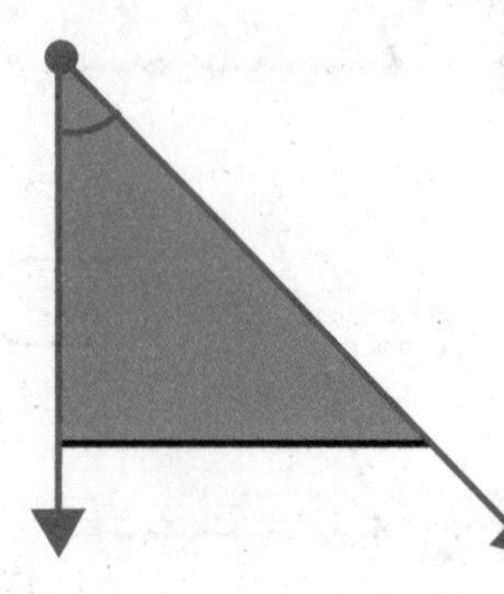

______

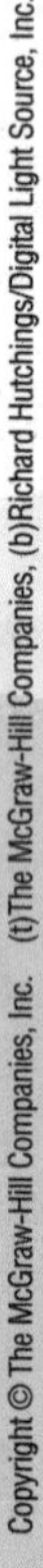

Name

Independent Practice

Use a protractor to measure each indicated angle.

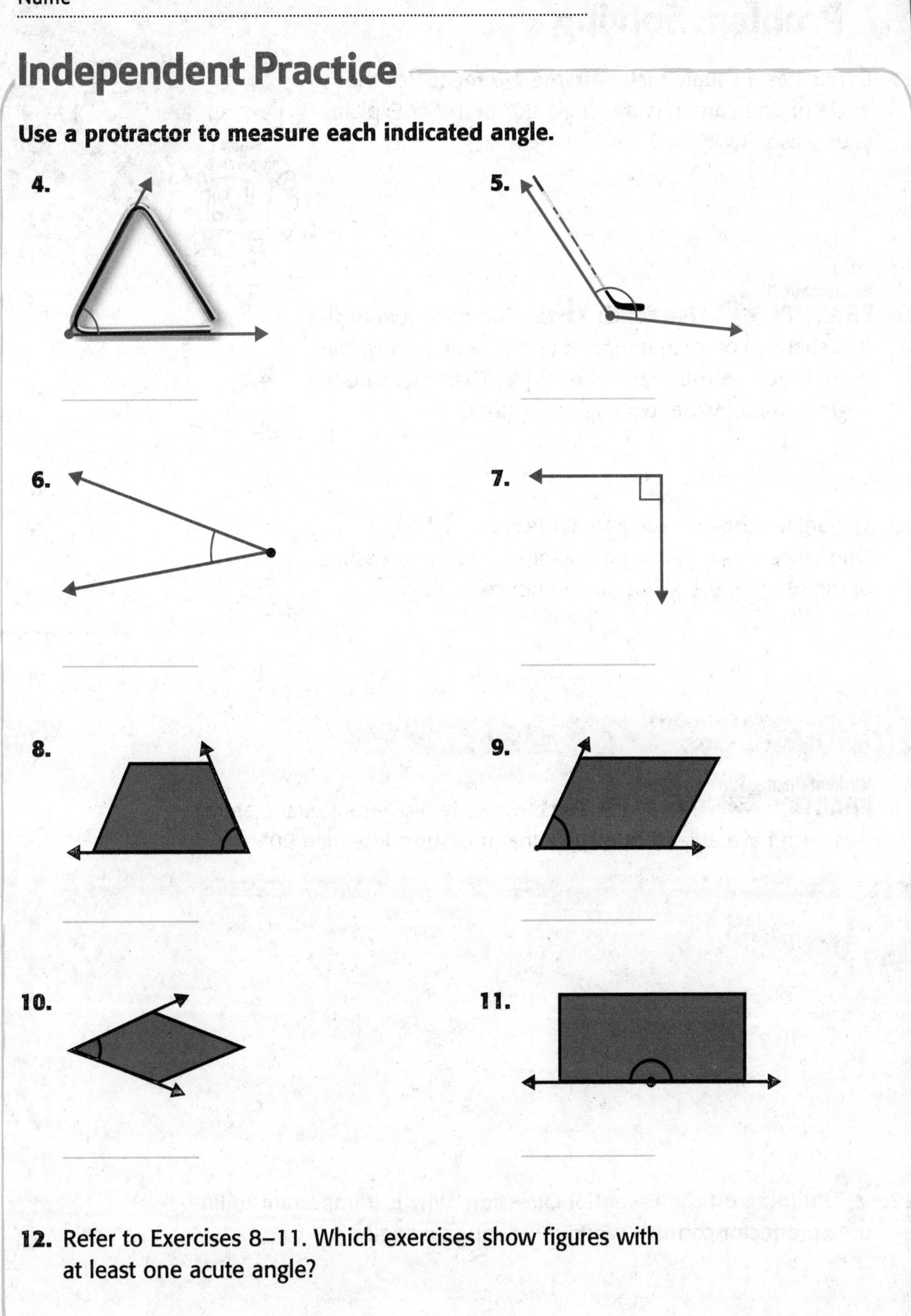

12. Refer to Exercises 8–11. Which exercises show figures with at least one acute angle?

Problem Solving

13. Darius has a square picture frame. He measures the angle of one corner. Is the angle 90° or 145°? Explain your reasoning.

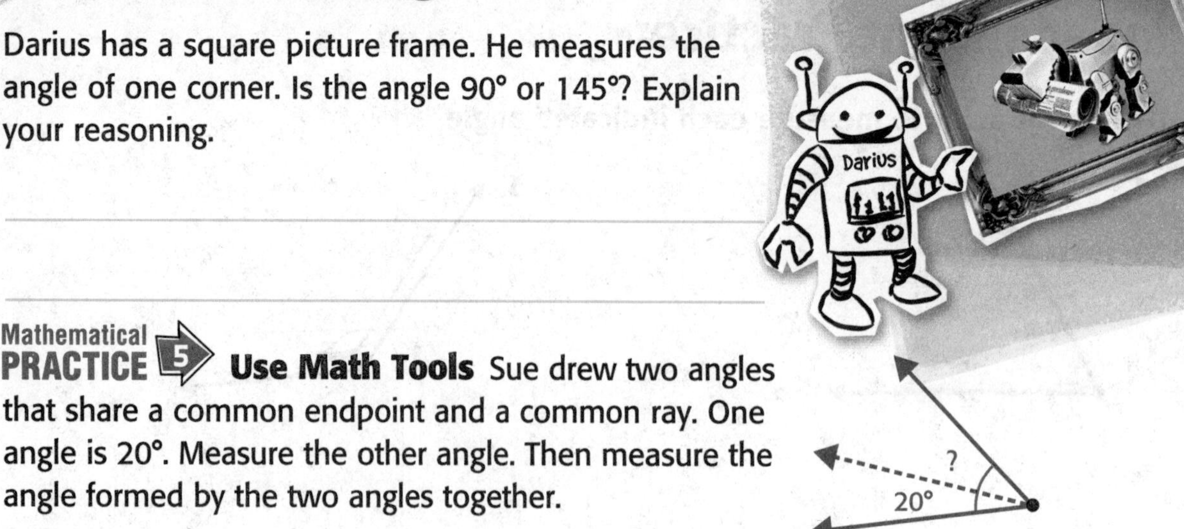

14. Mathematical PRACTICE 5 **Use Math Tools** Sue drew two angles that share a common endpoint and a common ray. One angle is 20°. Measure the other angle. Then measure the angle formed by the two angles together.

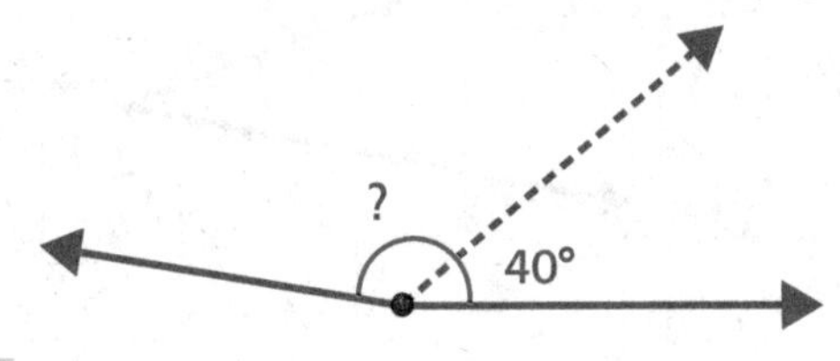

15. The angles shown have a total measure of 170°. One angle is 40°. Use a protractor to find the measure of the other angle. What do you notice?

? 40°

HOT Problems

16. Mathematical PRACTICE 5 **Use Math Tools** Draw two different quadrilaterals that each have at least one angle that measures less than 90°.

My Drawing!

17. **Building on the Essential Question** Why is it important to line up a protractor correctly when measuring an angle?

Name ______________________

MY Homework

Lesson 5
Measure Angles

Homework Helper

eHelp Need help? connectED.mcgraw-hill.com

Camille is making the pinwheel shown below. Measure the angle shown in red.

Use a protractor.

1. **Line up the protractor.**

2. **Line up the angle.**

3. **Measure the angle.**

So, the angle measures 100°.

Practice

Use a protractor to measure each angle.

1.

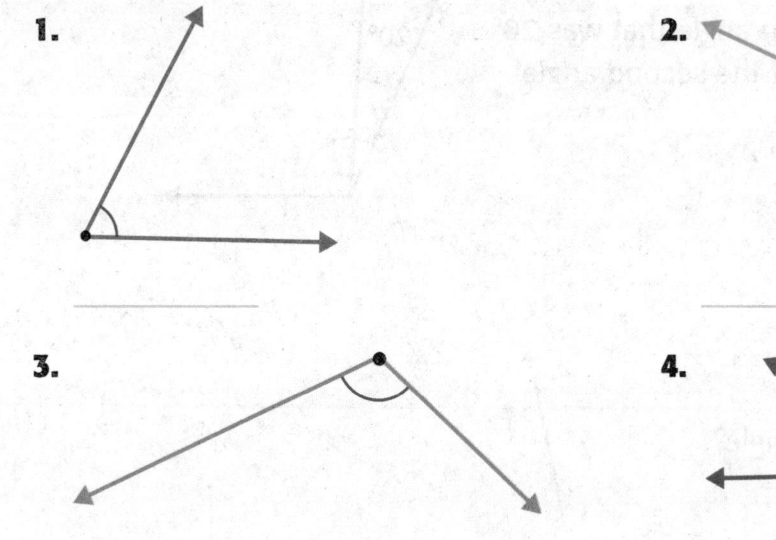

____

2.

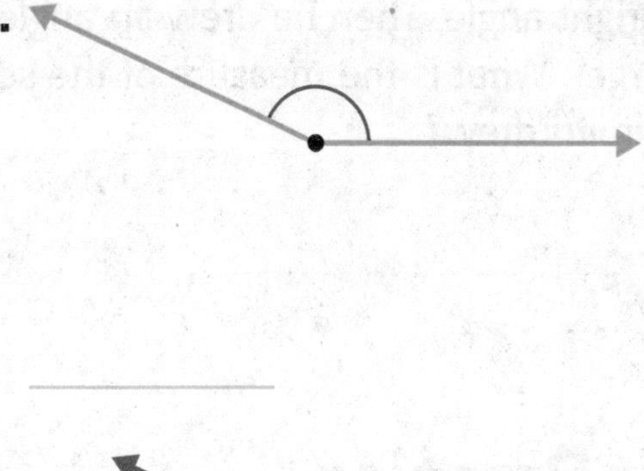

____

3.

____

4.

____

Use a protractor to measure each angle outlined in red.

5.

____

6.

____

7.

____

8.

____

Problem Solving

9. Heong cut a slice of birthday cake. The slice formed the angle shown. What is the measure of the angle shown?

____

10. Mathematical PRACTICE 5 **Use Math Tools** Dimitri drew a right angle. Then he drew an angle that was 20° larger. What is the measure of the second angle Dimitri drew?

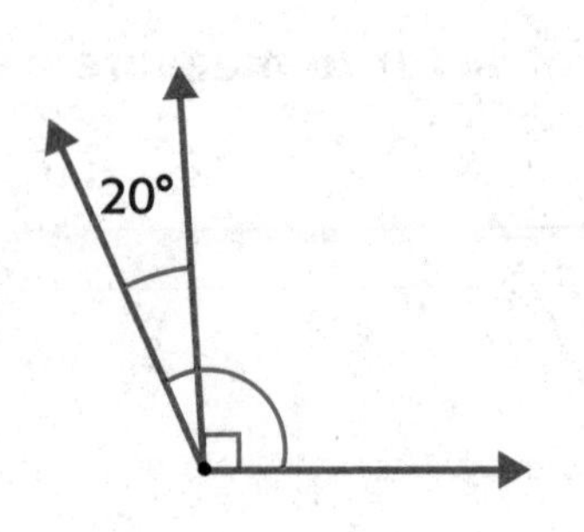

____

Test Practice

11. What is the measure of the angle?

Ⓐ 85° Ⓒ 75°

Ⓑ 80° Ⓓ 70°

Name

Draw Angles

Lesson 6

ESSENTIAL QUESTION
How are different ideas about geometry connected?

You have used a protractor to measure angles. A protractor can also be used to draw angles of a given measure.

Math in My World

Tools Watch Tutor

Example 1

A traffic sign shows an 80° angle. Draw an 80° angle.

1 Draw one ray of the angle.
Mark the endpoint and draw a ray.

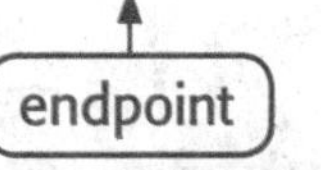

2 Measure the angle.
Place the protractor along the ray as you would to measure an angle. On the protractor, find 80°, and make a pencil mark.

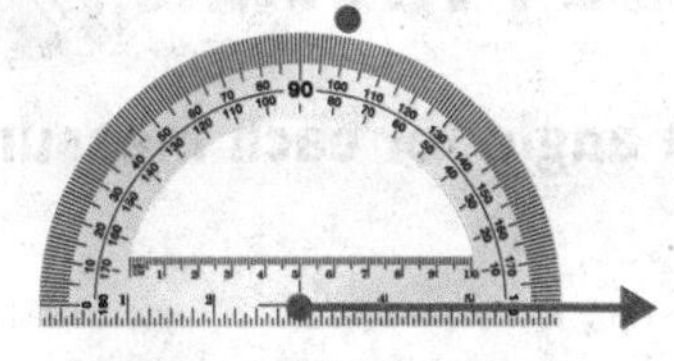

3 Draw the next ray of the angle.
Use a ruler to draw the ray that connects the endpoint and the pencil mark.

Draw your 80° angle below.

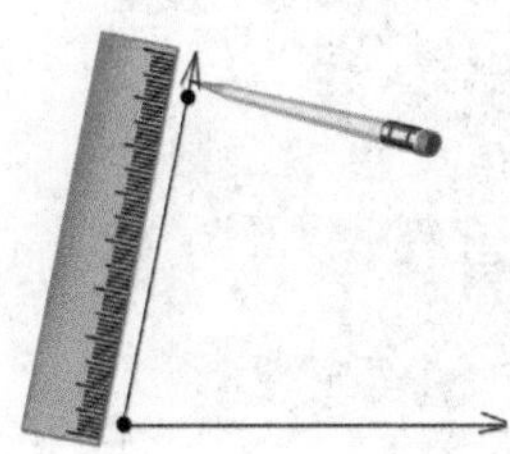

My Drawing!

Example 2

Tutor

Draw an angle that is greater than 10° and less than 30°. Measure and classify the angle.

Find 10° and 30° on the protractor. Draw an angle between the two measurements. Then measure the angle drawn.

My Drawing!

So, the angle drawn is ________. It is a(n) ________ angle.

Guided Practice

Check

Draw an angle for each measure.

1. 20°

2. 45°

3. 100°

Talk MATH

Describe how you would draw a 90° angle without using a protractor.

Name

Independent Practice

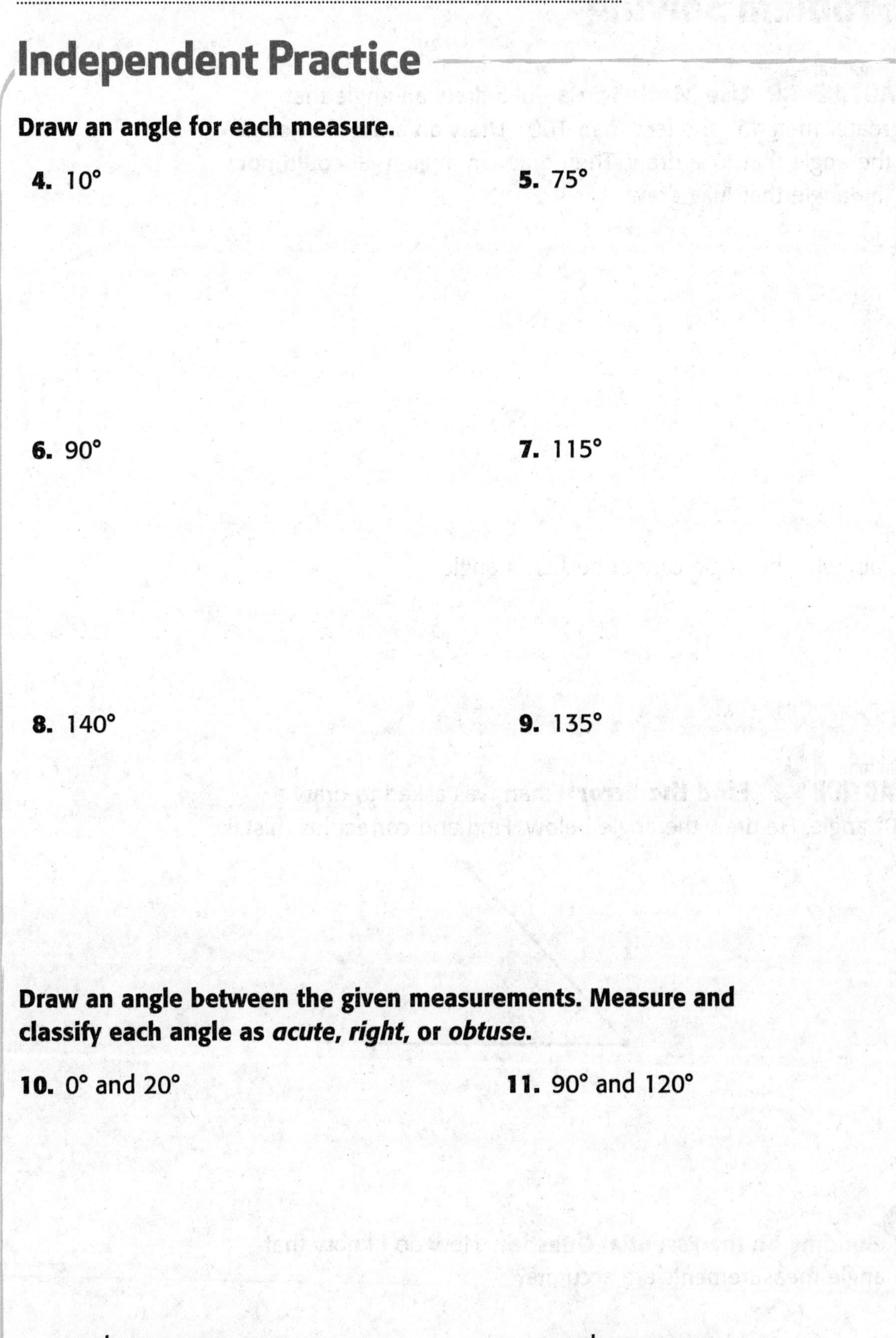

Draw an angle for each measure.

4. 10°

5. 75°

6. 90°

7. 115°

8. 140°

9. 135°

Draw an angle between the given measurements. Measure and classify each angle as *acute, right,* or *obtuse*.

10. 0° and 20°

angle measure: ______

type of angle: ______

11. 90° and 120°

angle measure: ______

type of angle: ______

Problem Solving

12. Mathematical PRACTICE 5 **Use Math Tools** Julia drew an angle that is greater than 45° and less than 100°. Draw an angle that could be the angle that Julia drew. Then draw an angle that could not be the angle that Julia drew.

My Drawing!

Explain why the angle cannot be Julia's angle.

HOT Problems

13. Mathematical PRACTICE 3 **Find the Error** Ethan was asked to draw a 130° angle. He drew the angle below. Find and correct his mistake.

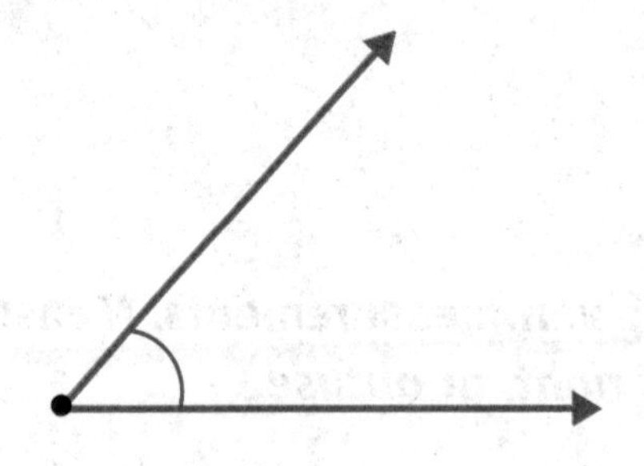

14. **Building on the Essential Question** How do I know that my angle measurements are accurate?

Name

MY Homework

Lesson 6
Draw Angles

Homework Helper

eHelp Need help? connectED.mcgraw-hill.com

Draw a 40° angle. Classify it as *acute, right,* or *obtuse*.

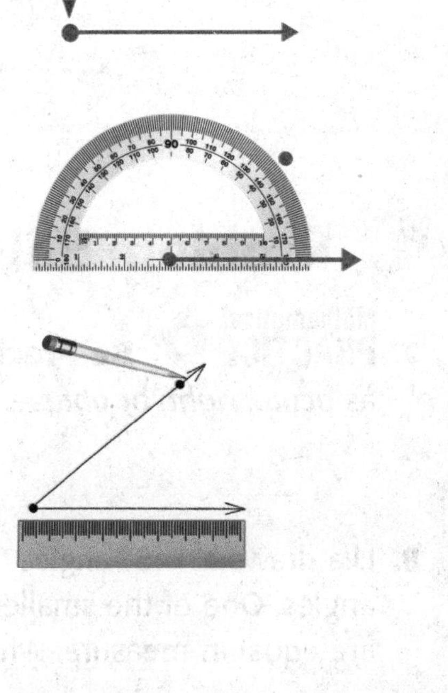

1 **Draw one ray of the angle.**
Mark the endpoint and draw a ray.

2 **Measure the angle.**
Place the protractor along the ray as you would to measure an angle. On the protractor, find 40°, and make a pencil mark.

3 **Draw the next ray of the angle.**
Use a ruler to draw the ray that connects the vertex and the pencil mark.

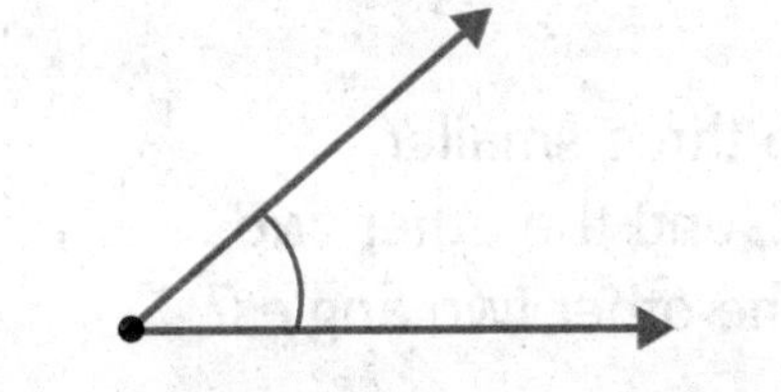

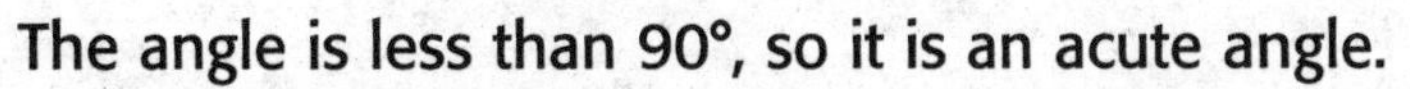

The angle is less than 90°, so it is an acute angle.

Practice

Draw an angle for each measure.

1. 65°

2. 140°

Draw an angle for each measure.

3. 80°

4. 35°

5. greater than 5° and less than 25°

6. greater than 90° and less than 120°

Problem Solving

7. Mathematical PRACTICE 6 **Be Precise** Classify the angle in Exercise 3 as *acute*, *right*, or *obtuse*.

____

8. Lila draws a 145° angle. Then she divides it into three smaller angles. One of the smaller angles measures 65°, and the other two are equal in measure. What is the measure of the other two angles?

____

Test Practice

9. Which is the correct drawing of a 160° angle?

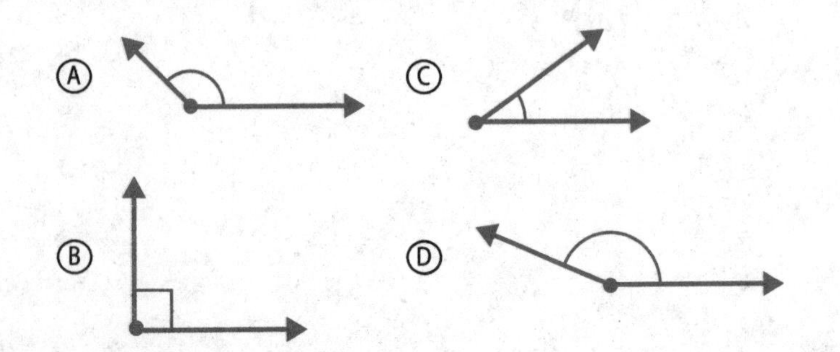

Name ____________

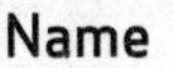

Solve Problems with Angles

Lesson 7

ESSENTIAL QUESTION
How are different ideas about geometry connected?

An angle can be decomposed, or broken, into non-overlapping parts. The angle measure of the whole is the sum of the angle measures of the parts.

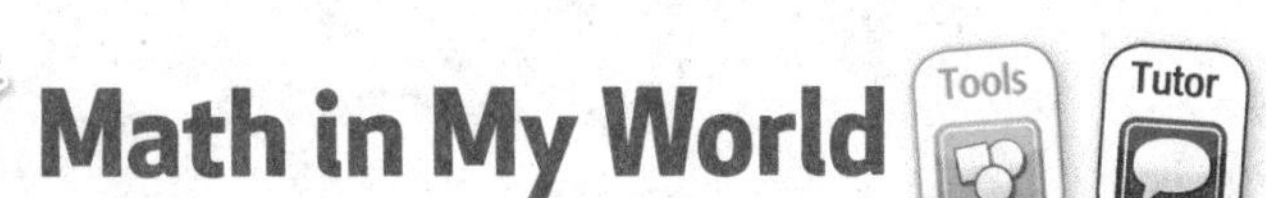

Example 1

Rachel and Dean made a sign out of fabric like the one shown to hang in the school gymnasium. The blue piece has a 35° angle. The red piece is attached to the longest side of the blue piece. Together, the pieces form a right angle. What is the angle shown on the red piece?

One Way Make a model.

Draw a 90° angle. Mark off a 35° angle. Measure the other angle.

?°
35°

The other angle has a measure of ____________.

Another Way Use an equation.

The 90° angle measure is the sum of two parts. One angle is 35°. Find the unknown angle measure.
Let r represent the unknown angle measure.

$35 + r = 90$

Since $35 + r = 90$, you know that $90 - 35 = r$. ← Addition and subtraction are inverse, or opposite, operations.

$r = 90 - 35$

$r =$ ____________

So, the angle shown on the red piece measures ____________.

Example 2

Find the combined measure of the angle shown.

One of the angles is 20°. The symbol on the other angle shows that it is a right angle. Therefore, it is 90°.

To find the combined measure of the angle, add the angle measures of the parts.

Let a represent the combined angle measure.

$a = 20° + 90°$

$a =$ ______

So, the combined measure of the angle is ______.

20°

Guided Practice

Check

Algebra Find each unknown.

1. The combined angle measure is 90°.

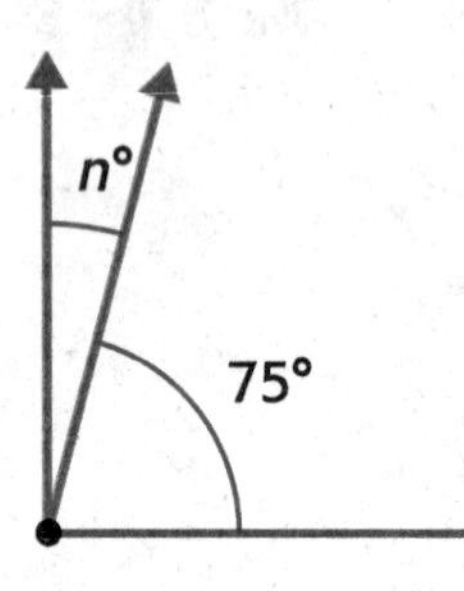

$n =$ ______

2. The combined angle measure is 130°.

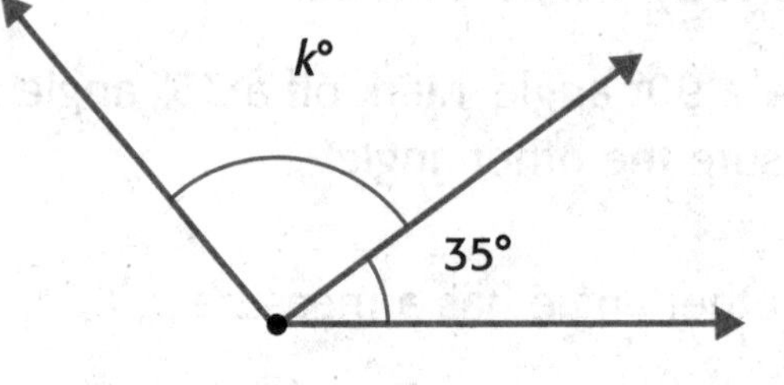

$k =$ ______

3. Find the combined angle measure.

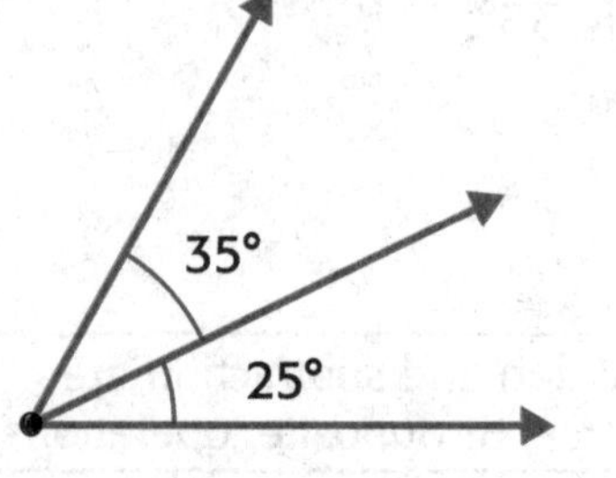

combined measure = ______

Talk MATH

How can the measures of parts of an angle be used to find the combined measure?

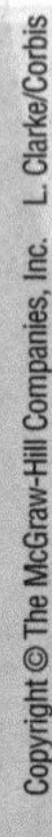

Name ....

CCSS

Independent Practice

Algebra Find each unknown.

4. The combined angle measure is 50°.

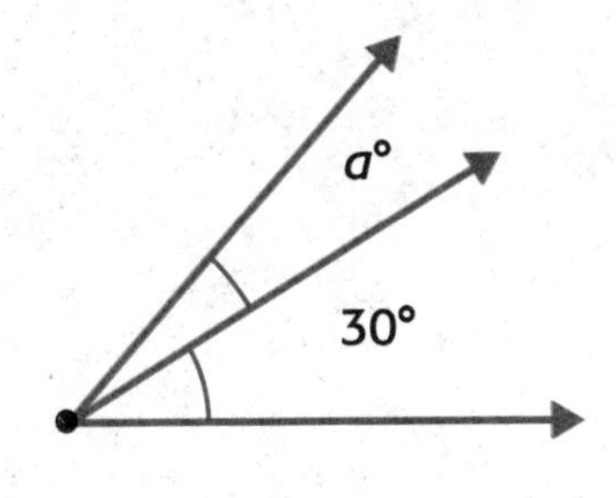

$a =$ ______

5. The combined angle measure is 90°.

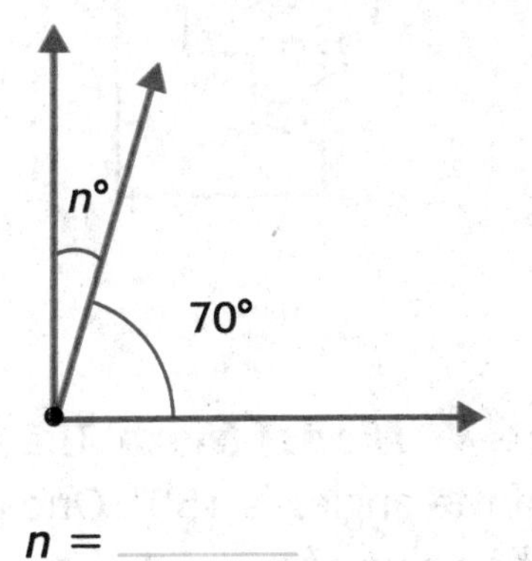

$n =$ ______

6. The combined angle measure is 125°.

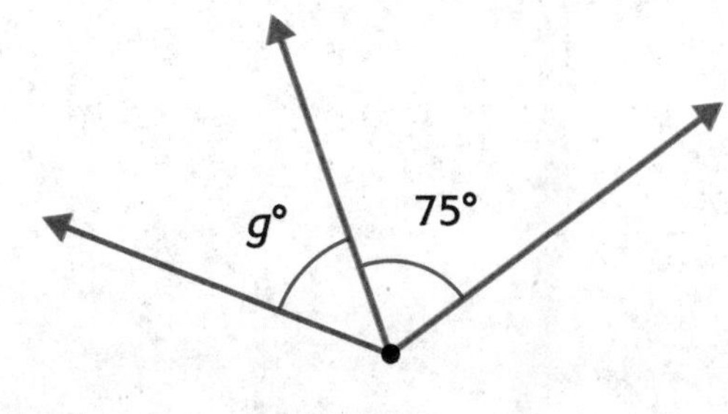

$g =$ ______

7. The combined angle measure is 150°.

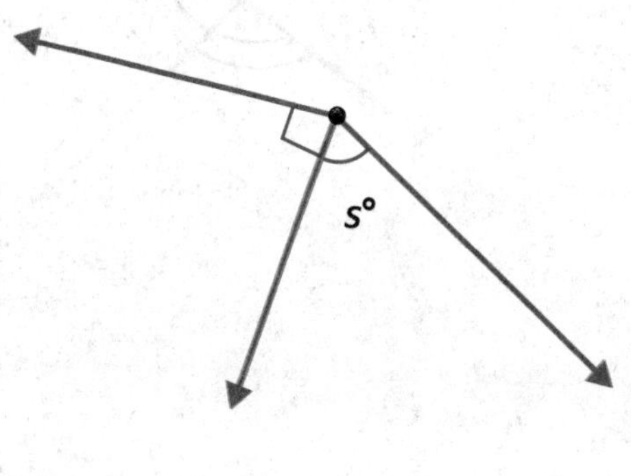

$s =$ ______

8. Draw a triangle with one right angle.

Find the combined measure of the three angles.

______

9. Draw a triangle with one obtuse angle.

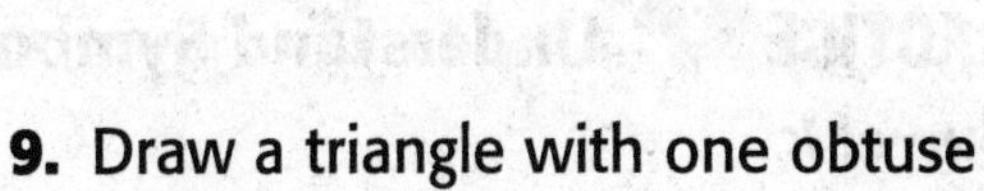

Find the combined measure of the three angles.

______

Problem Solving

10. The steps on a staircase should be 90°. One of the steps is crooked. The angle formed is 15° too large. What is the angle of that step?

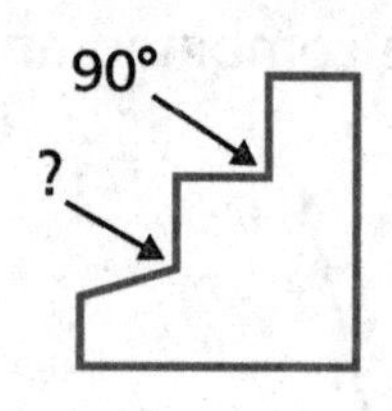

My Work!

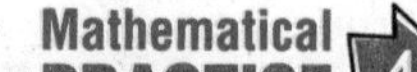

11. Mathematical PRACTICE 4 **Model Math** The combined measure of the angles is 150°. One angle measures 50°. Find the value of x.

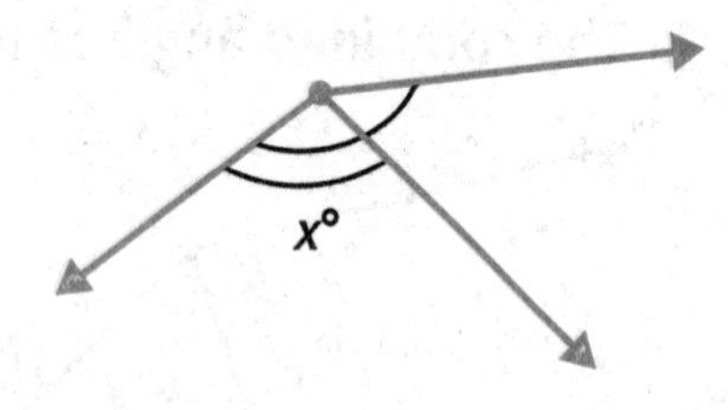

HOT Problems

12. Mathematical PRACTICE 2 **Understand Symbols** Find the value of k.

$k =$ ______

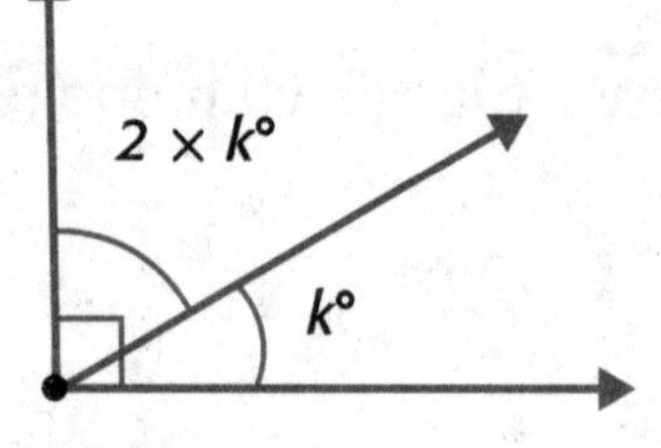

13. **Building on the Essential Question** How is addition related to angle measurement?

Name ....

Lesson 7

Solve Problems with Angles

Homework Helper

eHelp Need help? connectED.mcgraw-hill.com

Find the measure of the unknown angle.
The combined angle measure is 140°.

25°
$d°$

Use an equation.

You know the total measure is 140°. One angle is 25°.

Let d represent the unknown angle measure.

$$25 + d = 140$$

Since $25 + d = 140$, you know that $140 - 25 = d$. ← Addition and subtraction are inverse, or opposite operations.

$$d = 140 - 25$$

$$d = 115$$

So, the measure of the unknown angle is 115°.

Practice

Algebra Find each unknown.

1. The combined angle measure is 50°.

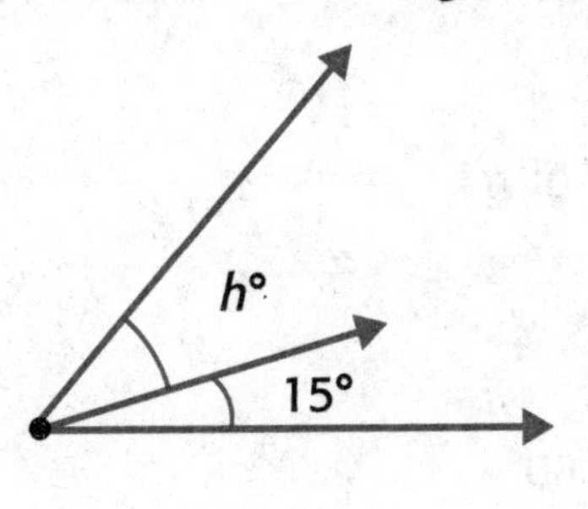

$h =$ ______

2. The combined angle measure is 135°.

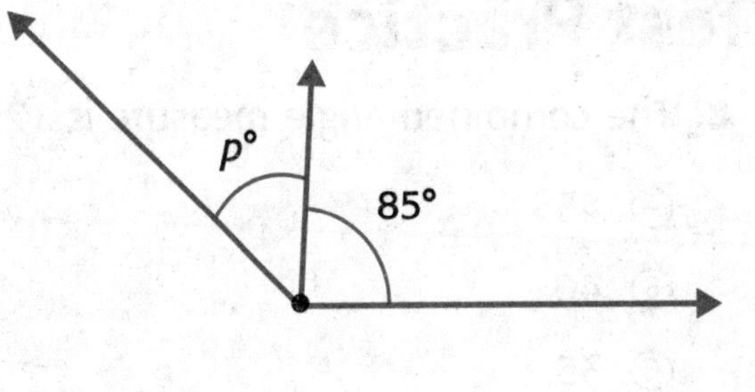

$p =$ ______

Algebra **Find each unknown.**

3. The combined angle measure is 70°.

45°

f°

$f =$ ______

4. The combined angle measure is 115°.

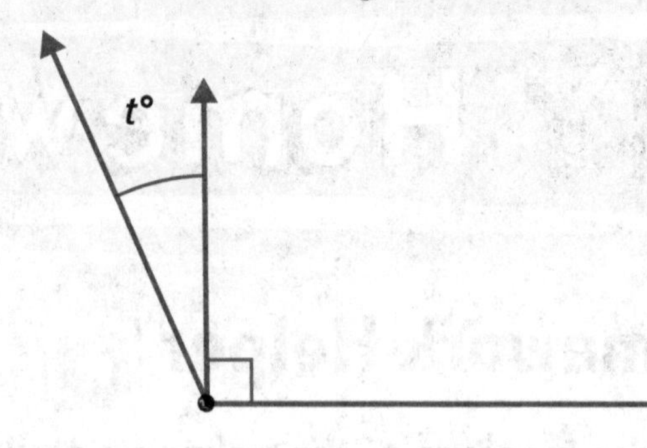

$t =$ ______

5. The combined angle measure is 180°.

$x =$ ______

6. Find the value of r.

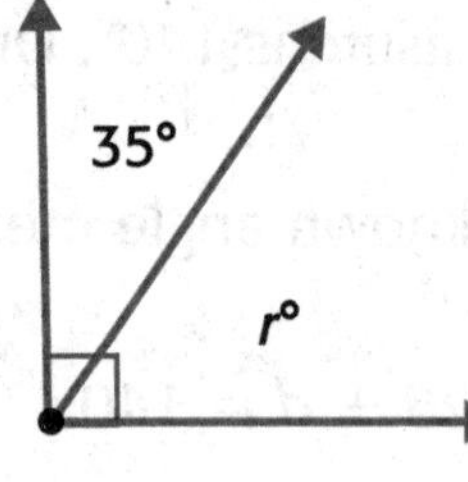

$r =$ ______

Problem Solving

7. Mathematical PRACTICE 1 **Make a Plan** Suppose you draw a line from the center of a clock face to the number 12. When the minute hand gets to 3 on the clock face, the line and minute hand form a 90° angle. What angle does the line and the minute hand make when the minute hand is on 2?

______

Test Practice

8. The combined angle measure is 120°. What is the value of n?

Ⓐ 45

Ⓑ 40

Ⓒ 35

Ⓓ 30

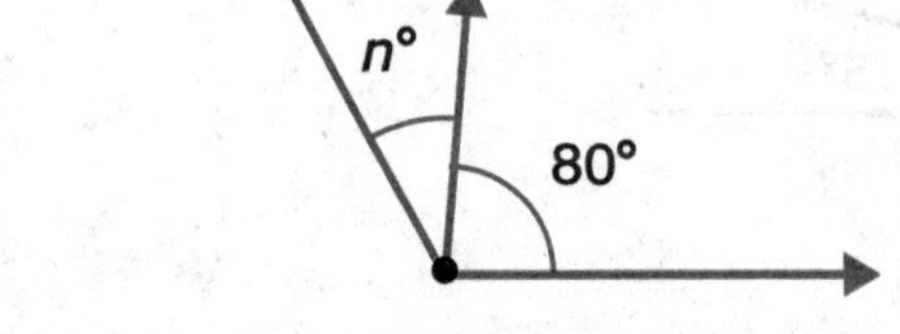

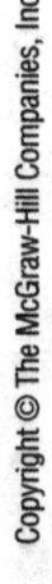

Check My Progress

Vocabulary Check

Vocab

Unscramble each word to complete each sentence. Use a word from the word bank.

acute angle **angle** **degree (°)** **obtuse angle** **right angle**

1. thrig lngae

   A ______________ measures 90°.

2. leagn

   An ______________ is a geometric shape that is formed when two rays have the same endpoint.

3. tacue aleng

   An ______________ measures greater than 0° and less than 90°.

4. geeedr

   The unit used to measure an angle is called a ______________.

5. oseutb egnla

   An ______________ measures greater than 90° but less than 180°.

Concept Check

Check

Classify each angle as *right, acute,* or *obtuse*.

6. ______________

7. ______________

8. ______________

Problem Solving

9. An angle on a grocery cart is made up of 100 one-degree angles. Find the measure of the angle.

____

10. The angle shown is made up of 112 one-degree angles. Find the measure of the angle.

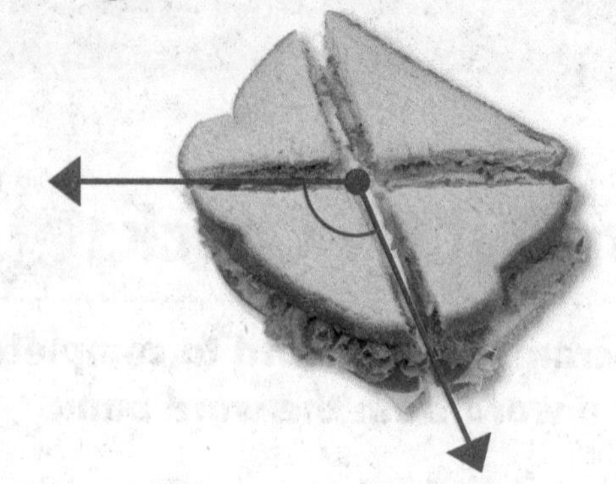

____

11. Draw an angle that measures between 40° and 50°.

My Drawing!

Test Practice

12. What is the measure of the angle in degrees and as a fraction of a full turn?

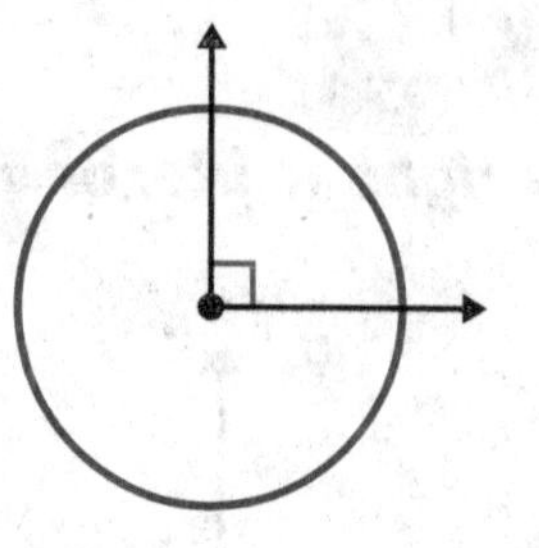

Ⓐ 90°; $\frac{3}{4}$ turn

Ⓑ 90°; $\frac{1}{360}$ turn

Ⓒ 90°; $\frac{1}{4}$ turn

Ⓓ 180°; $\frac{1}{2}$ turn

Name ....

Triangles

Lesson 8

ESSENTIAL QUESTION
How are different ideas about geometry connected?

There are many different kinds of triangles. You can classify triangles by the measure of their angles.

Lunch time!

Math in My World

Example 1

This sandwich is cut in half. Classify the triangle represented by the half sandwich as *right*, *acute*, or *obtuse*. Determine if any of the sides are perpendicular.

A **right triangle** has one right angle.

How many right angles are there? ______

The two sides that form the right angle are perpendicular.

An **acute triangle** has three acute angles.

How many acute angles are there? ______

An **obtuse triangle** has one obtuse angle.

How many obtuse angles are there? ______

So, the triangle is a(n) ______ triangle.

Key Concept Classify Triangles by Angles

An obtuse triangle has one obtuse angle.

An acute triangle has three acute angles.

A right triangle has one right angle. The two sides that form the right angle are perpendicular.

You can also identify vertices and line segments in triangles.

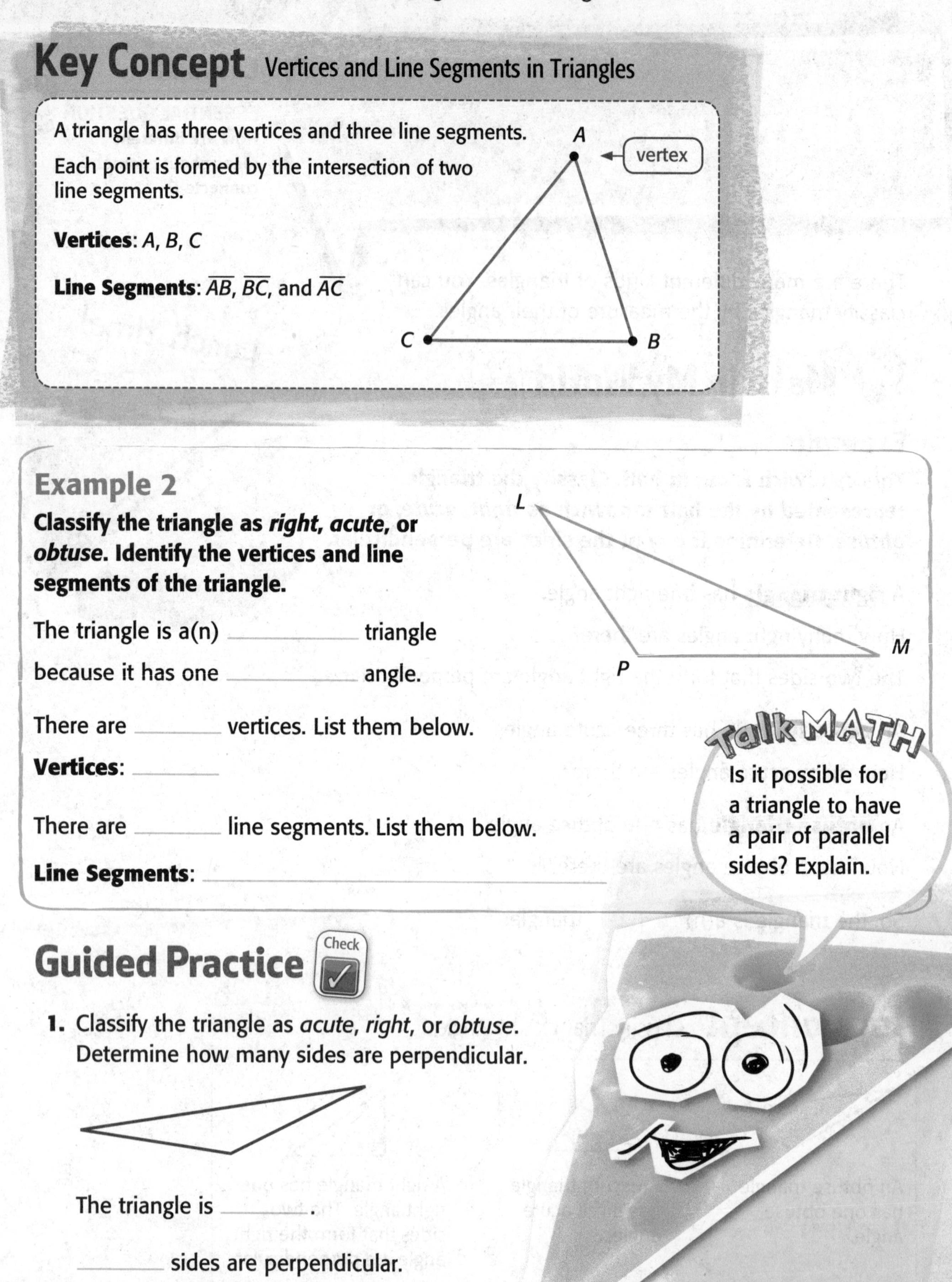

Key Concept Vertices and Line Segments in Triangles

A triangle has three vertices and three line segments.

Each point is formed by the intersection of two line segments.

Vertices: *A*, *B*, *C*

Line Segments: $\overline{AB}$, $\overline{BC}$, and $\overline{AC}$

Example 2

Classify the triangle as *right, acute,* or *obtuse*. Identify the vertices and line segments of the triangle.

The triangle is a(n) ________ triangle

because it has one ________ angle.

There are ______ vertices. List them below.

Vertices: ______

There are ______ line segments. List them below.

Line Segments: ________________________

Talk MATH

Is it possible for a triangle to have a pair of parallel sides? Explain.

Guided Practice Check

1. Classify the triangle as *acute*, *right*, or *obtuse*. Determine how many sides are perpendicular.

The triangle is ______.

______ sides are perpendicular.

Name

Independent Practice

Classify each triangle as *acute, right,* or *obtuse*.
Circle the triangles that have any perpendicular sides.

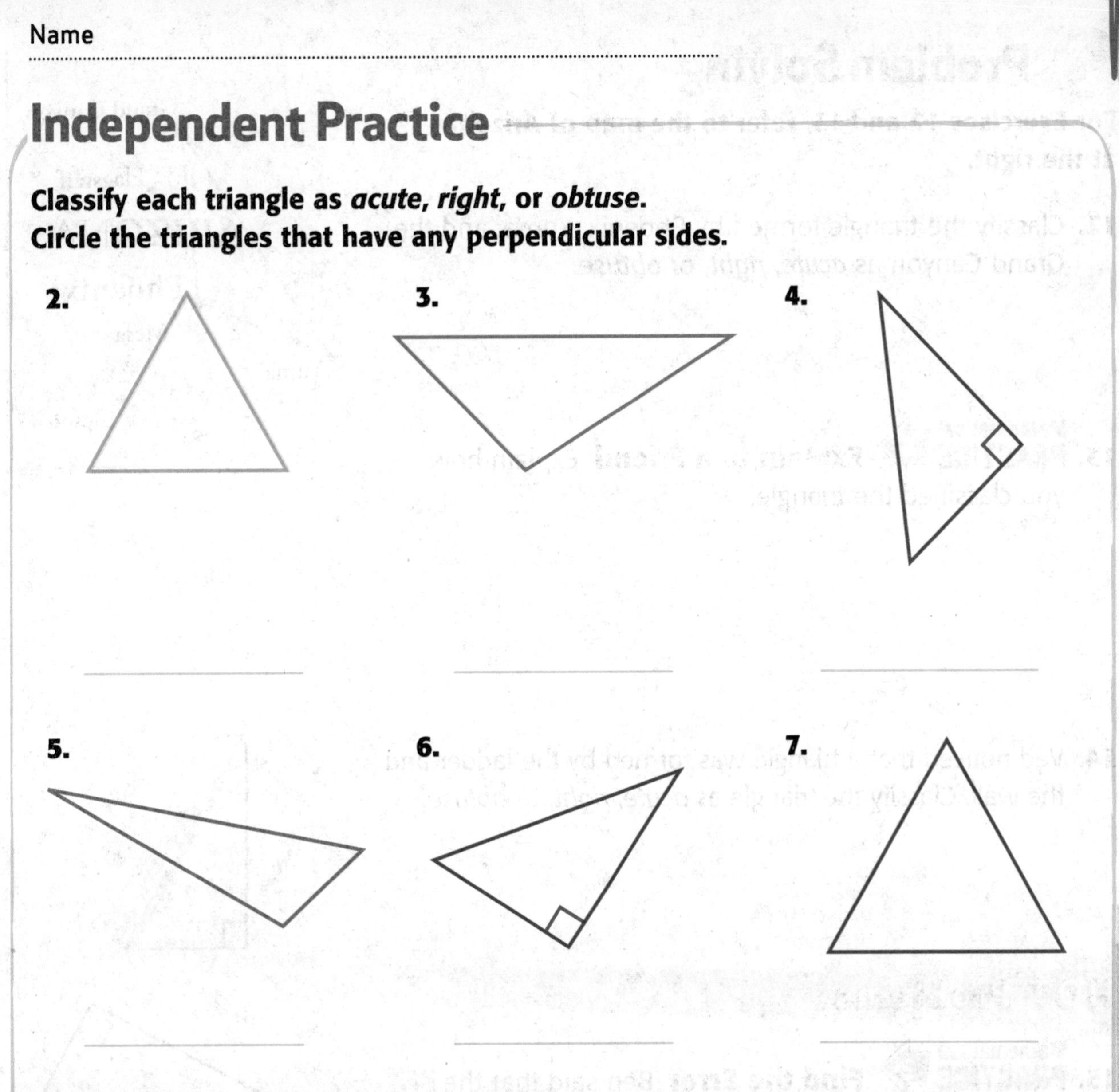

8. Draw three line segments that form a right triangle.

9. Draw three line segments that form an obtuse triangle.

10. Which exercises on this page show right triangles?

11. Which exercises on this page show figures with perpendicular line segments?

Problem Solving

For Exercises 12 and 13, refer to the map of Arizona at the right.

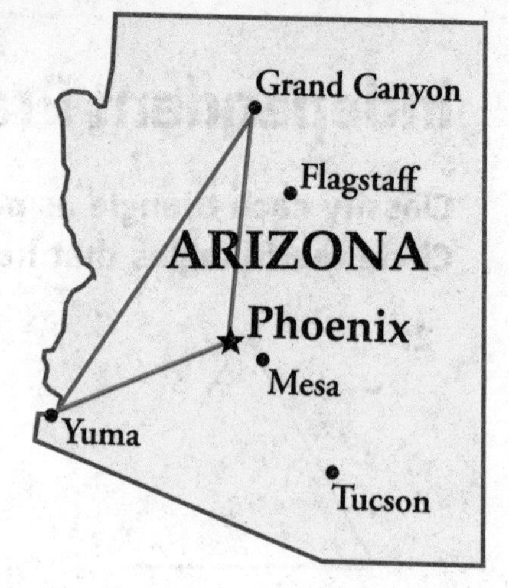

12. Classify the triangle formed by Phoenix, Yuma, and the Grand Canyon as *acute*, *right*, or *obtuse*.

13. Mathematical PRACTICE 6 **Explain to a Friend** Explain how you classified the triangle.

14. Ved noticed that a triangle was formed by the ladder and the wall. Classify the triangle as *acute*, *right*, or *obtuse*.

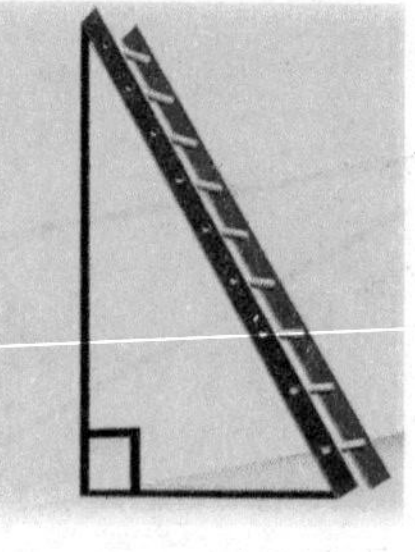

HOT Problems

15. Mathematical PRACTICE 3 **Find the Error** Ben said that the triangle shown is an acute triangle because the angle shown is acute. Find and correct his mistake.

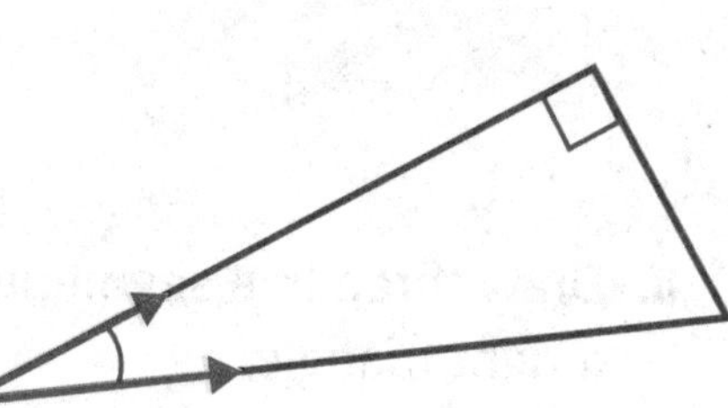

16. **Building on the Essential Question** Is it possible for a triangle to have two obtuse angles? Explain.

MY Homework

Lesson 8
Triangles

Homework Helper

eHelp Need help? connectED.mcgraw-hill.com

Identify the vertices and line segments of each triangle. Identify it as *right, acute,* or *obtuse*.

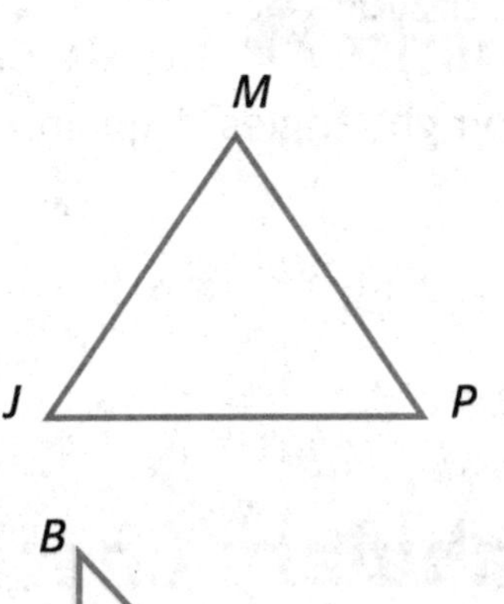

There are three vertices. They are *J, M,* and *P*.

There are three line segments. They are $\overline{JM}$, $\overline{MP}$, and $\overline{JP}$.

The triangle has 3 acute angles, so it is an acute triangle.

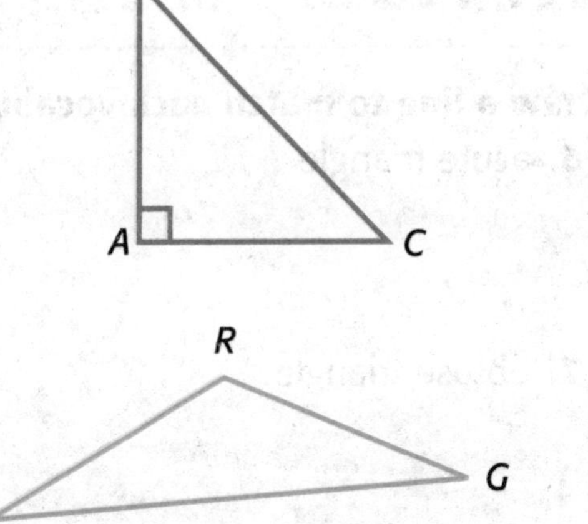

The vertices are *A, B,* and *C*.

The three line segments are $\overline{AB}$, $\overline{BC}$, and $\overline{AC}$.

The triangle has a right angle, so it is a right triangle.

The three vertices are *T, R,* and *G*.

The three line segments are $\overline{TR}$, $\overline{RG}$, and $\overline{TG}$.

The triangle has an obtuse angle, so it is an obtuse triangle.

Practice

Identify the perpendicular sides of each right triangle.

1. **2.**

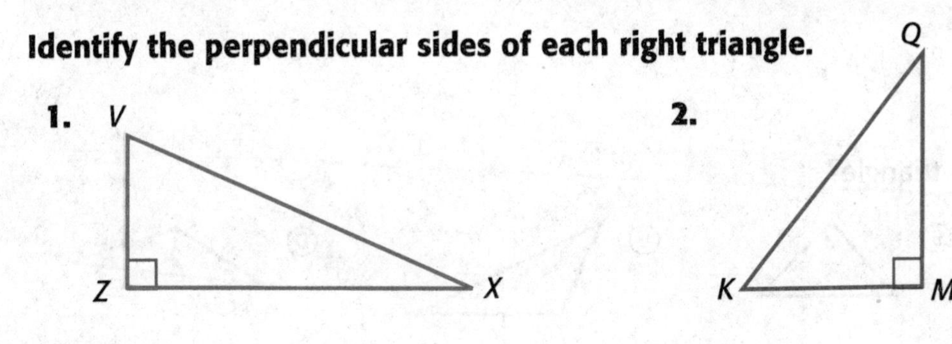

Classify each triangle as *acute, right,* or *obtuse*. Determine how many sides appear to be perpendicular.

3.

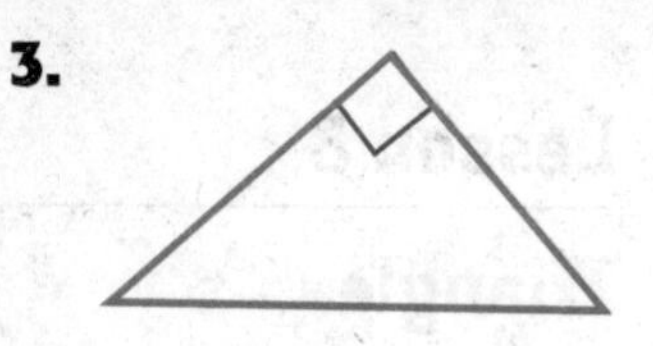

4.

______________________ ______________________

Problem Solving

5. Mathematical PRACTICE 3 **Justify Conclusions** Can a triangle have two right angles? Explain.

____

____

Vocabulary Check

Draw a line to match each vocabulary term to its example.

6. acute triangle •

7. obtuse triangle •

8. right triangle •

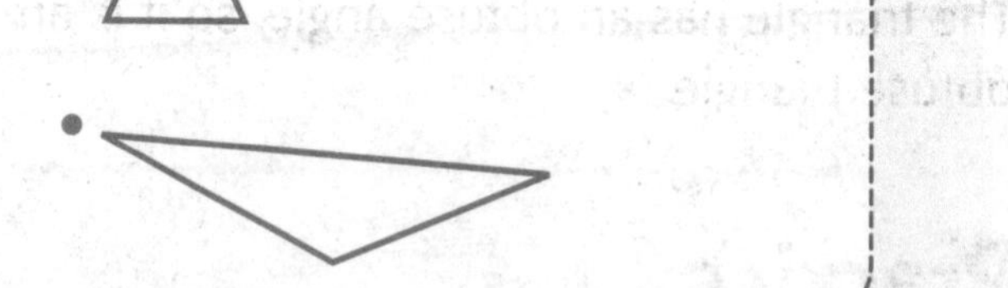

Test Practice

9. Which is an acute triangle?

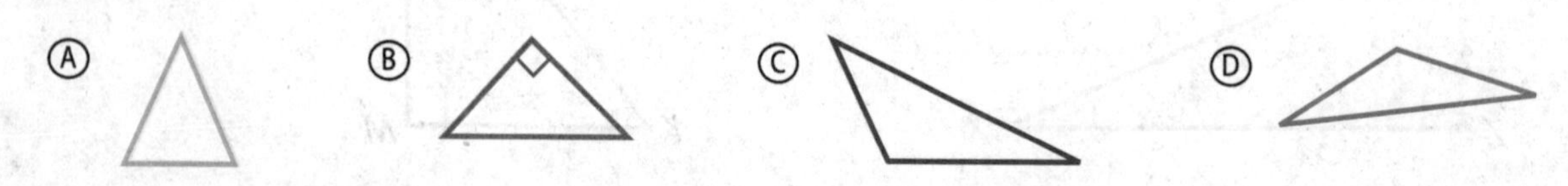

Ⓐ Ⓑ Ⓒ Ⓓ

Name 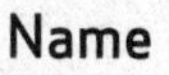

Quadrilaterals

Lesson 9

ESSENTIAL QUESTION
How are different ideas about geometry connected?

All quadrilaterals have 4 sides and 4 angles.
There are many different kinds of quadrilaterals.

Math in My World

SPEED
LIMIT
25

Example 1

The speed limit sign represents a quadrilateral. Classify the angles formed by the quadrilateral. Determine if any of the sides are parallel or perpendicular.

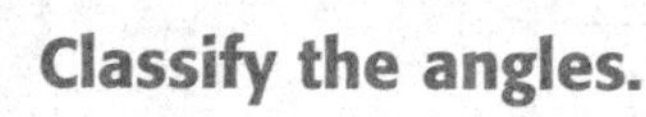

1 Classify the angles.

There are ________ right angles,

________ acute angles, and ________ obtuse angles.

2 Determine if there are any parallel or perpendicular sides.

The top and ________ sides are parallel.

The left and ________ sides are parallel.

Opposite sides are parallel.

Since there are 4 right angles, the sides that form each right angle are perpendicular.

So, there are ________ pairs of perpendicular sides.

Notice that opposite sides are also equal in length.

A quadrilateral with 4 right angles, opposite sides equal in length, and opposite sides parallel is a *rectangle*. A rectangle is a special kind of quadrilateral.

Key Concept Quadrilaterals

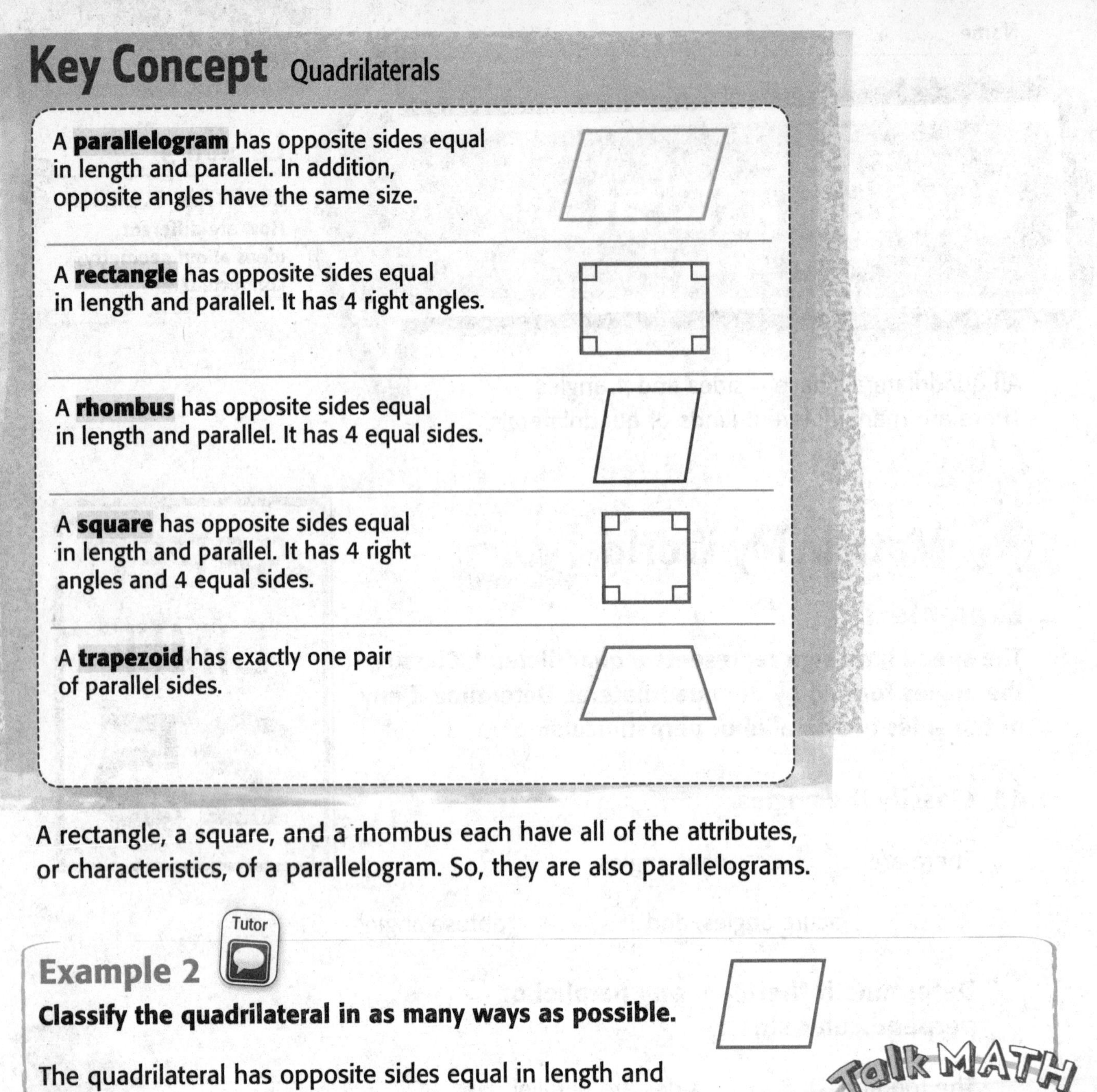

A **parallelogram** has opposite sides equal in length and parallel. In addition, opposite angles have the same size.

A **rectangle** has opposite sides equal in length and parallel. It has 4 right angles.

A **rhombus** has opposite sides equal in length and parallel. It has 4 equal sides.

A **square** has opposite sides equal in length and parallel. It has 4 right angles and 4 equal sides.

A **trapezoid** has exactly one pair of parallel sides.

A rectangle, a square, and a rhombus each have all of the attributes, or characteristics, of a parallelogram. So, they are also parallelograms.

Example 2

Tutor

Classify the quadrilateral in as many ways as possible.

The quadrilateral has opposite sides equal in length and opposite sides parallel. It also has ________ equal sides.

So, it is a ____________ and a ____________.

Talk MATH

Explain why a square is also a parallelogram.

Guided Practice

1. Classify the quadrilateral in as many ways as possible.

It is a ____________,

a ____________, a ____________,

and a ____________.

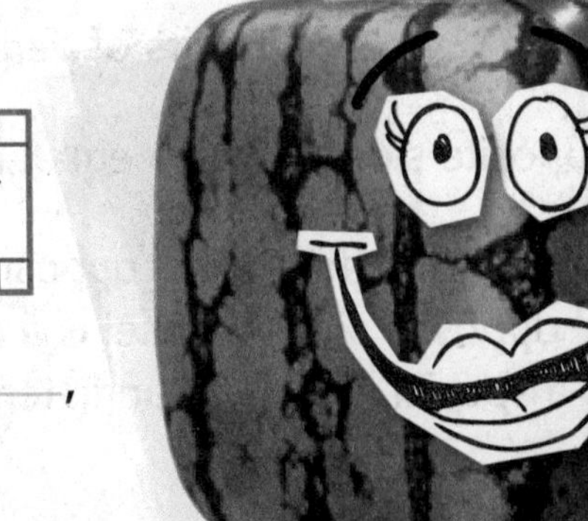

Name

Independent Practice

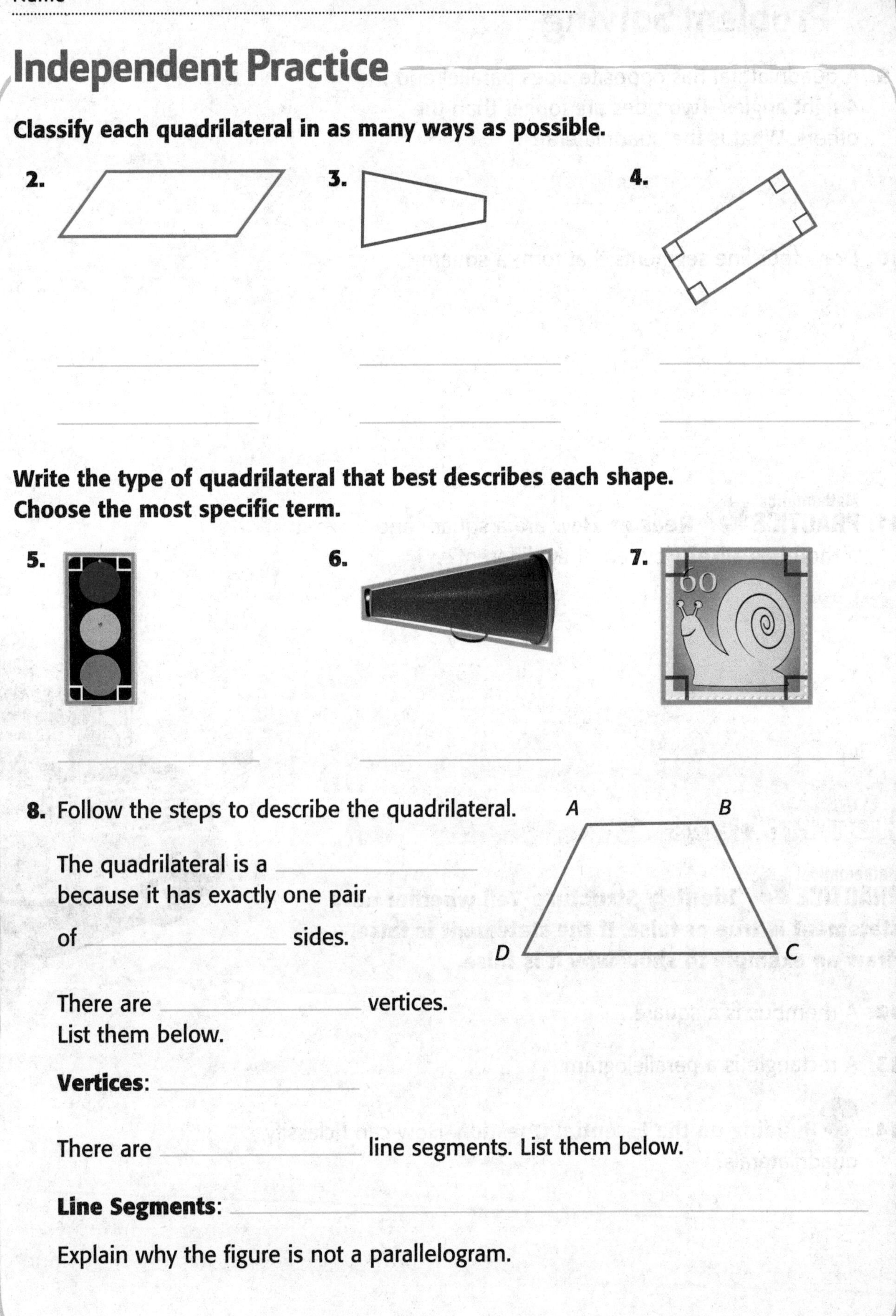

Classify each quadrilateral in as many ways as possible.

2.

3.

4.

Write the type of quadrilateral that best describes each shape. Choose the most specific term.

5.

6.

7.

8. Follow the steps to describe the quadrilateral.

The quadrilateral is a ______ because it has exactly one pair of ______ sides.

There are ______ vertices. List them below.

Vertices: ______

There are ______ line segments. List them below.

Line Segments: ______

Explain why the figure is not a parallelogram.

Real World!

Problem Solving

9. A quadrilateral has opposite sides parallel and 4 right angles. Two sides are longer than the others. What is the quadrilateral?

____

10. Draw four line segments that form a square.

My Work!

11. Mathematical PRACTICE 2 **Reason** How are a square and a rhombus alike? How are they different?

____

____

____

HOT Problems

Mathematical PRACTICE 7 **Identify Structure Tell whether each statement is true or false. If the statement is false, draw an example to show why it is false.**

12. A rhombus is a square. _______________

13. A rectangle is a parallelogram. _______________

14. **Building on the Essential Question** How can I classify quadrilaterals?

____

____

____

Name

MY Homework

Lesson 9

Quadrilaterals

Homework Helper

eHelp

Need help? connectED.mcgraw-hill.com

Classify each quadrilateral in as many ways as possible.

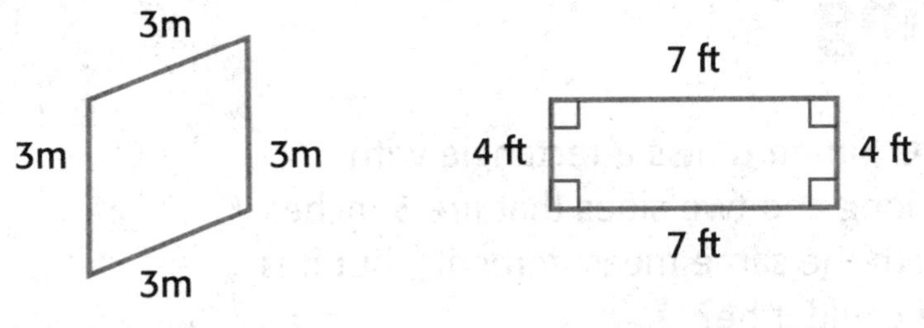

Quadrilateral 1 **Quadrilateral 2**

Quadrilateral 1 is a parallelogram because its opposite sides are equal in length and parallel.

It is also a rhombus because it has 4 sides equal in length.

Quadrilateral 2 is a parallelogram because its opposite sides are equal in length and parallel.

It is also a rectangle because it has 4 right angles.

Practice

Classify each quadrilateral in as many ways as possible.

1. ____________

2. ____________

____

____

____

Draw and classify a quadrilateral that fits each description.

3. 4 right angles, opposite sides equal in length and parallel

4. opposite sides equal in length and parallel

Problem Solving

5. Mathematical PRACTICE 6 **Be Precise** Simon draws a rectangle with two sides that are 2 inches long and two sides that are 3 inches long. Chaz draws a figure with the same measurements, but it is not a rectangle. What figure could it be?

Vocabulary Check

6. Explain the difference between a parallelogram and a rhombus.

7. Explain the difference between a square and a rectangle.

8. How is a trapezoid different from the other four types of quadrilaterals you have learned about?

Test Practice

9. Which quadrilateral does *not* have opposite sides equal in length?

Ⓐ rhombus Ⓑ trapezoid Ⓒ parallelogram Ⓓ square

Draw Lines of Symmetry

Lesson 10

ESSENTIAL QUESTION
How are different ideas about geometry connected?

A figure has **line symmetry** if it can be folded over a line so that one half of the figure matches the other half. This fold line is called the **line of symmetry**.

Math in My World

Watch Tutor

Example 1

Determine whether the sign at the right has line symmetry. If it does, draw the line(s) of symmetry on the figure.

1. **Determine if the figure has line symmetry.**
The figure can be folded in half vertically so that the left side matches the right side.

So, the figure has line symmetry.

2. **Draw the line of symmetry.**
Draw a vertical line through the center of the figure.

Some figures have more than one line of symmetry.

The pentagon at the right has five lines of symmetry.

Notice that the arrow ends that are not labeled are the other ends of the arrows that are labeled.

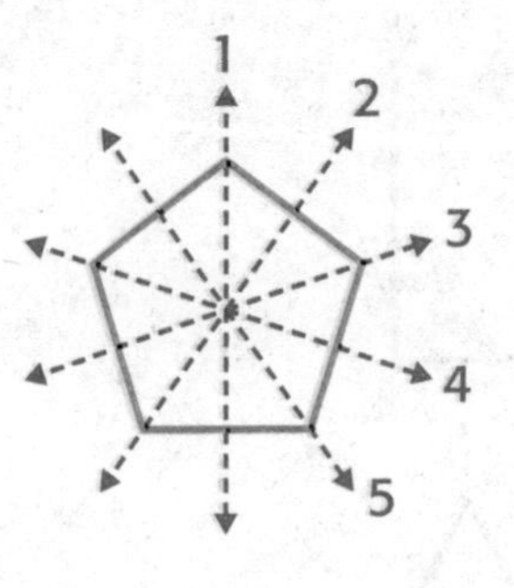

Some figures do not have any lines of symmetry.

The trapezoid at the right does not have any lines of symmetry.

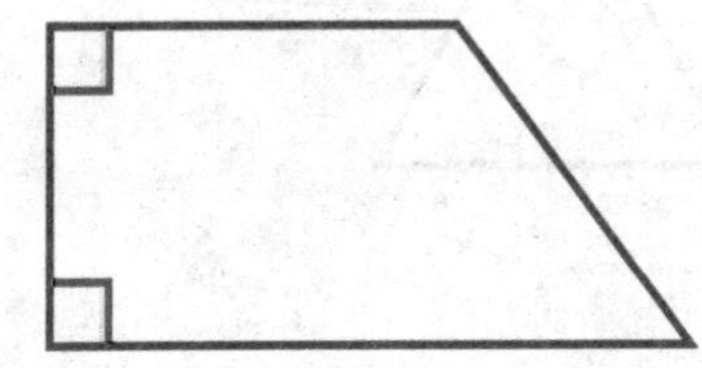

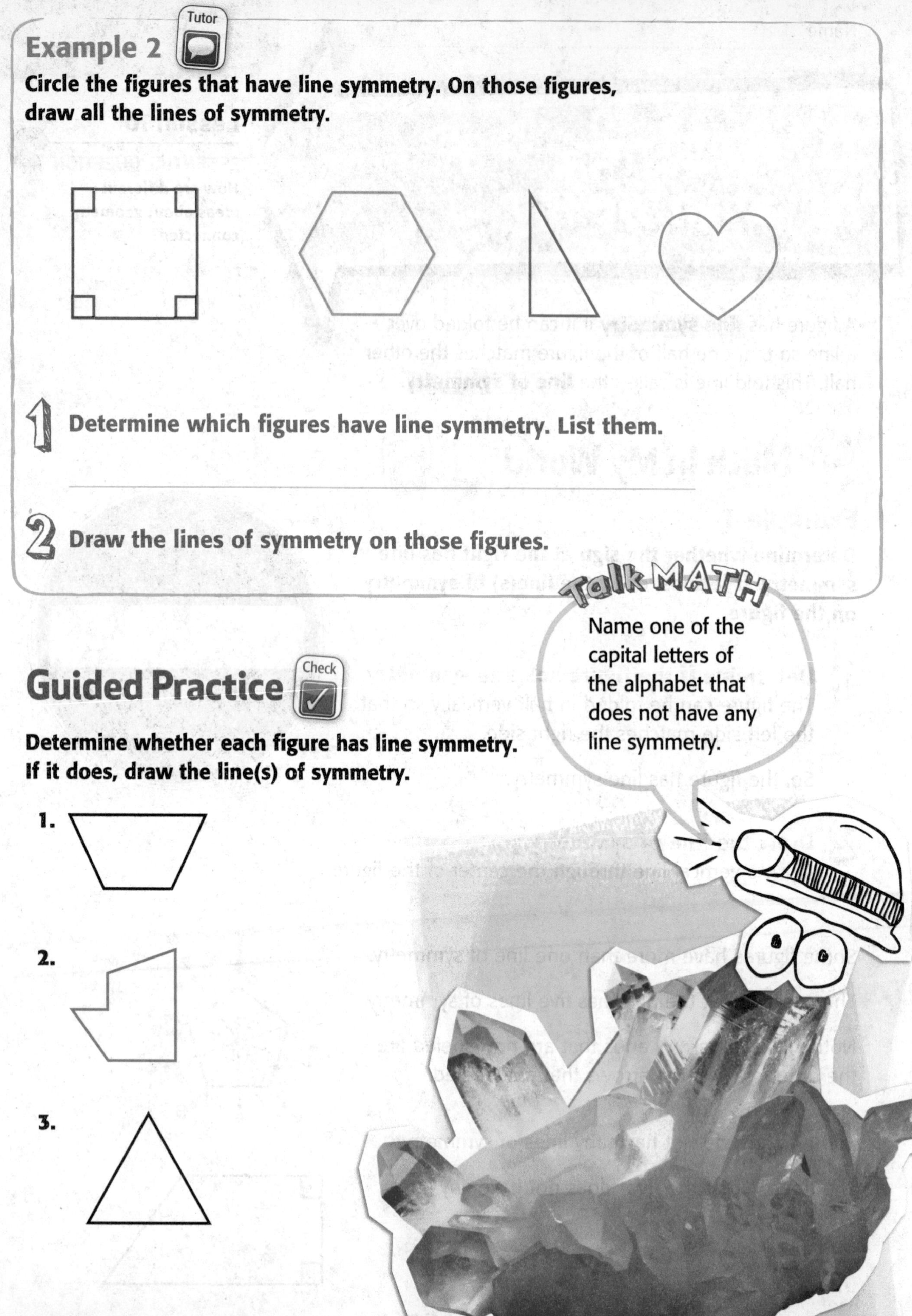

Example 2

Circle the figures that have line symmetry. On those figures, draw all the lines of symmetry.

1 **Determine which figures have line symmetry. List them.**

2 **Draw the lines of symmetry on those figures.**

Guided Practice

Determine whether each figure has line symmetry. If it does, draw the line(s) of symmetry.

1.

2.

3.

Name

Independent Practice

Determine whether each figure has line symmetry. Write *yes* or *no*. Draw the line(s) of symmetry on the figures that have line symmetry.

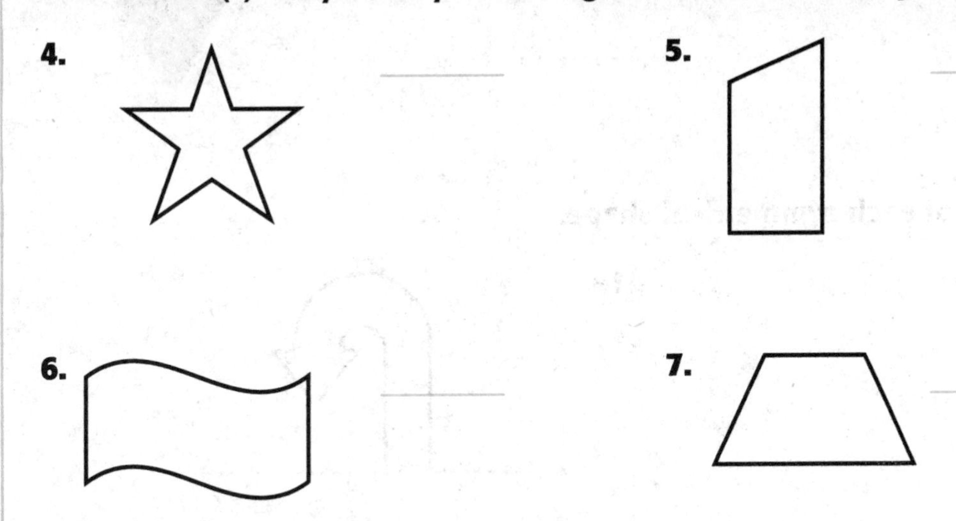

Circle the figures that have line symmetry. Cross out the figures that do not have line symmetry.

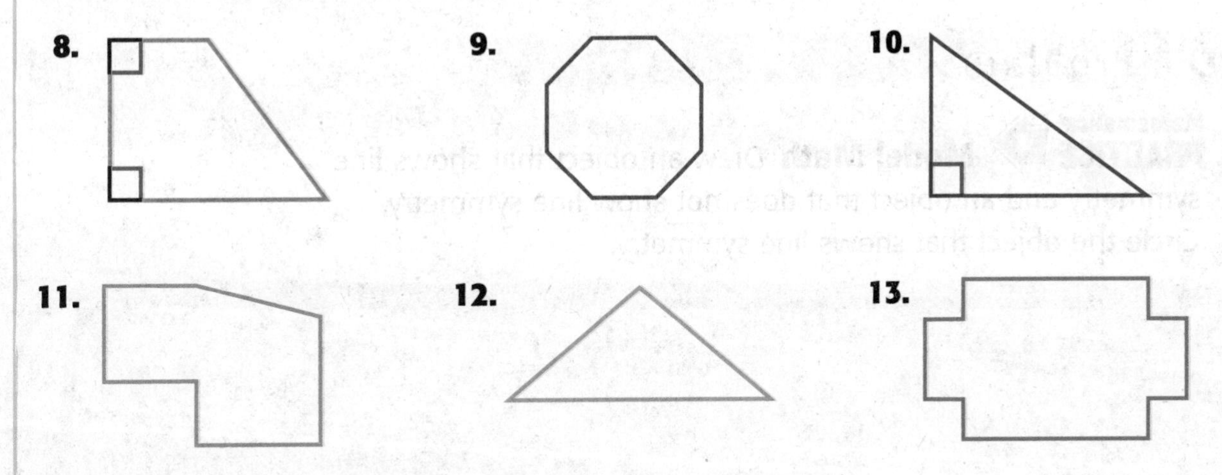

Determine whether the dotted line is a line of symmetry for each figure. Write *yes* or *no*.

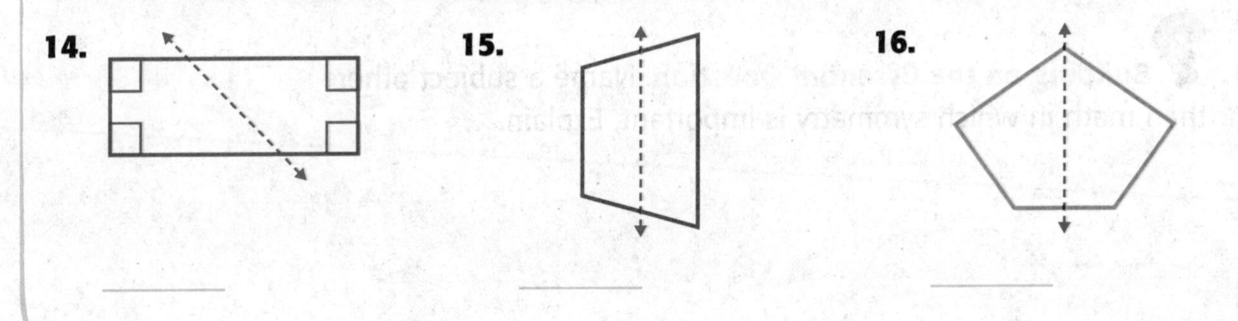

Problem Solving

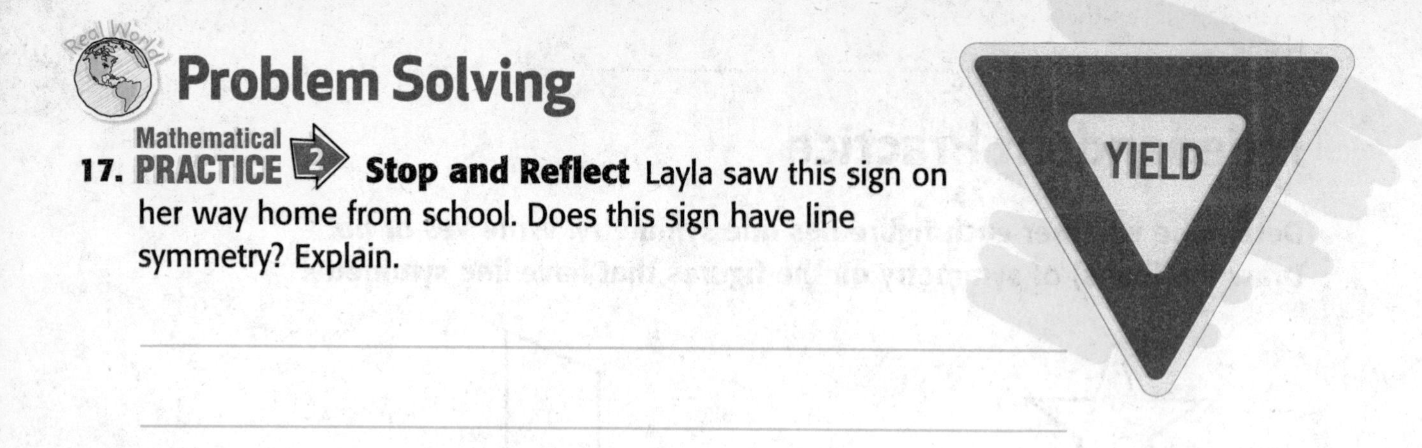

17. **Mathematical PRACTICE 2** **Stop and Reflect** Layla saw this sign on her way home from school. Does this sign have line symmetry? Explain.

Draw the other half of each symmetrical shape.

18. 19.

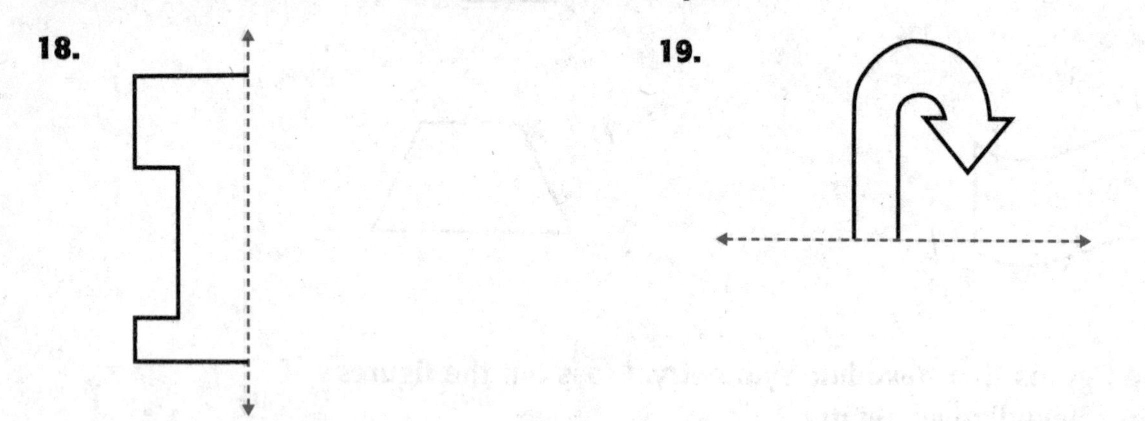

HOT Problems

20. **Mathematical PRACTICE 4** **Model Math** Draw an object that shows line symmetry and an object that does not show line symmetry. Circle the object that shows line symmetry.

My Drawing!

21. **Building on the Essential Question** Name a subject other than math in which symmetry is important. Explain.

Name

MY Homework

Lesson 10

Draw Lines of Symmetry

Homework Helper

eHelp

Need help? connectED.mcgraw-hill.com

Determine whether the hospital sign has line symmetry. If it does, draw the line(s) of symmetry on the figure.

1 **Determine if the figure has line symmetry.**
The figure can be folded in half vertically so that the left side matches the right side.

It can also be folded in half horizontally so that the top matches the bottom.

So, the figure has line symmetry.

2 **Draw the lines of symmetry.**
Draw a vertical line through the center of the figure.

Draw a horizontal line through the center of the figure.

There are 2 lines of symmetry.

Practice

Determine whether each figure has line symmetry. Write *yes* or *no*. Draw the line(s) of symmetry on the figures that have line symmetry.

1. ______

2. ______

3. ______

4. ______

Determine whether the dotted line is a line of symmetry for each figure. Write *yes* or *no*.

5. ______

6. ______

Draw the other half of each symmetrical shape.

7.

8.

Problem Solving

9. Mathematical PRACTICE 4 **Model Math** Vince wrote his name in all capital letters. How many of the letters have line symmetry? List them.

______

Vocabulary Check

Choose the correct word(s) to complete each sentence.

line of symmetry | line symmetry

10. If a figure can be folded into identical halves, it has ______.

11. The fold is the ______.

Test Practice

12. How many lines of symmetry does the sign have?

Ⓐ 3 Ⓒ 1

Ⓑ 2 Ⓓ 0

Name ....

Real World

Problem-Solving Investigation

STRATEGY: Make a Model

Lesson 11

ESSENTIAL QUESTION
How are different ideas about geometry connected?

Learn the Strategy

Watch Tutor

Maya has a square piece of paper. She folds it in half so there are two triangle-shaped parts. She folds it in half again so there are four triangle-shaped parts. When she unfolds the paper, how many right angles are shown?

1 Understand

What facts do you know?

Maya folds a piece of paper in half ______ times diagonally.

What do you need to find?

the number of right angles when the paper is unfolded

2 Plan

I will make a model to find the answer.

3 Solve

Use a square piece of paper. Follow the same steps that Maya followed.

Fold the paper.

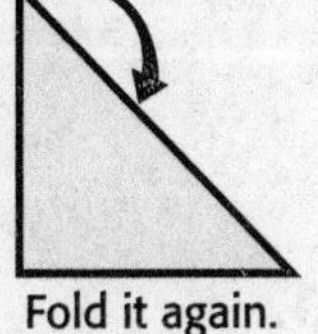
Fold it again.

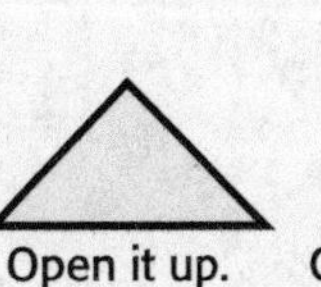
Open it up.

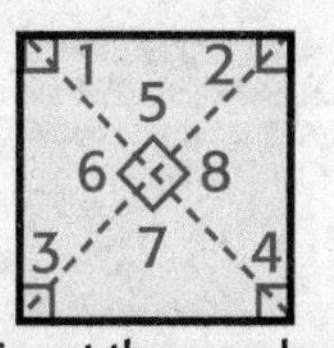

Count the number of right angles.

So, there are ______ right angles.

4 Check

Does your answer make sense? Explain.

____

Practice the Strategy

Sam has a card that is a quadrilateral. All sides are equal length. One of the angles measures 60°. Which quadrilateral is Sam's card?

1 Understand

What facts do you know?

____

____

What do you need to find?

____

____

2 Plan

____

3 Solve

4 Check

Does your answer make sense? Explain.

____

____

Name

CCSS

Apply the Strategy

Solve each problem by making a model.

My Work!

1. Mathematical PRACTICE 7 **Identify Structure** Mary Anne is making a pattern with quadrilaterals. She put squares in the first row, parallelograms in the second row, and trapezoids in the third row. She repeats this pattern four times. Which quadrilateral does she use in the tenth row?

2. Mathematical PRACTICE 4 **Model Math** Draw two lines on the square below so that three right triangles are formed.

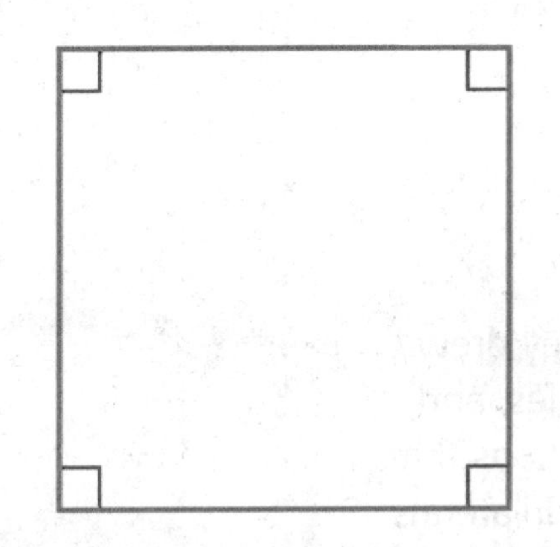

3. Mathematical PRACTICE 6 **Explain to a Friend** The first time a pizza is cut in half, there are 2 pieces. The second time each piece is cut in half, there are 4 pieces. The third time each piece is cut in half, there are 8 pieces. How many pieces will there be after each piece is cut the fourth time? Explain how you solved the problem.

Review the Strategies

Use any strategy to solve each problem.

- Make a model.
- Find a pattern.
- Make a table.
- Guess, check, and revise.

4. Mathematical PRACTICE 8 **Look for a Pattern** Mandy does one chore a day for her allowance. One day she washes dishes, the next she walks the dog, and the next she folds laundry. If she starts this cycle of chores on a Monday, which chore will she be doing the following Monday?

____

5. Kendra took photographs at the park. She photographed 20 dogs and owners in all. If there was a total of 64 legs, how many dogs and owners were there?

____

6. Mathematical PRACTICE 7 **Identify Structure** Corey drew 8 shapes. Four are squares, two are triangles, and the rest are parallelograms. Write two fractions that name the part of the shapes that are quadrilaterals.

____

7. Mathematical PRACTICE 1 **Make Sense of Problems** Mr. Lawrence used 100 tiles when he put a new floor in the kitchen. Thirty-four tiles are squares and 16 tiles are rectangles. Write the total amount of tiles that are squares and rectangles as a decimal and a fraction.

____

My Work!

MY Homework

Lesson 11
Problem Solving: Make a Model

Homework Helper

eHelp
Need help? connectED.mcgraw-hill.com

Bernie has three identical blocks. Each block has the shape of an acute triangle. If Bernie puts two of the blocks together, what quadrilateral does he form? If he joins the third block to the first two, what new quadrilateral does he form?

1 Understand

What facts do you know?

Bernie has three identical blocks shaped like acute triangles.

What do you need to find?

the quadrilaterals that Bernie will form by joining the blocks

2 Plan

Make a model to find the answer.

3 Solve

Draw an acute triangle. Join an identical acute triangle to the first one to form a quadrilateral.

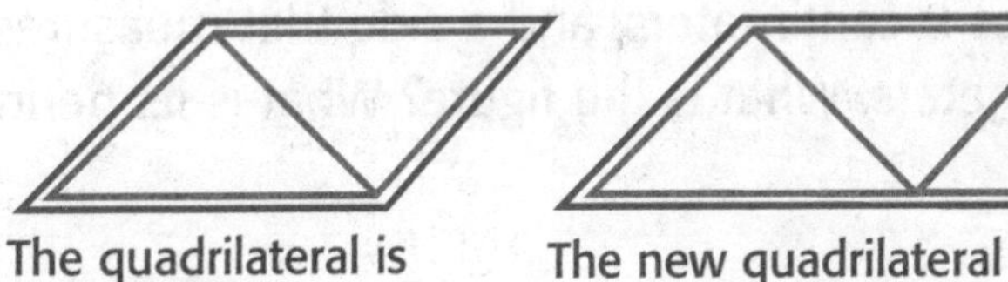

The quadrilateral is a parallelogram.

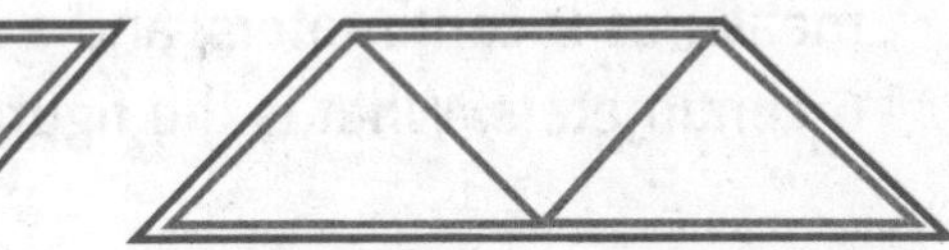

The new quadrilateral is a trapezoid.

Draw a third acute triangle joined to the first two triangles to form a new quadrilateral.

So, Bernie forms a parallelogram and a trapezoid.

4 Check

The figures match the information in the problem.

So, the answer makes sense.

Real World!

Problem Solving

Solve each problem by making a model.

My Work!

1. Mathematical PRACTICE 7 **Identify Structure** A quadrilateral has sides that measure 12 centimeters, 10 centimeters, 10 centimeters, and 5 centimeters. The side that measures 12 centimeters is parallel to the side that measures 5 centimeters. What is the figure?

2. Mathematical PRACTICE 6 **Explain to a Friend** Katerina draws a square. She wants to draw one line to divide the square into 2 triangles. Is it possible for her to divide the square into 2 obtuse triangles? Explain.

3. Is it possible for a quadrilateral to have only 2 right angles? If so, draw an example.

4. Mathematical PRACTICE 7 **Identify Structure** Kirby drew a quadrilateral with four right angles, a side that measures 9 centimeters, and a side that measures 6 centimeters. What is the figure? What is its perimeter?

5. Mathematical PRACTICE 6 **Be Precise** Is it possible for a trapezoid to have only 3 right angles? Explain.

Review

Chapter 14
Geometry

Vocabulary Check

Vocab

Draw an example of each vocabulary word.

1. acute angle	2. acute triangle	3. intersecting lines
4. line	5. line segment	6. obtuse angle
7. obtuse triangle	8. parallel lines	9. perpendicular lines
10. ray	11. right triangle	12. line symmetry

Concept Check

Describe each figure. Use *parallel, perpendicular,* or *intersecting.* Use the most specific term.

13.

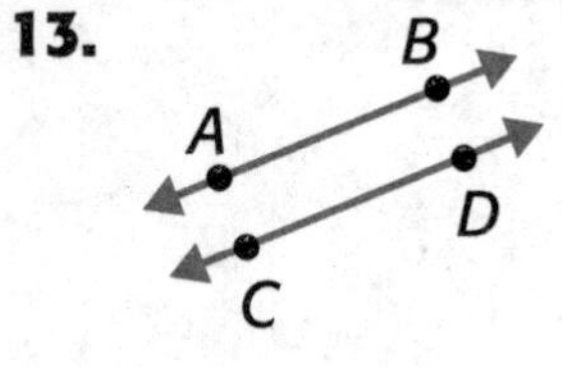

$\overleftrightarrow{AB}$ and $\overleftrightarrow{CD}$ are

____________________.

14.

U
T
S
V

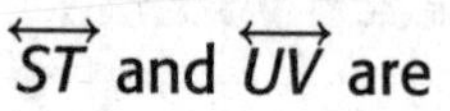

$\overleftrightarrow{ST}$ and $\overleftrightarrow{UV}$ are

____________________.

15. Draw an angle with a measure greater than $\frac{1}{4}$ turn.

16. Classify the angle as *right, acute,* or *obtuse.*

____

17. Classify the triangle as *acute, right,* or *obtuse.*

____

18. Write the type of quadrilateral that best describes the shape at the right.

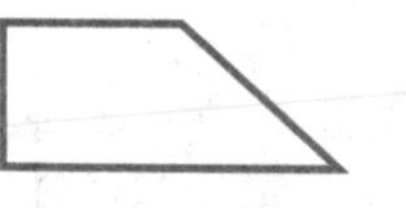

____

Draw the line(s) of symmetry in each figure.

19.

20.

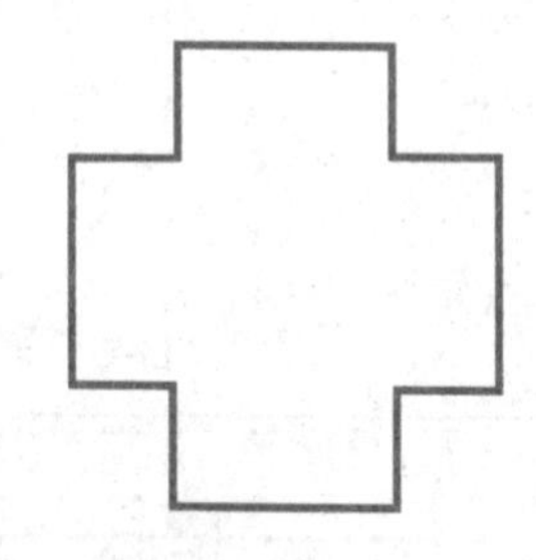

Name

Problem Solving

Real World

My Work!

21. Brent started his homework at 4:00 P.M. He completed it at the time shown. Write how far the minute hand has turned in degrees and as a fraction of a full turn.

22. Sam draws a quadrilateral. It has one pair of parallel sides. What figure did Sam draw?

23. Sergio saw an angle that measures greater than 0° and less than 90°. What type of angle did he see?

Test Practice

24. Levi's school is at the corner of High Street and Second Avenue. The corner forms a right angle. How might he describe the way the streets meet at his school?

Ⓐ The streets are parallel.

Ⓑ The streets are intersecting, but not perpendicular.

Ⓒ The streets never meet.

Ⓓ The streets are perpendicular.

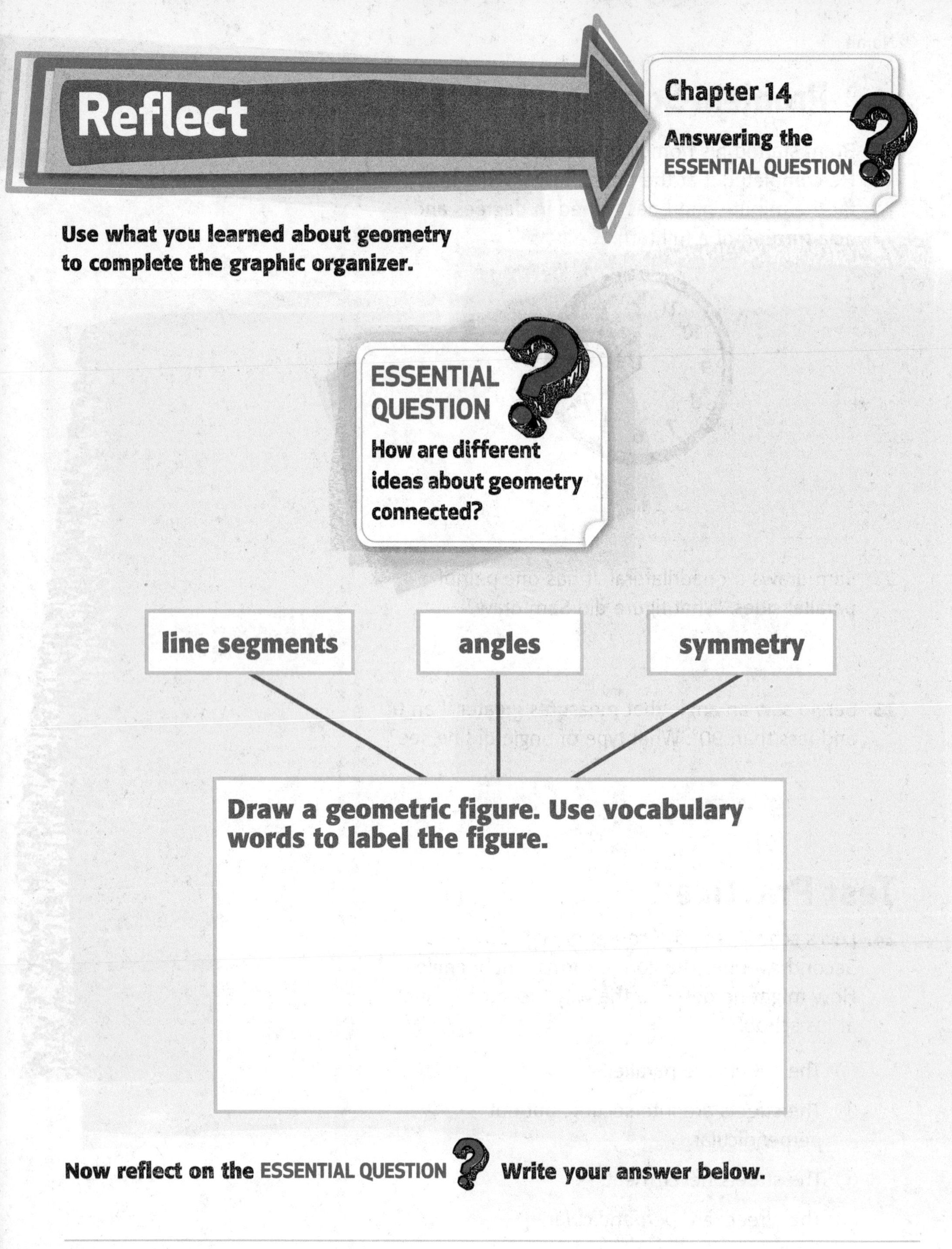

Use what you learned about geometry to complete the graphic organizer.

Now reflect on the ESSENTIAL QUESTION Write your answer below.

Glossary/Glosario

Vocab

Go online for the eGlossary.

Go to the **eGlossary** to find out more about these words in the following 13 languages:

Arabic • Bengali • Brazilian Portuguese • Cantonese • English • Haitian Creole
Hmong • Korean • Russian • Spanish • Tagalog • Urdu • Vietnamese

English | **Spanish/Español**

acute angle An *angle* with a measure greater than 0° and less than 90°.

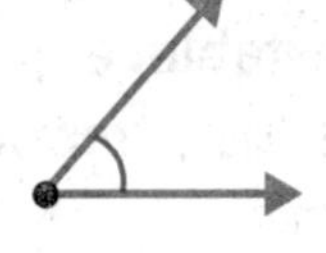

acute triangle A *triangle* with all three *angles* less than 90°.

add (adding, addition) An *operation* on two or more *addends* that results in a *sum*.

$$9 + 3 = 12$$

addend Any numbers being *added* together.

algebra A branch of mathematics that uses symbols, usually letters, to explore relationships between quantities.

ángulo agudo *Ángulo* que mide más de 0° y menos de 90°.

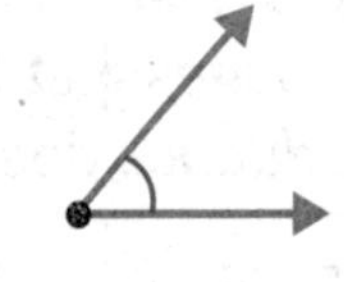

triángulo acutángulo *Triángulo* cuyos tres *ángulos* miden menos de 90°.

suma (sumar) *Operación* de dos o más *sumandos* que da como resultado una *suma*.

$$9 + 3 = 12$$

sumando Cualquier número que se *suma* a otro.

álgebra Rama de las matemáticas en la que se usan símbolos, generalmente letras, para explorar relaciones entre cantidades.

Aa

angle A figure that is formed by two *rays* with the same *endpoint*.

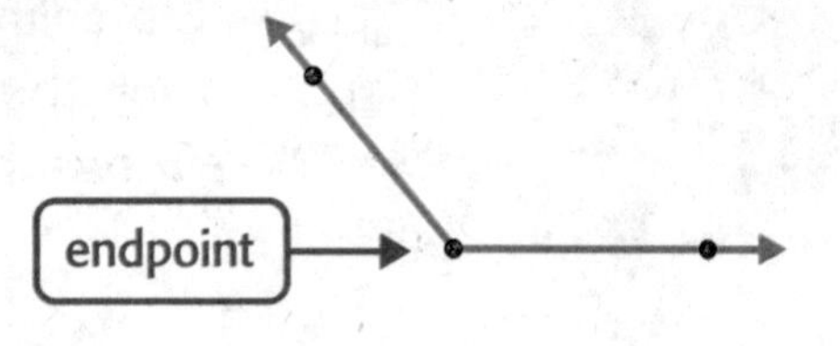

ángulo Figura formada por dos *semirrectas* con el mismo *extremo*.

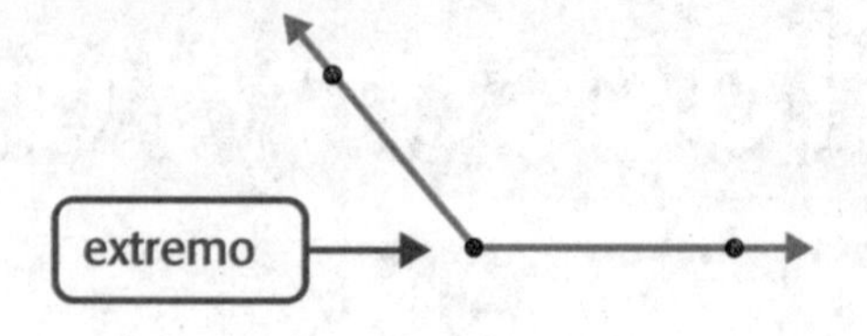

area The number of *square units* needed to cover the inside of a region or plane figure without any overlap.

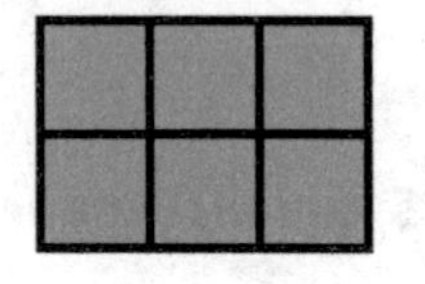

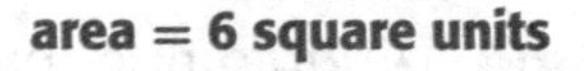
area = 6 square units

área Cantidad de *unidades cuadradas* necesarias para cubrir el interior de una región o figura plana sin superposiciones.

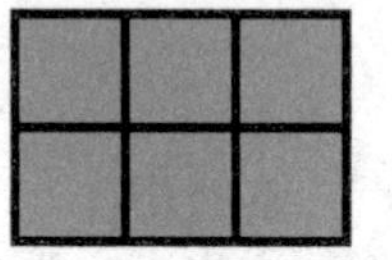

área = 6 unidades cuadradas

Associative Property of Addition The property that states that the grouping of the *addends* does not change the *sum*.

$$(4 + 5) + 2 = 4 + (5 + 2)$$

propiedad asociativa de la suma Propiedad que establece que la agrupación de los *sumandos* no altera la *suma*.

$$(4 + 5) + 2 = 4 + (5 + 2)$$

Associative Property of Multiplication The property that states that the grouping of the *factors* does not change the *product*.

$$3 \times (6 \times 2) = (3 \times 6) \times 2$$

propiedad asociativa de la multiplicación Propiedad que establece que la agrupación de los *factores* no altera el *producto*.

$$3 \times (6 \times 2) = (3 \times 6) \times 2$$

Bb

bar graph A graph that compares *data* by using bars of different *lengths* or heights to show the values.

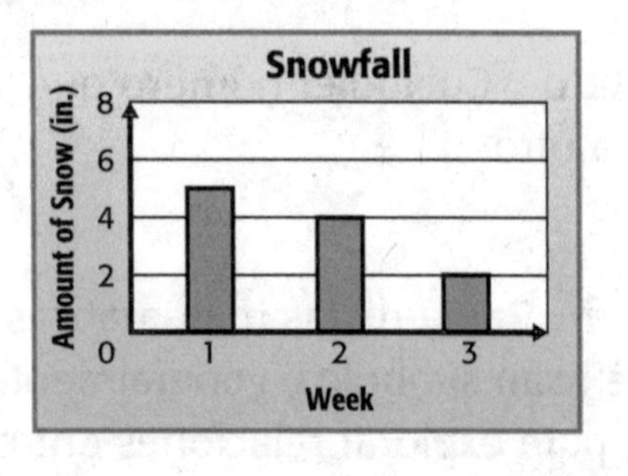

gráfica de barras Gráfica en la que se comparan los *datos* usando barras de distintas *longitudes* o alturas para mostrar los valores.

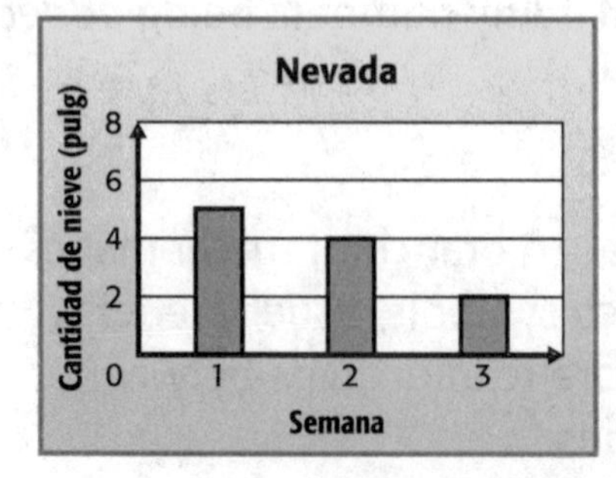

Cc

capacity The amount of liquid a container can hold.

capacidad Cantidad que puede contener un recipiente, medida en unidades de volumen.

centimeter (cm) A *metric* unit for measuring *length*.

100 centimeters = 1 meter

centímetro (cm) Unidad métrica de *longitud*.

100 centímetros = 1 metro

circle A closed figure in which all points are the same distance from a fixed point, called the center.

círculo Figura cerrada en la cual todos los puntos equidistan de un punto fijo llamado centro.

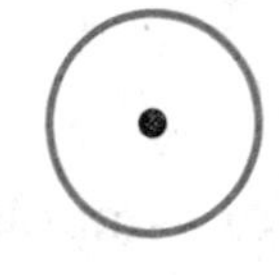

Commutative Property of Addition The property that states that the order in which two numbers are *added* does not change the *sum.*

12 + 15 = 15 + 12

propiedad conmutativa de la suma Propiedad que establece que el orden en el que se *suman* dos o más números no altera la *suma.*

12 + 15 = 15 + 12

Commutative Property of Multiplication The property that states that the order in which two numbers are *multiplied* does not change the *product.*

7 × 2 = 2 × 7

propiedad conmutativa de la multiplicación Propiedad que establece que el orden en el que se *multiplican* dos o más números no altera el *producto.*

7 × 2 = 2 × 7

compatible numbers Numbers in a problem or related numbers that are easy to work with mentally.

720 and 90 are compatible numbers for *division* because 72 ÷ 9 = 8.

números compatibles Números en un problema o números relacionados con los cuales es fácil trabajar mentalmente.

720 ÷ 90 es una división que usa números compatibles porque 72 ÷ 9 = 8.

composite number A whole number that has more than two factors.

12 has the factors 1, 2, 3, 4, 6, and 12.

número compuesto Número natural con más de dos factores.

12 tiene los factores 1, 2, 3, 4, 6 y 12.

Cc

congruent figures Two figures having the same size and the same shape.

figuras congruentes Dos figuras con la misma forma y el mismo tamaño.

convert To change one unit to another.

convertir Cambiar de una unidad a otra.

cup (c) A *customary* unit of *capacity* equal to 8 fluid ounces.

taza (tz) Unidad *usual* de *capacidad* que equivale a 8 onzas líquidas.

customary system The measurement system most often used in the United States. Units include *foot*, *pound*, and *quart*.

sistema usual Conjunto de unidades de medida de uso más frecuente en Estados Unidos. Incluye unidades como *el pie*, *la libra* y *el cuarto*.

Dd

data Numbers or symbols, sometimes collected from a *survey* or experiment, to show information. Datum is singular; data is plural.

datos Números o símbolos que muestran información, algunas veces recopilados a partir de una *encuesta* o un experimento.

decimal A number with one or more digits to the right of the decimal point, such as 8.37 or 0.05.

decimal Número con uno o más dígitos a la derecha del *punto decimal,* como 8.37 o 0.05.

decimal equivalents Decimals that represent the same number.

0.3 and 0.30

decimales equivalentes Decimales que representan el mismo número.

0.3 y 0.30

decimal point A period separating the ones and the *tenths* in a decimal number.

0.8 or $3.77

punto decimal Punto que separa las unidades de las *décimas* en un número decimal.

0.8 o $3.77

decompose To break a number into different parts.

descomponer Separar un número en diferentes partes.

degree (°) **a.** A unit for measuring angles. **b.** A unit of measure used to describe temperature.

grado (°) **a.** Unidad que se usa para medir ángulos. **b.** Unidad de medida que se usa para describir la temperatura.

denominator The bottom number in a *fraction.*

In $\frac{5}{6}$, 6 is the denominator.

denominador El número de abajo en una *fracción.*

En $\frac{5}{6}$, 6 es el denominador.

digit A symbol used to write numbers. The ten digits are 0, 1, 2, 3, 4, 5, 6, 7, 8, and 9.

dígito Símbolo que se usa para escribir los números. Los diez dígitos son 0, 1, 2, 3, 4, 5, 6, 7, 8 y 9.

Distributive Property To *multiply* a *sum* by a number, multiply each *addend* by the number and *add* the *products.*

$$4 \times (1 + 3) = (4 \times 1) + (4 \times 3)$$

propiedad distributiva Para *multiplicar* una *suma* por un número, multiplica cada *sumando* por ese número y luego *suma* los *productos.*

$$4 \times (1 + 3) = (4 \times 1) + (4 \times 3)$$

dividend A number that is being *divided.*

$3\overline{)19}$ **19 is the dividend.**

dividendo Número que se *divide.*

$3\overline{)19}$ **19 es el dividendo.**

division (divide) An *operation* on two numbers in which the first number is split into the same number of equal groups as the second number.

división (dividir) *Operación* entre dos números en la que el primer número se separa en tantos grupos iguales como indica el segundo número.

divisor The number by which the *dividend* is being *divided.*

$3\overline{)19}$ **3 is the divisor.**

divisor Número entre el cual se *divide* el *dividendo.*

$3\overline{)19}$ **3 es el divisor.**

Ee

elapsed time The amount of time that has passed from beginning to end.

tiempo transcurrido Cantidad de tiempo que ha pasado entre el principio y el fin de algo.

endpoint The point at either end of a *line segment* or the point at the beginning of a *ray.*

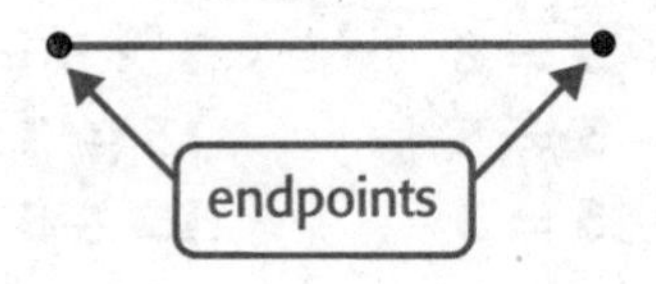

extremo Punto que se encuentra en cualquiera de los dos lados en que termina un *segmento de recta* o al principio de una *semirrecta.*

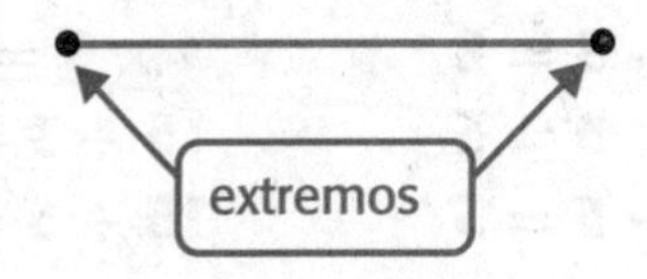

Ee

equation A sentence that contains an equals sign (=), showing that two *expressions* are equal.

ecuación Espresión matemática que contiene el signo igual, (=), que indica que las dos *expresiones* son iguales.

equilateral triangle A *triangle* with three *congruent* sides.

triángulo equilátero *Triángulo* con tres lados *congruentes*.

equivalent fractions *Fractions* that represent the same number.

$$\frac{3}{4} = \frac{6}{8}$$

fracciones equivalentes *Fracciones* que representan el mismo número.

$$\frac{3}{4} = \frac{6}{8}$$

estimate A number close to an exact value. An estimate indicates *about* how much.

47 + 22 is about
50 + 20 or 70.

estimación Número cercano a un valor exacto. Una estimación indica una cantidad approximada.

47 + 22 es aproximadamente
50 + 20 o 70.

expanded form/expanded notation The representation of a number as a *sum* that shows the value of each digit.

536 is written as 500 + 30 + 6.

forma desarrollada/notación desarrollada Representación de un número como la *suma* del valor de cada dígito.

536 puede escribirse como 500 + 30 + 6.

expression A combination of numbers, *variables,* and at least one *operation.*

expresión Combinación de números, *variables* y por lo menos una *operación.*

Ff

fact family A group of related facts using the same numbers.

$5 + 3 = 8$	$5 \times 3 = 15$
$3 + 5 = 8$	$3 \times 5 = 15$
$8 - 3 = 5$	$15 \div 3 = 5$
$8 - 5 = 3$	$15 \div 5 = 3$

familia de operaciones Grupo de operaciones relacionadas que usan los mismos números.

$5 + 3 = 8$	$5 \times 3 = 15$
$3 + 5 = 8$	$3 \times 5 = 15$
$8 - 3 = 5$	$15 \div 3 = 5$
$8 - 5 = 3$	$15 \div 5 = 3$

factor A number that *divides* a whole number evenly. Also a number that is *multiplied* by another number.

factor Número entre el que se *divide* otro número natural sin dejar residuo. También es cada uno de los números en una multiplicación.

factor pairs The two factors that are multiplied to find a product.

pares de factores Los dos factores que se multiplican para hallar un producto.

fluid ounce (fl oz) A *customary* unit of *capacity*.

onza líquida (oz líq) Unidad *usual* de *capacidad*.

foot (ft) A *customary* unit for measuring *length*. Plural is feet.

1 foot = 12 inches

pie (pie) Unidad *usual* de *longitud*.

1 pie = 12 pulgadas

formula An *equation* that shows the relationship between two or more quantities.

fórmula *Ecuación* que muestra la relación entre dos o más cantidades.

fraction A number that represents part of a whole or part of a set.

$\frac{1}{2}, \frac{1}{3}, \frac{1}{4}, \frac{3}{4}$

fracción Número que representa una parte de un todo o una parte de un conjunto.

$\frac{1}{2}, \frac{1}{3}, \frac{1}{4}, \frac{3}{4}$

frequency table A table for organizing a set of *data* that shows the number of times each result has occurred.

tabla de frecuencias Tabla para organizar un conjunto de *datos* que muestra el número de veces que se ha obtenido cada resultado.

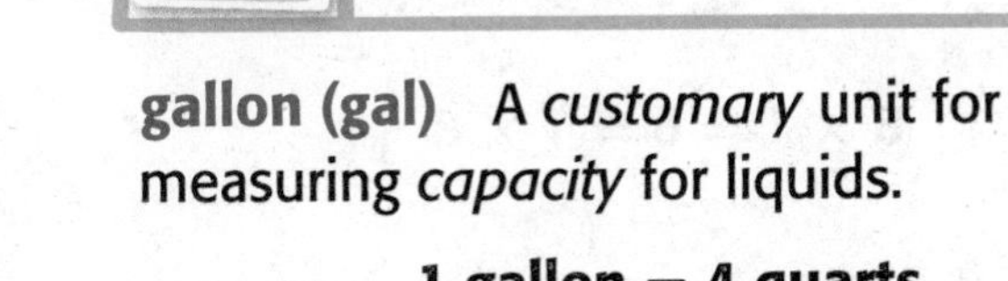

Gg

gallon (gal) A *customary* unit for measuring *capacity* for liquids.

1 gallon = 4 quarts

galón (gal) Unidad *usual* de *capacidad* de líquidos.

1 galón = 4 cuartos

gram (g) A *metric* unit for measuring *mass*.

gramo (g) Unidad *métrica* de *masa*.

Greatest Common Factor (GCF) The greatest of the common *factors* of two or more numbers.

The greatest common factor of 12, 18, and 30 is 6.

máximo común divisor (M.C.D.) El mayor de los *factores* comunes de dos o más números.

El máximo común divisor de 12, 18 y 30 es 6.

Hh

hexagon A *polygon* with six sides and six *angles.*

hexágono *Polígono* con seis lados y seis *ángulos.*

hundredth A *place-value* position. One of one hundred equal parts.

In the number 0.05, 5 is in the hundredths place.

centésima *Valor posicional.* Una de cien partes iguales.

En el número 4.57, 7 está en el lugar de las centésimas.

Ii

Identity Property of Addition For any number, zero plus that number is the number.

$3 + 0 = 3$ or $0 + 3 = 3$

propiedad de identidad de la suma Para todo número, cero más ese número da como resultado ese mismo número.

$3 + 0 = 3$ o $0 + 3 = 3$

Identity Property of Multiplication If you *multiply* a number by 1, the *product* is the same as the given number.

$8 \times 1 = 8 = 1 \times 8$

propiedad de identidad de la multiplicación Si *multiplicas* un número por 1, el *producto* es igual al número dado.

$8 \times 1 = 8 = 1 \times 8$

improper fraction A *fraction* with a *numerator* that is greater than or equal to the *denominator.*

$\frac{17}{3}$ or $\frac{5}{5}$

fracción impropia *Fracción* con un *numerador* igual al *denominador* o mayor que él.

$\frac{17}{3}$ o $\frac{5}{5}$

input A quantity that is changed to produce an output.

intersecting lines *Lines* that meet or cross at a point.

is equal to (=) Having the same value. The (=) sign is used to show two numbers or *expressions* are equal.

is greater than (>) An inequality relationship showing that the number on the left of the symbol is greater than the number on the right.

5 > 3 5 is greater than 3.

is less than (<) An inequality relationship showing that the number on the left side of the symbol is less than the number on the right side.

4 < 7 4 is less than 7.

isosceles triangle A *triangle* with at least 2 sides of the same *length*.

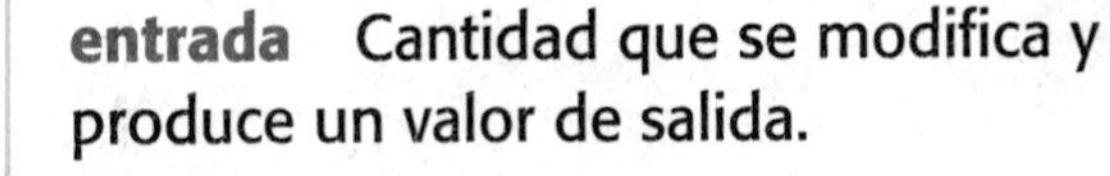

entrada Cantidad que se modifica y produce un valor de salida.

rectas secantes *Rectas* que se intersecan o se cruzan en un punto común.

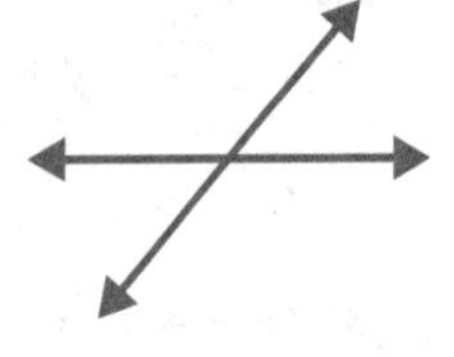

es igual a (=) Que tiene el mismo valor. Con el signo = se muestra que dos números o *expresiones* son iguales.

es mayor que (>) Relación de desigualdad que muestra que el número a la izquierda del signo es más grande que el número a la derecha.

5 > 3 5 es mayor que 3.

es menor que (<) Relación de desigualdad que muestra que el número a la izquierda del signo es más pequeño que el número a la derecha.

4 < 7 4 es menor que 7.

triángulo isósceles *Triángulo* que tiene por lo menos 2 lados del mismo *largo*.

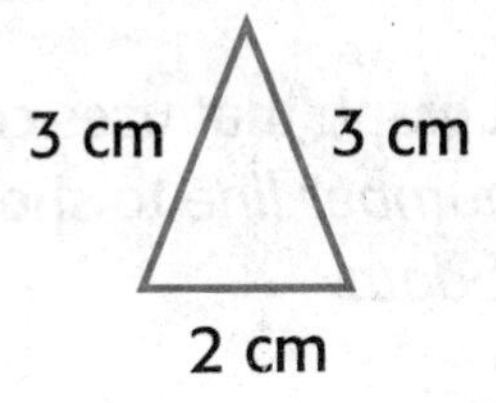

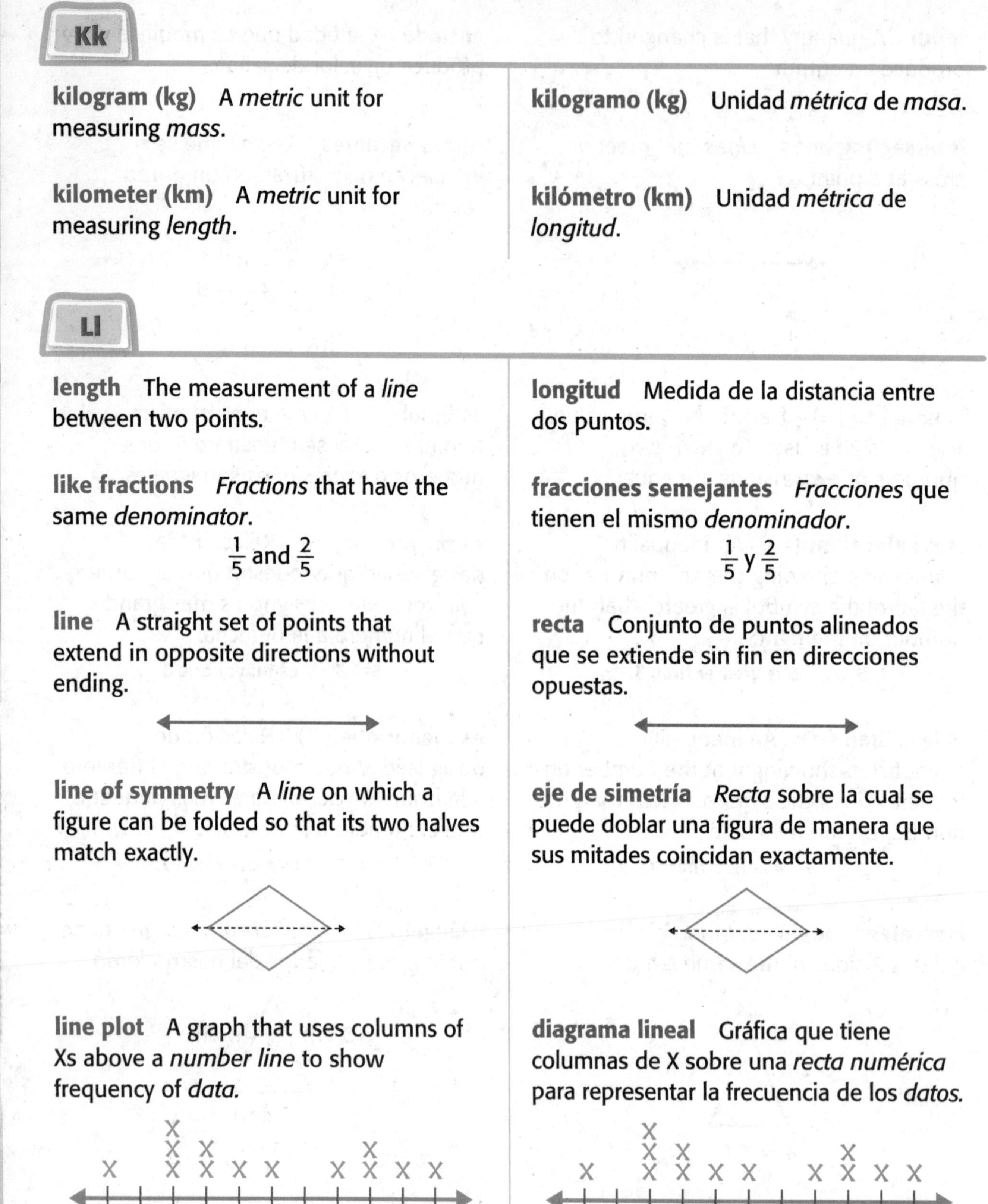

Kk

kilogram (kg) A *metric* unit for measuring *mass*.

kilogramo (kg) Unidad *métrica* de *masa*.

kilometer (km) A *metric* unit for measuring *length*.

kilómetro (km) Unidad *métrica* de *longitud*.

Ll

length The measurement of a *line* between two points.

longitud Medida de la distancia entre dos puntos.

like fractions *Fractions* that have the same *denominator*.

$\frac{1}{5}$ and $\frac{2}{5}$

fracciones semejantes *Fracciones* que tienen el mismo *denominador*.

$\frac{1}{5}$ y $\frac{2}{5}$

line A straight set of points that extend in opposite directions without ending.

recta Conjunto de puntos alineados que se extiende sin fin en direcciones opuestas.

line of symmetry A *line* on which a figure can be folded so that its two halves match exactly.

eje de simetría *Recta* sobre la cual se puede doblar una figura de manera que sus mitades coincidan exactamente.

line plot A graph that uses columns of Xs above a *number line* to show frequency of *data*.

diagrama lineal Gráfica que tiene columnas de X sobre una *recta numérica* para representar la frecuencia de los *datos*.

X X X X X X X X X X X X X
0 1 2 3 4 5 6 7 8 9 10

line segment A part of a *line* between two *endpoints.* The *length* of the line segment can be measured.

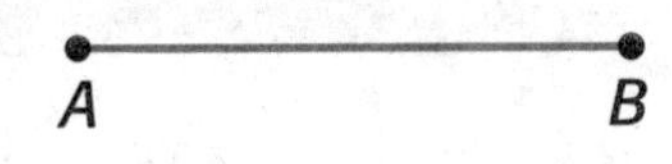

segmento de recta Parte de una *recta* entre dos *extremos.* La *longitud* de un segmento de recta se puede medir.

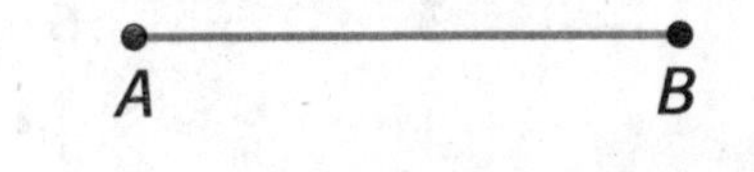

line symmetry A figure has *line symmetry* if it can be folded so that the two parts of the figure match, or are *congruent.*

simetría axial Una figura tiene *simetría axial* si puede doblarse de modo que las dos partes de la figura coincidan de manera exacta.

liter (L) A *metric* unit for measuring *volume* or *capacity.*

1 liter = 1,000 milliliters

litro (L) Unidad *métrica* de *volumen* o *capacidad.*

1 litro = 1,000 mililitros

Mm

mass The amount of matter in an object. Two examples of units of measure would be gram and kilogram.

masa Cantidad de materia en un cuerpo. El gramo y el kilogramo son dos ejemplos de unidades que se usan para medir la masa.

meter (m) A *metric* unit for measuring *length.*

metro (m) Unidad *métrica* de *longitud.*

metric system (SI) The decimal system of measurement. Includes units such as *meter, gram,* and *liter.*

sistema métrico (SI) Sistema decimal de medidas que se basa en potencias de 10 y que incluye unidades como *el metro, el gramo* y *el litro.*

mile (mi) A *customary* unit of measure for *length.*

1 mile = 5,280 feet

milla (mi) Unidad *usual* de *longitud.*

1 milla = 5,280 pies

milliliter (mL) A *metric* unit for measuring *capacity.*

1,000 milliliters = 1 liter

mililitro (mL) Unidad *métrica* de *capacidad.*

1,000 mililitros = 1 litro

millimeter (mm) A *metric* unit for measuring *length.*

1,000 millimeters = 1 meter

milímetro (mm) Unidad *métrica* de *longitud.*

1,000 milímetros = 1 metro

Mm

minuend The first number in a *subtraction* sentence from which a second number is to be subtracted.

8 – **3** = **5**

minuend **subtrahend** **difference**

mixed number A number that has a *whole number* part and a *fraction* part.

$6\frac{3}{4}$

multiple A multiple of a number is the *product* of that number and any whole number.

15 is a multiple of 5
because $3 \times 5 = 15$.

multiply (multiplication) An *operation* on two numbers to find their *product.* It can be thought of as repeated *addition.*

minuendo El primer número en un enunciado de *resta* del cual se restará un segundo número

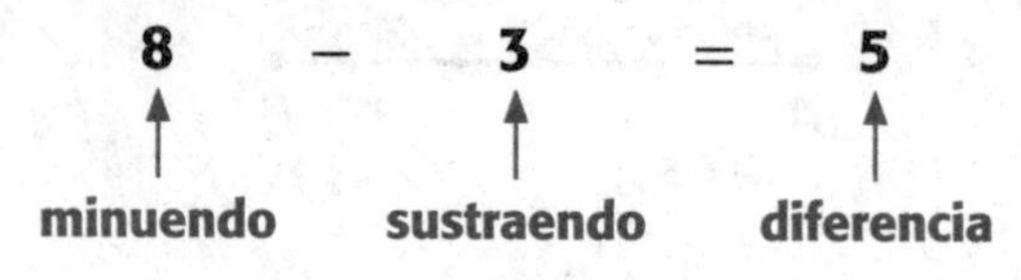

número mixto Número formado por un *número natural* y una parte *fraccionaria.*

$6\frac{3}{4}$

múltiplo Un múltiplo de un número es el *producto* de ese número y cualquier otro número natural.

15 es múltiplo de 5
porque $3 \times 5 = 15$.

multiplicar (multiplicación) *Operación* entre dos números para hallar su *producto.* También se puede interpretar como una *suma* repetida.

Nn

nonnumeric pattern Patterns that do not use numbers.

number line A *line* with numbers on it in order at regular intervals.

0 1 2 3 4 5 6 7 8 9 10

numerator The number above the bar in a *fraction*; the part of the fraction that tells how many of the equal parts are being used.

numeric pattern Patterns that use numbers.

patrón no numérico Patrón que no usa números.

recta numérica *Recta* con números ordenados a intervalos regulares.

0 1 2 3 4 5 6 7 8 9 10

numerador El número que está encima de la barra de *fracción*; la parte de la fracción que indica cuántas de las partes iguales en que se divide el entero se están usando.

patrón numérico Patrón que usa números.

obtuse angle An *angle* that measures greater than 90° but less than 180°.

ángulo obtuso *Ángulo* que mide más de 90° pero menos de 180°.

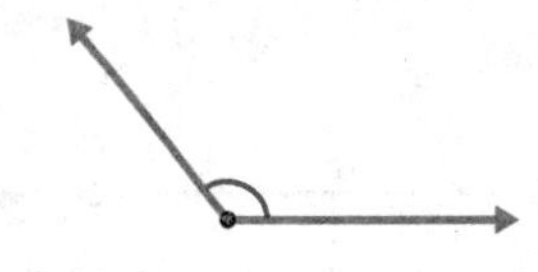

obtuse triangle A *triangle* with one *obtuse angle.*

triángulo obtusángulo *Triángulo* con un *ángulo obtuso.*

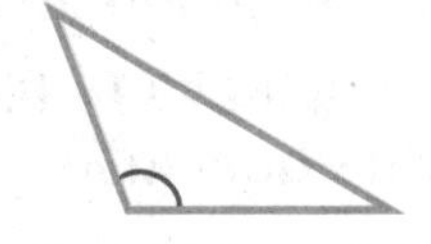

octagon A *polygon* with 8 sides and 8 *angles.*

octágono *Polígono* de 8 lados y 8 *ángulos.*

operation A mathematical process such as *addition* (+), *subtraction* (−), *multiplication* (×), or *division* (÷).

operación Proceso matemático como la *suma* (+), la *resta* (−), la *multiplicación* (×) o la *división* (÷).

order of operations Rules that tell what order to follow when evaluating an *expression:*
(1) Do the *operations* in *parentheses* first.
(2) *Multiply* and *divide* in order from left to right.
(3) *Add* and *subtract* in order from left to right.

orden de las operaciones Reglas que te indican qué orden seguir cuando evalúas una *expresión:*
(1) Resuelve primero las *operaciones* dentro de los *paréntesis.*
(2) *Multiplica* o *divide* en orden de izquierda a derecha.
(3) *Suma* o *resta* en orden de izquierda a derecha.

ounce (oz) A *customary* unit to measure *weight* or *capacity.*

onza (oz) Unidad *usual* de *peso* o *capacidad.*

output The result of an input quantity being changed.

salida Resultado que se obtiene al modificar un valor de entrada.

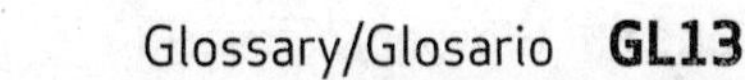

Pp

parallel lines *Lines* that are the same distance apart. Parallel lines do not meet.

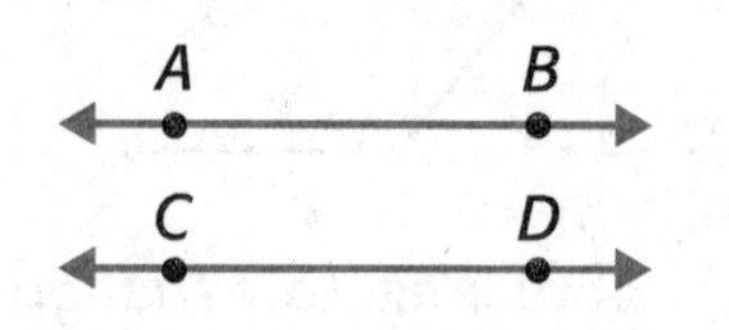

rectas paralelas *Rectas* separadas por la misma distancia en cualquier punto. Las rectas paralelas no se intersecan.

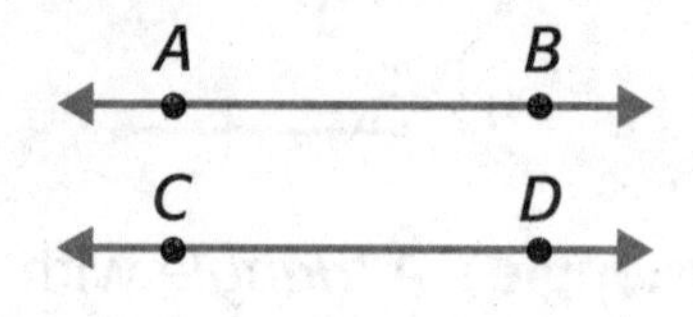

parallelogram A *quadrilateral* with four sides in which each pair of opposite sides are *parallel* and equal in *length.*

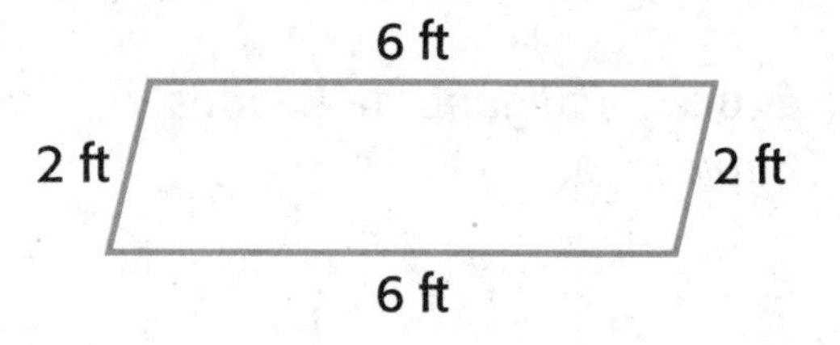

paralelogramo *Cuadrilátero* en el que cada par de lados opuestos son *paralelos* y tienen la misma *longitud.*

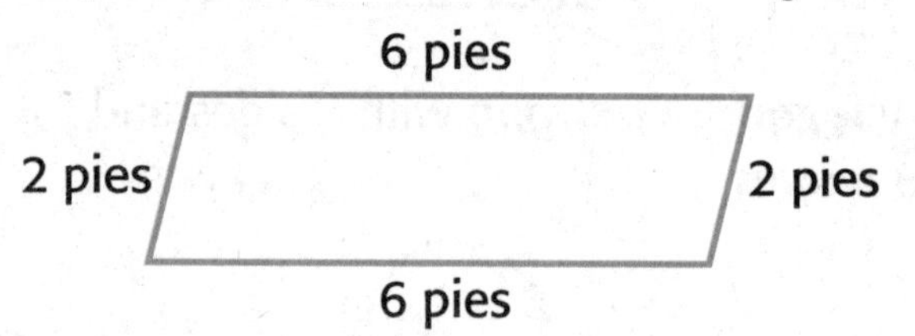

parentheses The enclosing symbols (), which indicate that the terms within are a unit.

paréntesis Los signos ( ) con que se encierran los términos, para indicar que cuando están adentro, forman una unidad.

partial products A multiplication method in which the products of each place value are found separately, and then added together.

productos parciales Método de multiplicación por el cual los productos de cada valor posicional se hallan por separado y luego se suman entre sí.

partial quotients A dividing method in which the dividend is separated into sections that are easy to divide.

cocientes parciales Método de división por el cual el dividendo se separa en secciones que son fáciles de dividir.

pattern A sequence of numbers, figures, or symbols that follows a rule or design.

2, 4, 6, 8, 10

patrón Sucesión de números, figuras o símbolos que sigue una regla o un diseño.

2, 4, 6, 8, 10

pentagon A *polygon* with five *sides* and five *angles.*

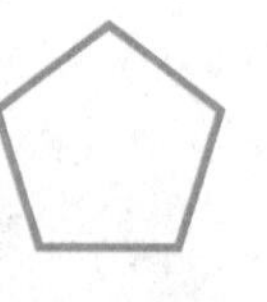

pentágono *Polígono* de cinco *lados* y cinco *ángulos.*

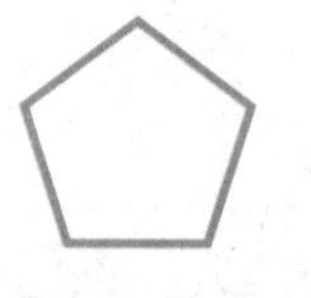

percent A ratio that compares a number to 100.

porcentaje Razón que compara un número con el 100.

perimeter The distance around a shape or region.

perímetro Distancia alrededor de una figura o región.

period The name given to each group of three digits on a place-value chart.

período Nombre dado a cada grupo de tres dígitos en una tabla de valor posicional.

perpendicular lines *Lines* that meet or cross each other to form *right angles.*

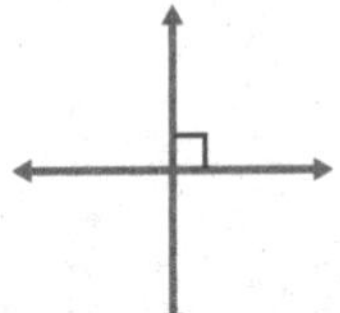

rectas perpendiculares *Rectas* que se intersecan o cruzan formando *ángulos rectos.*

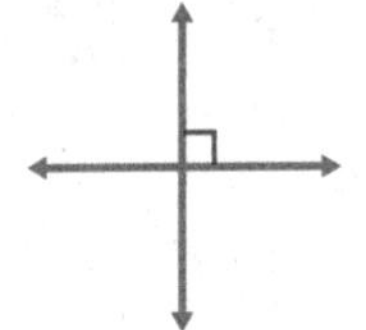

pint (pt) A *customary* unit for measuring *capacity*.

1 pint = 2 cups

pinta (pt) Unidad *usual* de *capacidad*.

1 pinta = 2 tazas

place value The value given to a *digit* by its position in a number.

valor posicional Valor dado a un *dígito* según su posición en un número.

point An exact location in space that is represented by a dot.

punto Ubicación exacta en el espacio que se representa con una marca puntual.

polygon A closed *plane figure* formed using *line segments* that meet only at their *endpoints.*

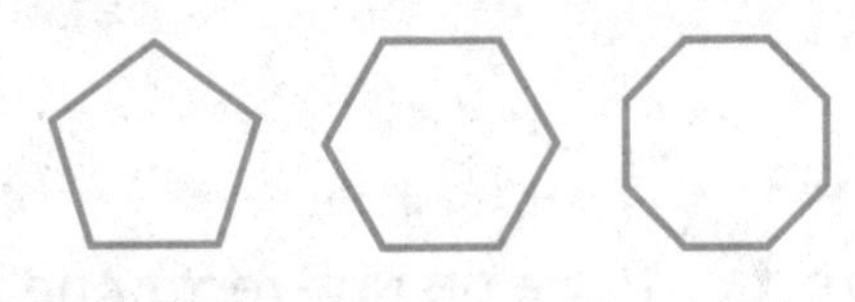

polígono *Figura plana* cerrada formada por *segmentos de recta* que solo se unen en sus *extremos*.

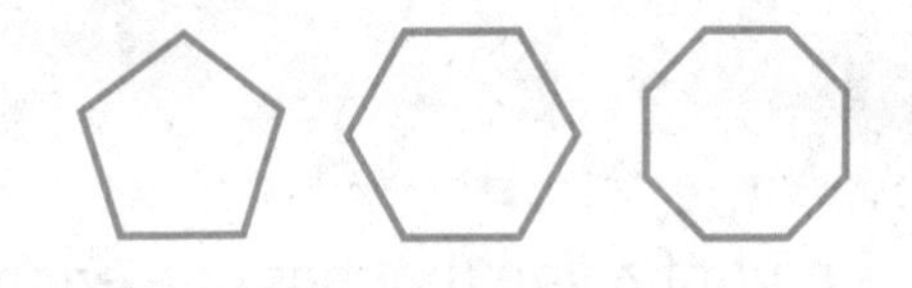

pound (lb) A *customary* unit to measure *weight* or *mass*.

1 pound = 16 ounces

libra (lb) Unidad *usual* de *peso* o *masa*.

1 libra = 16 onzas

Pp

prime number A whole number with exactly two *factors*, 1 and itself.

7, 13, and 19

número primo Número natural que tiene exactamente dos *factores*: 1 y sí mismo.

7, 13 y 19

product The answer or result of a *multiplication* problem. It also refers to expressing a number as the *product* of its *factors.*

producto Respuesta o resultado de un problema de *multiplicación.* Además, un número puede expresarse como el *producto* de sus *factores.*

protractor An instrument used to measure angles.

transportador Instrumento con el que se miden los ángulos.

Qq

quadrilateral A shape that has 4 sides and 4 *angles.*

square, rectangle, and *parallelogram*

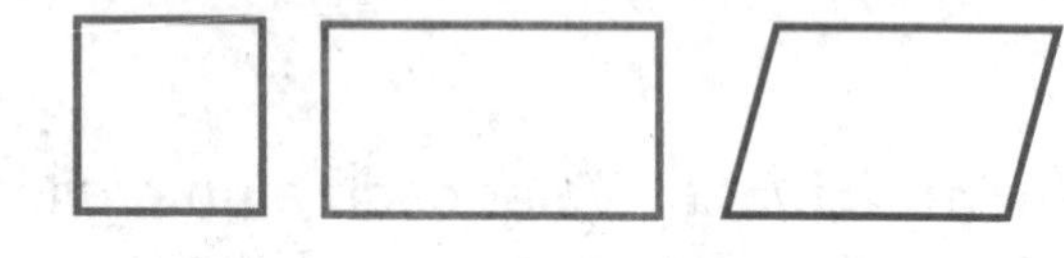

cuadrilátero Figura que tiene 4 lados y 4 *ángulos.*

cuadrado, rectángulo y *paralelogramo*

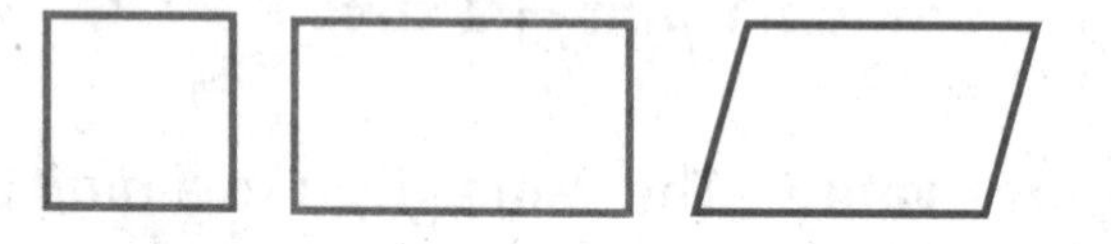

quart (qt) A *customary* unit for measuring *capacity.*

1 quart = 4 cups

cuarto (ct) Unidad *usual* de *capacidad.*

1 cuarto = 4 tazas

quotient The result of a *division* problem.

cociente Respuesta o resultado de un problema de *división.*

Rr

ray A part of a *line* that has one *endpoint* and extends in one direction without ending.

A *B*

semirrecta Parte de una *recta* que tiene un *extremo* y que se extiende sin fin en una dirección.

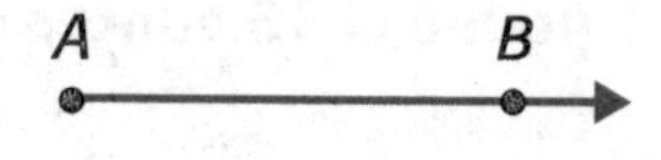

rectangle A *quadrilateral* with four *right angles*; opposite sides are equal and *parallel.*

regroup To use place value to exchange equal amounts when renaming a number.

remainder The number that is left after one whole number is *divided* by another.

repeated subtraction To subtract the same number over and over until you reach 0.

rhombus A *parallelogram* with four *congruent* sides.

right angle An *angle* with a measure of 90°.

right triangle A *triangle* with one *right angle.*

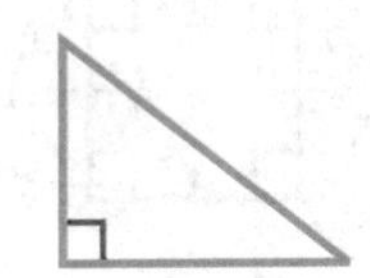
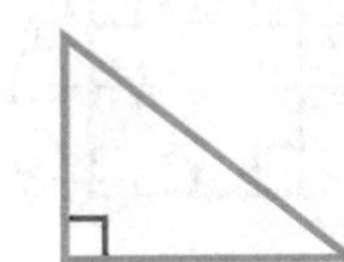

rectángulo *Cuadrilátero* con cuatro *ángulo rectos*; los lados opuestos son iguales y *paralelos.*

reagrupar Usar el valor posicional para expresar una cantidad de otra manera.

residuo Número que queda después de *dividir* un número natural entre otro.

resta repetida Procedimiento por el que se resta un número una y otra vez hasta llegar a 0.

rombo *Paralelogramo* con cuatro lados *congruentes.*

ángulo recto *Ángulo* que mide 90°.

triángulo rectángulo *Triángulo* con un *ángulo recto.*

Rr

round To change the value of a number to one that is easier to work with. To find the nearest value of a number based on a given *place value.*

redondear Cambiar el valor de un número a uno con el cual es más fácil trabajar. Hallar el valor más cercano a un número basándose en un *valor posicional* dado.

rule A statement that describes a relationship between numbers or objects.

regla Enunciado que describe una relación entre números u objetos.

Ss

second A unit of time.
60 seconds = 1 minute

segundo Unidad de tiempo.
60 segundos = 1 minuto

sequence The ordered arrangement of terms that make up a pattern.

secuencia Disposición ordenada de términos que forman un patrón.

simplest form A *fraction* in which the *numerator* and the *denominator* have no common *factor* greater than 1.

$\frac{3}{5}$ is the simplest form of $\frac{6}{10}$.

mínima expresión *Fracción* en la que el *numerador* y el *denominador* no tienen un *factor* común mayor que 1.

$\frac{3}{5}$ es la mínima expresión de $\frac{6}{10}$.

solve To replace a *variable* with a value that results in a true sentence.

resolver Despejar una *variable* y verdadera reemplazarla por un valor que haga que la ecuación sea.

square A *rectangle* with four *congruent* sides.

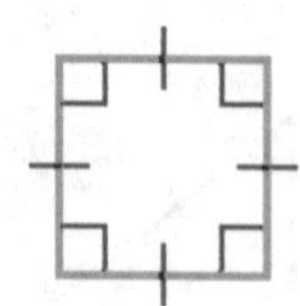

cuadrado *Rectángulo* de cuatro lados *congruentes.*

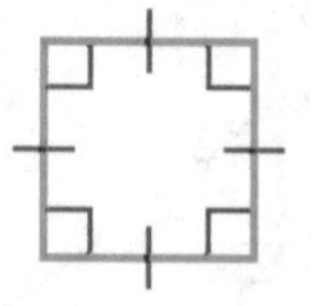

square unit A unit for measuring *area.*

unidad cuadrada Unidad para medir el *área.*

standard form/standard notation The usual way of writing a number that shows only its *digits,* no words.

537 89 1,642

forma estándar/notación estándar Manera habitual de escribir un número usando solo sus *dígitos,* sin usar palabras.

537 89 1,642

subtract (subtraction) An *operation* on two numbers that tells the *difference,* when some or all are taken away. Subtraction is also used to compare two numbers.

14 − 8 = 6

subtrahend A number that is *subtracted* from another number.

14 − 5 = 9
↑
subtrahend

sum The answer to an *addition* problem.

survey A method of collecting *data.*

restar (resta) *Operación* con dos números que indica la *diferencia,* entre ellos. Puede usarse para quitar una cantidad de otra o para comparar dos números.

14 − 8 = 6

sustraendo Un número que se *resta* de otro número.

14 − 5 = 9
↑
sustraendo

suma Respuesta o resultado que se obtiene al sumar.

encuesta Método para recopilar *datos.*

Tt

tally chart A way to keep track of *data* using *tally marks* to record the number of responses or occurrences.

What is Your Favorite Color?	
Color	**Tally**
Blue	卌 |||
Green	||||

tally mark(s) A mark made to keep track of and display *data* recorded from a *survey.*

tenth One of ten equal parts, or $\frac{1}{10}$.

term Each number in a numeric pattern.

tabla de conteo Manera de llevar la cuenta de los *datos* usando *marcas de conteo* para anotar el número de respuestas o sucesos.

¿Cuál es tu color favorito?	
Color	**Conteo**
azul	卌 |||
verde	||||

marca de conteo Marca que se hace para llevar un registro y representar *datos* recopilados en una *encuesta.*

décima Una de diez partes iguales o $\frac{1}{10}$.

término Cada número en un patrón numérico.

Tt

thousandth(s) One of a thousand equal parts, or $\frac{1}{1000}$. Also refers to a *place value* in a *decimal* number. In the *decimal* 0.789, the 9 is in the thousandths place.

milésima Una de mil partes iguales o $\frac{1}{1000}$. También se refiere a un *valor posicional* en un número *decimal.* En el *decimal* 0.789, el 9 está en el lugar de las milésimas.

three-dimensional figure A solid figure has three dimensions: *length,* width, and height.

figura tridimensional Figura sólida que tiene tres dimensiones: *largo,* ancho y alto.

ton (T) A *customary* unit to measure *weight.*

1 ton = 2,000 pounds

tonelada (T) Unidad *usual* de *peso.*

1 tonelada = 2,000 libras

trapezoid A *quadrilateral* with exactly one pair of *parallel* sides.

trapecio *Cuadrilátero* con exactamente un par de lados *paralelos.*

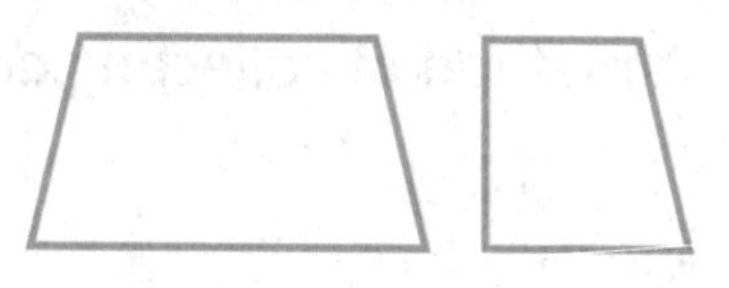

triangle A *polygon* with three sides and three *angles.*

triángulo *Polígono* con tres lados y tres *ángulos.*

two-dimensional figure A figure that lies entirely within one plane.

figura bidimensional Figura que puede representarse en un plano.

Uu

unit square A square with a side length of one unit.

cuadrado unitario Cuadrado cuyos lados miden una unidad de longitud.

unknown The amount that has not been identified.

incógnita La cantidad que no ha sido identificada.

Vv

variable A letter or symbol used to represent an unknown quantity.

variable Letra o símbolo que se usa para representar una cantidad desconocida.

Venn diagram A diagram that uses *circles* to display elements of different sets. Overlapping *circles* show common elements.

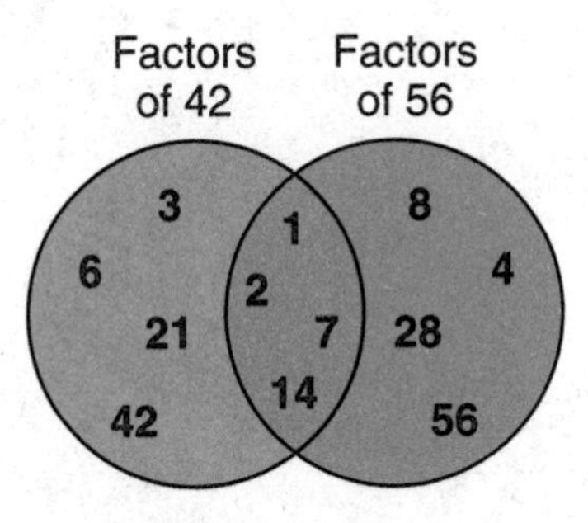

diagrama de Venn Diagrama con *círculos* para mostrar elementos de diferentes conjuntos. Los *círculos* sobrepuestos indican elementos comunes.

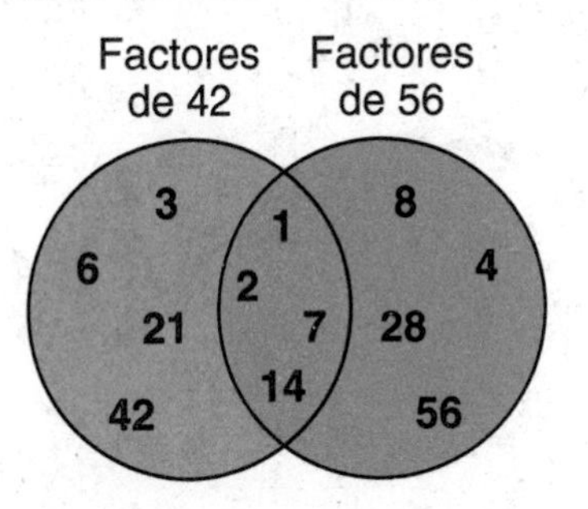

vertex The point where two *rays* meet in an *angle.*

vértice Punto donde se unen dos *semirrectas* formando un *ángulo.*

Ww

weight A measurement that tells how heavy an object is.

peso Medida que indica cuán pesado o liviano es un cuerpo.

word form/word notation The form of a number that uses written words.

forma verbal/notación verbal Forma de expresar un número usando palabras escritas.

Yy

yard (yd) A *customary* unit of *length* equal to 3 feet or 36 inches.

yarda (yd) Unidad *usual* de *longitud* igual a 3 pies o 36 pulgadas.

Zero Property of Multiplication
The property that states any number *multiplied* by zero is zero.

$0 \times 5 = 0 \qquad 5 \times 0 = 0$

propiedad del cero de la multiplicación
Propiedad que establece que cualquier número *multiplicado* por cero es igual a cero.

$0 \times 5 = 0 \qquad 5 \times 0 = 0$

Name

Work Mat 5: Tenths and Hundredths Models

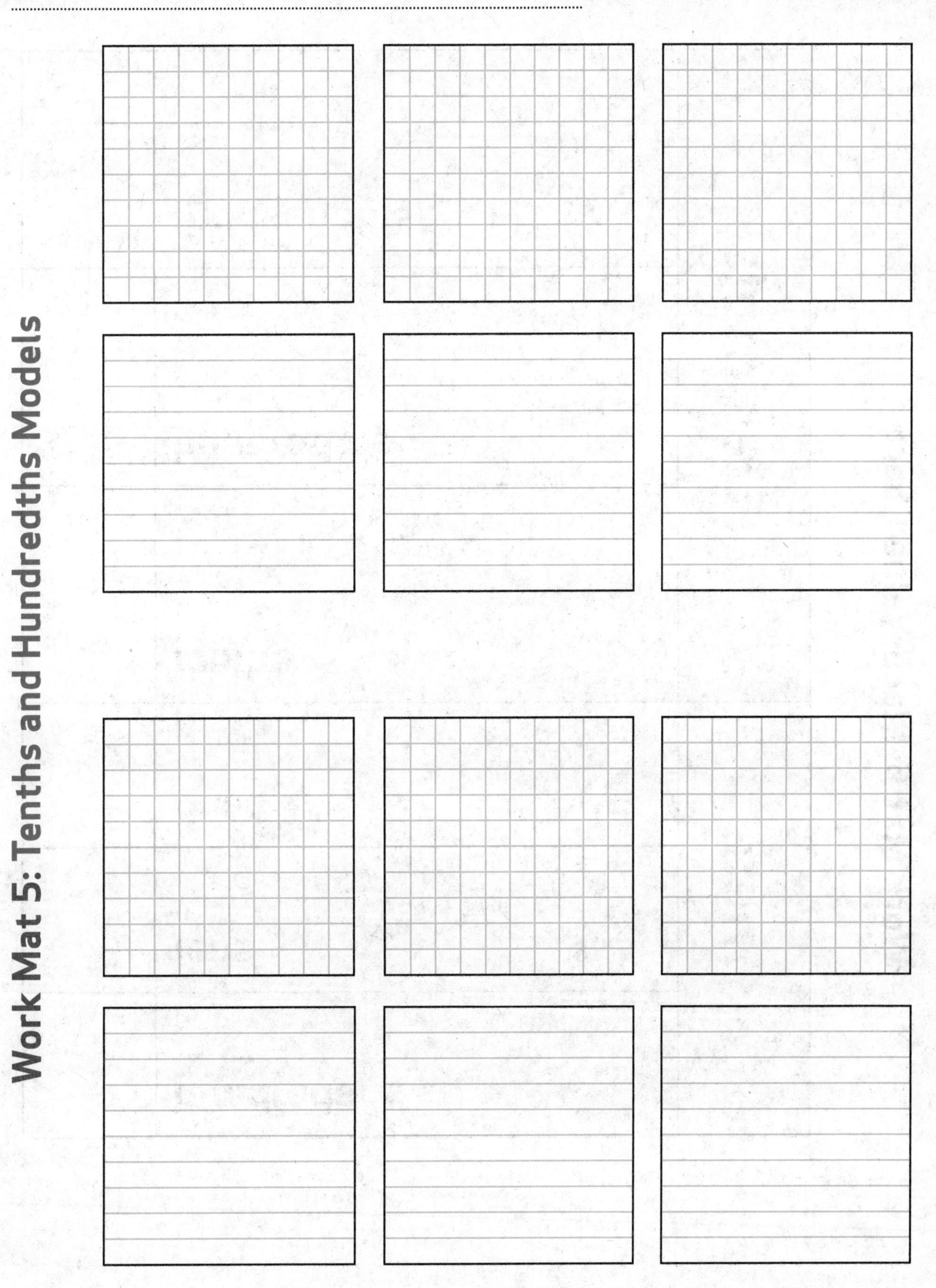

Work Mat 6: Place-Value Chart

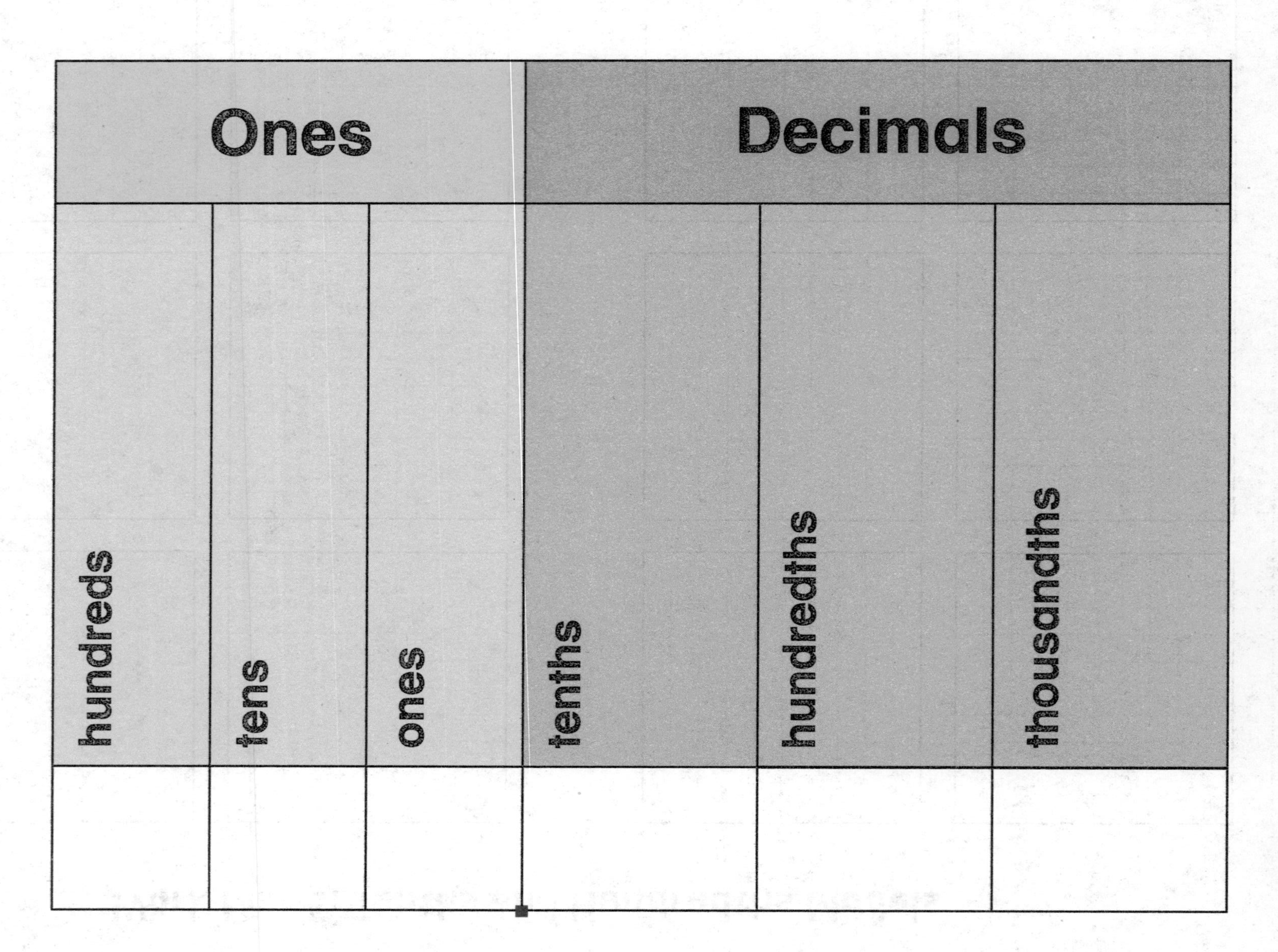

Name

Work Mat 7: Centimeter Grid

Work Mat 8: Number Lines

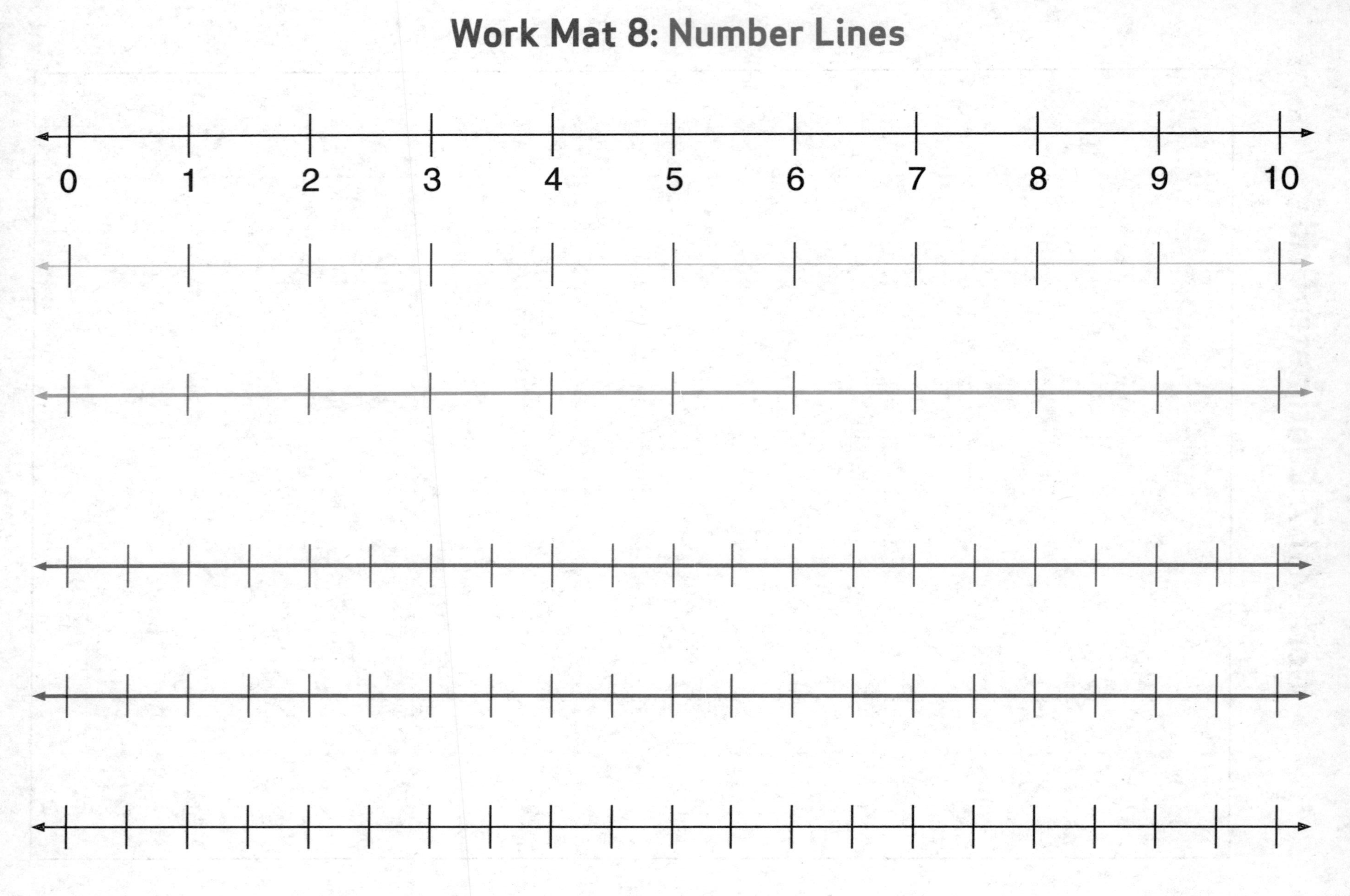